Advances in Experimental Medicine and Biology

Volume 1426

Advances in Experimental Medicine and Biology provides a platform for scientific contributions in the main disciplines of the biomedicine and the life sciences. This series publishes thematic volumes on contemporary research in the areas of microbiology, immunology, neurosciences, biochemistry, biomedical engineering, genetics, physiology, and cancer research. Covering emerging topics and techniques in basic and clinical science, it brings together clinicians and researchers from various fields.

Advances in Experimental Medicine and Biology has been publishing exceptional works in the field for over 40 years, and is indexed in SCOPUS, Medline (PubMed), EMBASE, BIOSIS, Reaxys, EMBiology, the Chemical Abstracts Service (CAS), and Pathway Studio.
2021 Impact Factor: 3.650 (no longer indexed in SCIE as of 2022)

Allan R. Brasier • Nizar N. Jarjour
Editors

Precision Approaches to Heterogeneity in Asthma

Springer

Editors
Allan R. Brasier
Department of Medicine, Division of Endocrinology Institute for Clinical and Translational Research
University of Wisconsin School of Medicine and Public Health
Madison, WI, USA

Nizar N. Jarjour
Department of Medicine, Division of Allergy, Pulmonary and Critical Care
University of Wisconsin School of Medicine and Public Health
Madison, WI, USA

ISSN 0065-2598 ISSN 2214-8019 (electronic)
Advances in Experimental Medicine and Biology
ISBN 978-3-031-32258-7 ISBN 978-3-031-32259-4 (eBook)
https://doi.org/10.1007/978-3-031-32259-4

This Springer imprint is published by the registered company Springer Nature Switzerland AG
The registered company address is: Gewerbestrasse 11, 6330 Cham, Switzerland

Introduction

Asthma is typically a life-long disease that has worldwide prevalence with substantial impact on quality of life and health-care utilization. Despite substantial advances in understanding its mechanisms and standardization of clinical practice guidelines, asthma continues to cause of substantial adverse health and economic impact. As recently as 2019, the Global Burden of Disease collaboration estimated that asthma affected ~262 million people worldwide. Asthma is the most common chronic disease in childhood, affecting 9–11% of children from ages 6–14, and one of the most common causes of school absenteeism. In adults, asthma was one of the major causes for years lived with disability. Finally, the World Health Organization (WHO) estimates that asthma is responsible for 455,000 deaths worldwide, principally in under-resourced environments.

The etiology of asthma remains obscure and thought to be the result of complex gene-environment and microbiome interactions. Accurate patient characterization is important to predict disease prognosis and guide therapy. Clinical studies have shown that several sub-phenotypes can be identified based on the age at disease onset, the presence of fixed airway function, and the inflammatory profile. Although the definition of phenotypes has been a step forward, there have been difficulties in directly linking phenotypes to effective therapeutic interventions. For example, in patients with non-severe asthma, T helper type 2 (T2) inflammation frequently resolves with glucocorticoid therapy; however, many patients with severe uncontrolled asthma have ongoing type 2 inflammation despite being treated with high-dose inhaled glucocorticoids. Fortunately, many of these patients with steroid-resistant persistent T2 inflammation can now benefit from approved biologic therapies that target cytokines (and cytokine receptors) that drive T2 inflammation.

Advances in molecular techniques and imaging have revealed the diverse biological pathways responsible for the persistent airway inflammation in asthma. The elucidation of biomarkers and cellular components specific to these inflammatory pathways allowed for categorization of asthma from generic phenotypes to specific endotype. Endotype refers to subtype of asthma driven by a distinct pathophysiological mechanism. The field is now making substantial advances on the understanding of endotypes to inform precision approaches in asthma therapy.

The first edition of this work, published in 2014 as *Heterogeneity in Asthma* addressed the emerging findings that asthma was heterogenous in etiology, genomics, and dynamic responses. The strong reception of this work by the community, with over 74,000 downloads as of 2022, indicated that the field is poised to apply precision approaches for the pathogenesis, diagnosis, and treatment of this disease. Since the publication of the first edition, substantial advances have been made identifying endotypes informing precision approaches. To embrace these advances in the field and link them to precision approaches, we have sought leading contributors to address epidemiology, mechanistic endotypes of disease, employing systems approaches using biomarker-informed clinical trials. It is our hope that this work will facilitate the current understanding of the spectrum of disease etiology, prognosis, and/or the likelihood of responding to the range of available therapeutic interventions.

To accomplish this goal, we have grouped the chapter content into four parts. Part I addresses the epidemiology of asthma, one of the most racially diverse respiratory illnesses, and ethnic minorities who tend to be the least responsive to asthma therapy, leading to highest morbidity and mortality. Emerging impacts of climate change on asthma are addressed. We consider how asthma is modified by social determinants of health, the different manifestations of asthma across the lifespan, and the sex effects of asthma. Furthermore, the effect of ongoing airway inflammation of asthma on other organs (e.g. brain, cardiovascular system) is also addressed.

Part II updates on studies related to the heterogeneous mechanistic pathways affecting the development and progression of asthma. We consider work in asthma phenotypes, emerging ideas on asthma endotypes, heterogeneity of treatment response and exciting work on discovery of regional airway changes manifesting are variation in ventilation, inflammation and remodeling. Amplifying asthma as a systemic disease, new discoveries on brain-airway interactions in asthma are described and its linkage with cognitive impairment. Part II concludes with latest work on integrative systems level studies, an approach incorporating clinical features, volatile organic compounds, proteomics, metabolomics, and transcriptomics to provide a more accurate and comprehensive understanding of this disease.

A subset of asthmatics suffers difficult-to-treat and exacerbation-prone asthma, incurring the greatest costs and health impacts of this disease. Part III discusses exacerbation-prone endotypes and the roles of innate and adaptive immunity in asthma exacerbations.

Part IV focuses on an update on how understanding heterogeneity can guide approaches to management. Current guidelines for stepped care and self-monitoring are presented including the Expert Panel Reports (EPR) published by United States National Heart, Lung, and Blood Institute (NHLBI) and the Global Strategy for Asthma Management and Prevention Report (GINA) published by the NHLBI and WHO. We review findings from biomarker-informed trials for severe asthma and advantages of master protocols and adaptive design methods for establishing efficacy of new interventions in prospectively defined subsets of patients and increased emphasis on effectiveness studies.

We hope that this work will help to advance the understanding of the spectrum of asthma and its heterogeneity, thus facilitating personalized approaches to its management from selecting treatments with greatest likelihood of patient response to paving the way for disease modifiers and therapeutics that have better chances to affect prognosis and/or achieve long-term disease remission. We anticipate that the ongoing exciting work built on biomarker-driven and effectiveness trials will lead to more streamlined and effective therapeutic interventions with reduced side effects, morbidities, and health-care utilization.

We are deeply indebted to our colleagues who have contributed to this volume, and thank you, the readers, for your input and feedback.

On behalf of the contributors,

Madison, WI, USA

Allan R. Brasier
Nizar N. Jarjour

Contents

Part I

The Spectrum of Asthma Epidemiology

Epidemiology of Asthma: Prevalence and Burden of Disease

Tamara Merhej and Joe G. Zein

Abstract

Asthma, a common airway disease, results in a significant burden to both patients and society worldwide. Yet, despite global political commitment backed by the United Nations, progress to reduce the burden of asthma remains inadequate. This is particularly true in low-income countries. To date, progress has been delayed by the lack of uniform data collection, imperfect surveillance methods, inadequate resources, poor access to effective therapies, substandard asthma education, ineffective governmental policies, rapid urbanization, progressive increase in asthma prevalence, increased life expectancy and obesity rates worldwide, asthma heterogeneity and disease complexity, smoking, and environmental exposures to allergens and pollution. A thorough understanding of the challenges facing the international community is essential to define future strategies to improve the burden of asthma.

Keywords

Epidemiology · Surveillance · Non-communicable disease · Incidence · Respiratory health · Global burden · Heterogeneity

T. Merhej
Brigham and Women's Hospital, Boston, MA, USA

J. G. Zein (✉)
Respiratory Institute. Cleveland Clinic, Cleveland, OH, USA
e-mail: zeinj@ccf.org

A. R. Brasier, N. N. Jarjour (eds.), *Precision Approaches to Heterogeneity in Asthma*, Advances in Experimental Medicine and Biology 1426,
https://doi.org/10.1007/978-3-031-32259-4_1

Introduction to Asthma Epidemiology

Asthma, which is characterized by inflammation and narrowing of the airways, is frequently manifested by cough, wheezing, dyspnea, and chest tightness [87]. It is frequently miss-diagnosed and suboptimally treated [1, 19, 24]. For example, in a prospective, multicenter Canadian study of 613 patients with physician-diagnosed asthma, 203 (33.1%) did not have asthma. This emphasizes the importance of confirming the diagnosis of asthma as the first step in the evaluation of patients presenting with asthma and severe asthma [41].

Asthma is known as a common chronic noncommunicable disease (NCD) and remains a major health problem globally. It is a cause of substantial health and economic burden to individuals and society and leads to premature death and reduced quality of life worldwide. In 2019, the Global Burden of Disease collaboration estimated that asthma affected as many as 262 million people worldwide and that asthma was one of the major causes for years lived with disability [28, 30]. Access to asthma care is usually limited, and effective therapies are frequently inaccessible and unaffordable. Although preventable with effective therapies and informative asthma education, about 1000 people die from asthma every day due to suboptimal management and reliance on rescue beta-agonists rather than controller medications.

Many governments have overlooked asthma in their plans to address NCDs and have made little progress in improving access to asthma management and medicines, especially inhaled corticosteroids (ICS), which are crucial for the long-term control of asthma. Evidence from the ongoing COVID-19 pandemic suggests that ICS are unlikely to have adverse effects on the outcome of SARSCoV2 infection among asthmatics, so their use should continue to be promoted, either alone or in combination with a long-acting β2-agonist inhaler [28].

Asthma Surveillance

The challenges in the diagnosis of asthma arise from disease heterogeneity and strict clinical and physiologic criteria. In the presence of characteristic symptoms and signs on physical exam, the diagnosis of asthma is supported by the presence of airway obstruction on spirometry and a significant bronchodilator response or a positive bronchoprovocation test [79]. However, the diagnosis of asthma in national and international surveillance programs is not so simple and depends on multiple interdependent criteria that vary between different countries and between different agencies within the same country. Internationally, asthma may be under or over-diagnosed depending on available resources (i.e., spirometry) and divergence in clinical practice. Invariably, bias is introduced by inadequate adjustment for confounding factors (e.g., smoking and biofuel exposure) and the lack of standardized definitions and surveillance questionnaires. Although this ambiguity and lack of consensus between surveillance programs have resulted in faulty estimates of disease incidence, prevalence, and burden, there is little doubt that asthma has significantly increased over the last 50 years.

Asthma Surveillance Programs

United States

In 1999, the United States Center for Disease Control and Prevention (CDC) established the National Asthma Control Program (NACP) to help improve asthma control, reduce asthma-related healthcare utilization, healthcare burden, and mortality [59]. Currently, the CDC funds state, regional, and local health programs to improve access to asthma therapy, promote quality measures, and inform policymakers about the burden of asthma. To date, NACP has resulted in significant improvement in asthma management, quality of care, patient education, and control across the United States, and fostered policies to help reduce air pollution. In Fig. 1 we detail several landmark events accomplished by the NACP since the establishment by the CDC in 1999.

In addition to the NACP, several public health programs have been established (Table 1). Among those, the CDC is tasked with a wide variety of surveillance activities to lower the burden of asthma, improve quality of life, reduce the number of days away from work or school, and improve access to care and affordable controller medications. Nationwide, state, regional, and local community data have been collected through the administration of surveys by the National Center for Health Statistics (NCHS) and the National Vital Statistics System (NVSS). Data collected by the CDC surveys are the main driver of public health policies to improve healthcare and resource allocation.

The prevalence of asthma in the US continues to increase. Adapted from the 2001–2020 National Health Interview Survey (NHIS) Data, Fig. 2 shows a steady increase in the lifetime prevalence of asthma in the United States. These trends have been increasing irrespective of sex (Fig. 1a), race and ethnicity (Fig. 1b), and income (Fig. 1c). The prevalence of lifetime asthma was higher in pubescent boys than girls and in adult women compared to men. It was also higher in non-Hispanic black than white individuals and inversely correlated with income. Among the different income categories, the highest lifetime asthma prevalence was seen among individuals living below the poverty line (Fig. 1c). Similarly, NHIS data on current asthma echoed that of lifetime asthma. In 2020, the prevalence of current asthma was higher in females (9.5%) than males (6.1%) and in non-Hispanic black (10.4) than white (7.7%) or Hispanic individuals (6.8%). Among Hispanics, current asthma prevalence was highest in Puerto Ricans (14.9%) and lowest in Mexican and Mexican-Americans (5.8%). Current asthma is also inversely correlated with income. It was higher in people living below the poverty threshold (11%) compared to those at 100–250% (8.7%), 250–450% (6.9%), and >450% (6.7%) of the poverty rate, respectively [61].

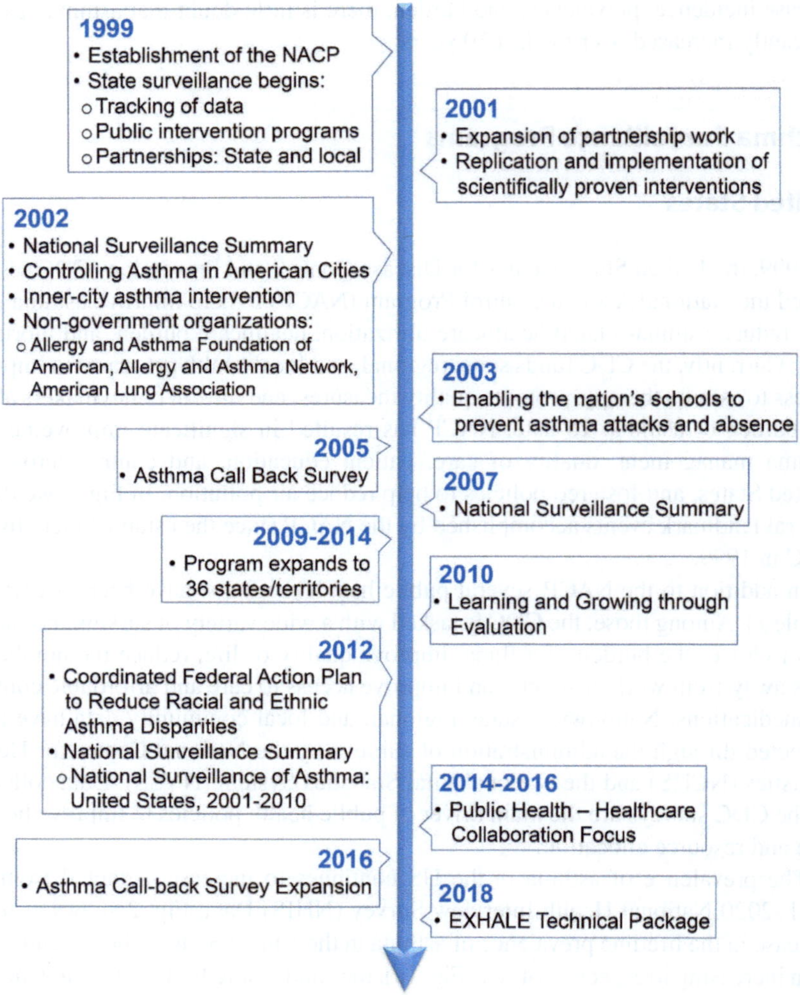

Fig. 1 Timeline of CDC's National Asthma Control Program (NACP) millstone achievements

Of the 25.1 million people who were diagnosed with current asthma in the US in 2019, 10.3 million (41.2%) had an asthma exacerbation, 1.8 million (7.3%) had an ER visit, 169,330 (0.7%) were hospitalized, and 3524 (0.014%) died from asthma. Of note, during that same period (i.e., 2019), the WHO estimated that asthma affected 260 million people worldwide and resulted in 455,000 (0.175% of those with current asthma) death [61]. These data clearly demonstrate that asthma-related mortality worldwide is more than tenfold higher than mortality in developed countries (such as the United States). Hence, >400,000 avoidable deaths could be saved globally every year with appropriate interventions.

Table 1 Method of surveillance in the United States

Abbreviation	Surveillance Program	Program Description
ACBS	Asthma Call-back Survey	Implemented in 2005 by the Centers for Disease Control and Prevention's (CDC) National Asthma Control Program (NACP), which funds states, cities, and schools to improve asthma surveillance, public awareness, and train and educate individuals with asthma, their families and healthcare professionals. The ACBS expanded quickly to 25 states in 2006 and 35 states in 2007.
BRFSS	Behavioral Risk Factor Surveillance System	Established in 1984, BRFSS is a system of health-related telephone surveys that collect state data about health-related risk behaviors, chronic health conditions, and use of preventive services. Being the largest health survey system in the world, BRFSS completes >400,000 adult interviews annually, in all 50 states, District of Columbia (DC), and 3 US territories.
HCUP	Healthcare Utilization Project	HCUP databases contain information on inpatient stays, ER visits, and ambulatory care, and uses weights to calculate national estimates of asthma outcomes, hospital mortality, and cost.
NAMCS	National Ambulatory Medical Care Survey	NAMCS has provided national data about the provision and use of ambulatory medical care services since 1989. Data are collected from national samples of visits to health centers and to nonfederally employed office-based physicians in the United States.
NCHS	National Center for Health Statistics	NCHS is part of the CDC and has the responsibility for producing vital and health statistics for the Nation.
NEPHTN	National Environmental Public Health Tracking Network	Part of the CDC, the NEPHTN combines health and environmental data from national, state, and city sources on environments and hazards, health effects, and population health.
NHAMCS	National Hospital Ambulatory Medical Care Survey	The NHAMCS collects data on the utilization and provision of ambulatory care services in hospital emergency and outpatient departments and ambulatory surgery locations of noninstitutional general and short-stay hospitals.
NHANES	National Health and Nutrition Examination Survey	NHANES is a program of the NCHS that is designed to assess the health and nutritional status of adults and children in the United States. The survey combines interviews and physical examinations. Enriched with minority populations, NHANES collects data on asthma exacerbations, asthma-related ER visits, and asthma medications. Like HCUP, it uses weights to calculate national estimates.
NHCS	National Hospital Care Survey	The NHCS integrates inpatient data collected by NHCS with data collected by NHAMCS to link information on inpatient and outpatient care provided to the same patient. NHCS data can also be merged with data from to the National Death Index, Medicaid, and Medicare to obtain a complete understanding of patient care.

(continued)

Table 1 (continued)

Abbreviation	Surveillance Program	Program Description
NHDS	National Hospital Discharge Survey	Conducted annually from 1965–2010, NHDS was designed to collect data on the characteristics of inpatients discharged from non-Federal short-stay hospitals and to examine important topics of interest in public health and for a variety of activities by scientific, academic, governmental, and commercial institutions.
NHIS	National Health Interview Survey	Part of the CDC, the NHIS, has tracked the health of the nation since 1957. NHIS uses personal household interviews to collect data on a broad range of health topics. Such data have been instrumental in monitoring health status, healthcare access, and progress toward accomplishing national health objectives.
NVSS	National Vital Statistics System	The NVSS is an inter-governmental system that provides the most complete data on births and deaths in the United States. It involves coordination between the NCHS, a division of the CDC, and different state health departments.

Global Surveillance

Although important, global asthma surveillance is frequently sporadic and limited. In fact, asthma prevalence was never studied in about half of the world's countries, and data collection about asthma-related hospital admissions is almost entirely restricted to high-income countries. This significantly limits the generalization of results to better understand the global burden of asthma. In an attempt to provide data on asthma incidence, prevalence, and burden, three major international studies have been involved in the surveillance of adults and children globally. Although important, these reports used a compilation of surveillance initiatives with diverse methodologies resulting in significant limitations to generalizability. Here, we will focus on three international studies.

The International Study of Asthma and Allergies in Children (ISAAC)

ISAAC is an epidemiological research program that was established in 1991 to study asthma, rhinitis, and eczema in children globally due to concern that the incidence and prevalence of these conditions were on the rise in western and developing countries. Involving more than 100 countries and nearly 2 million children, ISAAC aims to develop environmental measures and disease surveillance programs to reduce the burden of allergic and nonallergic diseases through targeted interventions. To date, findings from ISAAC's study have shown that these diseases are increasing, especially in developing countries. This highlights the urgent need for further population studies to better understand the underlying mechanisms of asthma, rhinitis, and eczema and the burden of these conditions.

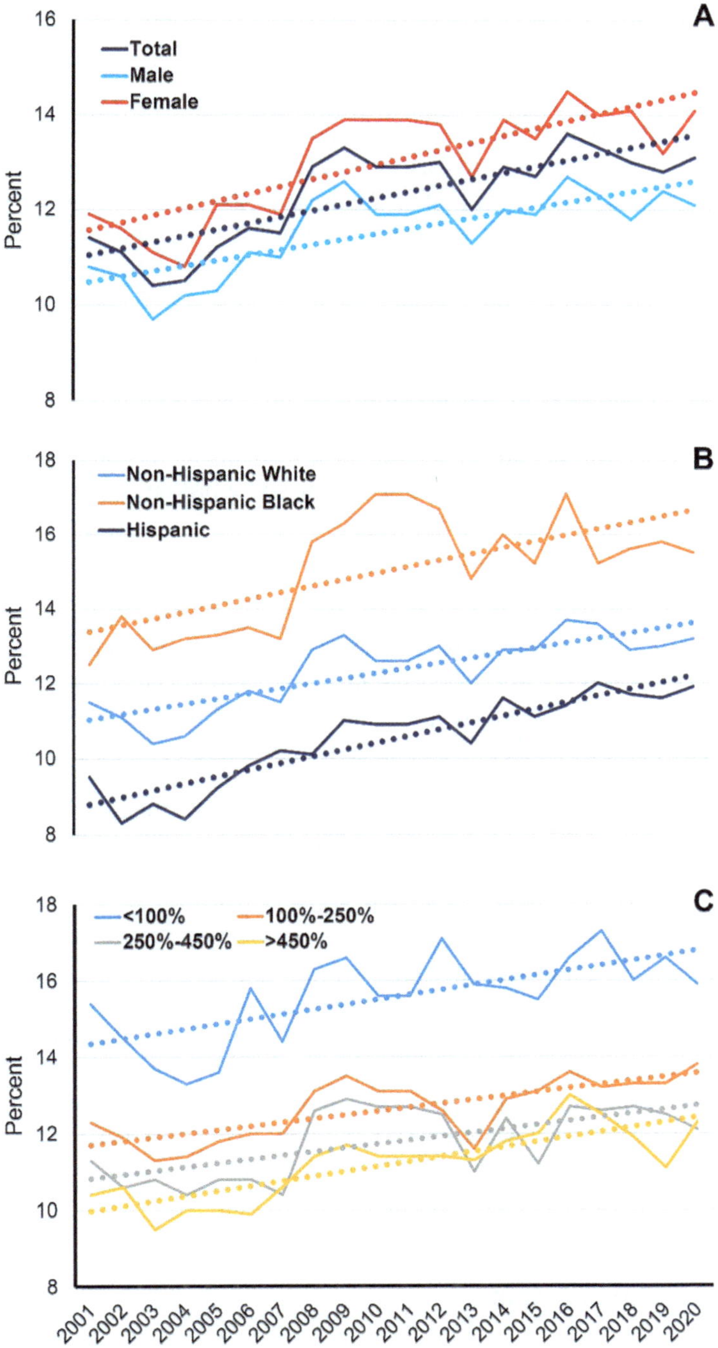

Fig. 2 Trends in lifetime asthma prevalence between 2001 and 2020, stratified by (**a**) sex, (**b**) race/ethnicity, and (**c**) poverty threshold. Adapted from 2001 to 2020 National Health Interview Survey (NHIS) Data

The European Community Respiratory Health Survey

Phase 1 of the European Community Respiratory Health Survey (ECRHS) was implemented in the 1980s in response to increases in asthma prevalence worldwide, in order to investigate the role of environmental factors in asthma pathobiology. The first stage of the ECRHS started in 1990 to collect data in 200,000 young adults from mainly European countries using standardized protocols. In contrast to the questionnaire stage 1, the clinical stage 2 enrolled 26,000 participants between 1998 and 2002. ECRHS was mostly funded by the European Commission, as part of the Quality of Life Program. Now ECRHS is entering stage 3, where previously enrolled participants are followed-up for a second time to assess changes in respiratory symptoms, (2) IgE sensitization, lung function decline, and development of chronic obstructive airway disease over a 20-year period (2002–2022). ECRHS 3 will also investigate the influence of changes in atopic status and obesity on exercise tolerance, lung function, lung function decline, and asthma prognosis of outcomes measures. ECRHS 3 will investigate the role of sex (including menopausal status) and environmental factors (e.g. occupation and air pollution) on observed effects and respiratory symptoms.

The World Health Surveys

Developed by the WHO in 2002, the World Health Surveys (WHS) aimed at standardizing data collection protocols to compare results within and between different countries [9]. The WHS was conducted in 178,215 individuals 18–45 years of age from 70 different countries. The prevalence of asthma was ascertained based on responses to questions related to self-reported asthma, doctor-diagnosed asthma, asthma therapy, and wheezing in the last 12 months, and prevalence varied significantly based on the question (4.3% for doctor-diagnosed asthma, 4.5% for clinical/treated asthma, and 8.6% for wheezing). The WHS highlights the importance of survey standardization and the effect of the survey question on asthma prevalence and burden, which may vary significantly based on the question.

The global prevalence rates varied by as much as 21-fold between the 70 countries and were highest in Australia (21.0% for doctor-diagnosed asthma, 21.5% for clinical/treated asthma, and 27.4% for wheezing). One in four individuals with clinical/treated asthma were current smokers, and one in five had never been treated for asthma. Data from the WHS highlights the impact of smoking worldwide and its global burden on asthma, especially in developing countries with limited resources. While asthma prevalence was highest in developed countries, nations with limited resources were also affected, especially with rising asthma prevalence caused by rapid urbanization in many developing countries [80].

Global Burden of Asthma

With more than 300 million people affected worldwide, asthma is associated with significant morbidity and cost to society [2]. It is the most common chronic disease in childhood, causing significant morbidity and mortality across the globe [29]. In a recent multinational study from 25 countries, asthma was found to affect 9.1% of children 6–7 years of age and 11.0% of children 13–14 years of age [29]. Among children, the worldwide prevalence of asthma negatively correlated with the gross national income [29]. Similarly, asthma prevalence has also been reported by the International Study of Asthma and Allergies in Childhood (ISAAC) after surveying about 1.2 million children seen in 233 centers in 98 countries [51]. While in developed countries, childhood asthma has been associated with allergic sensitization, obesity, environmental (maternal) tobacco smoke, and socioeconomic status [75], childhood asthma has been associated with biofuel and pesticide exposure in underdeveloped nations [40, 72, 90]. The burden of asthma was found to be similarly important in adults. A survey of 193,912 adults in 17 different countries showed that the prevalence of asthma and severe asthma positively correlated with gross national income, where high-income countries had the highest asthma (10.6%) and severe asthma (3.9%) prevalence [56]. Of note, countries with high adult asthma prevalence also had high asthma prevalence during childhood and adolescence [56]. This invariably increases the societal asthma burden and places great strains on the healthcare system in such countries.

Preventable asthma mortality is a real concern in developing countries. It has been frequently associated with suboptimal asthma care, poor access to effective and affordable therapies, and the lack of effective government policies and support [28]. Similar to mortality, severe asthma is a major concern in countries with limited resources. In fact, although the prevalence of asthma symptoms is greater in high-income countries than in low-income countries, the frequency of severe symptoms was similar in all age categories. These findings highlight the impact of pollution, smoking, and the risk of obesity in developing countries. This led the WHO to adopt several interventions to reduce the worldwide burden of asthma including tobacco control, provision of essential medicines and interventions, and the promotion of healthy diet and physical exercise. These data highlight the global burden of this treatable disease and the role we can all play, at both individual and community level, to improve asthma symptoms, disease control, and outcomes.

The Global Asthma Report

Similar to previously published reports in 2011, 2014, and 2018, the fourth Global Asthma Report (GAR) 2022 is focused on reducing the burden of asthma and improving asthma care. While the last three reports have been produced by the Global Asthma Network (GAN), [28] was a combined production by The International Union against Tuberculosis and Lung Disease (The Union) and the International Study of Asthma and Allergies in Childhood (ISAAC) [28]. With the contributions of 97 experts from 31 countries, [28] highlights several challenges relating to the prevention and management of asthma, especially during the COVID-19 pandemic. Unfortunately, progress in reducing the impact of chronic noncommunicable diseases such as asthma has been inadequate, especially in developing countries, although research findings and updated treatment recommendations have been successfully disseminated by the Global Asthma Network. Asthma outcomes continue to be negatively impacted by delays in diagnosis, poor access to high-quality care, and affordable medications. Analysis of the [28] data, which now includes information from the completed GAN Phase I survey of over 450,000 children and parents from 25 countries during the period 2015–2020, shows that asthma remained inadequately controlled in about 50% of treated children and adolescents and that one in five children and adolescents with asthma did not receive any effective therapy. Likewise, one-third of adults with asthma did not achieve adequate asthma control while on therapy. Information provided in [28] report may play a role in influencing policymakers to intervene in order to reduce the global burden of asthma.

Challenges to Asthma Surveillance, Therapy, and Control

The Heterogeneity of Asthma

Asthma is a complex heterogeneous disease that comprises several subphenotypes based on the clinical characteristics, the age at disease onset, and the inflammatory profile [45, 85]. Type 2-(T2-) high asthma is associated with eosinophilic airway inflammation, while the absence of T2 inflammation is T2-low [86]. Accurate patient characterization is important to predict disease prognosis and guiding therapy [41]. In the Severe Asthma Research Program (SARP III) cohort, about 34% of adults and 58% of children had eosinophilia (defined by ≥300 cells/μL): a biomarker of T2 inflammation [79]. In patients with nonsevere asthma, type 2 inflammation frequently resolves with glucocorticoid therapy [26, 41]. However, many patients with severe uncontrolled asthma have ongoing type 2 inflammation despite being treated with high-dose, inhaled glucocorticoids [21, 26, 41]. These patients with steroid-resistant persistent T2 inflammation can now benefit from approved biologic therapies that target cytokines (and cytokine receptors) that drive T2 inflammation including interleukin (IL-) 5 and thymic stromal lymphopoietin (TSLP) [10, 15, 41, 54, 64]. In addition to their glucocorticoid-sparing effect, these novel biologic therapeutics have been associated with reduced exacerbation rates and improved lung function, which translates into a better quality of life and decreased asthma burden [10, 15, 54, 64].

Until recently, very few therapeutic options were available for patients with T2-low asthma besides managing comorbidities such as the epidemic of obesity or diabetes [41]. In the United States, about 60% of the population meet diagnostic criteria for obesity [22]. Obesity is frequently associated with gastroesophageal reflux disease (GERD), obstructive sleep apnea (OSA), and diabetes, which have been associated with increased asthma severity and poor asthma control [22, 44, 99]. Even if patients do not meet the diagnostic criteria for diabetes, insulin resistance (IR) has been associated with poor outcomes in asthma [4]. IR, which is frequently seen in asthma and commonly exacerbated by corticosteroid therapy, has been associated with an accelerated decline in lung function and poor response to bronchodilator therapy [69]. Although the exact mechanisms underlying the pathobiology of severe asthma are uncertain, obesity-related metabolic dysfunction has been frequently implicated in the development of severe asthma [68]. In a cross-sectional data from patients enrolled in SARP, high serum IL-6 levels were associated with lower lung function and higher exacerbation rates in both obese and nonobese individuals [68]. The therapeutic role of anti-IL-6 inhibitors in asthma is currently being evaluated by the Precision Intervention for Severe Asthma (PrecISE), a biomarker driven NHLBI-sponsored adaptive-design clinical trial investigating therapeutic options in T2-low asthma [31] .

Phenotypes are also defined according to age at disease onset. Childhood-onset asthma occurs more frequently in young boys with atopy than in girls [82]. Although it resolves with adolescence in many males and female asthmatics, it continues into adulthood in a large proportion of patients [20, 74]. In such individuals, impaired lung function in childhood and male sex may impact future lung function and result in the development of fixed airway obstruction and chronic obstructive airway disease (COPD) in early adulthood [52]. These data emphasize the importance of serial lung function testing in children and young adults with persistent asthma to identify those at risk for chronic airflow obstruction in adulthood [52]. Asthma can also start at an older age, especially in women, and is associated with lower lung function, increased disease severity, and a higher rate of age-related comorbidities such as obesity, GERD, chronic rhinosinusitis, and obstructive sleep apnea [55, 94]. For all these reasons, elderly individuals are at highest risk for asthma-related morbidity and mortality [94]. They are also more likely to be underdiagnosed and undertreated [24]. These patients will benefit from a holistic approach tailored to their asthma phenotype that tackles medication compliance, age-related comorbidities, and disease severity.

Environmental Exposures

The remarkable increase in asthma prevalence in developed countries over time suggests that the environment plays an important role in asthma. In low and middle-income countries, urban residents are about 89% more likely to be diagnosed with asthma by a doctor and 77% more likely to self-report asthma than rural residents [71]. In the US, this rural-urban differentiation was not consistent, and county-level data on current asthma prevalence in adults and children were thought to be important for resource allocation [34]. In the United States, a lower prevalence of asthma among children in rural areas was only seen in 7 out of 27 states surveyed by the Center for Disease Control and Prevention (CDC) [34]. These differences might be related to farming conditions, traditional vs. industrialized, and their effect on the innate immune response. For example, children living on a traditional dairy farm in Central Europe might be subject to high microbial exposures as compared to children living on an industrialized farm in the United States [73]. Comparing environmental exposures, genetic ancestry, and immune profiles between Amish and Hutterite children, the asthma prevalence was four to six times lower in the Amish, and the house dust endotoxin levels were more than sixfold higher. These data emphasize the role of the microbiome and the regulation of the innate immune response on asthma pathogenesis and prevalence [73].

Ethnic and Racial Diversity

Asthma is one of the most racially diverse respiratory illnesses, and ethnic minorities tend to be the least responsive to asthma therapy and also have the highest morbidity and mortality [13, 89, 98]. Compared to non-Hispanic whites (12.9%), lifetime asthma prevalence is significantly higher in African Americans (15.6) and highest in Puerto Ricans (23%) [60]. Although the affordable care act (ACA) succeeded in reducing the rates of uninsured individuals in the United States, the persistence of cost barriers (i.e., medication and office visit costs) may be responsible for suboptimal improvement in asthma control at a population-level [77]. This creates a situation where asthma can appear to be more severe in racial and ethnic minorities and at higher risk for treatment failure resulting in increased healthcare utilization [84]. Furthermore, peripheral-blood profiles, which are used as clinical thresholds to determine the eligibility for asthma biologic therapies, differ by race and ethnicity [89]. Consequently, eligibility criteria for asthma biologic therapies may be different across racial groups. For instance, an eosinophil cutoff of 300 cells/μL was found to predict response to biologic therapies in a non-Hispanic White population but may poorly predict such response in a non-Hispanic Black or a Puerto Rican population [33]. These findings support the need for racial and ethnic-specific biomarker cutoffs in the era of precision medicine [33].

Sexual Dimorphism

Sex is a primary determinant of asthma severity. Clinical observations suggest that gender and sex hormones mediate asthma severity. Asthma varies in severity by sex across the lifespan [92, 93]. It is more prevalent and severe in pubescent boys compared to girls but greater in girls and women after puberty [36, 95, 98]. Girls with early menarche and multiparous women, who are exposed to higher estrogen levels and have greater cumulative sex hormones exposure, have a much higher risk for asthma and severe asthma [27, 35, 43]. Carrying a female fetus is associated with more severe asthma during pregnancy [23, 46, 65] that could be related to sex-specific alterations in the expression of placental genes of pregnant women with asthma. In fact, several genes linked to growth, inflammation, and immune pathways were reported to be differentially expressed in the placentae of asthmatic women carrying female vs. male fetuses [65]. Outside pregnancy, women of child-bearing age exhibit cyclical and diurnal variation in androgens and ovarian hormones that might have significant clinical implications [3, 7, 8, 12, 39]. Premenstrual worsening of asthma, which has been estimated to affect up to 40% of females with asthma, is associated with ~50% decrease in PC20 during the luteal phase, at peak estrogen and progesterone levels [78]. Asthmatic women receiving oral contraceptives have attenuated cyclical changes in airway reactivity in association with suppression of luteal upsurge in progesterone and estradiol [17, 43]. Likewise, asthma may improve with menopausal transition, when sex hormones wane [63, 81, 94].

Asthma also varies in severity across the lifespan [5, 48, 91, 93, 95]. In contrast to childhood asthma, asthma exacerbation, and healthcare utilization are higher among adult women as compared to men [67, 96, 98]. In a population-based study based on death certificates for US residents, we found that while asthma mortality in the US dropped by half between 1999 and 2015, the mortality in women consistently exceeded the mortality in men by twofold [67].

Although several population-based epidemiological studies suggest a gender discordance in asthma and severe asthma, the biologic mechanisms underlying the female bias in severity among adults with asthma are incompletely understood [36, 50, 53, 95, 98]. Sex hormone levels vary significantly across the lifespan between sexes and within the same sex. This variation in hormone levels could interact with pathogenic asthma genomic mechanisms to cause differences in airway inflammation, airway obstruction, and asthma severity [93]. The importance of the role of sex hormones in asthma has been recently demonstrated in two population-based studies from the US and the UK that revealed significant associations between sex hormones and asthma [37, 38]. Analysis of data from the UK Biobank and NHANES demonstrated that increased free testosterone levels were associated with lower odds of asthma and current wheeze in women and men, lower odds of asthma hospitalizations in women, and higher FEV_1 and FVC in men. Similarly, estradiol levels were associated with reduced odds of current asthma in obese women and nonobese men [37, 38] . These cumulative data suggest that sex hormones play an important role in asthma pathobiology. However, the role of hormone supplementation/modulation in personalized asthma care is yet to be determined.

The Burden of Asthma

In well-characterized studies such as the National Heart, Lung, and Blood Institute (NHLBI) - Severe Asthma Research Program (SARP), the diagnosis of severe asthma has been based on subjective, relatively arbitrary criteria such as those listed in the international European Respiratory Society (ERS) - American Thoracic Society (ATS) guidelines [18]. However, those guideline-based criteria are frequently missing from patient's electronic health records. Hence, severe asthma definition from real-world data is frequently associated with GINA step 4 and 5 therapy [83]. It is very important to identify severe asthma in daily practice to optimize patient care and resource allocation. Although severe asthma (SA) accounts for 10% of all asthma, it is responsible for 50% of its disease-associated costs and burden [76]. Left untreated, severe asthma is associated with significant morbidity and cost to both patients and society, especially in younger boys and adult women [19, 62, 95, 98].

Nevertheless, the burden of asthma cannot be solely correlated with the severe cases of the disease, since the current SA definitions (according to international [ERS-ATS] guidelines) are based on the dose of inhaled corticosteroids and the need for systemic steroids and a second controller medication [18]. These standardized international guidelines fell short of accounting for asthma control parameters such as asthma control tests (ACT). Recent GINA guidelines emphasize the importance of short-acting β_{2-} agonists (SABA) overuse as an indicator of poor asthma control [62] and recommend that patients using three or more canisters of SABA per year should be evaluated for possible treatment escalation and optimization due to the associated increased future risk for exacerbations, hospitalization and/or mortality [32]. Inhaler technique, errors, and adherence were also reported to be associated with worse asthma control [70]. Current guidelines also suggest stepping down asthma treatment once control has been attained and maintained for at least three months in order to minimize costs and potential adverse effects [49]. The burden of asthma may include a wide range of factors that negatively impact patient well-being. For example, patients with asthma may experience a pharmacologic burden related to the risk of complications caused by chronic corticosteroid use or the excessive use of beta-agonists [11]. They may also experience discrimination at work caused by asthma-related disability or face a psychological burden from having asthma or being in disadvantaged financial situations caused by any direct and indirect cost accrued from the price of asthma medications and time away from school and work. It is also important to understand the patient's perception of disease severity, treatment approaches and success, disease control, autonomy, and threats to self-identity [25, 58]. Although important, patient perception and preferences are frequently ignored. This may further amplify the burden felt by the patient [6]. Hence, it is important to recognize the different aspects of the burden of asthma to better understand its impact on individuals with asthma and on society.

Defining asthma remains a constant challenge, and new methods, tools, and scales are needed to recognize, diagnose and correctly classify the disease from its early stages [1]. With healthcare evolving toward a more patient-oriented approach, we should evaluate the burden of asthma from the patient's perspective [47, 58]. The decision-making should be shared between patients and clinicians to most effectively implement an optimal control regimen, a better quality of life, and reduce the asthma-associated health burden.

Asthma and the COVID-19 Pandemic

Over the last three years, the COVID-19 pandemic has caused broad disruption of healthcare services to chronic diseases such as asthma and amplified its burden worldwide. Overall, viral respiratory tract infections, most notably rhinovirus infections, have been associated with asthma exacerbations and poor asthma control [14]. This has also been seen with the previous H1N1 pandemic and with seasonal influenza [42]. Given these concerns, the CDC issued a warning associating moderate to severe and uncontrolled asthma with an increased risk for severe COVID-19 [16]. To further explore such association, we analyzed data from 72,064 individuals (9927 with asthma) infected with SARS-CoV-2 (i.e., the virus responsible for COVID-19) between April 1, 2020, and March 31, 2021. Furthermore, our analyses demonstrated that the association between asthma and COVID-19-related hospitalization and poor outcomes is influenced by asthma severity (where severe asthma is defined by the need for GINA treatment level steps 4 and 5) [97]. Similarly, openSAFELY, a large study of adults with COVID-19 ($n = 17{,}278{,}392$) from England, found that COVID-19-related death ($n = 10{,}926$) was associated with multiple risk factors such as male sex, older age, poverty, obesity, cardiovascular diseases, diabetes, and severe asthma [88]. These findings support the statement issued by the CDC and warrant close monitoring of patients with moderate to severe and uncontrolled asthma infected with SARS-CoV-2, as well as the consideration of early therapeutic interventions such as antiviral therapies or infusion of monoclonal antibodies.

Future Directions: The Role of Digital Health

Progress in reducing the burden of asthma has often been limited, especially in developing countries. Additionally, a greater proportion of patients with asthma fail to achieve asthma control with guideline-based therapies despite appropriate asthma education and effective therapies [66]. Although the reason underlying such outcomes is multifactorial, our current approach to asthma management has been far from perfect. Poor adherence to maintenance controller medication and suboptimal inhaler technique has been frequently associated with poor asthma control. To overcome such important barriers, digital health may play an important role in reducing these impairments. To be effective, digital health interventions for asthma monitoring should be personalized and easy to use in order to foster clinician and patient engagement [57]. Although digital health interventions show substantial promise for asthma disease monitoring and control, it may be restricted to high-income countries which can allocate the necessary resources for such technologies.

Summary

The prevalence of asthma has been increasing since the 1970s. This is of particular concern in poorer parts of the world. Although much has been done to lower the burden of asthma globally, progress has been very slow, especially in low-income countries where the burden of asthma continues to amplify. A multifaceted approach is needed to address all the challenges facing the international community and improve asthma burden globally.

References

1. Aaron SD, Vandemheen KL, FitzGerald JM, Ainslie M, Gupta S, Lemiere C, et al. Reevaluation of diagnosis in adults with physician-diagnosed asthma. JAMA. 2017;317(3):269–79. https://doi.org/10.1001/jama.2016.19627.
2. Akinbami LJ, Moorman JE, Liu X. Asthma prevalence, health care use, and mortality: United States, 2005-2009. Natl Health Stat Report. 2011;(32):1–14.
3. Albertsson-Wikland K, Rosberg S, Lannering B, Dunkel L, Selstam G, Norjavaara E. Twenty-four-hour profiles of luteinizing hormone, follicle-stimulating hormone, testosterone, and estradiol levels: a semilongitudinal study throughout puberty in healthy boys. J Clin Endocrinol Metab. 1997;82(2):541–9. https://doi.org/10.1210/jcem.82.2.3778.
4. Allinson JP, Patel PH, Donaldson GC. Obesity. Insulin Resistance and Asthma: Am J Respir Crit Care Med; 2022. https://doi.org/10.1164/rccm.202207-1271ED.
5. Almqvist C, Worm M, Leynaert B, working group of, G. A. L. E. N. W. P. G. Impact of gender on asthma in childhood and adolescence: a GA2LEN review. Allergy. 2008;63(1):47–57. https://doi.org/10.1111/j.1398-9995.2007.01524.x.
6. Baggott C, Chan A, Hurford S, Fingleton J, Beasley R, Harwood M, et al. Patient preferences for asthma management: a qualitative study. BMJ Open. 2020;10(8):e037491. https://doi.org/10.1136/bmjopen-2020-037491.
7. Bao AM, Liu RY, van Someren EJ, Hofman MA, Cao YX, Zhou JN. Diurnal rhythm of free estradiol during the menstrual cycle. Eur J Endocrinol. 2003;148(2):227–32. https://doi.org/10.1530/eje.0.1480227.
8. Barnes P, FitzGerald G, Brown M, Dollery C. Nocturnal asthma and changes in circulating epinephrine, histamine, and cortisol. N Engl J Med. 1980;303(5):263–7. https://doi.org/10.1056/NEJM198007313030506.
9. Bedirhan-Üstün. The World Health surveys, 2003. Retrieved from file:///C:/Users/zeinj/Downloads/9241562455.pdf. Accessed on 14 Nov 2022.
10. Bel EH, Wenzel SE, Thompson PJ, Prazma CM, Keene ON, Yancey SW, et al. Oral glucocorticoid-sparing effect of mepolizumab in eosinophilic asthma. N Engl J Med. 2014;371(13):1189–97. https://doi.org/10.1056/NEJMoa1403291.
11. Bleecker ER, Menzies-Gow AN, Price DB, Bourdin A, Sweet S, Martin AL, et al. Systematic literature review of systemic corticosteroid use for asthma management. Am J Respir Crit Care Med. 2020;201(3):276–93. https://doi.org/10.1164/rccm.201904-0903SO.
12. Brambilla DJ, Matsumoto AM, Araujo AB, McKinlay JB. The effect of diurnal variation on clinical measurement of serum testosterone and other sex hormone levels in men. J Clin Endocrinol Metab. 2009;94(3):907–13. https://doi.org/10.1210/jc.2008-1902.
13. Burchard EG, Avila PC, Nazario S, Casal J, Torres A, Rodriguez-Santana JR, et al. Lower bronchodilator responsiveness in Puerto Rican than in Mexican subjects with asthma. Am J Respir Crit Care Med. 2004;169(3):386–92. https://doi.org/10.1164/rccm.200309-1293OC.
14. Busse WW, Lemanske RF Jr, Gern JE. Role of viral respiratory infections in asthma and asthma exacerbations. Lancet. 2010;376(9743):826–34. https://doi.org/10.1016/S0140-6736(10)61380-3.
15. Castro M, Corren J, Pavord ID, Maspero J, Wenzel S, Rabe KF, et al. Dupilumab efficacy and safety in moderate-to-severe uncontrolled asthma. N Engl J Med. 2018;378(26):2486–96. https://doi.org/10.1056/NEJMoa1804092.
16. CDC. Risk of severe illness from COVID-19, 2022. Retrieved from https://www.cdc.gov/coronavirus/2019-ncov/need-extra-precautions/asthma.html. Accessed on 1 Nov 2022.
17. Chandler MH, Schuldheisz S, Phillips BA, Muse KN. Premenstrual asthma: the effect of estrogen on symptoms, pulmonary function, and beta 2-receptors. Pharmacotherapy. 1997;17(2):224–34.
18. Chung KF, Wenzel SE, Brozek JL, Bush A, Castro M, Sterk PJ, et al. International ERS/ATS guidelines on definition, evaluation and treatment of severe asthma. Eur Respir J. 2014;43(2):343–73. https://doi.org/10.1183/09031936.00202013.

19. Czira A, Turner M, Martin A, Hinds D, Birch H, Gardiner F, Zhang S. A systematic literature review of burden of illness in adults with uncontrolled moderate/severe asthma. Respir Med. 2022;191:106670. https://doi.org/10.1016/j.rmed.2021.106670.
20. DeBoer MD, Phillips BR, Mauger DT, Zein J, Erzurum SC, Fitzpatrick AM, et al. Effects of endogenous sex hormones on lung function and symptom control in adolescents with asthma. BMC Pulm Med. 2018;18(1):58. https://doi.org/10.1186/s12890-018-0612-x.
21. Denlinger LC, Phillips BR, Ramratnam S, Ross K, Bhakta NR, Cardet JC, et al. Inflammatory and comorbid features of patients with severe asthma and frequent exacerbations. Am J Respir Crit Care Med. 2017;195(3):302–13. https://doi.org/10.1164/rccm.201602-0419OC.
22. Dixon AE, Que LG. Obesity and asthma. Semin Respir Crit Care Med. 2022;43(5):662–74. https://doi.org/10.1055/s-0042-1742384.
23. Dodds L, Armson BA, Alexander S. Use of asthma drugs is less among women pregnant with boys rather than girls. BMJ. 1999;318(7189):1011.
24. Dunn RM, Busse PJ, Wechsler ME. Asthma in the elderly and late-onset adult asthma. Allergy. 2018;73(2):284–94. https://doi.org/10.1111/all.13258.
25. Eassey D, Reddel HK, Ryan K, Smith L. The impact of severe asthma on patients' autonomy: a qualitative study. Health Expect. 2019;22(3):528–36. https://doi.org/10.1111/hex.12879.
26. Fahy JV. Type 2 inflammation in asthma--present in most, absent in many. Nat Rev Immunol. 2015;15(1):57–65. https://doi.org/10.1038/nri3786.
27. Fida NG, Williams MA, Enquobahrie DA. Association of age at Menarche and menstrual characteristics with adult onset asthma among reproductive age women. Reprod Syst Sex Disord. 2012;1(3) https://doi.org/10.4172/2161-038X.1000111.
28. GAR. The global asthma report 2022, 2022. Retrieved from http://globalasthmareport.org/resources/Global_Asthma_Report_2022.pdf. Accessed on 1 Nov 2022.
29. Garcia-Marcos L, Asher MI, Pearce N, Ellwood E, Bissell K, Chiang CY, et al. The burden of asthma, hay fever and eczema in children in 25 countries: GAN phase I study. Eur Respir J. 2022;60(3) https://doi.org/10.1183/13993003.02866-2021.
30. GBD. Global burden of 369 diseases and injuries in 204 countries and territories, 1990-2019: a systematic analysis for the Global Burden of Disease Study 2019. Lancet. 2019;396(10258):1204–22. https://doi.org/10.1016/S0140-6736(20)30925-9.
31. Georas SN, Wright RJ, Ivanova A, Israel E, LaVange LM, Akuthota P, et al. The precision interventions for severe and/or exacerbation-prone (PrecISE) asthma network: an overview of network organization, procedures, and interventions. J Allergy Clin Immunol. 2022;149(2):488–516 e489. https://doi.org/10.1016/j.jaci.2021.10.035.
32. GINA. Global initiative for asthma, 2020. Retrieved from: https://ginasthma.org/wp-content/uploads/2021/05/GINA-Main-Report-2021-V2-WMS.pdf. Accessed 1 Aug 2022.
33. Gonzalez Burchard E, Borrell LN. Need for racial and ethnic diversity in asthma precision medicine. N Engl J Med. 2021;385(24):2297–8. https://doi.org/10.1056/NEJMe2114944.
34. Guo Z, Qin X, Pate CA, Zahran HS, Malilay J. Asthma among adults and children by urban-rural classification scheme, United States, 2016-2018. Public Health Rep. 2022;137(6):1100–6. https://doi.org/10.1177/00333549211047552.
35. Han CH, Chung JH, Kim SR. STROBE: the relationship between asthma and early menarche in Korean adolescents. Medicine (Baltimore). 2022;101(20):e29301. https://doi.org/10.1097/MD.0000000000029301.
36. Han MK, Arteaga-Solis E, Blenis J, Bourjeily G, Clegg DJ, DeMeo D, et al. Female sex and gender in lung/sleep health and disease. Increased understanding of basic biological, pathophysiological, and behavioral mechanisms leading to better health for female patients with lung disease. Am J Respir Crit Care Med. 2018;198(7):850–8. https://doi.org/10.1164/rccm.201801-0168WS.
37. Han YY, Forno E, Celedon JC. Sex steroid hormones and asthma in a nationwide study of U.S. Adults. Am J Respir Crit Care Med. 2020;201(2):158–66. https://doi.org/10.1164/rccm.201905-0996OC.

38. Han YY, Yan Q, Yang G, Chen W, Forno E, Celedon JC. Serum free testosterone and asthma, asthma hospitalisations and lung function in British adults. Thorax. 2020;75(10):849–54. https://doi.org/10.1136/thoraxjnl-2020-214875.
39. Hetzel MR, Clark TJ. Comparison of normal and asthmatic circadian rhythms in peak expiratory flow rate. Thorax. 1980;35(10):732–8. https://doi.org/10.1136/thx.35.10.732.
40. Hoppin JA, Umbach DM, London SJ, Alavanja MC, Sandler DP. Chemical predictors of wheeze among farmer pesticide applicators in the Agricultural Health Study. Am J Respir Crit Care Med. 2002;165(5):683–9. https://doi.org/10.1164/ajrccm.165.5.2106074.
41. Israel E, Reddel HK. Severe and difficult-to-treat asthma in adults. N Engl J Med. 2017;377(10):965–76. https://doi.org/10.1056/NEJMra1608969.
42. Jain S, Kamimoto L, Bramley AM, Schmitz AM, Benoit SR, Louie J, et al. Hospitalized patients with 2009 H1N1 influenza in the United States, April-June 2009. N Engl J Med. 2009;361(20):1935–44. https://doi.org/10.1056/NEJMoa0906695.
43. Jenkins MA, Dharmage SC, Flander LB, Douglass JA, Ugoni AM, Carlin JB, et al. Parity and decreased use of oral contraceptives as predictors of asthma in young women. Clin Exp Allergy. 2006;36(5):609–13. https://doi.org/10.1111/j.1365-2222.2006.02475.x.
44. Johnson O, Gerald LB, Harvey J, Roy G, Hazucha H, Large C, et al. An online weight loss intervention for people with obesity and poorly controlled asthma. J Allergy Clin Immunol Pract. 2022;10(6):1577–1586 e1573. https://doi.org/10.1016/j.jaip.2022.02.040.
45. Kuruvilla ME, Lee FE, Lee GB. Understanding asthma phenotypes, endotypes, and mechanisms of disease. Clin Rev Allergy Immunol. 2019;56(2):219–33. https://doi.org/10.1007/s12016-018-8712-1.
46. Kwon HL, Belanger K, Holford TR, Bracken MB. Effect of fetal sex on airway lability in pregnant women with asthma. Am J Epidemiol. 2006;163(3):217–21. https://doi.org/10.1093/aje/kwj032.
47. Lee LK, Obi E, Paknis B, Kavati A, Chipps B. Asthma control and disease burden in patients with asthma and allergic comorbidities. J Asthma. 2018;55(2):208–19. https://doi.org/10.1080/02770903.2017.1316394.
48. Leynaert B, Sunyer J, Garcia-Esteban R, Svanes C, Jarvis D, Cerveri I, et al. Gender differences in prevalence, diagnosis and incidence of allergic and non-allergic asthma: a population-based cohort. Thorax. 2012;67(7):625–31. https://doi.org/10.1136/thoraxjnl-2011-201249.
49. Lougheed MD, Lemiere C, Ducharme FM, Licskai C, Dell SD, Rowe BH, et al. Canadian Thoracic Society 2012 guideline update: diagnosis and management of asthma in preschoolers, children and adults. Can Respir J. 2012;19(2):127–64. https://doi.org/10.1155/2012/635624.
50. Mahmoud O, Granell R, Tilling K, Minelli C, Garcia-Aymerich J, Holloway JW, et al. Association of height growth in puberty with lung function: a longitudinal study. Am J Respir Crit Care Med. 2018; https://doi.org/10.1164/rccm.201802-0274OC.
51. Mallol J, Crane J, von Mutius E, Odhiambo J, Keil U, Stewart A, Group, I. P. T. S. The International Study of Asthma and Allergies in Childhood (ISAAC) phase three: a global synthesis. Allergol Immunopathol (Madr). 2013;41(2):73–85. https://doi.org/10.1016/j.aller.2012.03.001.
52. McGeachie MJ, Yates KP, Zhou X, Guo F, Sternberg AL, Van Natta ML, et al. Patterns of growth and decline in lung function in persistent childhood asthma. N Engl J Med. 2016;374(19):1842–52. https://doi.org/10.1056/NEJMoa1513737.
53. Mead J. Dysanapsis in normal lungs assessed by the relationship between maximal flow, static recoil, and vital capacity. Am Rev Respir Dis. 1980;121(2):339–42. https://doi.org/10.1164/arrd.1980.121.2.339.
54. Menzies-Gow A, Corren J, Bourdin A, Chupp G, Israel E, Wechsler ME, et al. Tezepelumab in adults and adolescents with severe, uncontrolled asthma. N Engl J Med. 2021;384(19):1800–9. https://doi.org/10.1056/NEJMoa2034975.
55. Moore WC, Meyers DA, Wenzel SE, Teague WG, Li H, Li X, et al. Identification of asthma phenotypes using cluster analysis in the Severe Asthma Research Program. Am J Respir Crit Care Med. 2010;181(4):315–23. https://doi.org/10.1164/rccm.200906-0896OC.

56. Mortimer K, Lesosky M, Garcia-Marcos L, Asher MI, Pearce N, Ellwood E, et al. The burden of asthma, hay fever and eczema in adults in 17 countries: GAN Phase I study. Eur Respir J. 2022;60(3) https://doi.org/10.1183/13993003.02865-2021.
57. Mosnaim G, Safioti G, Brown R, DePietro M, Szefler SJ, Lang DM, et al. Digital health technology in asthma: a comprehensive scoping review. J Allergy Clin Immunol Pract. 2021;9(6):2377–98. https://doi.org/10.1016/j.jaip.2021.02.028.
58. Mowrer JL, Tapp H, Ludden T, Kuhn L, Taylor Y, Courtlandt C, et al. Patients' and providers' perceptions of asthma and asthma care: a qualitative study. J Asthma. 2015;52(9):949–56. https://doi.org/10.3109/02770903.2015.1010731.
59. NACP. 2019. Retrieved from https://www.cdc.gov/asthma/nacp-20-years.htm. Accessed 14 Nov 2022.
60. NHIS. Lifetime asthma prevalence percents by age, United States: National Health Interview Survey, 2018. Available at: https://www.cdc.gov/asthma/nhis/2018/table2-1.htm. Accessed 17 Oct 2022.
61. NHIS. National Health Interview Survery Data 2001–2022, 2022. Retrieved from https://www.cdc.gov/asthma/nhis/default.htm. Accessed on 16 Nov 2022.
62. Nwaru BI, Ekstrom M, Hasvold P, Wiklund F, Telg G, Janson C. Overuse of short-acting beta2-agonists in asthma is associated with increased risk of exacerbation and mortality: a nationwide cohort study of the global SABINA programme. Eur Respir J. 2020;55(4) https://doi.org/10.1183/13993003.01872-2019.
63. O'Connor KA, Ferrell RJ, Brindle E, Shofer J, Holman DJ, Miller RC, et al. Total and unopposed estrogen exposure across stages of the transition to menopause. Cancer Epidemiol Biomark Prev. 2009;18(3):828–36. https://doi.org/10.1158/1055-9965.EPI-08-0996.
64. Ortega HG, Liu MC, Pavord ID, Brusselle GG, FitzGerald JM, Chetta A, et al. Mepolizumab treatment in patients with severe eosinophilic asthma. N Engl J Med. 2014;371(13):1198–207. https://doi.org/10.1056/NEJMoa1403290.
65. Osei-Kumah A, Smith R, Jurisica I, Caniggia I, Clifton VL. Sex-specific differences in placental global gene expression in pregnancies complicated by asthma. Placenta. 2011;32(8):570–8. https://doi.org/10.1016/j.placenta.2011.05.005.
66. Pavord ID, Mathieson N, Scowcroft A, Pedersini R, Isherwood G, Price D. The impact of poor asthma control among asthma patients treated with inhaled corticosteroids plus long-acting beta2-agonists in the United Kingdom: a cross-sectional analysis. NPJ Prim Care Respir Med. 2017;27(1):17. https://doi.org/10.1038/s41533-017-0014-1.
67. Pennington E, Yaqoob ZJ, Al-Kindi SG, Zein J. Trends in asthma mortality in the United States: 1999 to 2015. Am J Respir Crit Care Med. 2019;199(12):1575–7. https://doi.org/10.1164/rccm.201810-1844LE.
68. Peters MC, McGrath KW, Hawkins GA, Hastie AT, Levy BD, Israel E, et al. Plasma interleukin-6 concentrations, metabolic dysfunction, and asthma severity: a cross-sectional analysis of two cohorts. Lancet Respir Med. 2016;4(7):574–84. https://doi.org/10.1016/S2213-2600(16)30048-0.
69. Peters MC, Schiebler M, Cardet JC, Johansson MW, Sorkness R, DeBoer MD, et al. The impact of insulin resistance on loss of lung function and response to treatment in asthma. Am J Respir Crit Care Med. 2022; https://doi.org/10.1164/rccm.202112-2745OC.
70. Roche N, Aggarwal B, Boucot I, Mittal L, Martin A, Chrystyn H. The impact of inhaler technique on clinical outcomes in adolescents and adults with asthma: a systematic review. Respir Med. 2022;202:106949. https://doi.org/10.1016/j.rmed.2022.106949.
71. Rodriguez A, Brickley E, Rodrigues L, Normansell RA, Barreto M, Cooper PJ. Urbanisation and asthma in low-income and middle-income countries: a systematic review of the urban-rural differences in asthma prevalence. Thorax. 2019;74(11):1020–30. https://doi.org/10.1136/thoraxjnl-2018-211793.
72. Salameh P, Waked M, Baldi I, Brochard P, Saleh BA. Respiratory diseases and pesticide exposure: a case-control study in Lebanon. J Epidemiol Community Health. 2006;60(3):256–61. https://doi.org/10.1136/jech.2005.039677.

73. Stein MM, Hrusch CL, Gozdz J, Igartua C, Pivniouk V, Murray SE, et al. Innate immunity and asthma risk in Amish and Hutterite Farm children. N Engl J Med. 2016;375(5):411–21. https://doi.org/10.1056/NEJMoa1508749.
74. Strachan DP, Butland BK, Anderson HR. Incidence and prognosis of asthma and wheezing illness from early childhood to age 33 in a national British cohort. BMJ. 1996;312(7040):1195–9. https://doi.org/10.1136/bmj.312.7040.1195.
75. Subbarao P, Mandhane PJ, Sears MR. Asthma: epidemiology, etiology and risk factors. CMAJ. 2009;181(9):E181–90. https://doi.org/10.1503/cmaj.080612.
76. Sullivan SD, Rasouliyan L, Russo PA, Kamath T, Chipps BE, Group, T. S. Extent, patterns, and burden of uncontrolled disease in severe or difficult-to-treat asthma. Allergy. 2007;62(2):126–33. https://doi.org/10.1111/j.1398-9995.2006.01254.x.
77. Suri R, Macinko J, Inkelas M, Needleman J. The relationship between insurance status and the affordable care act on asthma outcomes among low-income US adults. Chest. 2022;161(6):1465–74. https://doi.org/10.1016/j.chest.2022.01.011.
78. Tan KS, McFarlane LC, Lipworth BJ. Loss of normal cyclical beta 2 adrenoceptor regulation and increased premenstrual responsiveness to adenosine monophosphate in stable female asthmatic patients. Thorax. 1997;52(7):608–11.
79. Teague WG, Phillips BR, Fahy JV, Wenzel SE, Fitzpatrick AM, Moore WC, et al. Baseline features of the Severe Asthma Research Program (SARP III) Cohort: differences with age. J Allergy Clin Immunol Pract. 2018;6(2):545–554 e544. https://doi.org/10.1016/j.jaip.2017.05.032.
80. To T, Stanojevic S, Moores G, Gershon AS, Bateman ED, Cruz AA, Boulet LP. Global asthma prevalence in adults: findings from the cross-sectional world health survey. BMC Public Health. 2012;12:204. https://doi.org/10.1186/1471-2458-12-204.
81. Troisi RJ, Speizer FE, Willett WC, Trichopoulos D, Rosner B. Menopause, postmenopausal estrogen preparations, and the risk of adult-onset asthma. A prospective cohort study. Am J Respir Crit Care Med. 1995;152(4 Pt 1):1183–8. https://doi.org/10.1164/ajrccm.152.4.7551368.
82. Turrin M, Rizzo M, Bonato M, Bazzan E, Cosio MG, Semenzato U, et al. Differences between early- and late-onset asthma: role of comorbidities in symptom control. J Allergy Clin Immunol Pract. 2022; https://doi.org/10.1016/j.jaip.2022.08.007.
83. Wang E, Wechsler ME, Tran TN, Heaney LG, Jones RC, Menzies-Gow AN, et al. Characterization of severe asthma worldwide: data from the International Severe Asthma Registry. Chest. 2020;157(4):790–804. https://doi.org/10.1016/j.chest.2019.10.053.
84. Wechsler ME, Castro M, Lehman E, Chinchilli VM, Sutherland ER, Denlinger L, et al. Impact of race on asthma treatment failures in the asthma clinical research network. Am J Respir Crit Care Med. 2011;184(11):1247–53. https://doi.org/10.1164/rccm.201103-0514OC.
85. Wenzel S. Severe asthma: from characteristics to phenotypes to endotypes. Clin Exp Allergy. 2012;42(5):650–8. https://doi.org/10.1111/j.1365-2222.2011.03929.x.
86. Wenzel SE, Schwartz LB, Langmack EL, Halliday JL, Trudeau JB, Gibbs RL, Chu HW. Evidence that severe asthma can be divided pathologically into two inflammatory subtypes with distinct physiologic and clinical characteristics. Am J Respir Crit Care Med. 1999;160(3):1001–8. https://doi.org/10.1164/ajrccm.160.3.9812110.
87. WHO. World Health Organization: asthma, 2022. Retrieved from: https://www.who.int/newsroom/fact-sheets/detail/asthma. Accessed on 7 Oct 2022.
88. Williamson EJ, Walker AJ, Bhaskaran K, Bacon S, Bates C, Morton CE, et al. Factors associated with COVID-19-related death using OpenSAFELY. Nature. 2020;584(7821):430–6. https://doi.org/10.1038/s41586-020-2521-4.
89. Wohlford EM, Huang PF, Elhawary JR, Millette LA, Contreras MG, Witonsky J, et al. Racial/ethnic differences in eligibility for asthma biologics among pediatric populations. J Allergy Clin Immunol. 2021;148(5):1324–1331 e1312. https://doi.org/10.1016/j.jaci.2021.09.005.
90. Wong GW, Brunekreef B, Ellwood P, Anderson HR, Asher MI, Crane J, et al. Cooking fuels and prevalence of asthma: a global analysis of phase three of the International Study of Asthma and Allergies in Childhood (ISAAC). Lancet Respir Med. 2013;1(5):386–94. https://doi.org/10.1016/S2213-2600(13)70073-0.

91. Wright AL, Stern DA, Kauffmann F, Martinez FD. Factors influencing gender differences in the diagnosis and treatment of asthma in childhood: the Tucson Children's Respiratory Study. Pediatr Pulmonol. 2006;41(4):318–25. https://doi.org/10.1002/ppul.20373.
92. Zein J, Gaston B. Sex differences in the pediatric lung: implications for lung disease. In: Tigno X, Silveyra P, editors. Sex-based differences in lung physiology. 1st ed. New York: Springer; 2021.
93. Zein JG, Denson JL, Wechsler ME. Asthma over the adult life course: gender and hormonal influences. Clin Chest Med. 2019;40(1):149–61. https://doi.org/10.1016/j.ccm.2018.10.009.
94. Zein JG, Dweik RA, Comhair SA, Bleecker ER, Moore WC, Peters SP, et al. Asthma is more severe in older adults. PLoS One. 2015;10(7):e0133490. https://doi.org/10.1371/journal.pone.0133490.
95. Zein JG, Erzurum SC. Asthma is different in women. Curr Allergy Asthma Rep. 2015;15(6):28. https://doi.org/10.1007/s11882-015-0528-y.
96. Zein JG, McManus JM, Sharifi N, Erzurum SC, Marozkina N, Lahm T, et al. Benefits of airway androgen receptor expression in human asthma. Am J Respir Crit Care Med. 2021;204(3):285–93. https://doi.org/10.1164/rccm.202009-3720OC.
97. Zein JG, Mitri J, Bell JM, Lopez D, Strauss R, Attaway AH. The relationship of asthma severity to COVID-19 outcomes. J Allergy Clin Immunol Pract. 2021; https://doi.org/10.1016/j.jaip.2021.10.041.
98. Zein JG, Udeh BL, Teague WG, Koroukian SM, Schlitz NK, Bleecker ER, et al. Impact of age and sex on outcomes and hospital cost of acute asthma in the United States, 2011-2012. PLoS One. 2016;11(6):e0157301. https://doi.org/10.1371/journal.pone.0157301.
99. Zhang P, Lopez R, Attaway AH, Georas SN, Khatri SB, Abi-Saleh S, Zein JG. Diabetes Mellitus is associated with worse outcome in patients hospitalized for asthma. J Allergy Clin Immunol Pract. 2020; https://doi.org/10.1016/j.jaip.2020.10.054.

The Role of Climate Change in Asthma

Anna Goshua, Vanitha Sampath, Jo Ann Efobi, and Kari Nadeau

Abstract

Human activity and increased use of fossil fuels have led to climate change. These changes are adversely affecting human health, including increasing the risk of developing asthma. Global temperatures are predicted to increase in the future. In 2019, asthma affected an estimated 262 million people and caused 455,000 deaths. These rates are expected to increase. Climate change by intensifying climate events such as drought, flooding, wildfires, sand storms, and thunderstorms has led to increases in air pollution, pollen season length, pollen and mold concentration, and allergenicity of pollen. These effects bear implications for the onset, exacerbation, and management of childhood asthma and are increasing health inequities. Global efforts to mitigate the effects of climate change are urgently needed with the goal of limiting global warming to between 1.5 and 2.0 °C of preindustrial times as per the 2015 Paris Agreement. Clinicians need to take an active role in these efforts in order to prevent further increases in asthma prevalence. There is a role for clinician advocacy in both the clinical setting as well as in local, regional, and national settings to install measures to

A. Goshua
Stanford School of Medicine, Stanford, CA, USA

V. Sampath · J. A. Efobi
Sean N. Parker Center for Allergy and Asthma Research at Stanford University, Palo Alto, CA, USA

K. Nadeau (✉)
Sean N. Parker Center for Allergy and Asthma Research at Stanford University, Palo Alto, CA, USA

Division of Pulmonary, Allergy, and Critical Care Medicine, Stanford University, Stanford, CA, USA

A. R. Brasier, N. N. Jarjour (eds.), *Precision Approaches to Heterogeneity in Asthma*, Advances in Experimental Medicine and Biology 1426,
https://doi.org/10.1007/978-3-031-32259-4_2

control and curb the escalating disease burden of childhood asthma in the setting of climate change.

Keywords

Climate Change · Air Pollution · Asthma · Wildfires · Pollen · Mold · Heterogeneity

Introduction

The increased prevalence of asthma in the last few decades [1] is associated with increased human activity and the burning of fossil fuels to meet the ever-growing energy needs of modern living. Since 1900, the Earth's surface temperature has risen 1 °C; approximately half of that warming has occurred since the 1970s [2]. The rate of warming has accelerated from 0.08 °C per decade in the 1880s to 0.18 °C since 1981. Increased fossil fuel use has led to increased release of heat-trapping greenhouse gases (GHGs) such as carbon dioxide (CO_2), methane, nitrous oxides (NO_x), ozone (O_3), and fluorinated gases. Further, deforestation and reduction of grasslands have decreased the Earth's ability to sequester carbon through absorption by the oceans and via plants through photosynthesis. Oceans absorb CO_2 and reduce atmospheric CO_2; however, over time, this has led to decreases in the ocean's pH and its ability to absorb CO_2. The acidity of the ocean today, on average, is about 25% greater than it was during preindustrial times [3]. The increased release of CO_2 and the decreased ability of the Earth to sequester it has led to increased concentrations of CO_2 in the atmosphere [4]. The concentration of atmospheric CO_2 has risen from about 280 parts per million (ppm) preindustrial revolution [5]—a number which remained stable for approximately 6000 years—to 421 ppm as of 2022 [6] (see Fig. 1).

GHGs and global temperatures are projected to increase further. In the best case scenario, global temperatures are projected to increase between 1.5 and 4.4 °C above preindustrial levels by 2100 [7]. These increases are based on predicted emission levels and take into account sustainability measures and socioeconomic factors [8]. Climate scientists warn that exceeding 1.5 °C will result in widespread catastrophic consequences to planetary health, which includes human health [9, 10].

Increases in global temperatures have led to climate change with increases in incidence and severity of climate events such as drought, flooding, heat waves, thunderstorms, sandstorms, and wildfires leading to increases in sea level rise, melting of glaciers, food and water scarcity, loss of biodiversity, and displacement and migration. Climate change adversely affects human health leading to heat strokes and other heat-related illnesses, respiratory diseases, waterborne diseases, vector-borne diseases, and mental health. Secondary effects of climate change that affect asthma include increases in air pollution (from wildfires and sandstorms) and increases in mold and allergenic pollen concentrations (global warming, thunderstorms, flooding, and alterations in weather patterns) (Fig. 2).

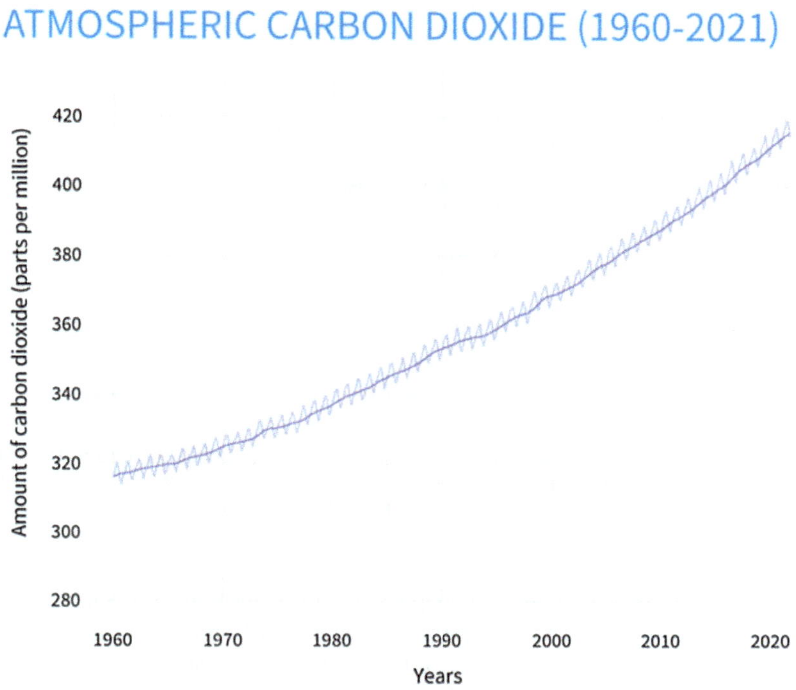

Fig. 1 This image demonstrates increases in carbon dioxide emissions since the Industrial Revolution, which has accelerated the Earth's warming. (Reproduced from https://www.climate.gov/news-features/understanding-climate/climate-change-atmospheric-carbon-dioxide)

Globally, asthma affected an estimated 262 million people in 2019 and caused 455,000 deaths [11]. Asthma is the most common chronic pediatric disease affecting about 7.8% of American children [12]. The prevalence varies by geography and other factors. A study of asthma prevalence in 70 countries found that prevalence varied by as much as 21-fold, with an average of 4.3% for doctor-diagnosed asthma and 4.5% for clinical/treated asthma [13]. Minoritized individuals, women, and those of low socioeconomic status are at higher risk of asthma [14].

Climate change is a health crisis, and health organizations are voicing their concern. The American Academy of Pediatrics, in their 2015 "Global Climate Change and Children's Health" policy statement, states that "climate change poses threats to human health, safety, and security. Children are uniquely vulnerable to these threats" [15]. The United Nations state that, "Climate Change is the defining issue of our time and we are at a defining moment" [16]. As global temperatures are expected to increase in the future, asthma prevalence as well as the severity of asthma exacerbations, are expected to increase [17].

In this chapter, we will discuss climate change and its effects on air pollution, pollen, and molds and their role in increasing risk and increasing severity of asthma exacerbations; the disproportionate impact of climate change on asthma burden in

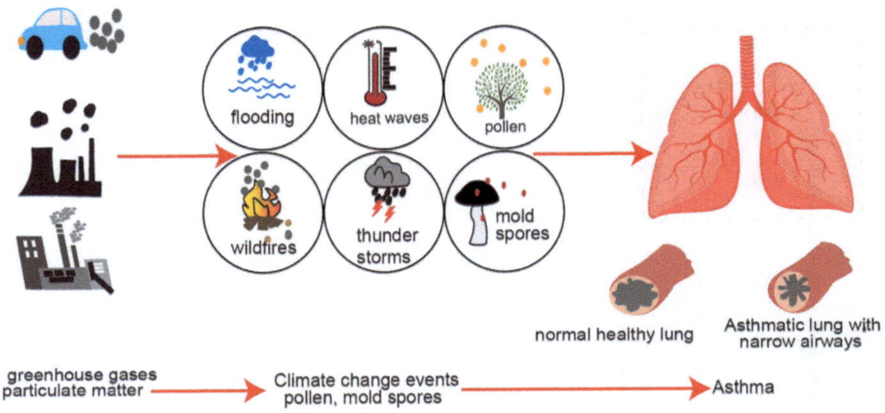

Fig. 2 Increased greenhouse gases have led to global warming and increased severity and frequency of climate change events (flooding, heat waves, wildfires, thunderstorms). Climate change has also led to increases in mold spores and pollen allergens, which increase the prevalence of allergic diseases

children in underserved areas; and adaptive and mitigation measures that clinicians can implement clinically and champion for change locally, regionally, and nationally.

Evidence also suggests that some climate change events such as heat and air pollution can act synergistically [18]. About 850 million children live in areas around the world where four or more climate change and environmental hazards, such as wildfires, flooding, air pollution, and resource scarcity, interact [19]. A 2009 study found that adolescent emergency room visits at similar levels of air pollution were 9.45% higher in the summer than in the winter [20]. Air pollution and extreme heat exposure during pregnancy are linked to increased risk of preterm delivery (<37 weeks) and small-for-gestational-age infants, especially in mothers who themselves have asthma [21]. Mitigation and adaptation strategies are needed to deal with future increases in asthma prevalence.

Air Pollution

Air pollution is strongly associated with asthma as well as other respiratory diseases. Climate change affects air quality in numerous ways. Increasing global temperatures and shifting wind patterns contribute to the proliferation of air pollutants like ground-level O_3 which has been shown to increase airway inflammation and can lead to increased asthma-related emergency room visits and hospitalizations [22]. Climate change is also affecting extreme weather events, such as the duration and intensity of wildfire seasons [23, 24] and sand and dust storms, which are both major contributors to air pollution [25]. In addition to air pollution caused by climate change-associated wildfires and sand and dust storms, pollutants released by human activity, such as urbanization, mining, deforestation, modern agricultural practices, and the manufacture of synthetic products for human use, are increasing the number and concentrations of air pollutants and further exacerbating asthma incidence and severity. Air pollution is a global problem.

The majority of the global population (99%) [26] live in areas where levels of air pollutants exceed World Health Organization (WHO) Global Air Quality Guidelines recommendations (see Fig. 3) [27]. Poor air quality is associated with both asthma development and worse asthma control, including lost school days and increased rates of emergency room visits and hospitalizations for acute exacerbations [28]. Infants, children, and young adolescents are particularly vulnerable to the effects of air pollution as they are still developing and growing rapidly and spend more time outdoors. Risks persist after birth and throughout childhood due to the unique anatomy and physiology of children. They breathe more quickly and inhale more air relative to their body weight than adults, which renders them more susceptible to the impacts of poor air quality [29]. Moreover, they have narrower airways prone to inflammation, obstruction, and collapse. Their immature immune systems are less able to fend off infections. The role of air pollutants in mediating asthma by skewing T naive cells toward a Th2-type allergic response is well characterized. During a Th2 allergic response, proinflammatory factors such as histamine, leukotrienes, and prostaglandins are released by mast cells and basophils. However, in addition to the classical Th2 response, other cell subsets that have been implicated in asthma pathogenesis include innate lymphoid cells type 2 (ILC2) and other T cells subsets, such as helper 17 (TH17), T follicular helper (Tfh), regulatory T (Treg), and gamma delta T (γδT) cells have been shown to mediate asthma. The allergic mechanisms are triggered through transmembrane receptors such as pattern recognition receptors (PRRs), toll like receptors (TLRs) as well as innate immune receptors such as NOD-like receptors (NLR), which recognize air pollutants, allergens, and bacterial and viral infections. O_3 mediates lung inflammation via reactive oxygen species. NLRP3 and TLRs have been shown to be activated by both O_3 and particulate matter (PM) [30, 31].

Environmental hazards to children's health begin in utero. Cross-sectional data from the National Survey of Children's Health (NSCH) collected in 2011 to 2012 found that children who were born preterm were 1.64 times more likely to develop asthma compared with those who were born term [32]. A 2011 study found that, in addition to increased asthma diagnoses, preterm infants may end up requiring more health services to manage their asthma, including increased corticosteroid use and emergency room visits [33]. Analysis of data from 3687 participants of the prospective Dutch PIAMA (Prevention and Incidence of Asthma and Mite Allergy) birth cohort found a higher incidence of asthma until the age of 20 years with higher exposure of nitrogen dioxide (a greenhouse gas), and particulate matter lesser than 10 μm [34].

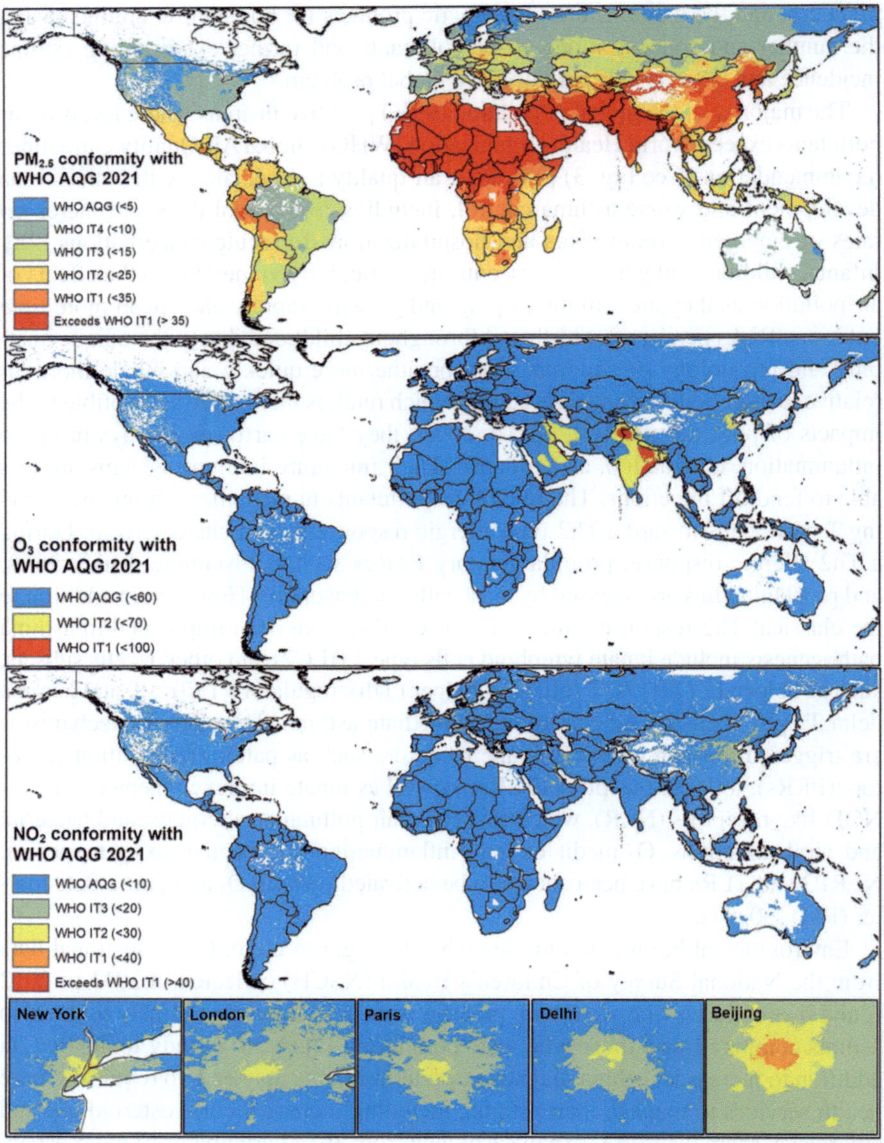

Fig. 3 Conformation of key pollutants worldwide with WHO 2021 Air Quality Guidelines. (Reproduced from Amini [100])

Wildfires

Wildfires are steadily causing more environmental damage. For example, only half-way through Europe's 2022 fire season, 600,731 hectares of land have been burned—the second-largest burn area on record [35]. By destroying trees and other vegetation, wildfires hamper the environment's capacity to absorb CO_2. Wildfires are themselves a significant source of carbon emissions. In 2021, 1.76 billion tons of carbon were released worldwide as a consequence of wildfires [36].

Wildfire smoke is a potent source of numerous air pollutants dangerous to human health, such as particulate matter (PM), carbon monoxide, methane, NO_x, formaldehyde, acrolein, polycyclic aromatic hydrocarbons, trace minerals, and various other compounds. PM with an aerodynamic diameter of 2.5 μm or less ($PM_{2.5}$) are known to be especially harmful because they have a large surface area for absorption but are small enough to penetrate deep into the lungs without being cleared by the mucociliary elevator, deposit into the terminal bronchioles and alveoli, damage alveolar macrophages, and even enter the bloodstream [37]. Accumulated $PM_{2.5}$ can lead to airway inflammation through oxidative stress and the release of proinflammatory epidermal growth factor receptor ligands, such as amphiregulin, transforming growth factor-alpha (TGFα) and -beta (TGFβ), that lead to the thickening of the bronchial walls and tissue fibrosis [38]. Even short-term exposure of 2 weeks or less to $PM_{2.5}$ has been linked to asthma exacerbation [39]. Studies indicate that wildfire smoke has more adverse health effects than ambient air pollution found in urban settings. This is thought to be due to the fact that wildfire smoke is more carbonaceous with more organic polar compounds, which have a greater potential to cause inflammation and oxidative stress in the lungs [40]. In addition, in wildfires that encompass structures, additional toxins from burnt buildings are present [41].

NO_x and certain organic compounds from wildfires can also react to form ground-level O_3. While elevated carbon monoxide levels tend to be restricted to the burned area, $PM_{2.5}$ and O_3 can spread over hundreds and even thousands of miles [42, 43]. The 2021 fires that burned large swathes of the West Coast resulted in air quality warnings as far east as Philadelphia and New York. Additionally, researchers are discovering that wildfire smoke may grow more toxic over time as it travels through the atmosphere and gets further oxidized into free radicals [44].

In the United States, approximately 7.4 million children are exposed to wildfire smoke every year, with some requiring hospitalization [45]. Over the course of the 2003 California wildfires, asthma-related hospital admissions increased by 25% during the fires and 56% in the aftermath [46].

Sand and Dust Storms

Much like wildfires, sand and dust storms also require three instigating conditions: wind, a dust source with minimal vegetation, and dry conditions [47]. Sand particles are defined as those 60 μm or larger, and dust particles are those that are smaller than 60 μm in diameter [48]. Climate change promotes droughts, desertification, changes in wind intensity, and vegetation degradation. These effects prime the environment for dust storms, especially in regions like the Middle East, where temperatures are increasing at twice the global average [49].

The primary air pollutant of concern produced by sandstorms is fine particulate matter ($PM_{2.5}$) [50]. These particles contain a variety of minerals and their oxides, organic content, pathogens, allergens (dust mites, pollen, and fungal spores), and anthropogenic pollutants in its matrix or on its surface. Some of the regions with the highest prevalence of asthma are found in regions of the Middle East with a high frequency of sand and dust storms. In Kuwait, for example, where dust storms blow through the region for almost a fifth of the year, 22.4% of children have asthma compared to 7.8% of American children [47, 51]. Other studies have shown that dust storm events are associated with an increased rate of hospital admissions for children with asthma [52, 53].

Global Warming, Flooding, and Thunderstorms

As described previously, climate change is influencing rainfall patterns and wind intensity. The warming of the Earth's surface causes more water to evaporate from the land and ocean, increasing moisture content in the air [54]. As such, in addition to stoking wildfires, climate change also contributes to the generation of powerful storm systems like hurricanes and cyclones. A 2020 analysis showed that the likelihood of a tropical storm developing into a Category 3 or higher hurricane, which involves winds of 110 miles per hour or higher, is increasing by 8% per decade. Moreover, ocean level rise worsens storm surges and flooding [55]. Storms also cause a myriad of indirect effects on children with asthma [56]. Disruptions to health care infrastructure disconnect children with asthma from critical health care services, such as prescription refills, to manage their asthma and acute care for severe exacerbations. In addition, the seasonality of aeroallergens like pollens and molds is influenced by changing temperatures.

Mold Spores

Molds are microscopic fungi characterized by the presence of multicellular filaments called hyphae. Multiple studies suggest that exposure to molds such as *Alternaria*, *Aspergillus*, *Cladosporium*, and *Penicillium* may contribute to the development of asthma [57]. Many molds contain proteases, which increase the permeability of the epithelial barrier and initiate a proinflammatory cascade that accelerate the onset the IgE-mediated sensitization and allergy [58]. Further, components of molds such as glucans, chitins, and proteases can also activate innate immune responses directly via pattern recognition receptors and pattern-associated molecular patterns [59].

Molds are aeroallergens that affect 5–10% of households in cold climates and 10–30% in moderate and warm climates [60]. They release spores that, via inhalation, can induce airway irritation and inflammation.

Flooding encourages the formation of mold. After Hurricane Katrina and Rita, high concentrations of molds were found; those homes with greater flood damage had higher levels of mold growth compared with homes with little or no flooding. Indoor mold exposure has previously been linked to asthma onset, exacerbations, and poorer asthma control in children [61]. Using data from the cross-sectional National Surveys of Children's Health (2017–2018) found that among 41,423 US children, the prevalence of asthma in children who were exposed to homes with mold (10.8%) was greater than those in homes without mold (7.2%) [62].

Pollen

Pollen exposure can precipitate asthma attacks. Pollen season has been lengthening all over the world over the past several decades, with many plants releasing pollen earlier in the season [63]. An analysis of pollen data from 1990 through 2018 found that pollen season starts 20 days earlier and features 21% more pollen volume [64]. A cross-sectional study of 29,257 patients with asthma found that the very early onset of spring was associated with an increased risk of asthma hospitalization [65]. This may, in part, be due to increasing global temperatures causing plants to alter their pollination season as an adaptive response to later frosts and earlier thaws [66]. Moreover, warmer temperatures and high CO_2 concentrations are conducive to more vigorous pollen growth and increased pollen production. For instance, a laboratory study of timothy grass found that the grass grown in an environment containing 800 ppm carbon dioxide produced almost twice as much pollen as grass grown in the presence of 400 ppm carbon dioxide [67]. The warmer temperatures are also enabling the geographic distribution of some allergens like ragweed to shift northward [68]. Projections show that asthma-related emergency room visits may increase by 8–14% by 2090, depending on global emissions [66].

Another factor that can trigger asthma during pollen season is thunderstorms, which can sweep up pollen and spores and fracture them into smaller fragments via electrical shock, enabling them to penetrate deeper into the lungs [69]. This phenomenon, known as thunderstorm asthma, tends to occur during the spring and summer periods when levels of aeroallergens are already high.

One of the most severe thunderstorm asthma events occurred in 2016 in Melbourne, Australia, overwhelming hospitals [70]. Individuals affected found TA to be associated with allergic rhinitis, ryegrass pollen sensitization, preexisting asthma, poor adherence to inhaled corticosteroid preventer therapy, hospital admission for asthma in the previous year, and outdoor location at the time of the storm. Individuals sensitized to allergens have a more robust Th2 response. Patients without a prior history of asthma were also affected.

Further, it has been suggested that the combination of high precipitation and low temperatures on thunderstorm days contributes to asthma attacks among people with asthma [71]. TA is relatively rare, with only 29 reported events since 1983; however, these events are likely to increase in frequency with global warming [72].

Health Disparities

Although the effects of climate change threaten the well-being of all populations, some are and will continue to be disproportionately affected. Most studies from North America, New Zealand, Asia, and Africa have shown that low socioeconomic status (SES) communities face higher concentrations of criteria air pollutants. The European research, on the other hand, is quite mixed [73]. Other studies have shown that children residing in low- and middle-income countries (LMICs) where air quality has continued to deteriorate in the setting of increased urbanization and economic development are disproportionately affected by increased asthma morbidity and mortality [74]. Additional challenges in LMICs include a lack of access to medications and health care services and health care infrastructure that is more susceptible to disruption from intensifying extreme weather events [75].

Disparities also persist in the United States along the lines of income, race, and ethnicity. Children who live just above, at, or below the poverty line have been shown to be more likely to have increased asthma morbidity [76]. A history of redlining has resulted in neighborhoods that are inherently less climate-resilient on account of heat-trapping asphalt, lack of greenery, and high-density buildings situated close to highways [77]. 71% of historically redlined neighborhoods are now considered low- or middle-income areas, and 64% are predominantly populated by Black and Hispanic people [78]. Rates of asthma are over twice as high among Black children compared to non-Hispanic White children [79]. Asthma-related mortality is also highest in Black children at 10.7% relative to 7.3% in Hispanics and 1.4% in non-Hispanic Whites [12]. In addition to environmental exposures throughout childhood that may account for some of these differences, the risks for Black children begin at birth—in 2020, the preterm birth rate among Black mothers was 14.4% compared to 9.1% and 9.8% among non-Hispanic White and Hispanic mothers [80]. Environmental factors such as air pollution and extreme heat can increase the rate of preterm birth [81, 82]. The earlier the preterm birth, the more immature a child's lungs are and the more susceptible they are to developing asthma and other respiratory problems.

Additionally, minority communities are 53–58% more likely to reside in neighborhoods with the highest predicted increases in asthma-related emergency room visits, per the Environmental Protection Agency's 2021 Climate Change and Social Vulnerability report [83]. Minority communities are also more likely to reside in neighborhoods disproportionately impacted by ozone and particulate matter pollution, extreme heat, and flooding.

Modeling projections at a 2 °C warming trajectory by 2100 show that individuals with low income and/or no high school diploma are 15% more likely to be living in areas projected to undergo the largest increases in childhood asthma diagnoses due to climate-related PM pollution [83].

Adaptation and Mitigation

All projections indicate continued warming of the Earth; however, the extent of the increase depends on global mitigation and adaptation efforts. Adaptation refers to making changes that help society adjust to the current and future effects of climate change, while mitigation involves tackling the root causes of these changes [84]. Children living in nonclimate-resilient communities will be at the highest risk of morbidity and mortality from asthma. A comprehensive asthma action plan must take into account disparities related to income, race, environment, neighborhood, and ethnicity to identify and address the needs of the most vulnerable children.

It is important that nations, organizations, and individuals take action to mitigate climate change to prevent risk of asthma. Clinicians can take action on several fronts to help their patients and to advocate for their broader communities.

Clinical Considerations

Clinicians must counsel patients and their families on risk-reduction strategies. Although the specifics of climate change risks vary depending on the region, there is information and best practice principles with which all pediatric clinicians should be familiar. In 2021, the American Board of Pediatrics launched a module entitled "Impact of Climate Change on Pediatric Health Care" to help clinicians familiarize themselves with the risks that climate change poses to children's health. The American College of Gynecologists has issued updated guidance with regard to the reproductive effects of environmental exposures like air pollution for clinicians to reference [85].

Screening for environmental risk factors can be integrated into the flow of the well-child visit. Patients and families should be counseled on both indoor and outdoor environmental exposures that can trigger asthma and provided with anticipatory guidance. They should be educated on safety measures that they can take during a wildfire such as appropriate use of safety masks, air filters, and air purifiers and taught to reference the Air Quality Index that indicates how safe the air outside is to breathe [86, 87]. There are also digital resources that enable patients to track pollen status across the country [88].

Clinicians can also counsel patients about the inhalers that they are using. Metered-dose inhalers are a potent source of hydrofluorocarbons, which are greenhouse gases that are thousands of times more potent than carbon dioxide [89]. Some inhalers are more environmentally friendly than others. For instance, a European study found that switching to dry-powder inhalers from a metered-dose inhaler resulted in a 422 kg decrease in a patient's annual carbon footprint [90]. Another study found that patients who switched to a dry-powder inhaler halved their carbon footprint without any changes to their asthma control [91].

Optimizing asthma control is key for both the patient's well-being and for reducing greenhouse gas emissions. A study based in the United Kingdom found that excess GHGs emissions per capita were eight times higher for children over the age of 12 with uncontrolled asthma as a result of increased inhaler use and health service utilization than children with controlled asthma [92].

Population Health

Clinicians can advocate for climate-friendly policy changes on local, regional, and national levels. To start, it is important to recognize that clinicians operate within systems that help to perpetuate climate change. The American health system is responsible for 8.5% of national carbon emissions and a quarter of all health-related emissions globally [93]. Clinicians can advocate for the decarbonization of their health systems, push for the use of renewable energy sources, and for retro-fitting old buildings to render them more energy efficient [94]. Moreover, they can encourage the implementation of environmentally friendly practices into their workspaces, such as recycling programs for emergency and operating rooms [95].

On a state and federal level, clinicians can look to existing policies and initiatives that have been implemented and positively impacted communities across the nation. For example, the Regional Greenhouse Gas Initiative in the northeastern United States, designed to reduce greenhouse gas emissions from the electrical power sector, prevented 537 new cases of asthma, 16,310 respiratory illnesses in children, and 112 preterm births from 2009 through 2014 [96].

Given the intersection of climate change with health disparities, there is also a role for clinicians to advocate for expanding social programs that help connect patients with housing, water, sanitation, health care, and other critical services. Climate change is making it difficult for people, especially those from low-income communities, to access these resources. For example, extreme weather events can destroy housing and important health service infrastructure and contaminate water systems [97].

Another possible area of intervention is spearheading initiatives to educate medical students, resident physicians, and attendings about the impacts of climate change on pediatric health.

Conclusion

Climate change is increasing the number and intensity of wildfires and heat waves, lengthening the fire, flooding, and pollen seasons, and potentiating aeroallergen production and distribution. All of these changes bear implications for the onset, exacerbation, and management of childhood asthma as well as disparities in health outcomes among children.

These effects will only escalate if nations fail to curb emissions to set us on track to meet a warming target of 1.5 °C by 2100. Mitigation and adaptation measures are needed to prepare for projected increases in asthma. However, A 2014 Global Asthma Network survey of 112 countries found that only 23% of nations had a national asthma plan in place for children to promote asthma education, early asthma detection, and access to timely treatment [98]. These countries that have implemented asthma plans observe a significant decrease in asthma-related morbidity and mortality. A program in Finland that focused on early diagnosis, active anti-inflammatory treatment, and networking between primary care providers and pharmacists reduced health costs by over 50% [99].

Clinicians should take action in their workplaces and advocate to ensure that their local, regional, and national communities have measures in place to prepare for and mitigate the rising disease burden of childhood asthma.

References

1. Dharmage SC, Perret JL, Custovic A. Epidemiology of asthma in children and adults. Front Pediatr. 2019;7:246.
2. Climate.gov. Climate change: global temperature. 2022. Available from: https://www.climate.gov/news-features/understanding-climate/climate-change-global-temperature.
3. EPA. Understanding the science of ocean and coastal acidification. Available from: https://www.epa.gov/ocean-acidification/understanding-science-ocean-and-coastal-acidification.
4. Ahmed F, Ali I, Kousar S, Ahmed S. The environmental impact of industrialization and foreign direct investment: empirical evidence from Asia-Pacific region. Environ Sci Pollut Res. 2022;29(20):29778–92.
5. Monastersky R. Global carbon dioxide levels near worrisome milestone. Nature. 2013;497(7447):13–4.
6. National Oceanic and Atmospheric Administration. Carbon dioxide now more than 50% higher than pre-industrial levels. 2022. Available from: https://www.noaa.gov/news-release/carbon-dioxide-now-more-than-50-higher-than-pre-industrial-levels#:~:text=Carbon%20dioxide%20measured%20at%20NOAA's,of%20California%20San%20Diego%20announced.
7. Januta A. Explainer: the U.N. climate report's five futures – decoded. 2021. Available from: https://www.reuters.com/business/environment/un-climate-reports-five-futures-decoded-2021-08-09/.
8. Agache I, Sampath V, Aguilera J, Akdis CA, Akdis M, Barry M, Bouagnon A, Chinthrajah S, Collins W, Dulitzki C, Erny B, Gomez J, Goshua A, Jutel M, Kizer KW, Kline O, LaBeaud AD, Pali-Schöll I, Perrett KP, Peters RL, Plaza MP, Prunicki M, Sack T, Salas RN, Sindher SB, Sokolow SH, Thiel C, Veidis E, Wray BD, Traidl-Hoffmann C, Witt C, Nadeau KC. Climate change and global health: a call to more research and more action. Allergy. 2022;77(5):1389–407.
9. NASA Global Climate Change. A degree of concern: why global temperatures matter. 2019. Available from: https://climate.nasa.gov/news/2865/a-degree-of-concern-why-global-temperatures-matter/.
10. Climate Action Tracker. Temperatures. Available from: https://climateactiontracker.org/global/temperatures/.
11. GBD 2019 Diseases and Injuries Collaborators. Global burden of 369 diseases and injuries in 204 countries and territories, 1990–2019: a systematic analysis for the Global Burden of Disease Study 2019. Lancet. 2020;396(10258):1204–22.
12. CDC. Most recent national asthma data. 2022. Available from: https://www.cdc.gov/asthma/most_recent_national_asthma_data.htm.
13. Asher MI, García-Marcos L, Pearce NE, Strachan DP. Trends in worldwide asthma prevalence. Eur Respir J. 2020;56(6):2002094.
14. Enilari O, Sinha S. The global impact of asthma in adult populations. Ann Glob Health. 2019;85(1):2.

15. Council on Environmental Health. Global climate change and children's health. Pediatrics. 2015;136(5):992–7.
16. United Nations. 2022. Available from: https://www.un.org/en/global-issues/climate-change.
17. Rorie A, Poole JA. The role of extreme weather and climate-related events on asthma outcomes. Immunol Allergy Clin N Am. 2021;41(1):73–84.
18. Grigorieva E, Lukyanets A. Combined effect of hot weather and outdoor air pollution on respiratory health: literature review. Atmosphere. 2021;12(6):790.
19. UNICEF. The climate crisis is a child rights crisis. 2021. Available from: https://www.unicef.org/reports/climate-crisis-child-rights-crisis.
20. Mohr LB, Luo S, Mathias E, Tobing R, Homan S, Sterling D. Influence of season and temperature on the relationship of elemental carbon air pollution to pediatric asthma emergency room visits. J Asthma. 2008;45(10):936–43.
21. Kemppainen M, Lahesmaa-Korpinen AM, Kauppi P, Virtanen M, Virtanen SM, Karikoski R, Gissler M, Kirjavainen T. Maternal asthma is associated with increased risk of perinatal mortality. PLoS One. 2018;13(5):e0197593.
22. EPA. Health effects of ozone in patients with asthma and other chronic respiratory disease. 2022. Available from: https://www.epa.gov/ozone-pollution-and-your-patients-health/health-effects-ozone-patients-asthma-and-other-chronic#:~:text=In%20these%20studies%2C%20ozone%20has,likely%20to%20indicate%20worsening%20asthma.
23. Rossiello MR, Szema A. Health effects of climate change-induced wildfires and heatwaves. Cureus. 2019;11(5):e4771.
24. McClure CD, Jaffe DA. US particulate matter air quality improves except in wildfire-prone areas. Proc Natl Acad Sci U S A. 2018;115(31):7901–6.
25. Han J, Dai H, Gu Z. Sandstorms and desertification in Mongolia, an example of future climate events: a review. Environ Chem Lett. 2021;19(6):4063–73.
26. WHO. Air pollution. Available from: https://www.who.int/health-topics/air-pollution#tab=tab_1.
27. World Health Organizations. New WHO global air quality guidelines aim to save millions of lives from air pollution. 2021. Available from: https://www.who.int/news/item/22-09-2021-new-who-global-air-quality-guidelines-aim-to-save-millions-of-lives-from-air-pollution.
28. Tiotiu AI, Novakova P, Nedeva D, Chong-Neto HJ, Novakova S, Steiropoulos P, Kowal K. Impact of air pollution on asthma outcomes. Int J Environ Res Public Health. 2020;17(17):6212.
29. Saikia D, Mahanta B. Cardiovascular and respiratory physiology in children. Indian J Anaesth. 2019;63(9):690–7.
30. Tuazon JA, Kilburg-Basnyat B, Oldfield LM, Wiscovitch-Russo R, Dunigan-Russell K, Fedulov AV, Oestreich KJ, Gowdy KM. Emerging insights into the impact of air pollution on immune-mediated asthma pathogenesis. Curr Allergy Asthma Rep. 2022;22(7):77–92.
31. Mumby S, Chung KF, Adcock IM. Transcriptional effects of ozone and impact on airway inflammation. Front Immunol. 2019;10:1610.
32. Zhang J, Ma C, Yang A, Zhang R, Gong J, Mo F. Is preterm birth associated with asthma among children from birth to 17 years old? -A study based on 2011–2012 US National Survey of Children's Health. Ital J Pediatr. 2018;44(1):151.
33. Goyal NK, Fiks AG, Lorch SA. Association of late-preterm birth with asthma in young children: practice-based study. Pediatrics. 2011;128(4):e830–8.
34. Gehring U, Wijga AH, Koppelman GH, Vonk JM, Smit HA, Brunekreef B. Air pollution and the development of asthma from birth until young adulthood. Eur Respir J. 2020;56(1):2000147.
35. Abnett K. Wildfires in Europe burn second-biggest area on record. 2022. Available from: https://www.reuters.com/world/europe/wildfires-europe-burn-second-biggest-area-record-2022-08-04/.
36. World Economic Forum. This is how much carbon wildfires have emitted this year. 2021. Available from: https://www.weforum.org/agenda/2021/12/siberia-america-wildfires-emissions-records-2021/.

37. Yang L, Li C, Tang X. The impact of PM(2.5) on the host defense of respiratory system. Front Cell Dev Biol. 2020;8:91.
38. Feng S, Gao D, Liao F, Zhou F, Wang X. The health effects of ambient PM2.5 and potential mechanisms. Ecotoxicol Environ Saf. 2016;128:67–74.
39. Wu J, Zhong T, Zhu Y, Ge D, Lin X, Li Q. Effects of particulate matter (PM) on childhood asthma exacerbation and control in Xiamen, China. BMC Pediatr. 2019;19(1):194.
40. Aguilera R, Corringham T, Gershunov A, Benmarhnia T. Wildfire smoke impacts respiratory health more than fine particles from other sources: observational evidence from Southern California. Nat Commun. 2021;12(1):1493.
41. Hauptman M, Balmes JR, Miller MD. The hazards of wildfire smoke exposure for children. Curr Probl Pediatr Adolesc Health Care. 2020;50(2):100756.
42. Tao Z, He H, Sun C, Tong D, Liang X-Z. Impact of fire emissions on U.S. air quality from 1997 to 2016–A modeling study in the satellite era. Remote Sens. 2020;12(6):913.
43. NPR. The western wildfires are affecting people 3,000 miles away. 2021. Available from: https://www.npr.org/2021/07/21/1018865569/the-western-wildfires-are-affecting-people-3-000-miles-away.
44. European Commission. 'Four times more toxic': how wildfire smoke ages over time. 2020. Available from: https://ec.europa.eu/research-and-innovation/en/horizon-magazine/four-times-more-toxic-how-wildfire-smoke-ages-over-time.
45. Rappold AG, Reyes J, Pouliot G, Cascio WE, Diaz-Sanchez D. Community vulnerability to health impacts of wildland fire smoke exposure. Environ Sci Technol. 2017;51(12):6674–82.
46. Delfino RJ, Brummel S, Wu J, Stern H, Ostro B, Lipsett M, Winer A, Street DH, Zhang L, Tjoa T, Gillen DL. The relationship of respiratory and cardiovascular hospital admissions to the southern California wildfires of 2003. Occup Environ Med. 2009;66(3):189–97.
47. Akpinar-Elci MMPHMD, Berumen-Flucker BMPH, Bayram HMDP, Al-Taiar AMDP. Climate change, dust storms, vulnerable populations, and health in the middle east: a review. J Environ Health. 2021;84(3):8–15.
48. Gross JE, Carlos WG, Dela Cruz CS, Harber P, Jamil S. Sand and dust storms: acute exposure and threats to respiratory health. Am J Respir Crit Care Med. 2018;198(7):P13–p4.
49. Finance Iaaoea. Without adaptation, Middle East and Central Asia face crippling climate losses. 2022. Available from: https://blogs.imf.org/2022/03/30/without-adaptation-middle-east-and-central-asia-face-crippling-climate-losses/.
50. Bell ML, Levy JK, Lin Z. The effect of sandstorms and air pollution on cause-specific hospital admissions in Taipei, Taiwan. Occup Environ Med. 2008;65(2):104.
51. Abal AT, Ayed A, Nair PC, Mosawi M, Behbehani N. Factors responsible for asthma and rhinitis among Kuwaiti school children. Med Princ Pract. 2010;19(4):295–8.
52. Thalib L, Al-Taiar A. Dust storms and the risk of asthma admissions to hospitals in Kuwait. Sci Total Environ. 2012;433:347–51.
53. Geravandi S, Sicard P, Khaniabadi YO, De Marco A, Ghomeishi A, Goudarzi G, Mahboubi M, Yari AR, Dobaradaran S, Hassani G, Mohammadi MJ, Sadeghi S. A comparative study of hospital admissions for respiratory diseases during normal and dusty days in Iran. Environ Sci Pollut Res Int. 2017;24(22):18152–9.
54. Change NGC. A force of nature: hurricanes in a changing climate. 2022. Available from: https://climate.nasa.gov/news/3184/a-force-of-nature-hurricanes-in-a-changing-climate/#:~:text=Moreover%2C%20according%20to%20Knutson%2C%20most,both%20of%20which%20fuel%20hurricanes.
55. Kossin JP, Knapp KR, Olander TL, Velden CS. Global increase in major tropical cyclone exceedance probability over the past four decades. Proc Natl Acad Sci U S A. 2020;117(22):11975–80.
56. Cowan KN, Pennington AF, Gregory T, Hsu J. Impact of hurricanes on children with asthma: a systematic literature review. Disaster Med Public Health Prep. 2022;16(2):777–82.
57. Kurup VP. Fungal allergens. Curr Allergy Asthma Rep. 2003;3(5):416–23.
58. D'Amato G, Chong-Neto HJ, Monge Ortega OP, Vitale C, Ansotegui I, Rosario N, Haahtela T, Galan C, Pawankar R, Murrieta-Aguttes M, Cecchi L, Bergmann C, Ridolo E, Ramon G,

Gonzalez Diaz S, D'Amato M, Annesi-Maesano I. The effects of climate change on respiratory allergy and asthma induced by pollen and mold allergens. Allergy. 2020;75(9):2219–28.
59. Sorrell TC, Chen SC. Fungal-derived immune modulating molecules. Adv Exp Med Biol. 2009;666:108–20.
60. Quansah R, Jaakkola MS, Hugg TT, Heikkinen SA, Jaakkola JJ. Residential dampness and molds and the risk of developing asthma: a systematic review and meta-analysis. PLoS One. 2012;7(11):e47526.
61. Caillaud D, Leynaert B, Keirsbulck M, Nadif R. Indoor mould exposure, asthma and rhinitis: findings from systematic reviews and recent longitudinal studies. Eur Respir Rev. 2018;27(148):170137.
62. Xiao S, Ngo AL, Mendola P, Bates MN, Barcellos AL, Ferrara A, Zhu Y. Household mold, pesticide use, and childhood asthma: a nationwide study in the U.S. Int J Hyg Environ Health. 2021;233:113694.
63. Paudel B, Chu T, Chen M, Sampath V, Prunicki M, Nadeau KC. Increased duration of pollen and mold exposure are linked to climate change. Sci Rep. 2021;11(1):12816.
64. Anderegg WRL, Abatzoglou JT, Anderegg LDL, Bielory L, Kinney PL, Ziska L. Anthropogenic climate change is worsening North American pollen seasons. Proc Natl Acad Sci U S A. 2021;118(7):e2013284118.
65. Sapkota A, Dong Y, Li L, Asrar G, Zhou Y, Li X, Coates F, Spanier AJ, Matz J, Bielory L, Breitenother AG, Mitchell C, Jiang C. Association between changes in timing of spring onset and asthma hospitalization in Maryland. JAMA Netw Open. 2020;3(7):e207551.
66. Neumann JE, Anenberg SC, Weinberger KR, Amend M, Gulati S, Crimmins A, Roman H, Fann N, Kinney PL. Estimates of present and future asthma emergency department visits associated with exposure to oak, birch, and grass pollen in the United States. GeoHealth. 2019;3(1):11–27.
67. Schmidt Charles W. Pollen overload: seasonal allergies in a changing climate. Environ Health Perspect. 2016;124(4):A70–A5.
68. Ziska LH. Climate, carbon dioxide, and plant-based aero-allergens: a deeper botanical perspective. Front Allergy. 2021;2:714724.
69. Kevat A. Thunderstorm asthma: looking back and looking forward. J Asthma Allergy. 2020;13:293–9.
70. Woodhead M. Hospitals overwhelmed with patients after "thunderstorm asthma" hits Melbourne. BMJ. 2016;355:i6391.
71. Park JH, Lee E, Fechter-Leggett ED, Williams E, Yadav S, Bakshi A, Ebelt S, Bell JE, Strosnider H, Chew GL. Associations of emergency department visits for asthma with precipitation and temperature on thunderstorm days: a time-series analysis of data from Louisiana, USA, 2010–2012. Environ Health Perspect. 2022;130(8):87003.
72. Stewart C, Young NL, Kim ND, Johnston DM, Turner R. Thunderstorm asthma: a review, risks for Aotearoa New Zealand, and health emergency management considerations. N Z Med J. 2022;135(1557):49–63.
73. Hajat A, Hsia C, O'Neill MS. Socioeconomic disparities and air pollution exposure: a global review. Curr Environ Health Rep. 2015;2(4):440–50.
74. Trikamjee T, Comberiati P, Peter J. Pediatric asthma in developing countries: challenges and future directions. Curr Opin Allergy Clin Immunol. 2022;22(2):80–5.
75. Mortimer K, Reddel HK, Pitrez PM, Bateman ED. Asthma management in low- and middle-income countries: case for change. Eur Respir J. 2022;60(3):2103179.
76. Akinbami LJ, Simon AE, Rossen LM. Changing trends in asthma prevalence among children. Pediatrics. 2016;137(1):1–7.
77. Gutschow B, Gray B, Ragavan MI, Sheffield PE, Philipsborn RP, Jee SH. The intersection of pediatrics, climate change, and structural racism: ensuring health equity through climate justice. Curr Probl Pediatr Adolesc Health Care. 2021;51(6):101028.
78. NCRC. HOLC "redlining" maps: the persistent structure of segregation and economic inequality. 2018. Available from: https://dataspace.princeton.edu/bitstream/88435/dsp01dj52w776n/1/NCRC-Research-HOLC-10.pdf.

79. HHS.gov. Asthma and African Americans. 2021. Available from: https://minorityhealth.hhs.gov/omh/browse.aspx?lvl=4&lvlid=15.
80. CDC. Preterm birth. 2021. Available from: https://www.cdc.gov/reproductivehealth/maternalinfanthealth/pretermbirth.htm.
81. Ju L, Li C, Yang M, Sun S, Zhang Q, Cao J, Ding R. Maternal air pollution exposure increases the risk of preterm birth: evidence from the meta-analysis of cohort studies. Environ Res. 2021;202:111654.
82. Chen X, Tan CM, Zhang X, Zhang X. The effects of prenatal exposure to temperature extremes on birth outcomes: the case of China. J Popul Econ. 2020;33(4):1263–302.
83. EPA. Climate change and social vulnerability in the United States: a focus on six impacts. 2021. Available from: https://www.epa.gov/system/files/documents/2021-09/climate-vulnerability_september-2021_508.pdf.
84. European Environment Agency. What is the difference between adaptation and mitigation? Available from: https://www.eea.europa.eu/help/faq/what-is-the-difference-between#:~:text=In%20essence%2C%20adaptation%20can%20be,(GHG)%20into%20the%20atmosphere.
85. ACOG. ACOG issues updated guidance on reducing patients' exposure to environmental toxins before and during pregnancy. 2021. Available from: https://www.acog.org/news/news-releases/2021/06/acog-updated-guidance-reducing-patients-exposure-to-environmental-toxins-before-and-during-pregnancy.
86. Airnow. Get air quality data where you live. 2022. Available from: https://www.airnow.gov.
87. California Air Resource Board. Protecting yourself from wildfire smoke. 2023. Available from: https://ww2.arb.ca.gov/protecting-yourself-wildfire-smoke.
88. Pollen.com. Pollen forecasts. Available from: https://www.pollen.com.
89. NDRC. Climate action: global transition away from HFCs. 2017. Available from: https://www.nrdc.org/experts/anjali-jaiswal/climate-action-global-transition-away-hfcs-moving-forw#:~:text=HFCs%20are%20incredibly%20potent%20greenhouse,or%20more%20of%20carbon%20dioxide.
90. Janson C, Henderson R, Löfdahl M, Hedberg M, Sharma R, Wilkinson AJK. Carbon footprint impact of the choice of inhalers for asthma and COPD. Thorax. 2020;75(1):82.
91. Woodcock A, Janson C, Rees J, Frith L, Löfdahl M, Moore A, Hedberg M, Leather D. Effects of switching from a metered dose inhaler to a dry powder inhaler on climate emissions and asthma control: post-hoc analysis. Thorax. 2022:thoraxjnl-2021-218088.
92. AstraZeneca. Carbon studies showed uncontrolled asthma is associated with an increased carbon footprint of asthma care. 2021. Available from: https://www.astrazeneca.com/media-centre/medical-releases/carbon-studies-showed-uncontrolled-asthma-is-associated-with-an-increased-carbon-footprint-of-asthma-care.html.
93. Dzau VJ, Levine R, Barrett G, Witty A. Decarbonizing the U.S. health sector – a call to action. N Engl J Med. 2021;385(23):2117–9.
94. The Climate Reality Project. Climate adaptation vs. mitigation: what's the fifference and why does it matter? 2019. Available from: https://www.climaterealityproject.org/blog/climate-adaptation-vs-mitigation-why-does-it-matter.
95. Hsu S, Thiel CL, Mello MJ, Slutzman JE. Dumpster diving in the emergency department. West J Emerg Med. 2020;21(5):1211–7.
96. Perera F, Cooley D, Berberian A, Mills D, Kinney P. Co-benefits to children's health of the U.S. regional greenhouse gas initiative. Environ Health Perspect. 2020;128(7):077006.
97. The World Bank. Social dimensions of climate change. Available from: https://www.worldbank.org/en/topic/social-dimensions-of-climate-change.
98. Asher I, Haahtela T, Selroos O, Ellwood P, Ellwood E. Global Asthma Network survey suggests more national asthma strategies could reduce burden of asthma. Allergol Immunopathol (Madr). 2017;45(2):105–14.
99. Serebrisky D, Wiznia A. Pediatric asthma: a global epidemic. Ann Glob Health. 2019;85(1):6.
100. Amini H. WHO air quality guidelines need to be adopted. Int J Public Health. 2021;66:1604483.

Life Course of Asthma

Sobia Farooq and Sumita Khatri

Abstract

Asthma is a heterogeneous chronic airway disease that can vary over a lifetime. Although broad categories of asthma by severity and type have been constructed, there remains a tremendous opportunity to discover an approach to managing asthma with additional factors in mind. Many in the field have suggested and are pursuing a novel paradigm shift in how asthma might be better managed, considering the life course of exposures, management priorities, and predicted trajectory of lung function growth. This approach will require a more holistic view of prenatal, postnatal, adolescence, hormonal and gender aspects, and the aging process. In addition, the environment, externally and internally, including in one's genetic code and epigenetic changes, are factors that affect how asthma progresses or becomes more stable in individuals. This chapter focuses on the various influences that may, to differing degrees, affect people with asthma, which can develop at any time in their lives. Shifting the paradigm of thought and strategies for care and advocating for public policies and health delivery that focus on this philosophy is paramount to advance asthma care for all.

Keywords

Prediction · Lung function · Trajectory · Family history · Prenatal exposure · Vitamin D deficiency · Heterogeneity

S. Farooq · S. Khatri (✉)
National Heart, Lung, and Blood Institute, CMO Division of Lung Diseases, Bethesda, MD, USA
e-mail: farooqs@ccf.org; sumita.khatri@NIH.gov; Sumita.Khatri@nih.gov

A. R. Brasier, N. N. Jarjour (eds.), *Precision Approaches to Heterogeneity in Asthma*, Advances in Experimental Medicine and Biology 1426,
https://doi.org/10.1007/978-3-031-32259-4_3

Introduction: Time for a Paradigm Shift

Asthma is a common chronic airway disease. Although many individuals have asthma that is relatively easy to manage, there are nuances to not only to an asthma diagnosis but also regarding its management and care. A particularly complex aspect is whether clinicians may be able to predict the course of asthma across a lifespan. Much of this depends upon various individual and environmental factors that may have initiated and promoted a predisposition toward asthma. To better understand the variables that promote an asthma state should be sought to reduce incidence in future generations, but it is imperative also to have a better understanding of the trajectory that can be expected across various subtypes of asthma.

Asthma symptomatology, although with many shared features, can also be unique based upon the types of inflammation and inciting factors. This chapter will summarize the current knowledge of these variables such that clinicians and researchers can adopt a diagnostic and therapeutic mindset that is open to varying presentations. This will also allow the medical community to customize and precisely care for (and manage with) our patients based upon the types of asthma they have. Even more importantly, as more research emerges, our charge is also to guide our patients with emerging knowledge available to counsel on how to maintain and improve their health-related quality of life.

One of the recent call to action for a paradigm shift was summarized by Szefler [1]. This perspective highlighted the commonly agreed-upon need: the ability to be more proactive with the management of asthma instead of reactive only to symptomatic episodes of acute care. Utilizing registries and large network trials, we must define differing trajectories of lung growth to help guide care [2, 3]. Prior studies and efforts includes NICE, NHLBI Childhood Asthma Management Program (CAMP), National Institute of Allergy and Infectious Diseases's Inner-City Asthma Consortium (CASINIAID) [4]. McGeachie et al. evaluated projections of patterns of future lung function, highlighting the need for ongoing monitoring, treatment, and counseling to patients and families [5].

Knowledge gleaned from these studies have the possibility to be life-altering from a lung health standpoint. If tracked over time, children and adolescents with asthma who are at risk may be able to manage other exposures and risk factors. With a proactive approach, modifiable risk factors can be mitigated to halt accelerated lung function decline or permit longer lung growth. This would be a shift in asthma management (more aligned with preventive care as in cardiovascular health and disease prevention) and a clinical tool to proactively counsel patients in a more precise manner.

The Prenatal Period and Asthma

It is well established that the timeline of asthma begins before birth. The mother's health and exposures have been known to be associated with the prevalence of wheezing as well as subsequent diagnosis of asthma in children. Exposures of concern are not only from air pollution or antibiotics. Prenatal maternal stress (MPS) has significant associations with the prevalence and development of asthma in children of mothers who have been subjected to MPS [6].

Exposures during certain periods of development contribute to the development of asthma; however, the genetic susceptibility of the mother and child also plays a role. The distinction of certain pollutants as the culprit is also fraught with confounders as often there are mixed pollutants in a pregnant mother's environment. However, the evidence below demonstrates that a significant amount of consideration be made for the prenatal and postnatal environment, which would also in turn help the mother's health.

Family History of Asthma

In this section, we will discuss some of the known associations that can predispose children to asthma in the prenatal period: Family history of asthma, and in particular maternal history of asthma, is a risk for development of asthma in their children. In a review of 33 international studies including the United States, Central/South America, Europe, Africa, and Australia, evaluating populations with asthma prevalence from 2% to 26%, a history of asthma in a first-degree relative was a consistent risk factor for asthma [7].

Fourteen international studies evaluated the family history of atopic disease using positive family history including asthma, allergic rhinitis, or atopic dermatitis. Relevant studies from Turkey, Saudi Arabia, India, Sweden, and Italy were included [8–12]. Overall, the risk associated with maternal, paternal, or sibling history of asthma was similar. Those studies found that a positive family history of atopy significantly increased the risk of asthma, with ORs ranging from 1.7 to 6.8.

Another group of ten studies with prevalence data was utilized to calculate a positive predictive value. Compiling these studies, the prevalence ranged from 11% to 37% with sensitivity from 4% to 43% [13–22]. This knowledge may allow individualized prevention efforts with known triggers and sensitizers, however, current evidence is not specific enough for precision intervention for all potentiating exposures. In this article, Burke also highlighted the few studies that evaluate interactions between additional risk factors to family history. Although studies have evaluated additional risks (maternal smoking or the mother history of death from asthma) [16, 23], not all studies observed such an interaction [24].

Prenatal Exposure to Tobacco Smoke

In utero, exposure to tobacco smoke is a risk factor for the development of asthma. The mechanisms include creating a more atopic Th2-rich milieu, polymorphisms of inflammatory cytokines, epigenetic changes, and ongoing postnatal Environmental Tobacco Smoke (ETS) exposure. The literature is rich in studies that have demonstrated that in utero exposure to ETS impairs lung function and increases bronchial hyperresponsiveness [25–30]. Structural changes also occur with the alteration of lung branching [31]. There is an increase in neonatal Th2 cytokine response to antigen exposure and a decreased production of IFN gamma when there is prenatal exposure to maternal tobacco smoking [32]. Furthermore, certain genetic polymorphisms such as GSTP1, TGF-beta, and IL1R are associated with a higher risk of asthma after exposure to ETS [27, 33, 34]. This supports the premise that both structural and inflammatory pathways contribute to the development of asthma with prenatal exposure to maternal tobacco smoking. Importantly, while it may be difficult to distinguish whether the incidence of asthma is an effect of prenatal vs. postnatal tobacco smoke as the pivotal exposure, it is likely there may be ongoing smoking after delivery, which significantly increases the prevalence of childhood asthma. However, it has been shown that prenatal exposure may contribute the greater risk for the development of wheeze or asthma [35, 36].

Prenatal Exposure to Pollution

It is well known that self-exposure to components of air pollution increases the incidence of childhood asthma. These postnatal exposures to the main components of air pollution (NO_2, fine particulate matter (PM) black carbon (BC), SO2) correlate with increased incidence of childhood asthma [37, 38]. However, a detailed study of prenatal exposure to polycyclic aromatic hydrocarbons (PAH) contained in diesel exhaust particles which are also emitted from other traffic sources, cigarette smoking, cooking, and space heating has been associated with childhood wheezing. A study of over 700 healthy women who were enrolled to evaluate their children for the association of PAH and ETS with asthma. While prenatal PAH exposure alone was not associated with asthma or IgE at age 5–6 years, the investigators found a positive interaction between prenatal PAH and prenatal ETS with asthma at age 5–6 years. It has already been noted that the prenatal period is a time of particular vulnerability from environmental exposures, including other pollutants such as ozone NO_2, CO, and PM [39, 40]. Evaluation of synergistic effects was also performed in a study of children exposed to ETS as well as pollutants, demonstrating that together, these children had increased odds of doctor-diagnosed asthma. It is also possible that the timing of the presentation and diagnosis of asthma is a feature related to prenatal exposure to pollutants, such that prenatal exposure to PAH can be associated with the onset wheeze in the first 2 years of life, but not necessarily later at age 3 or 4 years old [41].

Although the development of asthma is multifactorial (i.e., exposure, predisposition, and the functioning of certain innate adaptabilities), certain genes have demonstrated associations with asthma. Genes essential for antioxidant defense include cytosolic glutathione S-transferase (GST) genes. Genetic predisposition has been evaluated with GSTM1 polymorphisms, which play a role in air pollution and susceptibility to asthma (postnatal). It has been known that GSTP 1 is associated with airway hyperresponsiveness and atopy in adults, and variations in haplotypes play a role in the level of lung growth [42]. The haplotype Val^{105}/Val^{105} appears to be protective with reduced risk of asthma, atopy, and airway hyperresponsiveness [43]. Maternal, but not paternal, GSTP1 haplotypes were associated with asthma in their children. The genotype of mother and child had the most effect on airway hyperresponsiveness. A follow-up study in 2003 uncovered that in utero maternal environment, not genotype alone was driving whether asthma was to be present postnatally vs. not and found that a significant decrease in lung function was detected among children with mothers who smoked during their pregnancy and had a GSTM2 risk haplotype [27]. This finding was further supported by the additional risk of maternal smoking contributing to the risk of asthma, atopy, and airway hyperresponsiveness (AHR), demonstrating that mothers who smoked during pregnancy and had variation in GSTM2 were more likely to have children with lower lung function growth [27, 44].

This topic of exposome and asthma is a difficult line of research as many areas of exposure contribute to asthma, making the prediction of genetic risk complex. These studies can, however, be effective with genome-wide association studies (GWAS) or evaluating computer modeling when studying individuals with homogenous environments of exposure, such as children living on farms [33]. A notable conclusion from that study was that prediction of genetic predisposition of a complex disease such as asthma might improve when studied populations are in a relatively homogeneous milieu, as compared to children raised in an environment with more heterogeneous exposures.

Prenatal Exposure to Antibiotics

The microbiome plays a significant role in asthma, as it influences allergic sensitization, respiratory infections with wheezing, and the development of asthma. Several studies have shown an inverse relationship between rates of childhood asthma and exposure in infancy to microbe-rich environments [45, 46]. The robust development of innate immunity can be protective for asthma. The alteration of the microbiome may occur from both prenatal and postnatal exposure to antibiotics, and both types of exposure are associated with an increased risk of asthma [47]. Frequency of maternal use of antibiotics for urinary tract infections as well as infant antibiotic use in the first year of life, have been associated with increased risk of childhood asthma developing. In this Finland-based study, maternal use of any antibiotics during pregnancy was associated with an increased risk of asthma in the offspring (adjusted OR = 1.3 (95% CI 1.21–1.42)) and the strongest association observed for cephalosporins. Further postnatal use of antibiotics during the first year of life was associated with an increased risk of asthma OR 1.60 (1.48–1.73) [48].

A recent, large population-based cohort study of mother–child dyads of infants born between 1995 and 2003 in the United States was evaluated for exposure to antibiotics during the prenatal period [49]. In 84,214 mother–child dyads enrolled in the study with approximately 50–50 gender distribution and median gestational age at delivery of 39 weeks (IQR, 39–40), 54,141 of the 84,214 (64%) children were exposed to a total of 104,082 prenatal courses of antibiotics. In a significant dose-response relationship, the number of prenatal antibiotic courses was associated with childhood asthma. With an interquartile difference of two antibiotic courses (IQR, 0–2), the relative odds of childhood asthma increased by 33% (OR, 1.33 [95% CI, 1.26–1.39]) and 26% (aOR, 1.26 [95% CI, 1.20–1.33]) in the univariate and adjusted analyses, respectively. What is fascinating in this study is that the dose relationship was present in those children whose mothers did not have a history of asthma, demonstrating that exposure to these antibiotics affects the babies and predisposes them to asthma in a dose-dependent manner without a family history of asthma [49]. A meta-analysis of longitudinal studies on antibiotic use and subsequent development of wheeze or asthma demonstrated varying effects of antibiotic use on wheeze/asthma. When the methodology of the paper eliminated studies with reverse causation and confounding by indication, the risk of wheezing with antibiotic use was significantly reduced, suggesting an overestimation of the role of antibiotics in future wheeze and asthma. However, the potential mechanisms behind these associations may be the adverse long-term effect on gut microbiota of both mother and child as well as the vaginal microbiota of the mother. Furthermore, delayed colonization of child's gut microbiota may interfere with the maturation of child's immune system.

Prenatal Vitamin D Deficiency

Initial trials suggested reductions in asthma and recurrent wheeze with vitamin D intake; however, prenatal supplementation alone is not sufficient to prevent childhood asthma through mid-childhood. A study from Litonjua et al. demonstrated reduction in asthma and recurrent wheeze with vitamin D supplementation with an additional multivitamin supplementation had protective effects on asthma. In this trial, mothers had a daily vitamin D dose 40,000 IU + multivitamin or a daily placebo plus multivitamin with 400 IU. The primary outcomes were asthma, recurrent wheeze, or both [50]. By the age of 6 years, a total of 360 children met the criteria for asthma outcomes, and there was no significant effect from maternal supplementation with 4400 IU of vitamin D per day on the incidence of asthma or recurrent wheezing at the age of 6 years; However, among mothers with vitamin D levels less than 20 ng/mL, there was a higher risk of asthma among their children. Interestingly, children who were born to mothers in the vitamin D group had fewer positive specific IgE results (but the overall presence of sensitization to any allergen was not different), which suggests that although not fully protective, it may be possible to reduce the severity of asthma in their children. Therefore, while previous studies showed prenatal supplementation had significant effects in preventing asthma, this study demonstrated that prenatal supplementation alone is not sufficient to prevent childhood asthma through mid-childhood.

A study by Camargo et al. also evaluated a group of more than 1100 children and found a lower risk of recurrent wheezing if the child received prenatal vitamin D either from their mother's diet or from supplementation and evaluated them to the age of 3 years, revealing a lower risk of recurrent wheezing if the child had adequate vitamin D from their diet or from supplementation [51]. Further, a cohort study by Lu et al. used prospectively collected data to demonstrate a protective role of adequate vitamin D supplementation in pregnant women. Not only was there a lower risk of asthma by age 3 years, but a protective role also remained from 3 to 6 years old (both $p < 0.05$) for a trend of protection from asthma [52].

Therefore, the role of vitamin D is not clear, yet it does suggest that in situations where vitamin D deficiency is a potential risk, supplementation with multivitamin and vitamin D during pregnancy may have a beneficial effect.

Prenatal Diet and Asthma

There remains great interest in how an individual's dietary intake and habits affect asthma, not only in children and adults but also during the mother's diet while pregnant. The results are variable and depend upon the design of the study. The potential benefits of antioxidant and n-3 Polyunsaturated fatty acids (PUFA) have been hypothesized and comprise a Mediterranean diet; however, the potential for pulmonary and nonpulmonary benefits and the exact mechanisms and consistency of study outcomes are not clear [53]. There have been several interventional studies that have been equivocally related to antioxidant-rich diets, such as with high tocopherol or selenium ingestion, however, there may be more of a protective effect of vitamin E in earlier stages of development [54–56]. Other vitamins such as selenium may be protective against early wheezing [57]. The mechanism by which the Mediterranean diet may work is due to the antioxidant hypothesis [58]. This is based upon the premise that antioxidant deficiencies in diet, such as selenium and flavonoids, may have adverse effects, and in contrast, higher intake may have improvement with asthma, wheezing, and ventilatory function. Vulnerability to oxidative exposures is known to result in airway damage and reductions in airway compliance; however, no clear evidence of the benefits of oxidant/antioxidant supplementation exists. The optimal balance is uncertain as a high intake of vitamin C, and citrus has a counteractive function and may increase eczema and allergic sensitization [59].

In a study of higher prenatal exposure to a Mediterranean diet, a cohort of 460 children were evaluated after birth and up to the age of 6.5 years. High prenatal adherence to a Mediterranean diet demonstrated a protective effect during pregnancy on persistent wheeze, atopic wheeze, and atopy at the age of 6 ½ years old. High maternal Mediterranean diet score during pregnancy was also found to be protective against persistent wheeze, atopic wheeze, and atopy (OR 0.22 [0.08–0.58], 0.30 [0.10–0.90], and 0.55 [0.31–0.97]), respectively. A diet of vegetables more than eight times/week was also associated with less wheeze (OR 0.36 [0.14–0.92]) and atopy (OR 0.40 [0.22–0.72]) and was similarly protective as to fish intake more than two to three times per week [60]. Meanwhile, in another study of 1476 children, there were no associations with a Mediterranean diet during pregnancy being protective against asthma [61].

In a more recent review of maternal diet during pregnancy, seven studies explored this topic [62]. As dietary history and components were collected, outcomes of asthma or wheezing were evaluated. These children were seen at the age of 1, 3, or 4 years, and 5 years old. One study demonstrated a protective effect during the first year but not afterward [63], while another study of 690 children ages 7–18 years old demonstrated that high consumption of nuts was found to be inversely associated with wheezing (OR 0.46; 95% CI 0.20–0.98), whereas margarine increased the risk of both wheeze (OR 2.19; 95% CI 1.01–4.82) and allergic rhinitis (OR 2.10; 95% CI 1.31–3.37). This study demonstrated that adherence to the Mediterranean diet was protective against allergic rhinitis and provided modest protection for atopy and wheezing [64].

Although an aspect of prenatal health that is modifiable and personally important to many expectant mothers, the proper oxidant/antioxidant balance is elusive and difficult to quantify. Individual genetic and homeostatic factors as well as other environmental barriers and lifestyle, play a role. Diet studies can be difficult to evaluate with a single intervention, as there is likely variability in factors such as the amount of antioxidants/protective components, frequency, activity, and other exposures; thus, it is not surprising the protective or nonprotective effect is unclear.

Prenatal Maternal Obesity and Effects on Their Children

The impact of maternal obesity upon the health of their children has become a major concern, given the significant rise in the prevalence of obesity and its risk with the development of chronic health conditions. Associations that have been noted include both obstructive lung disease and asthma. The fetal environment may also affect lung development. Maternal over-nutrition can have an impact on metabolic, hormonal, and inflammatory factors that affect fetal lung development [65]. These factors may be from inflammatory mediators, impaired placental function, or obesity-related responses to fetal immune and respiratory systems, therefore increasing the risk of asthma or wheezing in the later years of life [66]. Many birth cohort and registry studies have been performed to assess maternal overweight/obese status and associated preschool wheezing or wheezing in school-aged children. Maternal fat mass, measured by skin fold thickness and body mass index (BMI), was associated with transient wheeze in the first 3 years of childhood. In another study of 14 European birth cohorts, maternal pregnancy overweight and obese status were both associated with increased risk of ever and recurrent wheezing in offspring, and associations were independent of other pregnancy complications or factors such as birth weight, gestational age, mode of delivery and breastfeeding status [67]. Similar findings were seen for school-aged children [68, 69].

Asthma at an early age can be difficult to diagnose; however, a study from 2008 was able to demonstrate a 1.5-fold increased risk of parent-reported asthma diagnosis in children at 3 years old who were born to obese mothers [70]. Maternal obesity (but not overweight status) was associated with wheezing in children 6–18 month olds in a study by Haberg et al. in 2009 [71]; however, other studies found obesity associated with a risk of preschool wheezing only among children with mothers with a history of asthma or atopy [72]. There are likely other factors, as some studies evaluated the association of maternal obesity increasing as a risk factor in proportion to weight, while others demonstrated an association of asthma with only obesity (BMI > 30) but not with overweight status (BMI 25–29.9) [72, 73]. In 2014, Forno et al. reported from a meta-analysis of 108,321 mother–child pairs with maternal obesity and outcomes of childhood ever-wheeze or asthma at an odds ratio of 1.49 [1.22–1.83] and current wheeze or asthma odds ratio of 1.35 [1.08–1.68] [74]. Interestingly, maternal overweight or obese status were associated with asthma onset by 12 years of age, but associated more significantly with asthma that was nonallergic. In addition, only boys developed nonallergic asthma, and allergic asthma appeared only in girls who developed asthma [75].

To better understand potential pathways, Rusconi and authors proposed the following mechanisms: maternal obesity as an inflammatory state, dietary patterns and nutrient intake in pregnant obese women, obesity and microbiome (including associated antibiotic therapy), obesity-related complications in pregnancy such as preeclampsia, and maternal and childhood obesity from epigenetic imprinting, environmental factors or habits [17].

Prenatal Maternal Stress (MPS)

Maternal stress during pregnancy and prenatal exposure to maternal stress have been associated with a higher risk of asthma [6]. Studies have examined how exposure to psychologic stress during pregnancy affects the child in utero, thus impacting postnatal and early childhood health. Critical developmental windows, such as when cell programming can result in changes to the stress response system, have a role in the development of inflammatory respiratory disorders including asthma. Oxidative stress, immune, autonomic and neuroendocrine factors such as the maturity of the fetus as well as the development process renders the fetus susceptible to stress-related outcomes [76]. The postnatal outcomes of allergy and asthma are also influenced by gender as well, with female children more adversely affected [77]. Lee et al. evaluated a prospective pregnancy cohort to evaluate the relative importance of exposure to prenatal and/or postnatal stress and its association with the development of asthma in their children. In an evaluation of self-reported negative life events (NLEs) – the recording of these NLE experiences resulted in physiologic disruptions. In addition, NLEs are compounded by prenatal exposure to traffic-related air pollution (TRAP), particularly black carbon (BC). Household allergies also play a role, with the odds of an asthma diagnosis increasing with each categorical increase in maternal stress. These models adjusted for sociodemographic factors, maternal atopy, neighborhood characteristics, and social and physical environmental factors. The results were consistent, demonstrating an exposure-response relationship between prenatal and early infancy as predictors of childhood asthma onset. Oxidative stress and the maternal-fetal interface of managing oxidative damage regulate oxidant balance. In addition, there was a relation between sex and timing-specific effects of early life stress, i.e., prenatal stress affected boys and postnatal stress affected girls. The mechanisms of these negative effects may be from prenatal stress altering maternal sympathetic and adrenomedullary systems as well as the hypothalamic-pituitary-adrenal axis (HPA), increasing placental CRH (pCHR) and fetal exposure to GCs [6]. As these have been studied in human and animal models, a set of biomarkers to assess risk would be useful to assess the regulation and functioning of the placenta, including epigenetic programming.

Epigenetics

Epigenetics is the mechanism of genetic alterations in response to exposures, and has relevance to how the life course of asthma develops. These factors may be from prenatal or postnatal exposures or experiences, as well as family history and maternal stress, as noted in the previous section.

Situations of high prenatal stress and increased wheeze include epigenetically deregulated neuroendocrine and neurotransmitter receptor pathways in both mothers and children. In children, calcium (Ca) signaling and Wingless integration site (Wnt) signaling were affected by regulating cell proliferation and differentiation in lung maturation. Stress-related effects can also influence the maternal and fetal microbiome. Further, the placenta manages the information between fetoplacental and maternal compartments, and the oxidative stress and maternal and fetal HPA axis of cortisol play key roles. Prenatal exposures also change other placental biomarkers including microRNAs (miRNAs-small noncoding RNA molecules that regulate numerous pathways).

In addition to the effects on asthma from maternal prenatal stress, additional inflammatory modifications occur through environmental exposures. These changes in genes are passed along over generations [78]. A more harmful respiratory environment with increased inflammation is promoted by various pathways and usual protective responses can be diminished. Epigenetic mechanisms can include DNA modification/methylation, histone modifications, and noncoding and transcriptional gene silencing, resulting in significant individual heterogeneity of asthma. DNA methylation is the transfer of methyl groups to form 5-methylcytosine, which is associated with changes in gene expression and repression. These changes are passed along over cell cycles and generations [78]. Histone modification also results in the repressed translation of proteins, resulting in changes in gene regulation and function. Nonprotein coding RNAs involved in the regulation of cellular functions also have a role in organ development and cell differentiation. Examples of critical asthma-related genes that have been modified include the beta-2 adrenergic receptor (ADRB2) as well as IL5RA, which can be modified by the air pollutant NO2 and cigarette smoking [79]. In addition to receptor modification, smooth muscle cell proliferation can be affected by microRNAs affecting Th1/Th2 inflammation from eosinophils, macrophages, and mast cells [80, 81]. The manner in which epigenetics functions is quite diverse and is an emerging science.

Childhood Exposures and Features of Asthma

Childhood Asthma Studies and Lung Function Projections

Original work from Speizer and Tager created a schematic of lung function trajectories over the first 30 years of life and found categories of normal growth, early decline, reduced growth, and reduced growth and early decline [82]. This potentially useful approach to predict asthma-related lung function over time warranted further prospective investigation.

To follow-up and evaluate these categories prospectively, the topic of lung function trajectory was examined in the Childhood Asthma Management Program (CAMP) study of children [5]. This study enrolled individuals with persistent asthma into trials of inhaled anti-inflammatory treatment from ages 5–12 years and then continued into phases of observational follow-up. Participants who had lung function testing at 23 years old were included, which comprised 805 of the original 1041 participants. This allowed assessment of their lung function trajectory, and patterns of lung function growth and decline. Evaluating 684 individuals, 175 (25%) were in the normal lung function without early decline category. However, 514 (75%) had abnormal patterns.

These individuals with patterns outside of normal were divided into categories based upon growth and decline (Fig. 1). (A) 160 (23%) classified as having '*reduced growth without early decline*,' demonstrating lower lung function without achieving maximum lung function and growth plateau at 21.9 years, (B) 178 (26%) demonstrated '*normal growth, early decline*' with max lung function at 20.6 years and the start of decline at 21.1 years, and (C) 176 (26%) demonstrated an early plateau with '*reduced growth and early decline*' with a mean max age of lung function at 20.6 years old, and the start of lung function decline at 21.3 years [5]. Risk factors for reduced growth included lower baseline FEV1, airway hyperresponsiveness, smaller response to bronchodilator, and male gender. The knowledge gleaned from this study is a hallmark of precision medicine for lung health.

Allergen Exposures During Childhood

In childhood, allergic sensitization and upper respiratory infections with wheezing are independent and synergistic risk factors in the development of asthma. Early bacterial diversity and exposure to endotoxin appear protective and a crucial element in the development of innate immunity [46]. Current factors related to urbanization, vaccination and sanitation have decreased microbial exposure in early life and reduced the diversity of the microbiome. Strains of certain bacteria are linked with the development of atopy, as decreased diversity has been shown to be predictive of allergic sensitization at age 6 years [83]. Gut microbiome colonization with *Bacteroides* and *Lactobacillus* species has been associated with decreased risk of allergy, while specific colonization in the gut with *Clostridia* species has been associated with wheezing and allergic sensitization [84].

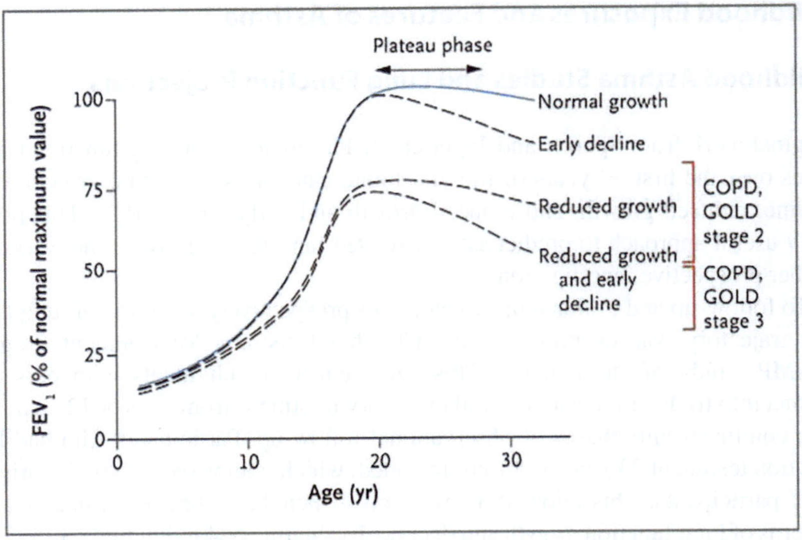

Fig. 1 Longitudinal lung function trajectories. (Illustrations: Permission granted from Clinics in Chest Medicine Figures, Biomedicines figures, Szefler JACI figures; New figure commissioned from Cleveland Clinic. McGeachie et al. [5])

Breastfeeding has been mostly favorable to the properties of the microbiome. A more protective profile of the microbiome community is noted with a predominance of *Corynebacterium* and *Colosigranulum*, which has been inversely associated with symptoms of wheezing and mild respiratory tract infections [85]. Conversely, oral sensitization to dust mite allergen can also be introduced through breastmilk. A study found the presence of dust mite allergen in breast milk correlating with risk of atopy and asthma at 5 years of age; however, this association was enhanced when mothers had asthma or atopy themselves, suggesting that certain combination of environmental factors and family history combine in the role of developing asthma [86].

Indoor environmental allergens include house dust mites (HDM), cockroaches, molds, rodents, and pets. Sensitization of such perennial allergens has contributed to asthma morbidity (number of asthma exacerbations, lost days from school/work, hospitalizations and emergency department (ED) visits, medication use, and asthma severity). Multimodal remediation of allergen exposures is the best approach to mitigating further atopy or asthma. Elimination of cockroach exposure can be effective in reducing asthma symptoms. The most effective approach includes professional extermination with insecticide, roach traps, and sealing food containers, as well as plugging holes, keeping kitchens clean, promptly washing dishes, and expeditious disposal of garbage. The role of cockroach sensitization and the development of asthma is worse when considered with other exposures [87–89]. Unfortunately, disparities in these exposures within areas affect individuals with lower socioeconomic status, with implications from an asthma standpoint well into adulthood.

In addition, pests and pets such as rodents (mice, rats, guinea pigs) are common sources of indoor allergens. Nearly ubiquitous, mouse protein can often be noted in house dust of homes in certain communities. Sensitization of mouse and allergen exposure appears to have a dose-dependent effect on the development of asthma [90, 91]. Similar to cockroach exposure, integrated pest management has been shown to reduce allergen burden in the home, a critical component of mitigation. This approach includes vacuuming, using low-toxicity pesticides, placing traps, and sealing holes [92].

Environmental Tobacco Smoke Around Children

Environmental Tobacco Smoke (ETS) is another major risk factor for asthma, and third-hand smoke (created by the off-gassing of tobacco smoke remnants on clothing and hair of those who smoke) that is easily placed into the respiratory environment of vulnerable nonsmoking individuals, is an ongoing threat that is infrequently addressed [93, 94]. Even without individuals actively smoking in the home, third-hand smoke exposure accumulates and off-gasses from clothes and hair, leaving residue for individual nonsmokers and causing indoor air pollution. ETS has been associated with poor asthma control, decreased health-related quality of life, and lower lung function. Often the severity of asthma is dependent solely on exposure to ETS, and compounded by genetic antioxidant capability [95].

In addition to ETS, indoor air quality can be significantly affected by chemical pollutants such as nitrogen dioxide and volatile organic compounds (VOCs). Nitrogen dioxide (NO2) is present indoors with high-temperature combustion and accumulates when there is inadequate ventilation/venting of combustion processes, such as from gas stoves, kerosene heaters, furnaces and fireplaces. The simple function of cooking can also increase the particulate matter concentration by 1.5 to 27-fold. In addition, VOCs are also emitted from construction materials, such as formaldehyde in particle board, and hobby and household materials such as oil-based paints, printer toners, and indoor plants [96]. These VOCs have been shown to increase oxidative stress and strain antioxidant mechanisms in airways [97].

Outdoor Air Pollution and Children

Many studies and reviews have addressed the role of environmental air pollution and childhood worsening of asthma [98–101]. The presence of high ozone levels, particulate matter, and composite particulate matter with traffic-related air pollution (TRAP) has deleterious effects on asthma. Much has been learned from the Children's Health Study, a longitudinal study in California that evaluates children with asthma, their symptoms, and lung function over time [101]. Lung growth and development is reduced in children who are exposed to higher levels of NO_2, and meanwhile improvement of air pollution and air quality reduces the proportion of children with clinically low FEV_1 [99, 100]. In addition, findings include that children who play outdoor three or more outdoor sports in high ozone season experience a higher incidence of asthma in childhood.

Additionally, a real-world study of pediatric asthma visits to acute urgent care/ Emergency Department demonstrated that during the Atlanta Olympics in 1996, when traffic was diverted away from central counties in Atlanta, the ozone levels were lower (due to less VOCs interacting with heat and sunlight thus producing less ozone) and during that period visits for urgent care for asthma were significantly reduced. For instance, peak daily ozone concentrations decreased by 27.9%, from 81.3 ppb during the baseline period to 58.6 ppb during the Olympic Games ($P < .001$). These correlated with the reduction in morning traffic counts. Unfortunately, but as expected, when the Olympics were over, this trend of air quality as well as acute asthma visits reverted back to pre-Olympic levels [102].

In children living within proximity to busy roads (within 400 m), TRAP also plays a significant role in the worsening of asthma/allergies. Janssen et al. demonstrated that children who attended schools closer to major roadways or highways with truck traffic demonstrated children were sensitized to pollen from truck traffic-related air pollution when the children were a sensitive subgroup, i.e., at baseline hide bronchial hyperreactivity and/or sensitization to common allergens experienced more allergy/sinus symptoms [103]. A study in Cleveland recently evaluated pediatric asthma visits during a period of localized and detailed monitoring of PM with speciation parameters, i.e., traffic-related vs industry/steel mill-related (CMAPS study). The frequency of ED visits for asthma in the area was associated with fine and coarse iron-related PM, which was associated with steel production. In addition, similar associations were seen near roadways and highways, speaking to the risk of PM related to motor vehicle emissions [104].

Changes in climate also worsen the pollution-related risk for children with asthma (as well as adults). There are direct and indirect threats to health as a result of changes in climate—pollution becomes worse with fewer areas of greenery, allergenicity of pollen become worse with longer spring and fall seasons and shorter winters. Indirect infrastructure challenges with the need to evaluate areas with extreme weather events, lack of electricity to manage devices such as nebulizers, and transportation to access healthcare are hampered [105].

Antibiotics, the Microbiome, and Asthma in Children

Prenatal and early life exposures significantly influence the development of asthma in childhood [98]. The role of the microbiome is recognized as significant in the development/inception of asthma. Early microbial exposures and the diversity of bacteria within the human microbiome are crucial to the development and sustaining of innate immunity [106]. Factors that affect the microbiome include environment, dietary, and antibiotic exposures. Urbanization, improved sanitation, and vaccination have reduced exposures early in life and reduced the biodiversity of the microbiome. For example, children who lived on farms had increased microbial exposure and more diversity and also had less asthma and atopy [106]. Further industrial farming, when compared with traditional farming techniques, was not as protective [107], as endotoxin exposure early in life appeared to be protective from an asthma standpoint.

Interestingly, the formation of the microbiome occurs through maternal transmission, and decreased diversity of the microbiome was predictive of allergic sensitization in early life, at age 6 years [83, 108]. In addition, the diversity of infant gut microbiota was protective from the risk of atopic sensitization [109]. Specifically, colonization in the gut with *Clostridia species* has been linked with wheezing and allergic sensitization. Conversely, colonization with *Bacteroides* and *Lactobacillus* species has been associated with a reduced risk of allergy [84]. This relation continues into early life, where microbial colonization of airways has played a role in the development of asthma. A study of airway bacterial colonization in asymptomatic neonates at 1 month of age revealed that early colonization with *S. pneumoniae*, *H. influenza*, and/or *M. catarrhalis* was associated with recurrent wheeze by age 5 years [110]. Meanwhile, antibiotic use in infancy was associated with decreased abundance of the more protective *Bifidobacterium* and *Bacteroides* species [111].

Exposure to Endotoxin

The presence of endotoxin has been implicated in various pulmonary effects, including airway inflammation and airflow obstruction. Endotoxin, the component of the outer membrane of gram-negative bacteria, is more prevalent in homes with dampness, mold, pets, or rodents. Not only in homes, endotoxin is prevalent in schools and workplaces, and personal endotoxin is associated with wheezing at 2 years old and with increased asthma symptoms and decreased FEV_1 [112, 113]. The risk of asthma has been inversely associated with increased microbial diversity [114]. Exposures to bacterial diversity in early life have been studied by comparing children living in farm vs. urban environments. As noted previously, in rural, farming environments, children had increased indoor and more diverse microbial exposures, and they experienced less asthma and atopy than children not living on farms [46]. Importantly, traditional farming environments such as in Amish communities, were environments in which asthma was noted to be four to six times lower [107]. The protective role of endotoxin, a component of gram-negative bacterial cell walls and a marker of microbial exposure, is believed to be the mechanism of this association.

Obesity and Asthma in Children

Both obesity and asthma are chronic diseases that are increasing in prevalence, although differing by area worldwide [115]. The interaction between obesity and asthma is very complex, including endocrine factors such as insulin resistance and dyslipidemia, physiologic factors such as fat distribution, as well as dietary habits

[116, 117]. Although fat mass distribution is more associated with health risk, simply evaluating BMI does not distinguish between fat mass vs. lean mass or subcutaneous from visceral adiposity [118, 119]. An overall categorization of the type of asthma can be loosely correlated with the T2-high endotype, associated with early onset and atopic asthma seen in childhood and adolescence, while the Th2-low endotype is more late onset and nonatopic, including many with obesity. Not only the characteristics but also the cytokines and their roles are different between those subtypes (Th2 is more IL5, IL-4 and thus eosinophil and IgE-related, and Th1 is more related to monocytes and neutrophils and CD8+ T cells, NK-cells related, with Th17 lymphocytes, driven by IL-17, IL-6, tumor necrosis factor (TNF), and TGF-β.)

From a mechanistic and pathophysiologic standpoint, factors that cause the interface and complications between obesity and asthma can be divided into three pathophysiological categories, per Sansone et al. 2020. These include mechanical impediment, adipokines, and hypoxic inflammation. Briefly, *mechanical impediment* reduces lung volume, airway caliber, and respiratory system compliance [120]. The limitations in respiratory excursion from obesity also lead to increasing airway resistance (from decreased stretch of airway smooth muscle (ASM) cells) and actin-myosin bridge detachment. Over time this can lead to remodeling and reduction in lung volume [121].

Secondly, *adipokine production*, including leptin and resistin, increases with increased BMI, while there is the reduction in circulating adiponectin. Leptin is a proinflammatory hormone that perpetuates oxidative stress, neutrophil chemotaxis, macrophage activation and Th1 Cytokines IL-6 and IFN gamma. Leptin also causes a shift from Th2 to Th1 cytokine production [122]. However, *adiponectin* appears to have a protective role against asthma, reducing airway hyper responsiveness and eosinophil/Th2 expression and recruitment. In both boys and girls, higher adiponectin concentration was associated with less exercise-induced bronchoconstriction and asthma symptoms/exacerbations [122].

Hypoxic inflammation is another mechanism promoted by the adipose tissue, which generates a hypoxic background that produces proinflammatory cytokines such as monocyte chemoattractant protein 1 (MCP-1), tumor necrosis factor (TNF), and IP-10 FN—inducible protein 10 (IP10), consistent with a Th1 inflammatory profile. Yet also adds to the Th2 profile with increased secretion of IL-6.

Interestingly, insulin has been shown to play a role in lung function and development, particularly as it relates to atopic response after allergen exposure. Type I Insulin resistance in rat models demonstrated a decreased atopic response after allergen exposure, and while reintroducing insulin treatment in these rats, the atopic response returned [123]. In addition, insulin may have contractile properties by binding to receptors on ASM surfaces or by increasing the contraction of ASM through inhibition of presynaptic M2 muscarinic receptors and thereby increasing acetylcholine release [124].

Gender, Puberty, and Asthma

Asthma is complex and can manifest differently in boys/men vs. girls/women. This next section highlights the changes in physiology and asthma overtime, taking into account various factors that are intrinsic and extrinsic (Figs. 2 and 3).

"It Starts Early": Effect of Female Fetus on Maternal Asthma

Developmentally asthma appears to be worse in boys as compared to girls until about puberty, after which it becomes more common in girls and women. Conversely, in utero the female sex of the fetus is linked with asthma severity. Sex-specific alterations in the expression of placental genes of pregnant women with asthma have been seen – higher number of genes with altered expression in placenta of asthmatic women carrying female vs male fetuses 59 vs 6 genes have been identified. These genes are linked to growth, inflammation, and immune pathways [125].

In 2011, a placental gene expression study by Osei-Kumah and colleagues showed that there are sex-specific alterations in pregnancies complicated by asthma using pathway analysis [125]. Additionally, maternal inhaled glucocorticoid use can

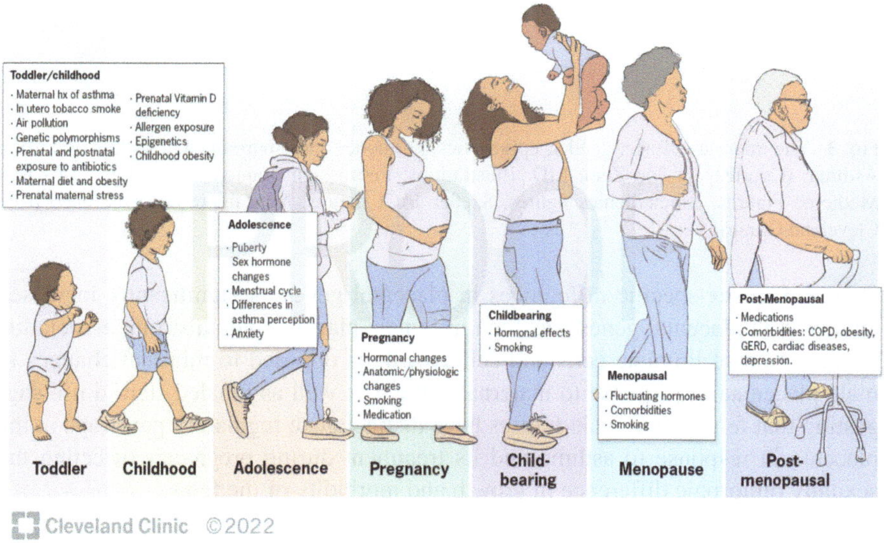

Fig. 2 Gender differences in asthma across the lifespan – unique features for women's health. Asthma is a chronic inflammatory airway disease with symptoms of cough, shortness of breath, chest tightness and or wheezing along with complex pathophysiology of numerous cell types and inflammatory mediators. Several epidemiological and clinical studies show that asthma may affect women differently than men. Genetic, environmental factors and asthma perception in women may impact healthcare utilization. The changes in immunity over the life span, the impact of sex hormones on the immune system, exposure, and environmental factors collectively influence the gene regulation. (Illustrations: Permission granted from Clinics in Chest Medicine Figures, Biomedicines figures, Szefler JACI figures; New figure commissioned from Cleveland Clinic)

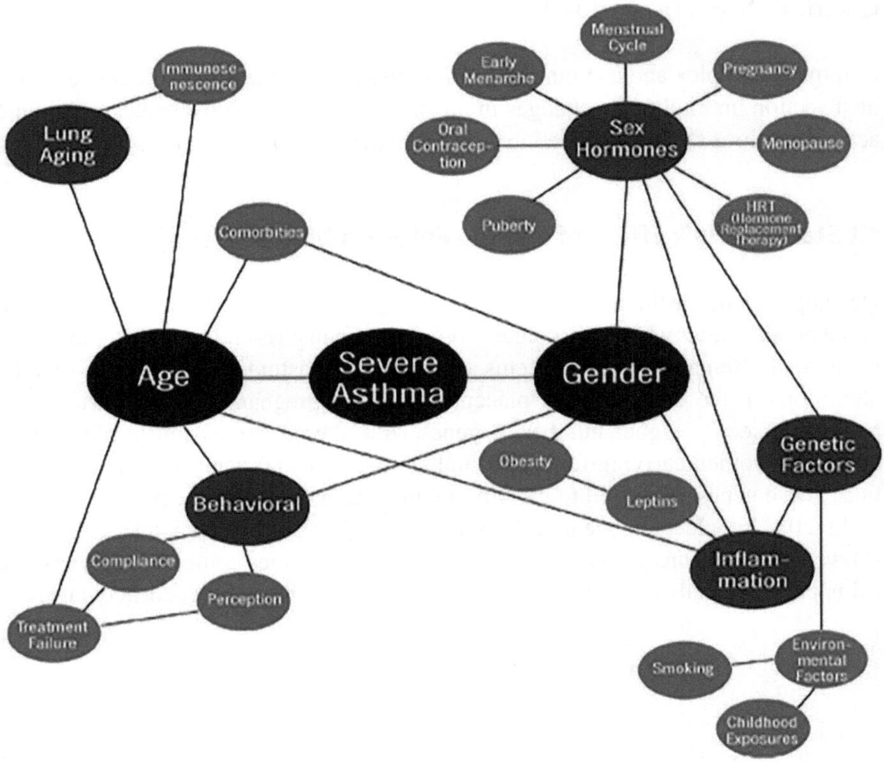

Fig. 3 The interplay of age, gender, epigenetics, hormones and environmental factors on Severe Asthma. (Courtesy of Joe Zein, MD. Illustrations: Permission granted from Clinics in Chest Medicine Figures, Biomedicines figures, Szefler JACI figures; New figure commissioned from Cleveland Clinic)

also lead to sex-specific differences in placental genes. A significantly increased alteration in placental genes was seen in female placenta with a strong association to mediation of immune cells and cell growth as opposed to minimal changes in male placentae in response to maternal asthma as well as inhaled steroid use suggesting that female and male fetuses have distinct gene regulatory pathways in the placenta in response to asthma and its treatment during pregnancy affecting the sexually dimorphic difference in growth and morbidity of the fetus.

Asthma Perception

Asthma in women is reported to be more severe in several studies. ENFUMOSA – a European Network for Understanding Mechanisms of Severe Asthma with 12 clinical specialist centers in nine European countries showed in a cross-sectional observational study showed that severe asthma was four times higher in women as compared to men [126]. Similar results were seen in an Indian (matched) cohort of

85 patients with 41 men and 44 women, with females having an increased perception of dyspnea, poor control of disease, along with the poor quality of life in comparison to the male counterparts with matched severity of asthma [127].

Similarly, other robust data and studies indicate that women, especially those of middle age, have more healthcare utilization with increased office and emergency room visits with increased medication use and mortality even after accounting for the severity of asthma and comorbid conditions [127–131]. Cydulka et al. showed in a multicenter, prospective cohort of 1291 patients presenting to the emergency department with an exacerbation, comparing peak expiratory flow rates PEFR along with reported symptom severity that women had higher hospital admissions rates and prolonged stay with similar lung function [128].

Similar data was seen in highly trained endurance Finnish swimmers – though the prevalence of asthma is comparable in female and male swimmers (19%), gender differences in asthma-like symptoms were noticed. Female swimmers had more coughing, while males had more allergies [132].

Another Nordic study validated the higher prevalence of anxiety with a preponderance to females with worse sleep quality and heightened anxiety [133]. This could perhaps imply gender differences in symptom perception, and reveal greater suffering and suboptimal coping skills in females rendering emphasis on different clinical and management approaches [127].

Medication Adherence Differences

Medication adherence, peak flow measurement, and behavior toward health care use differ between men and women. For example, although women are more likely to have unscheduled office visits and carry a rescue inhaler than men (61% vs. 30%), men were more likely to visit the emergency department because they ran out of their inhaled medications [134–136]. Recent data suggest that girls and women may benefit from anti-leukotriene therapy more than boys and men [137]. In both humans and mice, lipoxygenase–activating protein and novel type 5-lipoxygenase inhibitors were found to be more potent in girls and women, a response that was reversed by 5a-dihydrotestosterone. Combined, these data suggest that gender differences in asthma also may be related to differences in health behaviors, adherence to treatment guidelines, and response of controller therapy.

The work of Spaziano et al. [137] on the effect of anti-LT therapy differing in genders could, in theory, represent a different distribution of single nucleotide polymorphism (SNP) leading to better lung function, inflammation, and symptomatology. In the future, these findings could possibly help guide and craft personalized care.

Hormonal Differences in Asthma

Multifactorial etiologies have been associated with hormonal factors seen with puberty, during the menstrual cycle, pregnancy, and menopause. Both FEV1 and

FVC (forced vital capacity) are lower in women with and without asthma during luteal phase when levels of estradiol and progesterone are two to seven folds higher [138, 139]. However, asthmatic women had increased respiratory symptoms and lower lung function in the pre and menstrual period leading to increased health care utilization. Data from the Severe Asthma Research Program (SARP) was used to evaluate a cohort of more than 500 women and showed that peri-menstrual asthma (PMA) was more severe and had poor control. Aspirin sensitivity and lower FVC % predicted are associated with PMA after adjusting for multiple confounding factors, suggesting modifications in prostaglandins were contributing to this severe asthmatic phenotype [140].

By large, etiologies of PMA remain obscure, but the role of fluctuating hormones during menstrual cycle is one of the key mechanisms. The wavering of estrogens at ovulation and before menstruation and the progesterone secretion during the luteal phase followed by withdrawal seems to correlate with the deterioration of asthma during the luteal phase and/or during the first days of the menstrual cycle [141].

The role of oral contraceptives has been explored in PMA asthma and remains largely unanswered. Nwaru and colleagues studied this in a total of 3257 nonpregnant, 16- to 45-year-old women and found that the use of any kind of hormonal contraception was associated with a reduced risk of asthma diagnosis (odds ratio 0.68, 95% confidence interval 0.47–0.98) after adjustment of confounders [142]. Similarly, this approach was associated with a reduced risk of asthma exacerbation but with wide 95% confidence intervals. To date, evidence-based therapy for perimenstrual asthma is still lacking. A beneficial role of oral contraceptive pills (OCPs) was suggested in a small series – studies showing attenuation in a cyclical change in airway reactivity along with the reduction in diurnal PEFR variability with suppression of normal luteal phase rise in sex hormones [140].

Asthma of Mothers During Pregnancy

Thus far, we have discussed how prenatal exposures during pregnancy affect the child. In this section, we will briefly review the hormonal and physiologic effects of pregnancy on the expectant mother. It has been held that the clinical course follows a rule of thirds: about one-third of pregnant women experience improvement in the symptoms, one-third experience increased bronchospasm and worsening symptoms, one third remain the same; however, these changes are variable among women, and even among separate pregnancies [143, 144]. Worldwide, overall one in ten pregnant women have asthma and of these women, one in ten will have a severe asthma exacerbation during their pregnancy that requires oral corticosteroids [145]. Interestingly, prior pregnancies may be similar but do not necessarily predict the level of asthma control in future pregnancies [144, 146]. Postpartum, most women with asthma revert back to their prepartum level of asthma severity approximately at 3 months postpartum; however, approximately 25% can experience worsening asthma due to multiple influences.

Hormonal contributions to the course of asthma in pregnancy are related to cortisol, progesterone, estrogen, and prostaglandins. Cortisol levels rise during pregnancy and may account for the reduction in asthma symptoms nearer to the third trimester. While progesterone can demonstrate effects that alleviate asthma severity, such as cAMP-induced bronchodilation, direct smooth muscle relaxation, and increased sensitivity of central CO_2 receptors, it can also worsen asthma control via mechanisms of decreased beta-2 receptor responsiveness and increased airway inflammation and mucous production. Alterations in the maternal immune system can also be seen with increases in the Th2 inflammatory milieu with Th2 cytokines such as (IL-6, IL-8, ICAM-1, and regulated upon activation, normal T cell expressed and presumably secreted (RANTES) and eotaxin) relative to cells in nonpregnant women [145].

In addition to the hormonal effects on asthma, there is an obvious anatomic/physiologic contribution of symptoms during pregnancy which can be difficult to discern from asthma. Common physiologic changes in all pregnancies will likely cause dyspnea in most pregnant women due to the accommodation of the gravid uterus. Decreases in functional residual capacity (FRC) occur from both compartments of expiratory reserve volume (ERV) and residual volume (RV). In order to preserve the total lung capacity, the chest wall accommodates by increasing the subcostal angle as well as the transverse diameters and chest wall circumference. This partially accommodates the elevation of the diaphragm. In addition, increased diaphragm excursion and preserved diaphragm muscle strength help with the accommodation of increased tidal volumes [147, 148].

The known changes in lung function with pregnancy include a 400–700 mL (20–30%) drop in functional residual capacity (FRC), which decreases further by term, particularly in the supine position. A compensatory increase of inspiratory capacity is seen with a 200–350 mL (5–10%) increase during pregnancy. Overall, a relatively low 5% decrease in TLC occurs, with preserved vital capacity and increased tidal volume. These accommodations are created from a reduction in the functional residual capacity decreasing as much as 20% [149]. Understanding the known changes in pregnancy allows proper determination of the predominant source of dyspnea. Pulmonary function testing is particularly needed in situations when the cause is not clear. Certainly, in women with a history of asthma, asthma, and pulmonary function testing should be considered on a periodic basis throughout the pregnancy.

Importantly, there are differences in fetal growth by gender of the neonate if asthma is uncontrolled and untreated with inhaled corticosteroids. Female fetuses had reduced birthweight when their mothers had asthma, while male offspring grew normally. However with exacerbations, female fetuses continued growth, albeit at their lower trajectory, but males were more likely to be compromised to be growth restricted. Maternal asthma is associated with reduced female fetal growth [150, 151]. Fortunately, with active management of asthma, maternal as well as fetal complications can be reduced, and preterm delivery is reduced with active surveillance and management of asthma during pregnancy. Furthermore, this was possible without any association demonstrating that the use of bronchodilators and inhaled steroids was related to the risk of congenital malformations [152, 153].

Unfortunately there remains a tendency for expectant mothers with asthma to stop their asthma medications once they learn they are pregnant, likely from an unwarranted concern for harm for the fetus. Utilization of all classes of asthma medications decreased during pregnancy has been documented as follows: 23% decrease in ICS, 13% decrease in beta-agonist, and 54% decrease in rescue steroids [154].

Asthma and Menopause

Menopause is manifested with significantly fluctuating hormones and added comorbid conditions with resultant variable impact on prevalence and control of asthma [155]. The Respiratory Health in Northern Europe (RHINE) study described a new phenotype of asthma in a subgroup of women with disease onset after menopause [155]. Similarly, in the French E3N cohort, surgical menopause was associated with an increased risk of asthma onset. On the contrary the European Community Respiratory Healthy Survey I (ECRHS I) studied 884 menopausal women (aged 46–54 years old, of them 540 using hormone replacement therapy (HRT)) and did not find any association between self-reported asthma and HRT use in premenopausal or postmenopausal women [156]. Though in the ECRHS II trial showed that women transitioning to menopause with amenorrhea for 6 months had a reduction in lung function with increasing asthma symptomatology as compared to premenopausal women, particularly in leaner individuals [156, 157].

Although conflicting data exists in regards to asthma and menopause with discernable worsening characterized by severe and frequent exacerbations and poor response to therapy, in general, the impact of menopause on asthma has been a positive one. It has been studied in several cohorts, with the biggest data from Nurses' health Study on women who were hormonal replacement therapy (HRT) naïve had a lower incidence of asthma and was reversed by use of HRT [158]. Likewise, severe asthma improves as sex hormones wane in the menopausal transition [138, 159].

Zein et al. [160] studied a cross-sectional, prospective cohort of 1130 adults with severe asthma enrolled over 9 years span in Severe Asthma Research Program (SARP), showing not only a higher prevalence of asthma in women but also suggested a reversal of well-described gender differences at puberty when girls have severe asthma as compared to boys. This perhaps suggests a physiologic change in female lung at the time of menopause (Fig. 3).

Asthma in Transgender Individuals (Role of Hormonal Switch)

An inexplicable connection exists between transgender individuals and asthma. Dragon et al. showed in a systematic Medicare data review that transgender beneficiaries have a higher chronic disease burden; 29.6% of transgender as compared to

13.6% of cisgender patients [161, 162]. In general, asthma prevalence is higher in cisgender women as compared to cisgender men, 10.4% vs. 6.2%. The role of sex hormones remains unclear in cisgender women. Particularly in obese individuals with elevated testosterone and estradiol, there has been a decreased risk of asthma. However, the hormone replacement therapy in cisgender women, regardless of their obesity status, increases the risk of asthma.

Transgender individuals embody a distinctive population as they frequently pursue cross-sex hormone therapy. Very little is known about respiratory health in cross-sex hormone therapy, including asthma. This question was studied by Zein et al. in 56 million individuals, showing that transgender status is associated with a significantly higher risk of lifetime asthma, with the highest risk in male-to-female transgender individuals. This highlights the importance of both physicians and patients to be aware of the potential impact of estrogen on pulmonary outcomes in patients undergoing sex reassignment therapy [161].

Asthma in Elders

Asthma is often underdiagnosed in older adults, and current statistics on prevalence are likely underestimating asthma. Current literature suggests that the prevalence is between 4% and 13% in ages over 65 years. As of 2020, the Centers for Disease Control reports that 7.8% (±0.34%) of people 65 and older have asthma [163]. The numbers will continue to grow due to increases in our aging population. Asthma can have a higher prevalence in the elderly population, defined as 65 years old and older. Although often not recognized, asthma in elderly patients is associated with higher rates of mortality, hospitalization, and asthma with a near-fatal clinical picture, as compared to younger individuals with asthma [164, 165]. Mortality is fivefold higher in adults aged ≥65 years old vs. the younger group of 18–64 years.

Unfortunately, however, this is a population at higher risk for multiple reasons, including under-diagnosis, under-treatment or patient factors such as cognition, improper use, and polypharmacy [166, 167]. However, the unbelievably sobering statistic is that elder asthma patients are greater than five times more likely to die from their disease than younger patients with asthma (Fig. 1 deaths).

In contrast to long-standing asthma (from childhood or young adulthood), asthma in older populations has not fully been studied with similar depth. There has been limited research into how the lung changes with time, with respect to lung structure, the immunology, and cell biology within the organ and its biologic environment, the role of allergies in lungs over time, and more permanent changes in structure and physiologic functioning of the lung from airway remodeling.

A phenotype that can be noted regarding lung structure over time is the decrease in elastic recoil, more chest wall rigidity and kyphosis, and a reduction in respiratory muscle strength [168–170]. Up to a 31% reduction in respiratory muscle strength was noted in a group of individuals 24–75 years old. The reduced compliance was believed to be from calcification of costochondral joints, costal cartilage, and degenerative join disease, and that decreased elastic recoil occurs from a culmination of these various changes with an additional reduction of diaphragm and skeletal muscle strength [170]. Furthermore, other chronic conditions such as cardiac, neurologic, and other metabolic conditions such as malnutrition over time contribute to dyspnea from muscle weakness [165, 171]. In addition, in older patients, measurement of changes with age needs to account for natural changes in lung function, including obstructive and mixed patterns with increased residual volume (RV) and reductions in FEV_1, ratio, and FVC. While a decline in FEV_1 is expected and begins between 25 and 30 years old, individuals with asthma have lung function decline in a more accelerated pace [170, 172].

Comorbidities that have an enhanced effect on elderly individuals with asthma include COPD, chronic sinusitis, obesity, and depression. The coexistence of asthma with COPD is frequent, with rates of 40–60% in a study from Kauppi, and the overlap predicts low quality of life [173]. In addition, there is an increased prevalence of exacerbation in this population [174].

Whereas chronic rhinitis affects younger populations, older populations experience chronic sinusitis. In the SARP cohort, higher sinus disease was seen in older patients, both late onset and severe late-onset asthma [120]. Chronic rhinosinusitis, as well as gastroesophageal reflux disease (GERD), are more likely to be present in those with uncontrolled asthma (GERD: 30.7% vs. 22.3%, P 5 .006 and chronic sinusitis: 29.9% vs. 22.9%, $P = .02$). This is also similar to sinusitis with nasal polyposis. The underlying mechanism may also be related to Staphylococcus aureus super-antigens which may amplify IgE-related and eosinophilic inflammation [175].

Obesity in elderly individuals with airway disease imposes an added risk factor/symptom burden. Higher BMI has been associated with the prevalence of wheezing and physician-dx asthma. The risk of asthma increases in proportion to BMI or abdominal subcutaneous adiposity (without association with visceral adiposity) [176]. In addition to increased wheezing, a higher BMI increases the risk of corticosteroid insensitivity in the treatment of asthma [177]. Compounding the vicious cycle, the ability to function physically and be engaged in physical activity to lose weight also hamper means by which asthma may be better controlled [178].

Either as a result of their airways condition or by mechanisms of worsening asthma from depression itself, asthma and depression worsen asthma outcomes [179]. These may be related to behavioral reasons (lower quality of life (QOL) and inability for self-efficacy), which are predictors of future exacerbation. Without awareness, particularly in the elderly population, this may be an area to address that would be more counseling and interventions outside of asthma treatment itself.

Active smoking is associated with recurrent exacerbations in addition to known lung function decline, low QOL, hospitalizations, and increased mortality seen in all patients with asthma who smoke [167]. The opportunity to manage tobacco use with strategies for smoking cessation should also not be overlooked as seemingly futile, as it is a known and significant risk factor for exacerbation. Ongoing and consistent guidance and support for active smokers to quit will modify their risk.

Back to the Paradigm of Asthma

It has been rightfully proposed and a collective challenge to our profession to approach and manage the natural history of asthma in a proactively preventative and mitigating manner. Instead of episodic management, a more holistic and individualized management approach is needed. This requires awareness of inflection points from an exposure and risk standpoint to better predict the course of the disease over time, modify risks, and create a partnership with patients throughout their lifetime [1].

How do we do this? We in our profession must focus on the need to minimize risk and impairment with a goal to achieve optimal control. Guidelines include National Asthma Education and Prevention Program (NAEPP) and were updated most recently in 2021 and addressed intermittent ICS, LAMA, FeNO, BT, and indoor allergen reduction [3]. GINA guidelines are available and are revised annually, providing initial guidance such as single maintenance and reliever therapy (SMART) therapy. The NAEPP focused on the guidelines also embraced this concept [3].

Quality initiatives also provide guidelines for care, for instance, National Institute for Health and Care excellence [4]. This initiative addresses diagnosing and managing asthma in children to adults, focusing on diagnosis and control, reducing the risk of asthma attacks. True to current knowledge, personalized plans of asthma management is addressed, as well as the need to address the increasing threat of both indoor and outdoor air pollution.

Prediction of a trajectory is compelling in order to manage and counsel patients of all ages proactively. There is growing concern about poorly controlled asthma and irreversible pulmonary function (Fig. 1). McGeachie evaluated various trajectories from childhood to early adulthood (NHLBI CAMP), and a similar finding to what has been noted in COPD was noted in asthma. In addition, adding scores/metrics to avoid underestimating severity should be considered in the evaluation of asthma. The NIAID Inner-City Asthma Consortium (ICAC) CASI score correlates with levels of asthma severity, and Krouse et al. defined 'minimal important difference' reference values [180].

Certainly, a composite evaluation is what will be needed. Current studies from NIH/NHLBI such as SARP and PrecISE and other collaborative network trials are utilizing biomarkers evaluation and genetic predisposition. In a perfect world, hopefully soon, approaches to be considered by clinicians will be to: watch for lung function loss, check biomarkers to phenotype, prevent seasonal exacerbations proactively and with anticipatory step-up therapy, and a tailored approach of feedback-based technological adherence monitoring [181].

How Do We Engage with Our Patients Through the Lifespan of Asthma?

In order for our efforts to be successful, it will require partnership with patients in an authentic and proactive relationship. Health concerns are understandably associated with self-doubt and stigma, feeling that they may have brought it on

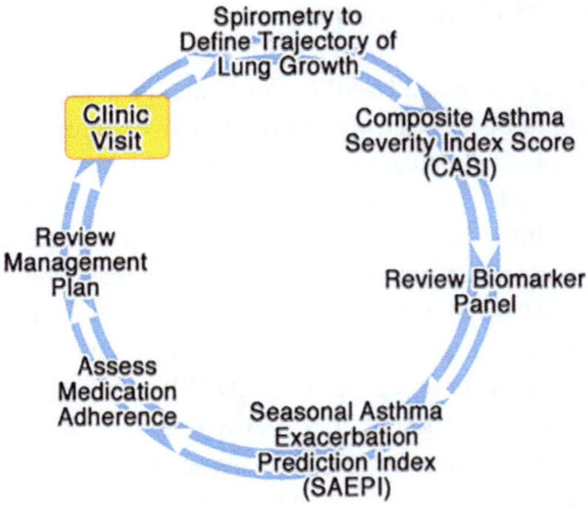

Fig. 4 Visits for asthma that assess and monitor risk. (Illustrations: Permission granted from Clinics in Chest Medicine Figures, Biomedicines figures, Szefler JACI figures; New figure commissioned from Cleveland Clinic)

themselves. Further, patients fear losing their independence and are concerned about becoming dependent on medications. As in all chronic conditions, with respect to asthma and lung health, success breeds success. Repeated exacerbations only make it more difficult to have a life where asthma is a barrier to a normal life. Adopting approaches that meet the need of our patients include shared decision-making and engaging children and their caregivers. Medication adherence and patient/parent satisfaction were higher. Effective self-management is particularly important for people with asthma, and empowering them to adjust to their symptoms and needs provides a patient-empowered approach [182].

As we care for patients and partner with them, various patient engagement practices and strategies have been proposed and evaluated. In one study, patient engagement increased when they perceived they were receiving education and receiving personalized attention [183]. Ultimately, what is needed is to ascertain what matters to and motivates our patients in their lives, and thereby make it personal and meaningful with shared goals for improvement of daily life and symptoms. That is what a motivational and self-management approach to care accomplishes.

Communication and conversations between patients and providers need to be a partnership and negotiation with agreed-upon timelines and goals. Such goals are set by determining specific personal goals and designing specific action plans to achieve those goals that are most important to them [184]. This also requires the reality of the patient's concerns, often hearing worries about longer-term medication use, costs of medications, and side effects, all of which can become a primary focus that needs to be addressed.

Figure 4 presents a model for conducting a clinic visit, including relating with motivational interviewing, shared goals, and personalized management and care [1].

Conclusion

Asthma is a protean respiratory condition comprising airway inflammation and episodic worsening of symptoms and airway inflammation. There is a growing awareness and appreciation of the "asthmas" with a variety of individually specific presentations, with a composite of prior etiologies from prenatal to elderly age, allowing for unique courses of comorbidities, lung function, and exacerbation vs. control. A means to predict the life course of asthma, if possible, may allow better proactive and preventive management strategies. Knowledge gained from large population health and network trials will add to this body of knowledge.

In the meantime, we can embrace individualized strategies for care and the setting of care, understand and reduce the impact of social disparities of health with local resources, and advocate for just public policies. In that way and in an ideal scenario, the partnership with healthcare providers and their patients will allow a more proactive than merely reactive approach to care. In that way, we hope that our patients and the communities we serve will be able to reach their highest potential.

References

1. Szefler SJ. Asthma across the lifespan: time for a paradigm shift. J Allergy Clin Immunol. 2018;142(3):773–80.
2. Asthma, G.I.f. Global strategy for asthma management and prevention, 2022. 2022.
3. Expert Panel Working Group of the National Heart, L., et al. 2020 Focused updates to the asthma management guidelines: a report from the National Asthma Education and Prevention Program Coordinating Committee Expert Panel Working Group. J Allergy Clin Immunol. 2020;146(6):1217–70.
4. Excellence, N.I.f.H.a.C. Asthma: diagnosis, monitoring and chronic asthma management. 2017. https://www.nice.org.uk/guidance/NG80/evidence.
5. McGeachie MJ, et al. Patterns of growth and decline in lung function in persistent childhood asthma. N Engl J Med. 2016;374(19):1842–52.
6. Lee A, et al. Prenatal and postnatal stress and asthma in children: temporal- and sex-specific associations. J Allergy Clin Immunol. 2016;138(3):740–7. e3
7. Burke W, et al. Family history as a predictor of asthma risk. Am J Prev Med. 2003;24(2):160–9.
8. Ronchetti R, et al. Is the increase in childhood asthma coming to an end? Findings from three surveys of schoolchildren in Rome, Italy. Eur Respir J. 2001;17(5):881–6.
9. Chhabra SK, et al. Risk factors for development of bronchial asthma in children in Delhi. Ann Allergy Asthma Immunol. 1999;83(5):385–90.
10. Kalyoncu AF, et al. Prevalence of asthma and allergic diseases in primary school children in Ankara, Turkey: two cross-sectional studies, five years apart. Pediatr Allergy Immunol. 1999;10(4):261–5.
11. al Frayh AR. Asthma patterns in Saudi Arabian children. J R Soc Health. 1990;110(3):98–100.
12. Norrman E, et al. Prevalence and incidence of asthma and rhinoconjunctivitis in Swedish teenagers. Allergy. 1998;53(1):28–35.
13. Frischer T, et al. Risk factors for childhood asthma and recurrent wheezy bronchitis. Eur J Pediatr. 1993;152(9):771–5.
14. Chhabra SK, et al. Prevalence of bronchial asthma in schoolchildren in Delhi. J Asthma. 1998;35(3):291–6.
15. Gray L, et al. Family patterns of asthma, atopy and airway hyperresponsiveness: an epidemiological study. Clin Exp Allergy. 2000;30(3):393–9.

16. London SJ, et al. Family history and the risk of early-onset persistent, early-onset transient, and late-onset asthma. Epidemiology. 2001;12(5):577–83.
17. Rusconi F, et al. Risk factors for early, persistent, and late-onset wheezing in young children. SIDRIA Collaborative Group. Am J Respir Crit Care Med. 1999;160(5 Pt 1):1617–22.
18. Sherman CB, et al. Early childhood predictors of asthma. Am J Epidemiol. 1990;132(1):83–95.
19. Takemura Y, et al. Relation between breastfeeding and the prevalence of asthma: the Tokorozawa Childhood Asthma and Pollinosis Study. Am J Epidemiol. 2001;154(2):115–9.
20. Ones U, Sapan N, Somer A, Dişçi R, Salman N, Güler N, Yalçin I. Prevalence of childhood asthma in Istanbul, Turkey. Allergy. 1997;52(5):570–5.
21. Oddy WH, Peat JK, de Klerk NH. Maternal asthma, infant feeding, and the risk of asthma in childhood. J Allergy Clin Immunol. 2002;110(1):65–7.
22. Abuekteish F, et al. Prevalence of asthma and wheeze in primary school children in northern Jordan. Ann Trop Paediatr. 1996;16(3):227–31.
23. Agabiti N, et al. The impact of parental smoking on asthma and wheezing. SIDRIA Collaborative Group. Studi Italiani sui Disturbi Respiratori nell'Infanzia e l'Ambiente. Epidemiology. 1999;10(6):692–8.
24. Oliveti JF, Kercsmar CM, Redline S. Pre- and perinatal risk factors for asthma in inner city African-American children. Am J Epidemiol. 1996;143(6):570–7.
25. Gilliland FD, et al. Effects of early onset asthma and in utero exposure to maternal smoking on childhood lung function. Am J Respir Crit Care Med. 2003;167(6):917–24.
26. Gilliland FD, et al. Maternal smoking during pregnancy, environmental tobacco smoke exposure and childhood lung function. Thorax. 2000;55(4):271–6.
27. Child F, et al. The association of maternal but not paternal genetic variation in GSTP1 with asthma phenotypes in children. Respir Med. 2003;97(12):1247–56.
28. Pattenden S, et al. Parental smoking and children's respiratory health: independent effects of prenatal and postnatal exposure. Tob Control. 2006;15(4):294–301.
29. Singh SP, et al. Prenatal cigarette smoke decreases lung cAMP and increases airway hyperresponsiveness. Am J Respir Crit Care Med. 2003;168(3):342–7.
30. Stick SM, et al. Effects of maternal smoking during pregnancy and a family history of asthma on respiratory function in newborn infants. Lancet. 1996;348(9034):1060–4.
31. Wongtrakool C, et al. Nicotine alters lung branching morphogenesis through the alpha7 nicotinic acetylcholine receptor. Am J Physiol Lung Cell Mol Physiol. 2007;293(3):L611–8.
32. Noakes PS, Holt PG, Prescott SL. Maternal smoking in pregnancy alters neonatal cytokine responses. Allergy. 2003;58(10):1053–8.
33. Krautenbacher N, et al. Asthma in farm children is more determined by genetic polymorphisms and in non-farm children by environmental factors. Pediatr Allergy Immunol. 2021;32(2):295–304.
34. Salam MT, et al. Transforming growth factor- 1 C-509T polymorphism, oxidant stress, and early-onset childhood asthma. Am J Respir Crit Care Med. 2007;176(12):1192–9.
35. Hugg TT, et al. Parental smoking behaviour and effects of tobacco smoke on children's health in Finland and Russia. Eur J Pub Health. 2008;18(1):55–62.
36. Jaakkola JJ, Gissler M. Are girls more susceptible to the effects of prenatal exposure to tobacco smoke on asthma? Epidemiology. 2007;18(5):573–6.
37. Lasso-Pirot A, Delgado-Villalta S, Spanier AJ. Early childhood wheezers: identifying asthma in later life. J Asthma Allergy. 2015;8:63–73.
38. Orellano P, et al. Effect of outdoor air pollution on asthma exacerbations in children and adults: systematic review and multilevel meta-analysis. PLoS One. 2017;12(3):e0174050.
39. Wu ZX, et al. Prenatal and early, but not late, postnatal exposure of mice to sidestream tobacco smoke increases airway hyperresponsiveness later in life. Environ Health Perspect. 2009;117(9):1434–40.
40. Slama R, et al. Traffic-related atmospheric pollutants levels during pregnancy and offspring's term birth weight: a study relying on a land-use regression exposure model. Environ Health Perspect. 2007;115(9):1283–92.

41. Jedrychowski WA, et al. Intrauterine exposure to polycyclic aromatic hydrocarbons, fine particulate matter and early wheeze. Prospective birth cohort study in 4-year olds. Pediatr Allergy Immunol. 2010;21(4 Pt 2):e723–32.
42. Cookson WO, et al. Maternal inheritance of atopic IgE responsiveness on chromosome 11q. Lancet. 1992;340(8816):381–4.
43. Fryer AA, et al. Polymorphism at the glutathione S-transferase GSTP1 locus. A new marker for bronchial hyperresponsiveness and asthma. Am J Respir Crit Care Med. 2000;161(5):1437–42.
44. Breton CV, et al. Variation in the GST mu locus and tobacco smoke exposure as determinants of childhood lung function. Am J Respir Crit Care Med. 2009;179(7):601–7.
45. Huang YJ, Boushey HA. The microbiome in asthma. J Allergy Clin Immunol. 2015;135(1):25–30.
46. Illi S, et al. Perennial allergen sensitisation early in life and chronic asthma in children: a birth cohort study. Lancet. 2006;368(9537):763–70.
47. Mueller NT, et al. The infant microbiome development: mom matters. Trends Mol Med. 2015;21(2):109–17.
48. Metsälä J, et al. Prenatal and post-natal exposure to antibiotics and risk of asthma in childhood. Clin Exp Allergy. 2015;45(1):137–45.
49. Turi KN, et al. Dose, timing, and spectrum of prenatal antibiotic exposure and risk of childhood asthma. Clin Infect Dis. 2021;72(3):455–62.
50. Litonjua AA, et al. Six-year follow-up of a trial of antenatal vitamin D for asthma reduction. N Engl J Med. 2020;382(6):525–33.
51. Camargo CA Jr, et al. Maternal intake of vitamin D during pregnancy and risk of recurrent wheeze in children at 3 y of age. Am J Clin Nutr. 2007;85(3):788–95.
52. Lu M, et al. Effect of early and late prenatal vitamin D and maternal asthma status on offspring asthma or recurrent wheeze. J Allergy Clin Immunol. 2021;147(4):1234–41. e3
53. Devereux G, Seaton A. Diet as a risk factor for atopy and asthma. J Allergy Clin Immunol. 2005;115(6):1109–17; quiz 1118.
54. Litonjua AA, Gold DR. Asthma and obesity: common early-life influences in the inception of disease. J Allergy Clin Immunol. 2008;121(5):1075–84. quiz 1085-6
55. Pearson PJ, et al. Vitamin E supplements in asthma: a parallel group randomised placebo controlled trial. Thorax. 2004;59(8):652–6.
56. Devereux G, et al. Early childhood wheezing symptoms in relation to plasma selenium in pregnant mothers and neonates. Clin Exp Allergy. 2007;37(7):1000–8.
57. Baiz N, et al. Prenatal exposure to selenium may protect against wheezing in children by the age of 3. Immun Inflamm Dis. 2017;5(1):37–44.
58. Seaton A, Godden DJ, Brown K. Increase in asthma: a more toxic environment or a more susceptible population? Thorax. 1994;49(2):171–4.
59. Martindale S, et al. Antioxidant intake in pregnancy in relation to wheeze and eczema in the first two years of life. Am J Respir Crit Care Med. 2005;171(2):121–8.
60. Chatzi L, et al. Mediterranean diet in pregnancy is protective for wheeze and atopy in childhood. Thorax. 2008;63(6):507–13.
61. de Batlle J, et al. Mediterranean diet is associated with reduced asthma and rhinitis in Mexican children. Allergy. 2008;63(10):1310–6.
62. Castro-Rodriguez JA, Garcia-Marcos L. What are the effects of a Mediterranean diet on allergies and asthma in children? Front Pediatr. 2017;5:72.
63. Pellegrini-Belinchón J, et al. Factors associated to recurrent wheezing in infants under one year of age in the province of Salamanca, Spain: is intervention possible? A predictive model. Allergol Immunopathol (Madr). 2016;44(5):393–9.
64. Chatzi L, et al. Protective effect of fruits, vegetables and the Mediterranean diet on asthma and allergies among children in Crete. Thorax. 2007;62(8):677–83.
65. Lock M, et al. Regulation of fetal lung development in response to maternal overnutrition. Clin Exp Pharmacol Physiol. 2013;40(11):803–16.

66. Martinez FD. The origins of asthma and chronic obstructive pulmonary disease in early life. Proc Am Thorac Soc. 2009;6(3):272–7.
67. Zugna D, et al. Maternal complications in pregnancy and wheezing in early childhood: a pooled analysis of 14 birth cohorts. Int J Epidemiol. 2015;44(1):199–208.
68. Caudri D, et al. Perinatal risk factors for wheezing phenotypes in the first 8 years of life. Clin Exp Allergy. 2013;43(12):1395–405.
69. Harskamp-van Ginkel MW, et al. A study on mediation by offspring BMI in the association between maternal obesity and child respiratory outcomes in the Amsterdam born and their development study cohort. PLoS One. 2015;10(10):e0140641.
70. Reichman NE, Nepomnyaschy L. Maternal pre-pregnancy obesity and diagnosis of asthma in offspring at age 3 years. Matern Child Health J. 2008;12(6):725–33.
71. Håberg SE, et al. Maternal obesity in pregnancy and respiratory health in early childhood. Paediatr Perinat Epidemiol. 2009;23(4):352–62.
72. Leermakers ET, et al. Maternal weight, gestational weight gain and preschool wheezing: the Generation R Study. Eur Respir J. 2013;42(5):1234–43.
73. Wright RJ, et al. Disrupted prenatal maternal cortisol, maternal obesity, and childhood wheeze. Insights into prenatal programming. Am J Respir Crit Care Med. 2013;187(11):1186–93.
74. Forno E, et al. Maternal obesity in pregnancy, gestational weight gain, and risk of childhood asthma. Pediatrics. 2014;134(2):e535–46.
75. Dumas O, et al. Longitudinal study of maternal body mass index, gestational weight gain, and offspring asthma. Allergy. 2016;71(9):1295–304.
76. Hsu HH, et al. Prenatal particulate air pollution and asthma onset in urban children. Identifying sensitive windows and sex differences. Am J Respir Crit Care Med. 2015;192(9):1052–9.
77. Keglowich LF, Borger P. The three A's in asthma – airway smooth muscle, airway remodeling & angiogenesis. Open Respir Med J. 2015;9:70–80.
78. Jones PA. Functions of DNA methylation: islands, start sites, gene bodies and beyond. Nat Rev Genet. 2012;13(7):484–92.
79. Arathimos R, et al. Epigenome-wide association study of asthma and wheeze in childhood and adolescence. Clin Epigenetics. 2017;9:112.
80. Chiba Y, et al. Down-regulation of miR-133a contributes to up-regulation of Rhoa in bronchial smooth muscle cells. Am J Respir Crit Care Med. 2009;180(8):713–9.
81. Mohamed JS, Lopez MA, Boriek AM. Mechanical stretch up-regulates microRNA-26a and induces human airway smooth muscle hypertrophy by suppressing glycogen synthase kinase-3β. J Biol Chem. 2010;285(38):29336–47.
82. Speizer FE, Tager IB. Epidemiology of chronic mucus hypersecretion and obstructive airways disease. Epidemiol Rev. 1979;1:124–42.
83. Abrahamsson TR, et al. Low diversity of the gut microbiota in infants with atopic eczema. J Allergy Clin Immunol. 2012;129(2):434–40, 440 e1–2.
84. van Nimwegen FA, et al. Mode and place of delivery, gastrointestinal microbiota, and their influence on asthma and atopy. J Allergy Clin Immunol. 2011;128(5):948–55 e1–3.
85. Biesbroek G, et al. The impact of breastfeeding on nasopharyngeal microbial communities in infants. Am J Respir Crit Care Med. 2014;190(3):298–308.
86. Baïz N, et al. Early oral exposure to house dust mite allergen through breast milk: a potential risk factor for allergic sensitization and respiratory allergies in children. J Allergy Clin Immunol. 2017;139(1):369–372.e10.
87. Cecchi L, D'Amato G, Annesi-Maesano I. External exposome and allergic respiratory and skin diseases. J Allergy Clin Immunol. 2018;141(3):846–57.
88. Dick S, et al. A systematic review of associations between environmental exposures and development of asthma in children aged up to 9 years. BMJ Open. 2014;4(11):e006554.
89. Gruchalla RS, et al. Inner City Asthma Study: relationships among sensitivity, allergen exposure, and asthma morbidity. J Allergy Clin Immunol. 2005;115(3):478–85.
90. Ahluwalia SK, et al. Mouse allergen is the major allergen of public health relevance in Baltimore City. J Allergy Clin Immunol. 2013;132(4):830–5.e1–2.

91. Torjusen EN, et al. Dose-response relationships between mouse allergen exposure and asthma morbidity among urban children and adolescents. Indoor Air. 2013;23(4):268–74.
92. Phipatanakul W, et al. Effect of environmental intervention on mouse allergen levels in homes of inner-city Boston children with asthma. Ann Allergy Asthma Immunol. 2004;92(4):420–5.
93. Hu FB, et al. Prevalence of asthma and wheezing in public schoolchildren: association with maternal smoking during pregnancy. Ann Allergy Asthma Immunol. 1997;79(1):80–4.
94. Lannerö E, et al. Maternal smoking during pregnancy increases the risk of recurrent wheezing during the first years of life (BAMSE). Respir Res. 2006;7(1):3.
95. Comhair SA, et al. Detrimental effects of environmental tobacco smoke in relation to asthma severity. PLoS One. 2011;6(5):e18574.
96. Weschler CJ, Shields HC. Indoor ozone/terpene reactions as a source of indoor particles. Atmos Environ. 1999;33(15):2301–12.
97. Kwon JW, et al. Exposure to volatile organic compounds and airway inflammation. Environ Health. 2018;17(1):65.
98. Subramanian A, Khatri SB. The exposome and asthma. Clin Chest Med. 2019;40(1):107–23.
99. Gauderman WJ, et al. The effect of air pollution on lung development from 10 to 18 years of age. N Engl J Med. 2004;351(11):1057–67.
100. Gauderman WJ, et al. Association of improved air quality with lung development in children. N Engl J Med. 2015;372(10):905–13.
101. McConnell R, et al. Asthma in exercising children exposed to ozone: a cohort study. Lancet. 2002;359(9304):386–91.
102. Friedman MS, et al. Impact of changes in transportation and commuting behaviors during the 1996 Summer Olympic Games in Atlanta on air quality and childhood asthma. JAMA. 2001;285(7):897–905.
103. Janssen NA, et al. The relationship between air pollution from heavy traffic and allergic sensitization, bronchial hyperresponsiveness, and respiratory symptoms in Dutch schoolchildren. Environ Health Perspect. 2003;111(12):1512–8.
104. Khatri SB, et al. Associations of air pollution and pediatric asthma in Cleveland, Ohio. ScientificWorldJournal. 2021;2021:8881390.
105. van Daalen KR, et al. The 2022 Europe report of the Lancet Countdown on health and climate change: towards a climate resilient future. Lancet Public Health. 2022;7(11):e942–65.
106. von Mutius E, et al. Exposure to endotoxin or other bacterial components might protect against the development of atopy. Clin Exp Allergy. 2000;30(9):1230–4.
107. Stein MM, et al. Innate immunity and asthma risk in Amish and Hutterite farm children. N Engl J Med. 2016;375(5):411–21.
108. Jimenez E, et al. Isolation of commensal bacteria from umbilical cord blood of healthy neonates born by cesarean section. Curr Microbiol. 2005;51(4):270–4.
109. Boyce JA, et al. Advances in mechanisms of asthma, allergy, and immunology in 2011. J Allergy Clin Immunol. 2012;129(2):335–41.
110. Bisgaard H, et al. Childhood asthma after bacterial colonization of the airway in neonates. N Engl J Med. 2007;357(15):1487–95.
111. Penders J, et al. Establishment of the intestinal microbiota and its role for atopic dermatitis in early childhood. J Allergy Clin Immunol. 2013;132(3):601–7. e8
112. Perzanowski MS, et al. Endotoxin in inner-city homes: associations with wheeze and eczema in early childhood. J Allergy Clin Immunol. 2006;117(5):1082–9.
113. Rabinovitch N, et al. Importance of the personal endotoxin cloud in school-age children with asthma. J Allergy Clin Immunol. 2005;116(5):1053–7.
114. Wlasiuk G, Vercelli D. The farm effect, or: when, what and how a farming environment protects from asthma and allergic disease. Curr Opin Allergy Clin Immunol. 2012;12(5):461–6.
115. Pearce N, et al. Worldwide trends in the prevalence of asthma symptoms: phase III of the International Study of Asthma and Allergies in Childhood (ISAAC). Thorax. 2007;62(9):758–66.
116. Arshi M, et al. Asthma and insulin resistance in children. Respirology. 2010;15(5):779–84.

117. Cottrell L, et al. Metabolic abnormalities in children with asthma. Am J Respir Crit Care Med. 2011;183(4):441–8.
118. Ashwell M, Hsieh SD. Six reasons why the waist-to-height ratio is a rapid and effective global indicator for health risks of obesity and how its use could simplify the international public health message on obesity. Int J Food Sci Nutr. 2005;56(5):303–7.
119. Guibas GV, et al. The obesity-asthma link in different ages and the role of body mass index in its investigation: findings from the Genesis and Healthy Growth Studies. Allergy. 2013;68(10):1298–305.
120. Moore WC, et al. Identification of asthma phenotypes using cluster analysis in the Severe Asthma Research Program. Am J Respir Crit Care Med. 2010;181(4):315–23.
121. Shore SA. Obesity and asthma: lessons from animal models. J Appl Physiol (1985). 2007;102(2):516–28.
122. Sood A, Shore SA. Adiponectin, leptin, and resistin in asthma: basic mechanisms through population studies. J Allergy (Cairo). 2013;2013:785835.
123. Cavalher-Machado SC, et al. Down-regulation of mast cell activation and airway reactivity in diabetic rats: role of insulin. Eur Respir J. 2004;24(4):552–8.
124. Karampatakis N, et al. Impaired glucose metabolism and bronchial hyperresponsiveness in obese prepubertal asthmatic children. Pediatr Pulmonol. 2017;52(2):160–6.
125. Osei-Kumah A, et al. Sex-specific differences in placental global gene expression in pregnancies complicated by asthma. Placenta. 2011;32(8):570–8.
126. The ENFUMOSA cross-sectional European multicentre study of the clinical phenotype of chronic severe asthma. Eur Respir J. 2003;22(3):470–7.
127. Chhabra SK, Chhabra P. Gender differences in perception of dyspnea, assessment of control, and quality of life in asthma. J Asthma. 2011;48(6):609–15.
128. Cydulka RK, et al. Differences between men and women in reporting of symptoms during an asthma exacerbation. Ann Emerg Med. 2001;38(2):123–8.
129. Serra-Batlles J, et al. Costs of asthma according to the degree of severity. Eur Respir J. 1998;12(6):1322–6.
130. Sinclair AH, Tolsma DD. Gender differences in asthma experience and disease care in a managed care organization. J Asthma. 2006;43(5):363–7.
131. Woods SE, Brown K, Engel A. The influence of gender on adults admitted for asthma. Gend Med. 2010;7(2):109–14.
132. Päivinen MK, Keskinen KL, Tikkanen HO. Swimming and asthma: differences between women and men. J Allergy (Cairo). 2013;2013:520913.
133. Sundberg R, et al. Asthma in men and women: treatment adherence, anxiety, and quality of sleep. Respir Med. 2010;104(3):337–44.
134. Axelsson M. Personality and reasons for not using asthma medication in young adults. Heart Lung. 2013;42(4):241–6.
135. Lindner PS, Lindner AJ. Gender differences in asthma inhaler compliance. Conn Med. 2014;78(4):207–10.
136. Zahran HS, et al. Predictors of asthma self-management education among children and adults--2006–2007 behavioral risk factor surveillance system asthma call-back survey. J Asthma. 2012;49(1):98–106.
137. Spaziano G, Esposito R, Mirra D, Piegari E, Rossi A, Roviezzo F, Gallelli L, Berrino L, D'Agostino B. SNPs in asthma patients: gender difference in anti-leukotriene therapy. Eur Respir J. 2020;56(Suppl. 64):2619.
138. Farha S, et al. Effects of the menstrual cycle on lung function variables in women with asthma. Am J Respir Crit Care Med. 2009;180(4):304–10.
139. Tan KS, McFarlane LC, Lipworth BJ. Loss of normal cyclical beta 2 adrenoceptor regulation and increased premenstrual responsiveness to adenosine monophosphate in stable female asthmatic patients. Thorax. 1997;52(7):608–11.
140. Rao CK, et al. Characteristics of perimenstrual asthma and its relation to asthma severity and control: data from the Severe Asthma Research Program. Chest. 2013;143(4):984–92.

141. Calcaterra V, et al. Perimenstrual asthma in adolescents: a shared condition in pediatric and gynecological endocrinology. Children (Basel). 2022;9(2):233.
142. Nwaru BI, Sheikh A. Hormonal contraceptives and asthma in women of reproductive age: analysis of data from serial national Scottish Health Surveys. J R Soc Med. 2015;108(9):358–71.
143. Kircher S, Schatz M, Long L. Variables affecting asthma course during pregnancy. Ann Allergy Asthma Immunol. 2002;89(5):463–6.
144. Schatz M, et al. The course of asthma during pregnancy, post partum, and with successive pregnancies: a prospective analysis. J Allergy Clin Immunol. 1988;81(3):509–17.
145. Murphy VE, Jensen ME, Gibson PG. Asthma during pregnancy: exacerbations, management, and health outcomes for mother and infant. Semin Respir Crit Care Med. 2017;38:160–73.
146. Gluck JC, Gluck PA. The effect of pregnancy on the course of asthma. Immunol Allergy Clin N Am. 2006;26(1):63–80.
147. Gee JB, et al. Pulmonary mechanics during pregnancy. J Clin Invest. 1967;46(6):945–52.
148. Gilroy RJ, Mangura BT, Lavietes MH. Rib cage and abdominal volume displacements during breathing in pregnancy. Am Rev Respir Dis. 1988;137(3):668–72.
149. Hegewald MJ, Crapo RO. Respiratory physiology in pregnancy. Clin Chest Med. 2011;32(1):1–13.
150. Murphy VE, et al. Maternal asthma is associated with reduced female fetal growth. Am J Respir Crit Care Med. 2003;168(11):1317–23.
151. Scott NM, et al. Placental cytokine expression covaries with maternal asthma severity and fetal sex. J Immunol. 2009;182(3):1411–20.
152. Murphy VE, et al. A meta-analysis of adverse perinatal outcomes in women with asthma. BJOG. 2011;118(11):1314–23.
153. Murphy VE, et al. The risk of congenital malformations, perinatal mortality and neonatal hospitalisation among pregnant women with asthma: a systematic review and meta-analysis. BJOG. 2013;120(7):812–22.
154. Enriquez R, et al. Cessation of asthma medication in early pregnancy. Am J Obstet Gynecol. 2006;195(1):149–53.
155. Balzano G, et al. Persistent airway inflammation and high exacerbation rate in asthma that starts at menopause. Monaldi Arch Chest Dis. 2007;67(3):135–41.
156. Gómez Real F, et al. Hormone replacement therapy, body mass index and asthma in perimenopausal women: a cross sectional survey. Thorax. 2006;61(1):34–40.
157. Real FG, et al. Lung function, respiratory symptoms, and the menopausal transition. J Allergy Clin Immunol. 2008;121(1):72–80.e3.
158. Troisi RJ, et al. Menopause, postmenopausal estrogen preparations, and the risk of adult-onset asthma. A prospective cohort study. Am J Respir Crit Care Med. 1995;152(4 Pt 1):1183–8.
159. O'Connor KA, et al. Total and unopposed estrogen exposure across stages of the transition to menopause. Cancer Epidemiol Biomark Prev. 2009;18(3):828–36.
160. Zein JG, Erzurum SC. Asthma is different in women. Curr Allergy Asthma Rep. 2015;15(6):28.
161. Dragon CN, et al. Transgender medicare beneficiaries and chronic conditions: exploring fee-for-service claims data. LGBT Health. 2017;4(6):404–11.
162. Han YY, Forno E, Celedón JC. Sex steroid hormones and asthma in a nationwide study of U.S. adults. Am J Respir Crit Care Med. 2020;201(2):158–66.
163. Prevention, C.f.D.C.a. Most recent national asthma data. [website] 2022 May 25, 2022 [cited 2022]. Available from: https://www.cdc.gov/asthma/most_recent_national_asthma_data.htm.
164. Ban GY, et al. Predictors of asthma control in elderly patients. Curr Opin Allergy Clin Immunol. 2016;16(3):237–43.
165. Yáñez A, et al. Asthma in the elderly: what we know and what we have yet to know. World Allergy Organ J. 2014;7(1):8.
166. Dunn RM, Busse PJ, Wechsler ME. Asthma in the elderly and late-onset adult asthma. Allergy. 2018;73(2):284–94.
167. Gibson PG, McDonald VM, Marks GB. Asthma in older adults. Lancet. 2010;376(9743):803–13.

168. Enright PL, et al. Respiratory muscle strength in the elderly. Correlates and reference values. Cardiovascular Health Study Research Group. Am J Respir Crit Care Med. 1994;149(2 Pt 1):430–8.
169. Estenne M, Yernault JC, De Troyer A. Rib cage and diaphragm-abdomen compliance in humans: effects of age and posture. J Appl Physiol (1985). 1985;59(6):1842–8.
170. Janssens JP. Aging of the respiratory system: impact on pulmonary function tests and adaptation to exertion. Clin Chest Med. 2005;26(3):469–84, vi–vii.
171. Lowery EM, et al. The aging lung. Clin Interv Aging. 2013;8:1489–96.
172. Lange P, et al. A 15-year follow-up study of ventilatory function in adults with asthma. N Engl J Med. 1998;339(17):1194–200.
173. Kauppi P, et al. Overlap syndrome of asthma and COPD predicts low quality of life. J Asthma. 2011;48(3):279–85.
174. Alshabanat A, et al. Asthma and COPD overlap syndrome (ACOS): a systematic review and meta analysis. PLoS One. 2015;10(9):e0136065.
175. Liu JN, et al. The prevalence of serum specific IgE to superantigens in asthma and allergic rhinitis patients. Allergy Asthma Immunol Res. 2014;6(3):263–6.
176. Song WJ, et al. Association between obesity and asthma in the elderly population: potential roles of abdominal subcutaneous adiposity and sarcopenia. Ann Allergy Asthma Immunol. 2012;109(4):243–8.
177. Kim SH, Sutherland ER, Gelfand EW. Is there a link between obesity and asthma? Allergy Asthma Immunol Res. 2014;6(3):189–95.
178. Dixon AE, et al. The challenge of addressing obesity in people with poorly controlled asthma. Obes Sci Pract. 2021;7(6):682–9.
179. Thomas M, et al. Asthma and psychological dysfunction. Prim Care Respir J. 2011;20(3):250–6.
180. Krouse RZ, et al. Minimally important differences and risk levels for the Composite Asthma Severity Index. J Allergy Clin Immunol. 2017;139(3):1052–5.
181. Esquivel A, et al. Effects of omalizumab on rhinovirus infections, illnesses, and exacerbations of asthma. Am J Respir Crit Care Med. 2017;196(8):985–92.
182. Kew KM, et al. Shared decision-making for people with asthma. Cochrane Database Syst Rev. 2017;10(10):CD012330.
183. Korwin A, et al. Exploring patient engagement: a qualitative analysis of low-income urban participants in asthma research. J Allergy Clin Immunol Pract. 2017;5(6):1625–1631.e2.
184. Liao Y, Gao G, Peng Y. The effect of goal setting in asthma self-management education: a systematic review. Int J Nurs Sci. 2019;6(3):334–42.

Asthma as a Systemic Disease: Cardiovascular Effects Associated with Asthma

Matthew C. Tattersall

Abstract

Asthma and cardiovascular disease (CVD) pose significant public health burdens. Airway inflammation is central to asthma pathophysiology and systemic inflammation, which occurs in asthma, is central to CVD pathophysiology. Numerous robust epidemiological studies have demonstrated deleterious systemic cardiovascular effects associated with the asthma syndrome. The cardiovascular effects associated with asthma include arterial injury, atherosclerotic CVD events, atrial fibrillation, and hypertension. Asthma is a heterogeneous disease, however, and the risk of CVD is not homogeneous across the various clinical phenotypes and molecular endotypes, highlighting prior inconsistent associations of asthma and its subtypes with various forms of CVD. The mechanistic underpinnings of the increased CVD risk in asthma remain multifactorial and undefined. Collectively, this supports the need for a precision approach in the identification of individuals with asthma who remain at elevated risk of development of cardiovascular diseases to guide both diagnostic and preventive interventions to decrease CVD risk among individuals living with asthma.

Keywords

Atrial fibrillation · Atherosclerotic Cardiovascular Disease · Body mass index · Inflammation · Interleukin-6 · Heterogeneity

M. C. Tattersall (✉)
University of Wisconsin School of Medicine and Public Health, Madison, WI, USA
e-mail: tattersall@wisc.edu

A. R. Brasier, N. N. Jarjour (eds.), *Precision Approaches to Heterogeneity in Asthma*, Advances in Experimental Medicine and Biology 1426,
https://doi.org/10.1007/978-3-031-32259-4_4

Abbreviations

AF	atrial fibrillation
ASCVD	Atherosclerotic Cardiovascular Disease
CVD	Cardiovascular Disease
IL	Interleukin

Background, Epidemiology of Asthma, and Cardiovascular Disease (CVD)

Asthma afflicts over 25 million individuals in the United States and 262 million people worldwide [1, 2]. Asthma was responsible for 455,000 deaths in 2019 [2]. The prevalence of asthma worldwide has increased over the past decade [2]. Worldwide, cardiovascular disease (CVD) is a leading cause of death. In 2019, there were 18.6 million deaths due to cardiovascular diseases and similar to asthma, the prevalence of cardiovascular disease worldwide continues to increase [2]. Asthma and CVD represent significant public health burdens worldwide. Airway inflammation is central to asthma pathophysiology, while systemic inflammation is central to the development and progression of several CVD diseases [3–9]. Emerging data have highlighted that airway inflammation in asthma may not occur in isolation, and there may be systemic interplay and consequences [10–13]. This chapter focuses on the cardiovascular effects of the asthma syndrome.

Asthma, Cardiovascular Disease, and Systemic Inflammation

Airway inflammation in asthma drives the clinical manifestations of the asthma syndrome. Airway inflammation has historically been characterized based on the presence or absence of T_2 inflammation [14]. T-helper 2 cells drive the T_2 inflammatory response by stimulating T_2 cytokines such as interleukin -4, -5, and -13, which in turn stimulate downstream T_2 cell type effectors such as eosinophils and mast cells clinically manifesting as atopic conditions and asthma [15, 16]. High T_2 inflammation counter-regulates T_1 inflammation [17]. Interleukin (IL) -4, a T_2-secreted cytokine, directly inhibits T_1 cells, whereas interferon gamma (IFN-γ), a pro-inflammatory, pro-atherogenic T_1 cytokine, directly inhibits T_2 cells [17, 18]. There are subsets of individuals with asthma who have a paucity of T_2 inflammation and do not respond to classic T_2-directed therapy [15]. Data from two United Kingdom cohorts and the United States Severe Asthma Research Program (SARP) cohort helped define distinct asthma endotypes based on predominant airway immune markers, such as high T_2 or low T_2, highlighting the heterogeneity of asthma endotypes within the asthma syndrome [14]. A major focus in asthma has been the characterization of inflammation in the airways; however, the coexistence of low-grade systemic inflammation in asthma has been less studied and is an emerging area of investigation [19]. Investigators from the Severe Asthma Research

Program (SARP) defined a clinical phenotype of asthma where asthmatic individuals with elevated serum IL-6 levels had lower lung function and more frequent asthma exacerbations [20]. Participants with elevated IL-6 in the SARP predominantly had elevated body-mass-index (BMI), however not all individuals with elevated BMI had elevated serum IL-6 levels. Indeed, only 38% of participants with a BMI in the obese range with asthma in SARP were characterized as high IL-6, highlighting the heterogeneity of the elevated IL-6 phenotype. A common clinical characteristic among the high IL-6 asthma phenotype was the presence of metabolic disturbance (diabetes mellitus) and hypertension. This finding has led to a hypothesis of an "outside-in" pathophysiology for asthmatic individuals with elevated IL-6, where chronic low grade systemic inflammation may have potential effects on the airways [13]. This hypothesis is being investigated in the Precision Interventions for Severe and/or Exacerbation-Prone Asthma (PrecISE) trial (NCT04129931), which is testing the effectiveness of systemic IL-6 inhibition among other therapies on asthma symptoms and lung function [19]. Systemic inflammation is also central to the initiation, progression, and resultant cardiovascular events in atherosclerotic cardiovascular disease (ASCVD) [3]. Prior studies have demonstrated that individuals with elevated inflammatory biomarkers and inflammatory diseases are at increased risk of developing ASCVD [3, 6, 21, 22]. Data from the Physicians Health Study demonstrated that elevated IL-6 levels in healthy men independently predicted future ASCVD events, supporting the idea that ASCVD is an inflammatory condition [23]. More recently, in the in the Antiinflammatory Therapy with Canakinumab for Atherosclerotic Disease (CANTOS) trial, treatment with a targeted anti-inflammatory agent to IL-1β reduced ASCVD events without any modification of traditional risk factors such as blood pressure or cholesterol, highlighting the role of low grade systemic inflammation on the pathogenesis of ASCVD [24]. Systemic inflammation affects the cardiovascular system beyond the development of ASCVD. Inflammation plays a central role in the pathogenesis of atrial fibrillation, the most common heart rhythm disorder in the United States [25, 26]. This shared systemic inflammation overlap of asthma and cardiovascular disease has prompted hypotheses that increased risks of cardiovascular disease may be a systemic manifestation of the asthma syndrome.

Early Atherosclerotic Process: Definitions of Arterial Injury in Asthma

The atherosclerotic process starts early in life occurring in response to ASCVD risk factors such as inflammation [27–29]. The term "arterial injury" encompasses the spectrum of pathologic changes to the artery including endothelial dysfunction, pathologic thickening of the intima-media complex, and inflammatory cell infiltration ultimately leading to atherosclerotic plaque deposition, structural changes in the arterial wall, and ASCVD events [27, 30]. ASCVD has a long latent period beginning in the earliest stages with arterial injury and culminating in ASCVD events [27]. Given the long latent period of ASCVD, methods to detect arterial injury, which is the precedent to ASCVD events can help define future ASCVD risk [31, 32]. Previous studies in the association of asthma, arterial injury, and ASCVD risk have utilized noninvasive imaging methods to identify and characterize the spectrum of arterial injury. We will first highlight brief descriptions of the imaging methods to define arterial injury, followed by a review of the data investigating arterial injury in individuals with asthma. An early stage in arterial injury is dysfunction of the endothelial lining of the artery. Brachial artery flow-mediated dilation (FMD) is the best noninvasive measure of endothelial function [33, 34]. Brachial FMD is an endothelium-dependent process which closely tracks with endothelial function in the coronary arteries [35]. Quantification of brachial FMD is a well-validated, standardized technique to measure endothelial function, which is predictive of future ASCVD events and has been used to assess the impact of ASCVD risk-lowering interventions and pharmacotherapies [34, 36]. Following endothelial dysfunction, there are structural changes to the arterial wall. The carotid intima-media thickness (IMT) is a measure of arterial injury which can be quantified on noninvasive imaging [27, 29]. The carotid IMT increases in thickness as part of the aging process, however in the presence of ASCVD risk factors such as inflammation and oxidative stress, arterial remodeling occurs more rapidly and is termed pathologic remodeling [30]. A thicker carotid IMT independently predicts future ASCVD events [37–41]. Arterial stiffening occurs in response to arterial structural changes [42]. These structural changes result from increased blood pressure, inflammation, oxidative stress, and other ASCVD risk factors [43]. Arterial stiffness can be measured noninvasively using measures of specific local vessel stiffness or proxy methods such as pulse wave velocity (PWV) to estimate the stiffness of the aorta, the major elastic arterial vessel [42]. These arterial stiffness measures also predict future ASCVD events and reflect subclinical arterial injury [44–46]. Carotid plaque, identified on vascular ultrasound, is a specific and late manifestation of the atherosclerotic process and is a strong and independent predictor of future ASCVD events [27, 29, 32, 47, 48] (Fig. 1). There have been numerous studies examining arterial injury among individuals with asthma compared with nonasthmatic individuals throughout a wide variety of ages (Table 1).

Asthma and Endothelial Dysfunction in Adult Asthma Populations

There are limited data on the association of asthma and brachial artery FMD, a noninvasive marker of endothelial dysfunction. In a small study of young adults, compared to controls, the asthma participants had worse endothelial dysfunction as measured on brachial artery FMD, despite having similar serum levels of C-reactive protein. The controls had an FMD of 10.4 ± 3.6% vs confirmed asthma FMD of 7.7 ± 3.6%.) [49]. Another small study of young adults with (n = 16) and without asthma (N = 16) did not find any difference in brachial FMD between the individuals with asthma compared with the controls; however, the study was powered to detect a 10% difference in FMD between asthma participants and controls. Prior

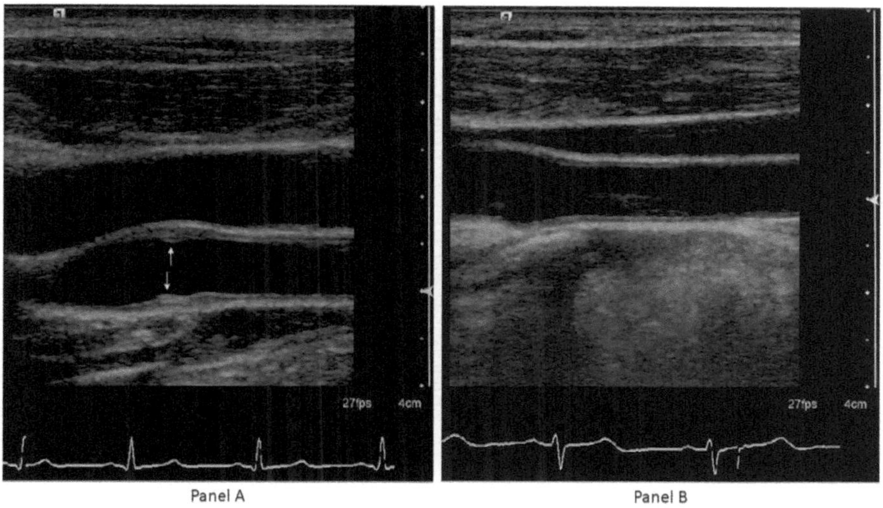

Fig. 1 Panel (**a**) Carotid plaque identification at the carotid bifurcation in a 45-year-old patient with asthma. The arrows indicated carotid plaque presence on the near and far walls of the carotid bifurcation. Panel (**b**) Carotid ultrasound of age-sex matched control without asthma showing no carotid plaque

studies indicate that FMD differences of 2% or more are typically regarded as clinically significant. Therefore, the study may have been underpowered to detect smaller but clinically significant FMD differences [33, 50, 51]. The association of asthma and endothelial dysfunction as measured by brachial FMD remains largely under-defined due to small study sizes on limited asthma phenotypes. Larger population-based studies including various asthma phenotypes are needed to better define this association.

Asthma and Carotid IMT in Adult Asthma Populations

Several large studies have investigated the association between carotid IMT and asthma in an adult population. In an early report from the Atherosclerosis Risk Factors in Male Youngsters (ARMY) and Bruneck studies, allergic conditions such as allergic rhinitis and asthma were associated with more carotid plaque and/or thicker carotid IMT [52]. In this study, both the presence of allergic conditions (allergic rhinitis and asthma) and serum IgE levels were associated with more carotid injury [52]. Inherent to the design of the cohort, the ARMY study excluded asthmatic participants at enrollment, and the prevalence of asthma among participants in the Buneck study was 2.4% ($N = 20$ asthmatic participants), which limited the study's ability to investigate the contribution of asthma alone in this association [52]. In a study from the Atherosclerosis Risk in Communities (ARIC) study, which included 13,627 men and women with carotid ultrasound data, the investigators found that women with self-reported adult-onset asthma (age ≥21 years old) but not

Table 1 Association of asthma and arterial injury in selected studies

	Study	Participants	Arterial Injury measure	Results
Pediatric population	Cakmak et al. [57]	$N = 84$ with asthma $N = 194$ controls	Carotid IMT	Asthma participants had ↑ Carotid IMT: 0.48 ± 0.06 mm asthma vs 0.42 ± 9.05, $p < 0.0001$
	Dratva et al. [58]	$N = 28$ with asthma $N = 229$ without asthma	Carotid IMT	Male asthma participants had ↑ Carotid IMT compared to controls: difference 0.023 mm 95% CI 0.003–0.043)
	Winder et al. [107]	$N = 22$ nonallergic asthma $N = 58$ allergic asthma $N = 268$ inhaled allergies $N = 1158$ without asthma/allergy	Carotid IMT	Compared to Controls: (411.7 μm) Nonallergic asthma: 411.7 μm, $p = 0.932$ Inhaled allergies 420.0 μm, $p = 0.118$ Allergic asthma: 430.8 μm, $p = 0.004$
	Tattersall et al. [60]	$N = 12$ no asthma, no atopy $N = 15$ Atopy, no asthma $N = 34$ Asthma, no atopy $N = 28$ Both asthma and atopy	Carotid IMT	Compared to Controls: (589 μm) Atopy alone: 653 μm, $p = 0.07$ Asthma alone: 649 μm, $p = 0.05$ Both asthma and atopy: 677 μm, $p = 0.005$
	Endothelial dysfunction (brachial FMD)			
Adult population	Henry et al. [49]	$N = 26$ asthma $N = 26$ controls $N = 15$ unconfirmed asthma	Brachial FMD PWV (radial-carotid)	FMD asthma compared to controls ($p < 0.05$): FMD Controls: 10.4 ± 3.6 FMD unconfirmed asthma: 7.3 ± 3.5 FMD confirmed asthma: 7.7 ± 3.6, PWV asthma compared to controls ($p = 0.58$) PWV controls: 7.9 ± 1.8 m/s PWV unconfirmed asthma: 8.5 ± 1.0 m/s PWV confirmed asthma 8.2 ± 2.5 m/s
	Moore et al. [51]	$N = 16$ asthma $N = 16$ controls	Brachial FMD PWV (radial-carotid)	No difference in FMD, asthma worse PWV
	Carotid IMT			

	Knoflach et al. [52]	Bruneck: $N = 12$ allergic rhinitis alone $N = 11$ asthma alone $N = 9$ both allergic rhinitis and asthma ARMY: $N = 34$ allergic rhinitis ($N = 2$ with asthma)	Carotid IMT	Allergic disorder/asthma ↑ carotid IMT compared to control: Bruneck: odds ratio (OR) 3.8 (95% CI 1.4–10.2) ARMY: odds ratio 2.5 (95% CI 1.1–5.5)
	Onufrak et al. [53]	$N = 390$ adult-onset asthma $N = 369$ childhood-onset asthma $N = 12{,}868$ no asthma	Carotid IMT	Women, not men, with adult-onset asthma ↑ carotid IMT adjusted difference: Nonasthma adjusted carotid IMT: 0.687 mm Adult-onset asthma adjusted carotid IMT (women): 0.713 mm, $p = 0.008$
	Arterial stiffness measures			
	Sun et al. [54]	$N = 169$ asthma $N = 1577$ no asthma	PWV (aortafemoral)	Baseline: Asthma 6.78 m/s, No asthma 6.13 m/s, $p = 0.048$ PWV progression: asthma 8.99 m/s, No asthma 2.95 m/s, $p = 0.043$
	Carotid plaque			
	Tattersall et al. [56]	$N = 109$ persistent asthma $N = 388$ intermittent asthma $N = 4532$ no asthma	Carotid plaque presence and score (range 0–12)	Persistent asthma compared with control had: ↑ presence of carotid plaque (67% vs 51%, $p = 0.003$), ↑ carotid plaque burden (mean ([SD] total plaque score: 2.1 [2.4] vs. 1.3 [1.8], $p = 0.002$ ↑ serum IL-6 mean (SD) 1.89 (1.61) vs. 1.52 (1.21), $p < 0.05$

Abbreviations: *FMD* Flow mediated dilation, *IL-6* Interluekin-6, *IMT* intima-media thickness; *PWV* Pulse wave velocity childhood-onset asthma had an adjusted 1.79 (95% confidence intervals (CI) 1.00–3.22) higher odds of having a thickened carotid IMT ≥1 mm [53]. In this study, there was no association between either childhood or adult-onset asthma and carotid atherosclerosis in men [53].

Asthma and Arterial Stiffness in Adult Asthma Populations

Arterial stiffness measures have been investigated in adults with asthma. As mentioned above, a small study of young adults with and without asthma found no difference in endothelial function but did find differences in arterial stiffness measured by the radial-carotid PWV. A report from the Bogalusa Heart Study investigated the association of childhood asthma and aorta-femoral PWV in a large cohort study. The investigators found at baseline, participants with childhood asthma had higher PWV (more stiff vessels) compared with nonasthmatic participants [54]. In addition, the participants with childhood asthma also had a more rapid progression of stiffness measures over a median 11-year follow-up period [54]. There was evidence of an interaction between asthma and systolic blood pressure and asthma and body mass index (BMI) [54]. Asthmatic participants with systolic blood pressures ≥120 mmHg or BMI ≥25 kg/m^2 had more rapid progression of PWV over the observational period [54]. These findings of increased baseline arterial stiffness and more rapid progression of arterial stiffness are significant as arterial stiffness measures not only predict incident ASCVD events but have strong associations with extra-cardiac target organ damage of the brain and kidney due to increased transmission of pulsatile flow in the end-organs [55].

Asthma and Carotid Plaque in Adult Asthma Populations

A report from the Multi-Ethnic Study of Atherosclerosis (MESA) examined a different measure of arterial injury, the association of asthma, and carotid plaque, a late manifestation of the carotid atherosclerotic process [56]. In the MESA, persistent asthmatic participants, defined as asthma requiring the need of daily controller medications were more likely to have the presence of carotid plaque (67% for persistent asthma participants versus 51% for nonasthma participants, $p = 0.003$) and the persistent asthma participants had a higher total mean (SD) plaque score (range: 0–12), (persistent asthma, 2.08 [2.35] versus nonasthmatic participants, 1.29 [1.80], $p \leq 0.002$) [56]. In models adjusting for biologic confounders, persistent asthma was associated with carotid plaque presence: odds ratio (OR), 1.83 (95% CI, 1.21–2.76), and burden measured by the total plaque score ($\beta = 0.66$, $p < 0.01$) in the MESA. In this study, participants with persistent asthma also had higher levels of serum inflammatory markers such as interleukin-6 (IL-6), compared to nonasthma participants; however, adjustment for these serum inflammatory markers (IL-6 and C-reactive protein) did not attenuate the association between persistent asthma and carotid plaque, highlighting a potential multifactorial aspect of this association [56].

Asthma and Arterial Injury in Pediatric Populations

Studies investigating arterial injury in the pediatric asthma population have focused on the carotid IMT as an early marker of arterial injury, as the presence of carotid plaque in this population is rare. In a study of asthmatic children, mean (SD) age 9.05 (3.22) years old compared to age-matched controls, the children with asthma had a thicker (worse) right common carotid IMT measurement (Asthma group: 0.48 ± 0.06 mm vs Control group: 0.42 ± 0.05 mm, $p < 0.0001$) [57]. The authors found similar results for the left common carotid artery [57]. Investigators in the Swiss Study on Air Pollution and Lung and Heart Disease in Adults (SAPALDIA) study found increased carotid IMT in male asthma participants compared to male nonasthma participants (difference 0.023 mm [95% CI 0.003–0.043]), but no difference in female asthma participants compared to female nonasthma participants [58]. The number of asthma participants in this study was small ($N = 28$), with 68% of asthma participants being male. Therefore, the results may have been underpowered to detect differences in female participants with and without asthma [58]. The Early Vascular Aging-Tyrol (EVA-Tyrol) cohort study investigated carotid IMT among allergic asthma, nonallergic asthma, and nonasthma participants. The mean (SD) age of participants was 17.8 (0.9) years. The investigators compared the 58 participants with allergic asthma, 22 with nonallergic asthma, and 268 with a history of inhaled allergies confirmed by a positive radioallergosorbent test (RAST) allergy to the 1158 participants without asthma or allergies. The investigators found no difference in the carotid IMT thickness between participants with either nonallergic asthma or inhaled allergies and nonasthma, nonallergic participants. The investigators did, however, observe a significantly higher carotid IMT among allergic asthma participants compared with the nonasthma, nonallergic participants (430.8 μm vs 411.7 μm, $p = 0.004$). A study from the National Institutes of Health funded Childhood Origins of Asthma (COAST) cohort investigated carotid IMT in a longitudinal asthma birth cohort [59]. Newborn infants were enrolled in the cohort based on parenteral history of asthma or respiratory allergies. Children were longitudinally assessed for the development of asthma and allergic conditions [59]. Carotid IMT was measured in a subsample of 89 participants at mean (SD) age of 15.3 (0.6) years old. A total of 12 participants never developed asthma or atopic conditions, 15 participants developed atopic conditions without asthma, 34 participants developed asthma alone, and 28 participants developed both asthma and atopic conditions [60]. In fully adjusted models, there was a progressive thickening of the right carotid bulb segment among asthma/atopic disease groups, with participants having both atopic conditions and asthma having the worst IMT ($p_{trend} = 0.011$) [60]. Compared to controls, the carotid IMT was thicker among those with atopic disease ($\beta = 77.4$, $p = 0.02$), asthma alone ($\beta = 64.2$, $p = 0.03$), and both atopic disease and asthma ($\beta = 85.8$, $p = 0.007$) [60]. There was no effect modification by sex

($p = 0.72$) and no interactive effects of serum IgE or lung function in the carotid IMT and asthma/atopic condition association [60]. Collectively, these pediatric studies consistently identified early arterial injury in individuals with asthma. A significant strength of these pediatric studies is the identification of early arterial injury in asthmatic children, as their physiology is less confounded by the comorbidities of aging.

Studies of Asthma and Atherosclerotic Cardiovascular Disease (ASCVD) Events

There have been numerous reports investigating the association of increased ASCVD events in individuals with asthma over the past several decades. These reports have varied considerably in the strength and direction of the associations, varying by specific ASCVD event subtypes, and effect modification by sex, age, asthma severity, and timing of asthma onset. (Table 1). In an early report of severe asthma patients from the early 1960s, 273 asthmatic patients had an increased risk of ischemic heart disease death compared to nonasthmatics (relative risk [RR] 1.9, 95% confidence intervals [CI] 1.4–2.4) [61]. This finding was driven by a higher risk among women with asthma (women RR: 2.5 [95% CI 1.7–3.3]) compared to men (RR: 1.4 [95% CI 0.8–2.0]) [61]. In a cross-sectional analysis of the Cardiovascular Health Study (CHS), a large cohort of older United States adults, there was a higher prevalence of coronary heart disease among nonsmoking men with asthma compared to nonasthma participants, but not women [62]. Asthma was an independent risk factor for incident stroke (hazard ratio [HR] 1.50, 95% CI 1.04–2.15) but not coronary heart disease (CHD) in the Atherosclerosis Risk in Communities (ARIC) study of middle-aged adults [63]. A subsequent report from ARIC showed that women but not men with adult-onset asthma had a higher risk of stroke (HR 2.08, 95% CI 1.30–3.32) and coronary heart disease (HR 1.78, 95% CI 1.18–2.67) [64] A National Health and Nutrition Examination Survey report identified increased coronary heart disease risk (but not stroke) among participants with adult (late) onset asthma (odds ratio [OR] 2.94, 95% CI 1.37–6.33) [65]. Administrative data from Kaiser Permanente showed an increased risk of incident coronary heart disease (HR, 1.41 95% CI 1.36–1.45) and cerebrovascular disease (HR 1.20, 95% CI 1.15–1.25) among patients with asthma; this association was stronger among women [66]. In a report from the Multi-Ethnic Study of Atherosclerosis (MESA), participants with persistent asthma were more likely than those without asthma to experience a first ASCVD event over 10 years in models adjusted for biologic confounders (HR 1.59, 95% CI 1.01–2.50) [11]. There was no evidence of effect modification by sex in this study [11]. A report from the Wisconsin Sleep Cohort, a longitudinal cohort study with over 1500 participants, found that late-onset asthma (onset >18 years old), but not early-onset asthma was associated with increased risk of future ASCVD events (HR 1.57 [95% CI 1.01–2.45]) [67–69]. Similar findings were demonstrated between a history of asthma and ASCVD risk among asthmatic participants compared to nonasthmatic participants in the

Framingham Offspring study (HR 1.28, 95% CI 1.07–1.54) [70]. In a large Tawain medical program, patients with current asthma had a higher risk of ASCVD mortality (HR 1.32, 95% CI 1.08–1.62), however this finding was driven due to an increased risk among men (HR 1.63, 95% CI 1.30–2.04) but not women (HR 0.65, 95% CI 0.40–1.07, $p_{interaction} = 0.012$) [71]. Despite heterogeneity in the definitions of asthma and ASCVD endpoints, most of these observational studies demonstrate an association of asthma with both prevalent and incident ASCVD. These data highlight additional considerations of effect modification of this association based on sex, asthma severity, and age of onset of asthma (Table 2).

Asthma and Atherosclerotic Cardiovascular Disease: Mechanisms

Mechanistically there are several putative shared inflammatory pathways that may drive both airway inflammation and result in arterial injury leading to ASCVD. There are candidate inflammatory pathways present in both T_2 high and T_2 low asthma endotypes. One potential shared inflammatory mechanism between asthma and ASCVD is the 5-Lipo-oxygenase (5-LO) pathway. In this pathway, arachidonic acid undergoes enzymatic conversion by the enzyme 5-LO into an intermediate epoxide which eventually becomes a leukotriene, which is a paracrine inflammatory substance produced by immune cells which induces inflammation [72]. In the airway, leukotrienes are potent mediators of the clinical manifestations of asthma by constricting smooth muscle, inducing tissue edema, and serving as a chemoattractant for additional inflammatory cell types [73]. The 5-LO enzyme is also expressed in high levels in human atherosclerotic plaques [74]. In human carotid plaques, the levels of 5-LO expressed directly correlated to the stability of the atherosclerotic plaque and recent clinical symptoms of plaque rupture, providing further support that the levels of 5-LO within the atherosclerotic plaque provide insight into the stage and vulnerability of the plaque [74]. In animal models, blockade of the leukotriene receptors results in less inflammatory cell recruitment and reduces atheroma progression [75]. Therefore, 5-LO and downstream leukotrienes are potential candidates in the overlap of asthma and ASCVD. In T_2 low asthma, there are several potential mediators of the ASCVD process. ASCVD has typically been characterized as a T_1 (low T_2) disease state. Interferon-gamma (IFN-γ), is a key T_1 inflammatory mediator and has many proatherogenic effects including activation of monocytes, macrophages, and dendritic cells and stimulation of matrix metalloproteinases which induce plaque rupture [76, 77]. In animal models, the administration of IFN-γ promotes atherogenesis [78–80]. Another prominent cytokine in low T_2 endotype is interleukin 6 (IL-6) [20, 81]. Asthmatic individuals with elevated IL-6 levels have more severe asthma and greater asthma morbidity [13, 20]. IL-6 exerts proatherogenic properties in part through, directly inhibiting T-Regulatory cells (T_{Reg} cells) which permits the T_1 response to proceed [82]. IL-6 also induces arterial injury by causing endothelial dysfunction and platelet activation [83]. Prior longitudinal cohort studies have demonstrated that IL-6 is an independent predictor of

Table 2 Association of asthma and CVD events in selected studies

Study	Participants	Data source	Study location (years)	Asthma Subtype/ definition	Endpoint	Results
Enright et al. [62]	5201	CHS Cohort	United States (1989–1990)	Asthma *Defined as:* Physician diagnosed	Prevalence of CVD (CHD, MI, CHF, stroke, LV hypertrophy, ECG abnormality) among asthma participants	No difference in MI, CHF, stroke, Left Ventricular (LV) hypertrophy prevalence
Iribarren et al. [108]	151,620	Administrative Claims/Medical Records Data	United States (1964–2000)	Asthma *Defined as either:* Physician diagnosed or Asthma hospitalization	ICD-8, ICD-9 codes for CHD	Adjusted HR (95% CI), Men: 0.99 (0.93–1.05) Women: 1.22 (1.14–1.31)
Schanen et al. [63]	13,501	ARIC Cohort	United States (1987–2000)	Asthma *Defined as* Physician diagnosed	Incident CHD or Stroke	Adjusted HR (95% CI) CHD: 0.87 (0.66–1.14) Stroke: 1.50 (1.04–2.15)
Pollevick et al. [70]	3612	Framingham Offspring Cohort	United States (1979–2013)	Asthma *Defined as* Physician diagnosed	Incident CVD event: MI, angina, coronary insufficiency, stroke, TIA, CHF	Adjusted HR (95% CI) CVD: 1.28 (1.07–1.54)

Late-onset/adult asthma and CVD						
Onufrak et al. [64]	15, 792	ARIC Cohort	United States (1987–2001)	Adult- Onset-Asthma *Defined as* Physician diagnosed and age of onset (>18 years)	Incident CHD or Stroke	Adjusted RR (95% CI) Adult-onset asthma, women: 1.78 (1.18–2.67). No association for child-onset asthma or men with asthma
Lee et al. [65]	16,943	NHANES Study	United States (1996–2006)	Adult- Onset-Asthma *Defined as* Physician diagnosed and age of onset (>18 years)	Prevalent Stroke, CHF, CHD, CVD	Adjusted OR (95% CI) Adult-Onset Asthma: 2.07 (1.2–3.7) for CVD; 2.26 (1.2–4.2) for CHD
Iribarren et al. [66]	203, 595	Administrative Claims/Medical Records Data	United States (1996–2008)	Adult- Onset-Asthma *Defined as* ICD-9 codes for asthma with age of onset (>18 years)	Incident CHD, stroke, CHF	Adjusted HR (95% CI) CHD: 1.40 (1.35–1.45) Stroke: 1.20 (1.15–1.25) CHF: 2.14 (2.06–2.22) CHD risk higher in women $p_{interaction}$, <0.0001
Tattersall et al. [67]	1267	WSC	United States (1989–2013)	Late (adult) Onset Asthma *Defined as* Physician diagnosed, age of onset (>18 years) on ≥2 interviews	Incident CVD Events: coronary death, MI, angina, stroke, coronary revascularization, CHF, or CVD death	Adjusted HR (95% CI) Late-onset asthma 1.57 (95% CI 1.01–2.45)
Current asthma, asthma severity, and CVD						
Torén et al. [61]	262	Medical Records	Sweden (1962–1986)	Severe Asthma *Defined as:* Asthma treated with daily oral corticosteroids >1 year	ICD-8 codes for all-cause and CVD causes of death	Unadjusted RR (95% CI) Ischemic Heart Disease: 1.9 (1.4–2.4)

(continued)

Table 2 (continued)

Study	Participants	Data source	Study location (years)	Asthma Subtype/ definition	Endpoint	Results
Tattersall et al. [11]	6792	MESA Cohort	United States (2000–2013)	Persistent Asthma *Defined as:* Physician diagnosed and Current use of daily controller medications	MI, cardiac arrest, angina, stroke, CHD death, CVD, and stroke death	Adjusted HR (95% CI) 1.59 (1.01–2.50)
Strand et al. [71]	446,346	Taiwan Health Program	Taiwan (1994–2011)	Current Asthma *Defined as:* Physician diagnosed and Current use of daily controller medications	ICD-9, ICD-10 for CVD, CHD, and stroke death	Adjusted HR (95% CI) CVD mortality: 1.32 (1.08–1.62) CHD mortality: 1.16 (0.78–1.73) Stroke mortality: 1.23 (0.86–1.74) Effect modification by sex for men and CVD mortality $p_{interaction} = 0.012$, and stroke $p_{interaction} = 0.043$

Abbreviations: *CHD* Coronary heart disease, *CHF* Congestive heart Failure, *CHS* Cardiovascular Health Study, *CVD* Cardiovascular disease, *CIMT* carotid intima-media thickness, *CHD* Coronary Heart Disease, *MESA* Multi-Ethnic Study of Atherosclerosis, *MI* Myocardial infarction, *RR* Relative Risk, *TRF* Traditional Risk Factors, *CI* Confidence Intervals, *HR* Hazard Ratio, *MI* Myocardial Infarction, *CVD* Cardiovascular Disease incident ASCVD events [23]. Additional evidence of a shared inflammatory basis has been demonstrated in studies of C-reactive protein (CRP), a nonspecific marker of systemic inflammation which independently predicts ASCVD [4–6]. CRP is elevated in asthmatic individuals compared with nonasthmatic individuals and elevates further during exacerbations, highlighting the systemic inflammation which occurs in response to airway inflammation [11, 84].

Beyond ASCVD: Associations of Asthma with Atrial Fibrillation and Hypertension

While the association of asthma and ASCVD has been heavily investigated, recent studies have investigated additional impacts of asthma on the cardiovascular system. These findings highlight other specific manifestations of the asthma syndrome on cardiovascular health and have significant overlap with the mechanisms leading to the increased risk of ASCVD among asthmatic individuals.

Asthma and Incident Atrial Fibrillation

Atrial fibrillation (AF) is the most common heart arrhythmia afflicting more than 59.7 million individuals worldwide [85]. Inflammation is central to the underlying pathophysiology of AF, supporting the biologic plausibility of the studies investigating the association of asthma and incident AF. These studies differ in the definition of asthma and reported varying results in the association of asthma and incident AF. A nested case–control study using administrative claims from Taiwan reported 1.2-fold higher odds (95% CI 1.1–1.3) of developing AF among asthma individuals currently being treated with controller medications. The Norwegian population-based Nord-Trøndelag Health Study (HUNT) investigated the association of asthma and asthma control on incident AF. The investigators found associations between a history of physician-diagnosed asthma and an incident AF (HR 1.38, 95% CI 1.18–1.61) and a higher risk of incident AF in those with uncontrolled asthma (HR 1.74, 95% CI 1.26–2.42) [86]. There was no evidence of effect modification of this association based on sex or body mass index [86]. In addition, the investigators performed mediation analyses including adjusting for CRP levels and beta-agonist therapy, which did not change the strength of the association, highlighting a potential multifactorial mechanistic underpinning [86]. Another study investigating administrative claims data from the National Health Services hospitals in England found no association of asthma with AF (odds ratio, 1.02 [95% CI 0.97–1.07]) [87]. Finally, a report from the MESA reported an association between persistent asthma and incident AF in risk-factor adjusted models (HR 1.49, [95% CI 1.03–2.14]) [88]. In sensitivity analyses, despite the persistent asthma participants having higher levels of serum inflammatory markers (CRP, IL-6, and fibrinogen), there was no attenuation of persistent asthma and incident AF association by adjustment for these inflammatory markers, consistent with the findings from the HUNT study [86, 88].

Adjustment for autonomic measures (heart rate variability and sleep apnea also did not attenuate this association [88]. The MESA study did, however, identify effect modification by sex, where women with persistent asthma had a higher risk of incident AF compared to women without asthma (HR 1.97, [9% CI 1.26–3.07]), $p_{interaction} = 0.058$. Taken together, these findings highlight an increased risk for AF development among individuals with asthma, with increasing risk based on severity and control of asthma.

Asthma and Atrial Fibrillation Mechanisms

Similar to asthma and ASCVD, a clear underlying mechanism driving increased AF risk among individuals with asthma has not yet been defined. There are several candidate mechanistic underpinnings. Changes in autonomic function are central in the pathophysiology of both asthma and AF and could plausibly account for the increased risk [89, 90]. While the MESA study did not find a change in association of persistent asthma and incident AF with adjustment for heart rate variability, autonomic function is dynamic and one measure of autonomic function may not account for rapid changes that occur in response to asthma exacerbations, medications, or control [88, 90]. Another possibility is the aforementioned eicosanoid pathway and leukotriene signaling. Leukotriene receptors are expressed on the myocardium and cardiac conduction system and pharmacologic inhibition leads to decreased heart rate, improved heart rate variability and improved heart rate control in permanent atrial fibrillation [91, 92]. Cytokine mediators such as IL-6, also elevated in specific asthma endotypes have been reported to increase risk of incident AF [93]. Finally, the contribution of asthma medications is a consideration. While the association of asthma, asthma control, and incident AF was not modified by adjustment for short-acting beta-agonists in the HUNT study, the use of asthma medications such as beta-agonists and oral glucocorticoids may have an impact on AF risk [94, 95].

Asthma and Hypertension: A Combination with Systemic Implications

Hypertension is highly prevalent among asthmatic patients compared to nonasthmatic patients, and this association is independent of traditional risk factors [96–98]. Several studies have highlighted the association between asthma and hypertension. In a study of 74,342 participants in the Canadian Community Health Survey, participants with asthma were more likely than nonasthmatic participants to have hypertension even after adjustment for traditional risk factors (odds ratio 1.36 [95% CI 1.21–1.53]) [96]. In a large administrative claim database of 117,922 patients, the patients with asthma were more likely than nonasthmatic patients to have a diagnosis of hypertension. The investigators also found that patients with more severe asthma were more likely to have a diagnosis of hypertension, highlighting the heterogeneity in the prevalence of hypertension among various asthma subtypes and severity [98]. In a large medical record study from the National Health Service hospitals in England a total of 60,424 patients with asthma were compared to 302,120 patients without asthma, and the investigators found that asthma was associated with hypertension (odds ratio 1.66 [1.61–1.70]) in risk-factor adjusted models [87]. The association between asthma and hypertension is likely multifactorial including a common inflammatory basis [97]. In a network analysis of shared gene signatures between asthma and hypertension, the investigators identified six functional modules of overlap. Among the overlap areas of genes were drug-response genes, cytokine signaling such as TH17 and IFN-γ and oxidative stress [99]. These studies highlighting the comorbid conditions of asthma and hypertension fit well with the aforementioned data on increased arterial stiffness in asthmatic participants from the Bogalusa study [54]. Arterial stiffness measures predict incident hypertension, and as vessels become stiffer, the blood pressure becomes more difficult to control [100, 101]. From a pathophysiologic perspective, increased distending pressures on elastic arteries lead to more rapid elastin degradation and pathologic vascular remodeling, resulting in less compliant or stiffer arterial vessels which further drives elevations in systolic pressure [101]. These findings of increased arterial stiffness measures and prevalent hypertension not only impact the cardiovascular system but may also play a role in the risk of noncardiovascular conditions among asthmatic individuals. In a report from the Framingham Offspring study, higher arterial stiffness independently predicted future mild cognitive impairment, all-cause dementia and Alzheimer disease, highlighting the vast implications of arterial injury, remodeling, and overt hypertension in asthma [102].

Clinical Implications of Asthma and Cardiovascular Diseases

There have been many epidemiological studies highlighting the increased risk of cardiovascular disease, namely ASCVD, among asthmatic individuals. It is imperative for clinicians to be aware of the increased ASCVD risk and higher prevalence of comorbid conditions such as hypertension among asthmatic patients. In a subset of the SARP, hypertension was prevalent among all asthma participants; however, over 52% of high IL-6 asthma participants had hypertension [20, 97]. Diabetes mellitus was also prevalent in the SARP, especially among high IL-6 asthma participants [20]. Addressing modifiable risk factors at clinic visits can work to mitigate risk among all patients. One method to address modifiable risk factors while engaging in health promotion is the use of the American Heart Association's "Life's Essential 8" construct [103]. The components of this construct include important modifiable risk factors proven to improve cardiometabolic health. These components may also improve asthma and asthma symptoms by addressing shared comorbid and overlapping disease states [104]. For adult patients, the American College of Cardiology/American Heart Association 2019 Guidelines on the Primary Prevention of Cardiovascular Disease recommend the application of a 10-year ASCVD risk score to define ASCVD risk [105]. In cases where 10-year ASCVD risk is borderline (5% to <7.5%) or intermediate (≥7.5% to <20%), the use of risk-enhancing conditions can help refine risk [105]. Inflammatory conditions such as rheumatoid arthritis, lupus, or Human Immunodeficiency Virus infection are included as risk-enhancing conditions as are patients with persistently elevated serum inflammatory biomarkers [105]. Clinicians should be aware in patients with asthma, there may be some clinical phenotypes/molecular endotypes which may have laboratory evidence of persistent systemic inflammation, highlighting an ASCVD risk-enhancing feature. Finally, previous articles have highlighted the importance of addressing the comorbid condition of hypertension in asthma care [97]. Treatment includes nuanced approaches in avoiding potential agents that may complicate or worsen asthma symptoms while controlling systemic blood pressures [97]. Addressing cardiovascular risk factors through lifestyle and health promotion early in life may help change the trajectory of disease development later in life [106].

References

1. Akinbami LJ, Moorman JE, Bailey C, Zahran HS, King M, Johnson CA, et al. Trends in asthma prevalence, health care use, and mortality in the United States, 2001-2010. NCHS Data Brief. 2012;(94):1–8. PubMed PMID: 22617340. Epub 2012/05/24. eng.
2. Collaborators GDaI. Global burden of 369 diseases and injuries in 204 countries and territories, 1990-2019: a systematic analysis for the Global Burden of Disease Study 2019. Lancet. 2020;396(10258):1204–22. PubMed PMID: 33069326. PMCID: PMC7567026. eng.
3. Hansson GK. Inflammation, atherosclerosis, and coronary artery disease. N Engl J Med. 2005;352(16):1685–95. PubMed PMID: 15843671. Epub 2005/04/22. eng.
4. Ridker PM, Hennekens CH, Buring JE, Rifai N. C-reactive protein and other markers of inflammation in the prediction of cardiovascular disease in women. N Engl J Med. 2000;342(12):836–43. PubMed PMID: 10733371. Epub 2000/03/25. eng.
5. Ridker PM, Cushman M, Stampfer MJ, Tracy RP, Hennekens CH. Inflammation, aspirin, and the risk of cardiovascular disease in apparently healthy men. N Engl J Med. 1997;336(14):973–9. PubMed PMID: 9077376. Epub 1997/04/03. eng.
6. Ridker PM, Danielson E, Fonseca FA, Genest J, Gotto AM Jr, Kastelein JJ, et al. Rosuvastatin to prevent vascular events in men and women with elevated C-reactive protein. N Engl J Med. 2008;359(21):2195–207. PubMed PMID: 18997196. Epub 2008/11/11. eng.
7. Go AS, Mozaffarian D, Roger VL, Benjamin EJ, Berry JD, Blaha MJ, et al. Heart disease and stroke statistics--2014 update: a report from the American Heart Association. Circulation. 2014;129(3):e28–e292. PubMed PMID: 24352519. Epub 2013/12/20. eng.
8. Djukanovic R. Airway inflammation in asthma and its consequences: implications for treatment in children and adults. J Allergy Clin Immunol. 2002;109(6 Suppl):S539–48. PubMed PMID: 12063510. eng.
9. Jarjour NN, Kelly EA. Pathogenesis of asthma. Med Clin North Am. 2002;86(5):925–36. PubMed PMID: 12428538. eng.
10. Ash SY, Rahaghi FN, Come CE, Ross JC, Colon AG, Cardet-Guisasola JC, et al. Pruning of the pulmonary vasculature in asthma. The Severe Asthma Research Program (SARP) Cohort. Am J Respir Crit Care Med. 2018;198(1):39–50. PubMed PMID: 29672122. PMCID: PMC6034125. eng.
11. Tattersall MC, Guo M, Korcarz CE, Gepner AD, Kaufman JD, Liu KJ, et al. Asthma predicts cardiovascular disease events: the multi-ethnic study of atherosclerosis. Arterioscler Thromb Vasc Biol. 2015;35(6):1520–5. PubMed PMID: 25908767. PMCID: Pmc4441553. Epub 2015/04/25. eng.

12. Teague WG, Phillips BR, Fahy JV, Wenzel SE, Fitzpatrick AM, Moore WC, et al. Baseline features of the Severe Asthma Research Program (SARP III) Cohort: differences with age. J Allergy Clin Immunol Pract. 2018;6(2):545–54.e4. PubMed PMID: 28866107. PMCID: PMC5832534. Epub 2017/08/31. eng.
13. Peters MC, Fahy JV. Metabolic consequences of obesity as an "outside in" mechanism of disease severity in asthma. Eur Respir J. 2016;48(2):291–3. PubMed PMID: 27478182. PMCID: PMC6020152. eng.
14. Wenzel SE. Asthma phenotypes: the evolution from clinical to molecular approaches. Nat Med. 2012;18(5):716–25. PubMed PMID: 22561835. Epub 2012/05/09. eng.
15. Fahy JV. Type 2 inflammation in asthma--present in most, absent in many. Nat Rev Immunol. 2015;15(1):57–65. PubMed PMID: 25534623. PMCID: PMC4390063. eng.
16. Locksley RM. Asthma and allergic inflammation. Cell. 2010;140(6):777–83. PubMed PMID: 20303868. PMCID: PMC3134388. eng.
17. Mosmann TR, Coffman RL. TH1 and TH2 cells: different patterns of lymphokine secretion lead to different functional properties. Annu Rev Immunol. 1989;7:145–73. PubMed PMID: 2523712. Epub 1989/01/01. eng.
18. Kaiko GE, Horvat JC, Beagley KW, Hansbro PM. Immunological decision-making: how does the immune system decide to mount a helper T-cell response? Immunology. 2008;123(3):326–38. PubMed PMID: 17983439. PMCID: 2433332. eng.
19. Israel E, Denlinger LC, Bacharier LB, LaVange LM, Moore WC, Peters MC, et al. PrecISE: precision medicine in severe asthma: an adaptive platform trial with biomarker ascertainment. J Allergy Clin Immunol. 2021;147(5):1594–601. PubMed PMID: 33667479. PMCID: PMC8113113. Epub 20210302. eng.
20. Peters MC, McGrath KW, Hawkins GA, Hastie AT, Levy BD, Israel E, et al. Plasma interleukin-6 concentrations, metabolic dysfunction, and asthma severity: a cross-sectional analysis of two cohorts. Lancet Respir Med. 2016;4(7):574–84. PubMed PMID: 27283230. PMCID: PMC5007068. Epub 2016/06/11. eng.
21. Solomon DH, Karlson EW, Rimm EB, Cannuscio CC, Mandl LA, Manson JE, et al. Cardiovascular morbidity and mortality in women diagnosed with rheumatoid arthritis. Circulation. 2003;107(9):1303–7. PubMed PMID: 12628952. Epub 2003/03/12. eng.
22. Stein JH, Hsue PY. Inflammation and arterial injury in individuals with human immunodeficiency virus infection. JAMA Cardiol. 2016;1(4):481–2. PubMed PMID: 27438326. Epub 2016/07/22. eng.
23. Ridker PM, Rifai N, Stampfer MJ, Hennekens CH. Plasma concentration of interleukin-6 and the risk of future myocardial infarction among apparently healthy men. Circulation. 2000;101(15):1767–72. PubMed PMID: 10769275. Epub 2000/04/19. eng.
24. Ridker PM, Everett BM, Thuren T, MacFadyen JG, Chang WH, Ballantyne C, et al. Antiinflammatory therapy with Canakinumab for atherosclerotic disease. N Engl J Med. 2017;377:1119. PubMed PMID: 28845751. Epub 2017/08/29. eng.
25. Conen D, Ridker PM, Everett BM, Tedrow UB, Rose L, Cook NR, et al. A multimarker approach to assess the influence of inflammation on the incidence of atrial fibrillation in women. Eur Heart J. 2010;31(14):1730–6. PubMed PMID: 20501475. PMCID: PMC2903714. Epub 2010/05/27. eng.
26. Kornej J, Börschel CS, Benjamin EJ, Schnabel RB. Epidemiology of atrial fibrillation in the 21st century: novel methods and new insights. Circ Res. 2020;127(1):4–20. PubMed PMID: 32716709. PMCID: PMC7577553. Epub 20200618. eng.
27. Tattersall MC, Stein JH. Carotid intima-media thickness and plaque assessment. In: Shapiro MD, editor. Cardiovascular risk assessment in primary prevention. Cham: Springer International Publishing; 2022. p. 487–503.
28. Nagai Y, Metter EJ, Earley CJ, Kemper MK, Becker LC, Lakatta EG, et al. Increased carotid artery intimal-medial thickness in asymptomatic older subjects with exercise-induced myocardial ischemia. Circulation. 1998;98(15):1504–9. PubMed PMID: 9769303. Epub 1998/10/14. eng.

29. Stein JH, Korcarz CE, Hurst RT, Lonn E, Kendall CB, Mohler ER, et al. Use of carotid ultrasound to identify subclinical vascular disease and evaluate cardiovascular disease risk: a consensus statement from the American Society of Echocardiography Carotid Intima-Media Thickness Task Force. Endorsed by the Society for Vascular Medicine. J Am Soc Echocardiogr. 2008;21(2):93–111; quiz 89-90. PubMed PMID: 18261694. Epub 2008/02/12. eng.
30. Ungvari Z, Tarantini S, Donato AJ, Galvan V, Csiszar A. Mechanisms of vascular aging. Circ Res. 2018;123(7):849–67. PubMed PMID: 30355080. PMCID: PMC6248882. Epub 2018/10/26. eng.
31. Gepner AD, Young R, Delaney JA, Budoff MJ, Polak JF, Blaha MJ, et al. Comparison of carotid plaque score and coronary artery calcium score for predicting cardiovascular disease events: the Multi-Ethnic Study of Atherosclerosis. J Am Heart Assoc. 2017;6(2):e005179. PubMed PMID: 28196817. PMCID: PMC5523788. Epub 20170214. eng.
32. Gepner AD, Young R, Delaney JA, Tattersall MC, Blaha MJ, Post WS, et al. Comparison of coronary artery calcium presence, carotid plaque presence, and carotid intima-media thickness for cardiovascular disease prediction in the Multi-Ethnic Study of Atherosclerosis. Circ Cardiovasc Imaging. 2015;8(1):e002262. PubMed PMID: 25596139. PMCID: PMC4299916. Epub 2015/01/18. eng.
33. Corretti MC, Anderson TJ, Benjamin EJ, Celermajer D, Charbonneau F, Creager MA, et al. Guidelines for the ultrasound assessment of endothelial-dependent flow-mediated vasodilation of the brachial artery: a report of the International Brachial Artery Reactivity Task Force. J Am Coll Cardiol. 2002;39(2):257–65. PubMed PMID: 11788217. Epub 2002/01/15. eng.
34. Stein JH, Klein MA, Bellehumeur JL, McBride PE, Wiebe DA, Otvos JD, et al. Use of human immunodeficiency virus-1 protease inhibitors is associated with atherogenic lipoprotein changes and endothelial dysfunction. Circulation. 2001;104(3):257–62. PubMed PMID: 11457741. Epub 2001/07/18. eng.
35. Anderson TJ, Uehata A, Gerhard MD, Meredith IT, Knab S, Delagrange D, et al. Close relation of endothelial function in the human coronary and peripheral circulations. J Am Coll Cardiol. 1995;26(5):1235–41. PubMed PMID: 7594037. Epub 1995/11/01. eng.
36. Deanfield JE, Halcox JP, Rabelink TJ. Endothelial function and dysfunction: testing and clinical relevance. Circulation. 2007;115(10):1285–95. PubMed PMID: 17353456. Epub 2007/03/14. eng.
37. Prati P, Tosetto A, Vanuzzo D, Bader G, Casaroli M, Canciani L, et al. Carotid intima media thickness and plaques can predict the occurrence of ischemic cerebrovascular events. Stroke. 2008;39(9):2470–6. PubMed PMID: 18617662. Epub 2008/07/10. eng.
38. Johnsen SH, Mathiesen EB, Joakimsen O, Stensland E, Wilsgaard T, Løchen ML, et al. Carotid atherosclerosis is a stronger predictor of myocardial infarction in women than in men: a 6-year follow-up study of 6226 persons: the Tromsø Study. Stroke. 2007;38(11):2873–80. PubMed PMID: 17901390. Epub 2007/09/27. eng.
39. Chambless LE, Folsom AR, Clegg LX, Sharrett AR, Shahar E, Nieto FJ, et al. Carotid wall thickness is predictive of incident clinical stroke: the Atherosclerosis Risk in Communities (ARIC) study. Am J Epidemiol. 2000;151(5):478–87. PubMed PMID: 10707916. eng.
40. Chambless LE, Heiss G, Folsom AR, Rosamond W, Szklo M, Sharrett AR, et al. Association of coronary heart disease incidence with carotid arterial wall thickness and major risk factors: the Atherosclerosis Risk in Communities (ARIC) Study, 1987-1993. Am J Epidemiol. 1997;146(6):483–94. PubMed PMID: 9290509. Epub 1997/09/18. eng.
41. Lorenz MW, von Kegler S, Steinmetz H, Markus HS, Sitzer M. Carotid intima-media thickening indicates a higher vascular risk across a wide age range: prospective data from the Carotid Atherosclerosis Progression Study (CAPS). Stroke. 2006;37(1):87–92. PubMed PMID: 16339465. Epub 2005/12/08. eng.
42. Segers P, Rietzschel ER, Chirinos JA. How to measure arterial stiffness in humans. Arterioscler Thromb Vasc Biol. 2020;40(5):1034–43. PubMed PMID: 31875700. PMCID: PMC7180118. Epub 20191226. eng.

43. Lacolley P, Regnault V, Laurent S. Mechanisms of arterial stiffening: from Mechanotransduction to epigenetics. Arterioscler Thromb Vasc Biol. 2020;40(5):1055–62. PubMed PMID: 32075419. Epub 20200220. eng.
44. Yang EY, Chambless L, Sharrett AR, Virani SS, Liu X, Tang Z, et al. Carotid arterial wall characteristics are associated with incident ischemic stroke but not coronary heart disease in the Atherosclerosis Risk in Communities (ARIC) study. Stroke. 2012;43(1):103–8. PubMed PMID: 22033999. PMCID: PMC3246524. Epub 2011/10/29. eng.
45. Gepner AD, Korcarz CE, Colangelo LA, Hom EK, Tattersall MC, Astor BC, et al. Longitudinal effects of a decade of aging on carotid artery stiffness: the multiethnic study of atherosclerosis. Stroke. 2014;45(1):48–53. PubMed PMID: 24253542. PMCID: PMC3888489. Epub 2013/11/21. eng.
46. Stern R, Tattersall MC, Gepner AD, Korcarz CE, Kaufman J, Colangelo LA, et al. Sex differences in predictors of longitudinal changes in carotid artery stiffness: the Multi-Ethnic Study of Atherosclerosis. Arterioscler Thromb Vasc Biol. 2015;35(2):478–84. PubMed PMID: 25477347. PMCID: PMC4304990. Epub 2014/12/06. eng.
47. Inaba Y, Chen JA, Bergmann SR. Carotid plaque, compared with carotid intima-media thickness, more accurately predicts coronary artery disease events: a meta-analysis. Atherosclerosis. 2012;220(1):128–33. PubMed PMID: 21764060. Epub 2011/07/19. eng.
48. Nambi V, Chambless L, Folsom AR, He M, Hu Y, Mosley T, et al. Carotid intima-media thickness and presence or absence of plaque improves prediction of coronary heart disease risk: the ARIC (Atherosclerosis Risk In Communities) study. J Am Coll Cardiol. 2010;55(15):1600–7. PubMed PMID: 20378078. PMCID: PMC2862308. eng.
49. Henry SL, Moore LE, Brotto AR, Rowland S, Fuhr D, Stickland MK. Systemic vascular health is compromised in both confirmed and unconfirmed asthma. Respir Med. 2022;200:106932. PubMed PMID: 35870436. Epub 20220714. eng.
50. Korcarz CE, Benca R, Barnet JH, Stein JH. Treatment of obstructive sleep apnea in young and middle-aged adults: effects of positive airway pressure and compliance on arterial stiffness, endothelial function, and cardiac hemodynamics. J Am Heart Assoc. 2016;5(4):e002930. PubMed PMID: 27039928. PMCID: PMC4859282. Epub 2016/04/05. eng.
51. Moore LE, Bhutani M, Petersen SR, McMurtry MS, Byers BW, Tedjasaputra V, et al. Physical activity, fitness, and vascular health in patients with asthma. J Allergy Clin Immunol. 2015;136(3):809–11.e3. PubMed PMID: 25913103. Epub 20150424. eng.
52. Knoflach M, Kiechl S, Mayr A, Willeit J, Poewe W, Wick G. Allergic rhinitis, asthma, and atherosclerosis in the Bruneck and ARMY studies. Arch Intern Med. 2005;165(21):2521–6. PubMed PMID: 16314550. Epub 2005/11/30. eng.
53. Onufrak S, Abramson J, Vaccarino V. Adult-onset asthma is associated with increased carotid atherosclerosis among women in the Atherosclerosis Risk in Communities (ARIC) study. Atherosclerosis. 2007;195(1):129–37. PubMed PMID: 17045272. PMCID: Pmc2128256. Epub 2006/10/19. eng.
54. Sun D, Li X, Heianza Y, Nisa H, Shang X, Rabito F, et al. History of asthma from childhood and arterial stiffness in asymptomatic young adults: the Bogalusa heart study. Hypertension (Dallas, Tex : 1979). 2018;71(5):928. PubMed PMID: 29632104.
55. Vasan RS, Short MI, Niiranen TJ, Xanthakis V, DeCarli C, Cheng S, et al. Interrelations between arterial stiffness, target organ damage, and cardiovascular disease outcomes. J Am Heart Assoc. 2019;8(14):e012141. PubMed PMID: 31303106. PMCID: PMC6662123. Epub 20190713. eng.
56. Tattersall MC, Dasiewicz AS, McClelland RL, Jarjour NN, Korcarz CE, Mitchell CC, et al. Persistent asthma is associated with carotid plaque in MESA. J Am Heart Assoc. 2022;11(23):e026644.
57. Cakmak A, Zeyrek D, Cece H, Erel O. The relationship between carotid intima media thickness and oxidative stress in asthmatic children. Asian Pac J Allergy Immunol. 2010;28(4):256–61. PubMed PMID: 21337909. Epub 2011/02/23. eng.

58. Dratva J, Caviezel S, Schaffner E, Stolz D, Rothe T, Kuenzli N, et al. Is there a gender-specific association between asthma and carotid intima media thickness in Swiss adolescents? Eur J Pediatr. 2018;177(5):699–707. PubMed PMID: 29411143. Epub 2018/02/08. eng.
59. Lemanske RF. The childhood origins of asthma (COAST) study. Pediatr Allergy Immunol. 2002;13(s15):38–43. PubMed PMID: 12688623. eng.
60. Tattersall MC, Evans MD, Korcarz CE, Mitchell C, Anderson E, DaSilva DF, et al. Asthma is associated with carotid arterial injury in children: the Childhood Origins of Asthma (COAST) Cohort. PLoS One. 2018;13(9):e0204708. PubMed PMID: 30261051. PMCID: PMC6160166. Epub 2018/09/28. eng.
61. Toren K, Lindholm NB. Do patients with severe asthma run an increased risk from ischaemic heart disease? Int J Epidemiol. 1996;25(3):617–20. PubMed PMID: 8671564. Epub 1996/06/01. eng.
62. Enright PL, Ward BJ, Tracy RP, Lasser EC. Asthma and its association with cardiovascular disease in the elderly. The Cardiovascular Health Study Research Group. J Asthma. 1996;33(1):45–53. PubMed PMID: 8621370. Epub 1996/01/01. eng.
63. Schanen JG, Iribarren C, Shahar E, Punjabi NM, Rich SS, Sorlie PD, et al. Asthma and incident cardiovascular disease: the Atherosclerosis Risk in Communities Study. Thorax. 2005;60(8):633–8. PubMed PMID: 16061703. PMCID: Pmc1747501. Epub 2005/08/03. eng.
64. Onufrak SJ, Abramson JL, Austin HD, Holguin F, McClellan WM, Vaccarino LV. Relation of adult-onset asthma to coronary heart disease and stroke. Am J Cardiol. 2008;101(9):1247–52. PubMed PMID: 18435952. PMCID: Pmc2396340. Epub 2008/04/26. eng.
65. Lee HM, Truong ST, Wong ND. Association of adult-onset asthma with specific cardiovascular conditions. Respir Med. 2012;106(7):948–53. PubMed PMID: 22445771. Epub 2012/03/27. eng.
66. Iribarren C, Tolstykh IV, Miller MK, Sobel E, Eisner MD. Adult asthma and risk of coronary heart disease, cerebrovascular disease, and heart failure: a prospective study of 2 matched cohorts. Am J Epidemiol. 2012;176(11):1014–24. PubMed PMID: 23139248. Epub 2012/11/10. eng.
67. Tattersall MC, Barnet JH, Korcarz CE, Hagen EW, Peppard PE, Stein JH. Late-Onset asthma predicts cardiovascular disease events: the Wisconsin Sleep Cohort. J Am Heart Assoc. 2016;5(9):e003448. PubMed PMID: 27559071. Epub 2016/08/26. eng.
68. Young T, Palta M, Dempsey J, Skatrud J, Weber S, Badr S. The occurrence of sleep-disordered breathing among middle-aged adults. N Engl J Med. 1993;328(17):1230–5. PubMed PMID: 8464434. Epub 1993/04/29. eng.
69. Hla KM, Young T, Hagen EW, Stein JH, Finn LA, Nieto FJ, et al. Coronary heart disease incidence in sleep disordered breathing: the Wisconsin Sleep Cohort Study. Sleep. 2015;38(5):677–84. PubMed PMID: 25515104. PMCID: Pmc4402672. Epub 2014/12/18. eng.
70. Pollevick ME, Xu KY, Mhango G, Federmann EG, Vedanthan R, Busse P, et al. The relationship between asthma and cardiovascular disease: an examination of the Framingham Offspring Study. Chest. 2021;159(4):1338–45. PubMed PMID: 33316236. Epub 20201211. eng.
71. Strand LB, Tsai MK, Wen CP, Chang SS, Brumpton BM. Is having asthma associated with an increased risk of dying from cardiovascular disease? A prospective cohort study of 446346 Taiwanese adults. BMJ Open. 2018;8(5):e019992. PubMed PMID: 29858410. PMCID: PMC5988076. Epub 20180531. eng.
72. Di Gennaro A, Haeggstrom JZ. The leukotrienes: immune-modulating lipid mediators of disease. Adv Immunol. 2012;116:51–92. PubMed PMID: 23063073. Epub 2012/10/16. eng.
73. Drazen JM. Leukotrienes as mediators of airway obstruction. Am J Respir Crit Care Med. 1998;158(5 Pt 3):S193–200. PubMed PMID: 9817745. eng.
74. Spanbroek R, Grabner R, Lotzer K, Hildner M, Urbach A, Ruhling K, et al. Expanding expression of the 5-lipoxygenase pathway within the arterial wall during human atherogenesis. Proc Natl Acad Sci U S A. 2003;100(3):1238–43. PubMed PMID: 12552108. PMCID: Pmc298757. Epub 2003/01/29. eng.

75. Aiello RJ, Bourassa PA, Lindsey S, Weng W, Freeman A, Showell HJ. Leukotriene B4 receptor antagonism reduces monocytic foam cells in mice. Arterioscler Thromb Vasc Biol. 2002;22(3):443–9. PubMed PMID: 11884288. Epub 2002/03/09. eng.
76. Voloshyna I, Littlefield MJ, Reiss AB. Atherosclerosis and interferon-gamma: new insights and therapeutic targets. Trends Cardiovasc Med. 2014;24(1):45–51. PubMed PMID: 23916809. PMCID: PMC3844070. Epub 2013/08/07. eng.
77. Springall R, Amezcua-Guerra LM, Gonzalez-Pacheco H, Furuzawa-Carballeda J, Gomez-Garcia L, Marquez-Velasco R, et al. Interferon-gamma increases the ratio of matrix metalloproteinase-9/tissue inhibitor of metalloproteinase-1 in peripheral monocytes from patients with coronary artery disease. PLoS One. 2013;8(8):e72291. PubMed PMID: 23951304. PMCID: PMC3741271. Epub 2013/08/21. eng.
78. Whitman SC, Ravisankar P, Elam H, Daugherty A. Exogenous interferon-gamma enhances atherosclerosis in apolipoprotein E−/− mice. Am J Pathol. 2000;157(6):1819–24. PubMed PMID: 11106554. PMCID: PMC1885762. Epub 2000/12/07. eng.
79. Gupta S, Pablo AM, Jiang X, Wang N, Tall AR, Schindler C. IFN-gamma potentiates atherosclerosis in ApoE knock-out mice. J Clin Invest. 1997;99(11):2752–61. PubMed PMID: 9169506. PMCID: PMC508122. Epub 1997/06/01. eng.
80. Liuzzo G, Goronzy JJ, Yang H, Kopecky SL, Holmes DR, Frye RL, et al. Monoclonal T-cell proliferation and plaque instability in acute coronary syndromes. Circulation. 2000;101(25):2883–8. PubMed PMID: 10869258. Epub 2000/06/28. eng.
81. Kuruvilla ME, Lee FE, Lee GB. Understanding asthma phenotypes, endotypes, and mechanisms of disease. Clin Rev Allergy Immunol. 2019;56(2):219–33. PubMed PMID: 30206782. PMCID: PMC6411459. eng.
82. Kimura A, Kishimoto T. IL-6: regulator of Treg/Th17 balance. Eur J Immunol. 2010;40(7):1830–5. PubMed PMID: 20583029. Epub 2010/06/29. eng.
83. Yudkin JS, Kumari M, Humphries SE, Mohamed-Ali V. Inflammation, obesity, stress and coronary heart disease: is interleukin-6 the link? Atherosclerosis. 2000;148(2):209–14. PubMed PMID: 10657556. Epub 2000/02/05. eng.
84. Fujita M, Ueki S, Ito W, Chiba T, Takeda M, Saito N, et al. C-reactive protein levels in the serum of asthmatic patients. Ann Allergy Asthma Immunol. 2007;99(1):48–53. PubMed PMID: 17650829. Epub 2007/07/27. eng.
85. Roth GA, Mensah GA, Johnson CO, Addolorato G, Ammirati E, Baddour LM, et al. Global burden of cardiovascular diseases and risk factors, 1990-2019: update from the GBD 2019 Study. J Am Coll Cardiol. 2020;76(25):2982–3021. PubMed PMID: 33309175. PMCID: PMC7755038. eng.
86. Cepelis A, Brumpton BM, Malmo V, Laugsand LE, Loennechen JP, Ellekjaer H, et al. Associations of asthma and asthma control with atrial fibrillation risk: results from the Nord-Trondelag Health Study (HUNT). JAMA Cardiol. 2018;3(8):721–8. PubMed PMID: 29998294. PMCID: PMC6143075. Epub 2018/07/13. eng.
87. Carter P, Lagan J, Fortune C, Bhatt DL, Vestbo J, Niven R, et al. Association of cardiovascular disease with respiratory disease. J Am Coll Cardiol. 2019;73(17):2166–77. PubMed PMID: 30846341. Epub 2019/03/09. eng.
88. Tattersall MC, Dasiewicz AS, McClelland RL, Gepner AD, Kalscheur MM, Field ME, et al. Persistent asthma is associated with increased risk for incident atrial fibrillation in the MESA. Circ Arrhythm Electrophysiol. 2020;13(2):e007685–e. PubMed PMID: 32013555. Epub 2020/02/04. eng.
89. Lemanske RF, Kaliner MA. Autonomic nervous system abnormalities and asthma. Am Rev Respir Dis. 1990;141(3 Pt 2):S157–61. PubMed PMID: 2155565. eng.
90. Lewis MJ, Short AL, Lewis KE. Autonomic nervous system control of the cardiovascular and respiratory systems in asthma. Respir Med. 2006;100(10):1688–705. PubMed PMID: 16524708. Epub 20060309. eng.
91. Sanak M, Dropinski J, Sokolowska B, Faber J, Rzeszutko M, Szczeklik A. Pharmacological inhibition of leukotriene biosynthesis: effects on the heart conductance. J Physiol Pharmacol. 2010;61(1):53–8. PubMed PMID: 20228415. Epub 2010/03/17. eng.

92. Sokolowska B, Dropinski J, Rzeszutko M, Szczeklik W, Sanak M, Szczeklik A. Influence of leukotriene biosynthesis inhibition on heart rate in patients with atrial fibrillation. Int J Cardiol. 2010;145. Netherlands:625–6.
93. Hu YF, Chen YJ, Lin YJ, Chen SA. Inflammation and the pathogenesis of atrial fibrillation. Nat Rev Cardiol. 2015;12(4):230–43. PubMed PMID: 25622848. Epub 2015/01/28. eng.
94. Karthikeyan R, Krishnamoorthy S, Maamidi S, Kaza AM, Balasubramanian N. Effect of inhaled corticosteroids on systemic inflammation in asthma. Perspect Clin Res. 2014;5(2):75–9. PubMed PMID: 24741484. PMCID: Pmc3980548. Epub 2014/04/18. eng.
95. van der Hooft CS, Heeringa J, Brusselle GG, Hofman A, Witteman JC, Kingma JH, et al. Corticosteroids and the risk of atrial fibrillation. Arch Intern Med. 2006;166(9):1016–20. PubMed PMID: 16682576. Epub 2006/05/10. eng.
96. Dogra S, Ardern CI, Baker J. The relationship between age of asthma onset and cardiovascular disease in Canadians. J Asthma. 2007;44(10):849–54. PubMed PMID: 18097862. eng.
97. Christiansen SC, Zuraw BL. Treatment of hypertension in patients with asthma. N Engl J Med. 2019;381(11):1046–57. PubMed PMID: 31509675. eng.
98. Christiansen SC, Schatz M, Yang SJ, Ngor E, Chen W, Zuraw BL. Hypertension and asthma: a comorbid relationship. J Allergy Clin Immunol Pract. 2016;4(1):76–81. PubMed PMID: 26342745. Epub 20150903. eng.
99. Zolotareva O, Saik OV, Königs C, Bragina EY, Goncharova IA, Freidin MB, et al. Comorbidity of asthma and hypertension may be mediated by shared genetic dysregulation and drug side effects. Sci Rep. 2019;9(1):16302. PubMed PMID: 31705029. PMCID: PMC6841742. Epub 20191108. eng.
100. Kaess BM, Rong J, Larson MG, Hamburg NM, Vita JA, Levy D, et al. Aortic stiffness, blood pressure progression, and incident hypertension. JAMA. 2012;308(9):875–81. PubMed PMID: 22948697. PMCID: PMC3594687. eng.
101. Mitchell GF. Arterial stiffness and hypertension: chicken or egg? Hypertension. 2014;64(2):210–4. PubMed PMID: 24799614. PMCID: PMC4185002. eng.
102. Pase MP, Beiser A, Himali JJ, Tsao C, Satizabal CL, Vasan RS, Seshadri S, Mitchell GF, et al. Aortic stiffness and the risk of incident mild cognitive impairment and dementia. Stroke. 2016;47(9):2256. PubMed PMID: 27491735.
103. Lloyd-Jones DM, Allen NB, Anderson CAM, Black T, Brewer LC, Foraker RE, et al. Life's essential 8: updating and enhancing the American Heart Association's construct of cardiovascular health: a presidential advisory from the American Heart Association. Circulation. 2022;146(5):e18–e43. PubMed PMID: 35766027. Epub 20220629. eng.
104. Juel CTB, Ali Z, Nilas L, Ulrik CS. Asthma and obesity: does weight loss improve asthma control? A systematic review. J Asthma Allergy. 2012;5:21–6.
105. Arnett DK, Blumenthal RS, Albert MA, Buroker AB, Goldberger ZD, Hahn EJ, et al. 2019 ACC/AHA guideline on the primary prevention of cardiovascular disease: a report of the American College of Cardiology/American Heart Association Task Force on Clinical Practice Guidelines. Circulation. 2019;140(11):e596–646. PubMed PMID: 30879355. Epub 2019/03/19. eng.
106. Lloyd-Jones DM, Lewis CE, Schreiner PJ, Shikany JM, Sidney S, Reis JP. The Coronary Artery Risk Development In Young Adults (CARDIA) Study: JACC focus seminar 8/8. J Am Coll Cardiol. 2021;78(3):260–77. PubMed PMID: 34266580. PMCID: PMC8285563. eng.
107. Winder B, Kiechl SJ, Gruber NM, Bernar B, Gande N, Staudt A, et al. The association of allergic asthma and carotid intima-media thickness in adolescence: data of the prospective early vascular ageing (EVA)-Tyrol cohort study. BMC Cardiovasc Disord. 2022;22(1):11. PubMed PMID: 35042472. PMCID: PMC8764773. Epub 20220118. eng.
108. Iribarren C, Tolstykh IV, Eisner MD. Are patients with asthma at increased risk of coronary heart disease? Int J Epidemiol. 2004;33(4):743–8. PubMed PMID: 15131088. Epub 2004/05/08. eng.

Structural Racism and the Social Determinants of Health in Asthma

Adali Martinez and Neeta Thakur

Abstract

Asthma prevalence and morbidity are disproportionately higher among minoritized communities in the United States. Racial and ethnic disparities in asthma result from complex interactions across biological, environmental, and social factors. Asthma is considered a complex heterogeneous disease consisting of different phenotypes, some of which may be more common in individuals impacted by the downstream effects of structural racism and lack of access to the social determinants of health. Structural racism across generations has created and reinforced inequitable systems through policies and practices which are embedded in the economic, educational, health care, and justice systems (Bailey et al., N Engl J Med 384(8):768–773, 2021; Bailey et al., Lancet 389:1453–1463, 2017; Williams et al., Annu Rev Public Health 40:105–125, 2019). This manifests in an inequitable distribution of resources and the social determinants of health affecting an individual's physical and social environment (Bailey et al., Lancet 389:1453–1463, 2017; Thakur et al., Am J Respir Crit Care Med 202:943–949, 2020; Martinez et al., J Allergy Clin Immunol 148(5):1112–1120, 2021). In this chapter, we outline how inequity in housing, zoning laws, urban planning, education, employment, healthcare access, and healthcare delivery is linked to higher asthma prevalence and morbidity. We also describe the role that chronic physiologic stress has on asthma by enhancing neuroimmune and immunologic responses to environmental exposures. Interventions aimed at addressing the physical or social environment of an individual or community have been shown to improve asthma outcomes in patients at higher risk of severe disease.

A. Martinez · N. Thakur (✉)
Division of Pulmonary, Allergy, Critical Care and Sleep, Department of Medicine, University of California San Francisco, San Francisco, CA, USA
e-mail: neeta.thakur@ucsf.edu

A. R. Brasier, N. N. Jarjour (eds.), *Precision Approaches to Heterogeneity in Asthma*, Advances in Experimental Medicine and Biology 1426,
https://doi.org/10.1007/978-3-031-32259-4_5

Keywords

Racism · Social determinants of health · Health equity · Physiological stress · Neighborhood inner-city · Environment · Intervention · Heterogeneity

Overview

Asthma affects over 4.2 million children and 21 million adults in the United States (7.8% of the population) [1], with a disproportionate disease burden among minoritized populations. A higher prevalence of asthma exists in children and adults living below the federal poverty level [1], highlighting disparities across socioeconomic status. Asthma prevalence is also substantially higher in Black and Indigenous populations, 10.3% and 11.3%, respectively, compared to the non-Hispanic White population [1]. In addition to higher prevalence, Black and Latinx populations are more likely to have severe or difficult-to-control asthma with higher rates of healthcare utilization compared to non-Hispanic White populations [1–4].

Asthma is considered a complex heterogeneous disease consisting of different phenotypes. These phenotypes are thought to reflect particular biologic pathways and help predict drug response; they may also help categorize individuals at risk for poor outcomes [5]. This precision medicine approach to asthma has broadened our understanding of asthma and opened up novel therapeutic targets that have otherwise stalled for over two decades. To advance further and address asthma morbidity across populations, in addition to investigating the phenotypic and genetic variations in populations with asthma, it is exceedingly important to recognize the contextual factors contributing to asthma disparities. Racial and ethnic disparities in asthma result from complex interactions across biological, environmental, and social factors, including structural racism and the social determinants of health [6].

Structural racism is the totality of ways in which society fosters discrimination by creating and reinforcing inequitable systems through intentional policies and practices sanctioned by various levels of government and institutions [7–9]. It is embedded in the economic, educational, health care, and justice systems, which manifests in an inequitable distribution of resources and the social determinants of health [8, 10]. Structural racism leads to proximal pathways of inequity affecting the physical and social environments [6]. In this chapter, we outline how socioeconomic, political constructs, and structural racism impact the social determinants of health and asthma disparities in the United States (Schematically shown in Fig. 1).

Neighborhood and Housing

Perhaps the most important social determinant of health impacting asthma outcomes is housing and the physical environment. Disparities in housing and neighborhoods are deeply rooted in structural racism, resulting from several key policies and practices, most notably the practice of redlining, which sanctioned existing racist segregation and exclusion practices, including racial covenants (i.e., discriminatory clauses put into recorded documents, such as deeds, to prohibit Black and other ethnic minorities from buying or living in properties) [11]. In the 1930s, as an effort to recover from the Great Depression and promote home ownership, the federally sponsored Home Owners' Loan Corporation (HOLC) provided low-interest, long-term loans to prevent foreclosure. Due to popularity, HOLC created maps of at least 239 US cities ranking communities on mortgage worthiness [7, 12]. Communities with large Black or immigrant populations were graded and outlined in red, flagging them as hazardous investment areas [7, 12, 13]. As a result, residents from Black and immigrant communities were barred from receiving HOLC loans, which had lasting effects over generations [11].

Redlining and other exclusionary housing practices have led to an inequitable distribution of wealth and the concentration of poverty, resulting in a higher prevalence of substandard housing in Black and immigrant communities [12–14]. Systematic disinvestment in these communities has also led to under-resourced public services, including schools, healthcare facilities, and public parks [7, 15, 16]. Over generations, historically, redlines census tracts have been subjected to downstream policies, such as eminent domain and discriminatory zoning laws, which affected where highways and toxic hazard sites were built, often co-located in Black and immigrant communities [12, 14, 17]. The proximity to hazardous exposures and

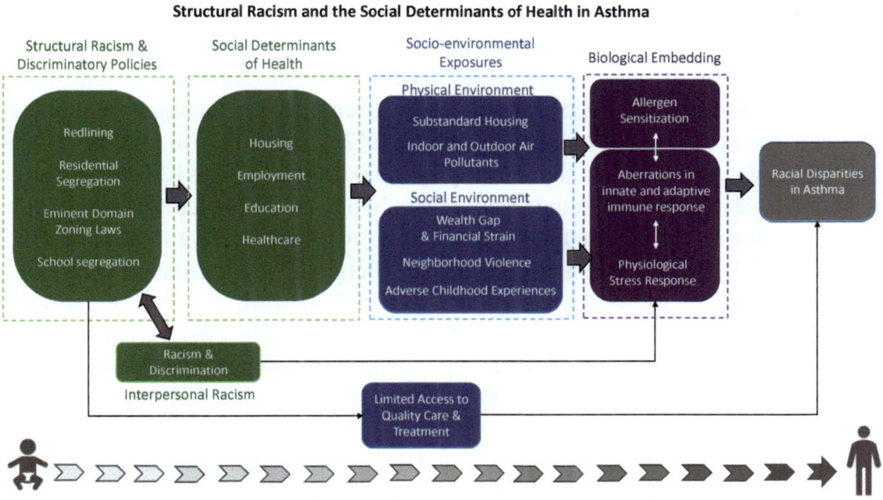

Fig. 1 Structural racism in the form of discriminatory policies and practices has created and reinforced inequitable systems and disparities in the social determinants of health. This results in disproportionately higher risk of exposures associated with asthma prevalence and morbidity in both the physical and social environments. These exposures are biologically embedded, having both a direct and indirect impact on asthma by enhancing immunologic responses. Structural racism and discriminatory practices have a direct impact on limiting access to quality asthma care and advanced treatments. Interpersonal racism and discrimination may also be biologically embedded as chronic stress, contributing to racial disparities in asthma. (Figure adapted from Martinez et al. [6])

reduced access to healthcare and health-promoting factors, such as green space, is the likely mechanism for the observed increase in asthma morbidity in these neighborhoods [18]. Historically redlined census tracts have 2.4 times higher rates of asthma-related emergency department visits than census tracts deemed good investments [14].

Indoor Allergens

A large body of literature documenting the association between substandard housing and asthma morbidity exists [19–21]. Early childhood exposure and sensitization to allergens found in substandard housing results in increased asthma incidence, and continued exposure to these allergens may also exacerbate asthma symptoms [13, 22]. These allergens include cockroaches, house dust mites, animals (cats, dogs, rodents), and molds [22–25]. Poor housing conditions and shared walls of multifamily swellings provide an optimal environment for pests, and studies have found that cockroach allergens are detectable in 85% of inner-city US homes, and approximately two-thirds of children living in these communities demonstrate skin test sensitivity to cockroaches [23]. As compared to other allergens, sensitization and exposure to cockroaches and dust mite allergens are associated with an increased rate of hospital admissions, unscheduled healthcare visits, wheezing, increased asthma symptom days, and missed school days [22, 24, 25].

Socio-environmental factors may contribute to type 2 inflammation through early-life exposure to specific allergens, leading to sensitization and increasing the risk for asthma. Low socioeconomic status has been linked to increases in total IgE, allergen sensitization, and levels of specific IgE [26]. Sensitization to cockroaches, animals, and ragweed differs significantly by socioeconomic status [26] Sensitization to allergens results in an immune response involving IgE antibodies but also includes IgG, IgA antibodies [27]. Individuals sensitized to allergens have more robust T_h2 responses [27]. T cells from individuals with allergy are more prone to make cytokines typical of T_h2 responses, including IL-4 and IL-5, while T cells from individuals without allergy generally produce IL-2 and IFN-y [27]. This may manifest in phenotypic different asthma presentations and differences in asthma outcomes with the highest morbidity among children with high wheeze and allergic sensitization [28].

Longitudinal home-based environmental interventions that entail targeted education and guided environmental modifications have been shown to effectively reduce indoor allergen levels and decrease the number of days with asthma symptoms and unscheduled healthcare visits [25, 29]. This intervention was most particularly effective in reducing exposure to cockroaches and dust mite allergens [29]. However, given the heterogeneity of asthma, these interventions are most effective when preceded by allergen testing and/or a thorough history. The 2020 Focused Updates to the Asthma Management Guidelines: A Report from the National Asthma Education and Prevention Program Coordinating Committee Expert Panel Working Group recommends a multicomponent allergen-specific mitigation intervention in individuals with asthma who have symptoms related to identified indoor allergens [30]. Interventions include air filtration, integrated pest management, mold mitigation, and carpet removal [30] Households in marginalized communities and low-income households face additional challenges in eliminating asthma allergens from home, limited by cost and lack of autonomy when renting. For patients with limited resources, engaging with local public health and housing programs can be helpful in identifying and addressing indoor allergen triggers [31].

Exposure to Pollutants

Residential segregation has shaped the distribution of resources at wealth at the individual and community levels [32]. Discriminatory policies such as redlining, eminent domain, and zoning laws have concentrated Black and immigrant communities in areas with high traffic density and higher levels of ambient air pollution [12, 14, 17, 32]. Historically redlined census tracts and communities with higher concentrations of Black and immigrant populations have nearly twice as high diesel exhaust particulate (DEP) emissions and increased levels of NO_2 and $PM_{2.5}$ compared to non-redlined, predominantly White neighborhoods [14, 33–35]. There is a large body of literature demonstrating significant associations between exposure to DEP, $PM_{2.5}$, NO_2 and right of asthma development in childhood and asthma-related morbidity [36–39]. In marginalized communities, many of these pollutants are higher indoors than outdoors, and studies have shown that exposure to indoor pollutants is independently associated with asthma symptoms [23].

Ambient air pollutants are associated with various asthma phenotypes. Ambient air pollutants have been demonstrated to elicit an immune response via airway inflammation and activation of antigen-presenting cells that interact with CD4+ T lymphocytes and polarize these cells toward the Th2 cytokine pathway [40]. There is also emerging research noting the role of air pollution in the upregulation of non-Th2 subsets also involved in airway disease, including Th17, TFh, and Treg [40]. Long-term $PM_{2.5}$ is also associated with lower lung function levels and slower lung function growth in children [41].

On the individual level, there are a few strategies to mitigate the impact of air pollution such as using air quality alert systems, limiting outdoor exercise on high-air pollution days or near sources of air pollutants, and wearing facemasks in certain circumstances [42]. The Particulate Reduction Education in City Homes study, for instance, explored the utility of HEPA air cleaners in mitigating the effects of indoor air pollution on asthma symptoms [43, 44]. In this randomized control trial, HEPA air cleaners led to a significant reduction in indoor PM concentrations; however, these levels continued to be above the EPA standards for outdoor air quality [43, 44]. Children in the HEPA air cleaner arms of the study had an increase in symptom-free days compared to the control group, but no difference was seen in symptom free-nights or acute asthma healthcare events [44]. This is a promising intervention worth further research and possibly at a larger scale involving multi-unit housing or public buildings such as schools. There are several limitations to these individual-level behavioral interventions, particularly for patients with limited resources; thus, public policy, such as federal enforcement of the Environmental Protection Agency (EPA) National Ambient Air Quality Standards, plays a crucial role in reducing air pollution impacting marginalized populations [42, 45].

Socioeconomic Position

Socioeconomic position (SEP) measures differential access to resources and encapsulates social class and social stratification [46]. Residential segregation and redlining have shaped the distribution of resources and wealth at the individual and community levels [32]. This has resulted in racial and ethnic differences in the quality of education, which limits employment and income potential, all of which are factors contributing to socioeconomic position [46–48]. Asthma prevalence is approximately 1.6 times higher among individuals with low family income than individuals with higher family income [1]. Lower SES is also associated with worse asthma control, greater emergency health service use, and worse asthma self-efficacy [49].

Access to quality education is highly influenced by where an individual lives, both within the United States and globally [47, 50]. In the United States, highly segregated schools have resulted in poorer access to education due to a lack of community resources and funding from low property tax revenue [47]. The relationship between educational attainment and asthma outcomes has been studied with various populations worldwide, exploring the impact of both individual's own education on their asthma symptoms and parental educational attainment on their children's asthma.

Individuals with adult-onset asthma with lower educational attainment are an increased risk of uncontrolled asthma [51] and are more likely to have asthma-related emergency health service visits [49]. Parental educational attainment is also associated with increased asthma prevalence, severe persistent asthma, and higher rescue medication use in children [52–54].

SEP is thought to impact asthma prevalence and severity through both extrinsic and intrinsic stressors. Persons with low SEP are more likely to live in substandard housing in close proximity to highways and other hazardous air pollutants, but housing only partially explains the relationship between low SEP and asthma morbidity [53]. Low SEP can also impact an individual's access to health resources, insurance status, ability to pay for medications, and health literacy [53, 54]. Lower SEP can also have a profound effect on an individual's intrinsic stress response [55, 56]. In children with asthma, lower parental SEP is associated with higher expression of inflammatory profiles, including higher eosinophil count, larger Th2 and Th1 cytokine responses, and marginally lower morning cortisol values compared with the high SEP group [55, 56], suggesting a mechanistic pathway of contributing different asthma phenotypes. Above, we discussed the role of sensitization to indoor allergens leading to increased asthma risk. Independent of allergic sensitization, SEP has been demonstrated to activate both type 2 and non-type 2 inflammation and enhance airway inflammation [56, 57], impacting lung function and drug response [58, 59].

Interventions aimed at addressing the adverse circumstances of low SEP have been shown to be effective in improving asthma symptoms [46, 60–62]. Several of these interventions enlist the support of community health workers, which are members of the community, often with cultural and language affinity, trained in health advocacy and community resources [60–62]. One of the best evidence interventions for addressing low-income environments is asthma home visiting programs, which address both environmental exposures but also provide linkages for community resources [29, 63, 64]. A promising intervention is medical-legal partnerships, which entails an attorney embedded in a healthcare setting to address social determinants of health [65].

Healthcare

Significant racial and ethnic disparities exist in access to health care throughout the United States [15]. Despite living in dense urban cities, minoritized individuals have less access to high-quality preventative healthcare, often relying on emergency healthcare services [15]. Several reasons for this disparity are explored in the literature, including higher uninsurance rates, high prescription costs, primary care physician shortages, fewer referrals to specialty care, and communication barriers [15, 66–68].

From 1997 to 2018, the uninsurance rate among adults with asthma in the United States decreased from 19.4% to 9.6%, coinciding with the passing of the Affordable Care Act [66]. However, despite this reduction in uninsurance rates, the proportion of individuals foregoing medical care or medications due to costs has not improved. Racial and ethnic minorities with asthma have higher rates of inadequate healthcare coverage compared to White adults [66]. Rising drug prices, particularly of inhalers after their reformulation, has impacted affordability and access to necessary asthma treatment [66]

Access to quality asthma care includes access to both primary care and specialty care. Within major cities, highly segregated zip codes are more likely to have a shortage of primary care physicians [15], resulting in an overreliance on inconsistent emergency services. As a result, patients from marginalized communities are less likely to receive appropriate asthma medications for preventative care or post-emergency care [69]. There are also significant racial and ethnic disparities in the prescription of ICS therapy among children and adults with asthma [67, 70, 71], likely contributing to higher rates of poorly controlled asthma.

Interventions to mitigate differential access to quality asthma care include pediatric asthma pathways and emergency service referrals to specialized asthma care. Pediatric asthma pathways have demonstrated improvements in the quality of care and reductions in inpatient length of stays by prompting clinicians to adopt evidence-based guidelines [72, 73]. Referral programs for high-risk children, such as the "Yes We Can" clinic, could also aid in reducing hospitalizations by providing comprehensive and holistic asthma care to children with high emergency healthcare utilization and high rescue inhaler use [74].

Advancements in precision medicine for asthma have led to the development of novel therapeutic targets and biologic treatment. Unfortunately, there are significant disparities in access to these novel therapies, with restricted access for publicly insured patients and fewer referrals for racial and ethnic minorities [67]. Due to the complexity and heterogeneity of asthma, certain patients would benefit from referral to an asthma specialist and consideration of biologic treatments. Yet, less than a quarter of patients with uncontrolled asthma are referred to an asthma specialist for management [75]. Patients that are publicly insured are less likely than those that are privately insured to be referred for an asthma treatment or biologic treatment visit [67]. Within the publicly insured population seen for specialized asthma care, only 60% of the population in asthma treatment visits were White, yet they accounted for up to 80% of patients evaluated for biologic treatment visits [67], highlighting a second layer of racial and ethnic disparities in access to advanced targeted therapy.

Individuals from racial and ethnic minorities with non-English language preferences face additional barriers to quality asthma care. For instance, parents of children with asthma whose preferred language is Spanish have been noted to have lower asthma self-management knowledge scores than parents whose preferred language is English [68]. Physician factors that contribute to impaired communication with patients include the underuse of interpreter services, unintentional racial and ethnic biases influencing the interpretation of patient symptoms, and poor provider understanding of the patient's cultural disease model [76]. Video-based education has been shown to increase asthma knowledge among Latinx parents with non-English language preferences and also among parents with low health literacy [68].

Biological Embedding of the Social Context

Environmental exposures, such as allergens and air pollution, have been well established as risk factors for asthma prevalence and morbidity [36–39]. There is growing evidence demonstrating the role of chronic psychosocial stress as a significant contributor to asthma [77], having an independent association with asthma and also playing a role in enhancing neuroimmune and immunologic responses to environmental exposures [78]. The relationship between chronic stress and asthma has been best described for financial strain (*see* section "Socioeconomic Position"); there is also increasing evidence for other sources of psychosocial stress, including adverse childhood experiences (ACEs), community violence, racial discrimination, and acculturation.

ACEs are stressful or traumatic experiences that occur before the age of 18 [79]. In various populations of different racial and ethnic backgrounds, higher exposure to adverse childhood experiences early in life has been associated with a higher prevalence of asthma in children and adults, regardless of phenotype [80–84]. In addition to ACEs, exposure to violence in adulthood has also been linked to asthma morbidity. In an individual level, intimate partner violence has been associated with increased odds of asthma exacerbations, higher symptom burden, and lack of asthma control [85]. In a community level, exposure to neighborhood violence has been associated with increased hospitalizations and emergency care for asthma [86]. The relationship between community violence and asthma risk persists when controlling for individual and neighborhood level confounders; however, neither community violence nor other individual level factors fully account for the excess asthma burden among Black individuals [87]. Exposure to violence may induce epigenetic changes that increase susceptibility to asthma. For instance, methylation of ADCYAP1R1 has been associated with both exposure to violence and increased asthma odds in a study of Puerto Rican children [88].

We previously discussed the mechanisms by which discriminatory policies disproportionally placed Black individuals in communities with higher exposure to environmental exposures known to trigger asthma [12, 14, 17, 32]. Additionally, Black individuals are subjected to individualized racism and discrimination, which can also have an impact on health and the physiologic stress response. Perceived discrimination has been associated with increased odds of asthma and poor asthma control among Black youth [89]. While the association between discrimination and asthma among Latinx youth is less robust, higher acculturation has been shown to be associated with a higher prevalence of asthma in Mexican Americans and other Latinx subgroups [90].

Stressful events early in life have been found to have a wide range of physiological effects on neural, endocrine, and immune responses [91]. Activation of the hypothalamic–pituitary–adrenal (HPA) and the sympathetic nervous system leads to the secretion of glucocorticoids (cortisol) and catecholamines (epinephrine and norepinephrine) [91, 92]. Chronic glucocorticoid and β-adrenergic receptors receptor activation in lymphocytes leads to desensitization and decreased responsiveness to corticosteroid and β-agonist treatment [93]. Chronic stress has also been shown to result in changes in the expression of β-adrenergic receptors (B2AR) and glucocorticoid receptor (GR-alpha) genes [77]. Glucocorticoids and catecholamines may also medicate T_h2 polarization by inhibiting the production of Type 1 cytokines in T_h1 and promoting T_h2 phenotype [94, 95]. Similarly, glucocorticoids promote inflammation by decreasing the number of T_{regs} and promoting T_h17 cell differentiation [57, 96].

Universal screening of ACEs and current unmet social needs in clinical settings can help further identify and address the impacts of early life and current adversity on asthma morbidity. Screening for ACEs and current unmet social needs can help identify children and adults that would benefit from the additional support provided through medical-social care models such as case management, community health workers, and asthma home visiting programs which have all been shown to improve outcomes for children with poorly controlled asthma [60–63].

Conclusion

Precision medicine in asthma has broadened our understanding of the biological pathways responsible for the development of various asthma phenotypes, leading to novel therapeutic targets. Independent of this research, disparities in the social determinants of health, specifically housing, socioeconomic position, healthcare access, and delivery, have been linked to higher asthma prevalence and morbidity [19–21, 46, 49, 53, 67, 69]. There is also a growing interest in defining the role of structural racism in influencing the social determinants of health and asthma disparities [6–8, 14]. Structural racism has resulted in an inequitable distribution of resources, leading to inequity in housing, zoning laws, urban planning, education, employment, healthcare access, and healthcare delivery [6, 8, 10, 32]. This has led to a disproportionately higher risk of exposure to indoor allergens and air pollutants known to trigger asthma symptoms, particularly within racial and ethnic minority communities [12–14, 17, 22, 23, 32, 33]. Socioeconomic barriers, trauma, community violence, discrimination, and acculturation may also increase the risk of asthma morbidity through the biological embedding of chronic stress, enhancing neuroimmune and immunologic responses to environmental exposures [77–84, 89–91].

Emerging research explores the role of socio-environmental factors on inflammatory pathways involved in developing specific asthma phenotypes. However, while social and environmental interventions have been shown to improve asthma outcomes, less is known about the impact of these interventions on mitigating specific inflammatory pathways. Further research is needed to explore the role of targeted asthma therapy, specifically in racial and economically disadvantaged groups, with less access to specialty asthma care and biologic treatments. Lastly, to have a meaningful impact on asthma disparities, there is a need for research in this field to translate into reform in public policy in the form of environmental regulations, housing standards, and healthcare access.

References

1. National Center for Environmental Health. Most recent national asthma data. Center for Disease Control and Prevention. 2020.
2. Zahran HS, Bailey CM, Damon SA, Garbe PL, Breysse PN. Morbidity and mortality weekly report vital signs: asthma in children-United States, 2001–2016 [Internet]. Available from: https://www.cdc.gov/nchs/nhis/index.htm.
3. Lowe AA, Bender B, Liu AH, Solomon T, Kobernick A, Morgan W, et al. Environmental concerns for children with asthma on the Navajo Nation. Ann Am Thorac Soc. 2018;15(6):745–53.
4. Keet CA, Matsui EC, McCormack MC, Peng RD. Urban residence, neighborhood poverty, race/ethnicity and asthma morbidity among children in Medicaid. J Allergy Clin Immunol. 2017;140(3):822–7.
5. Gauthier M, Ray A, Wenzel SE. Evolving concepts of asthma. Am J Respir Crit Care Med. 2015;192(6):660–8.
6. Martinez A, de la Rosa R, Mujahid M, Thakur N. Structural racism and its pathways to asthma and atopic dermatitis. J Allergy Clin Immunol. Elsevier Inc. 2021;148(5):1112–20.

7. Bailey ZD, Feldman JM, Bassett MT. How structural racism works-racist policies as a root cause of U.S. racial health inequities. N Engl J Med. 2021;384(8):768–73.
8. Bailey ZD, Krieger N, Agénor M, Graves J, Linos N, Bassett MT. Structural racism and health inequities in the USA: evidence and interventions. Lancet. Lancet Publishing Group. 2017;389:1453–63.
9. Williams DR, Lawrence JA, Davis BA. Racism and health: evidence and needed research. Annu Rev Public Health [Internet]. 2019;40:105–25. Available from: https://doi.org/10.1146/annurev-publhealth-.
10. Thakur N, Lovinsky-Desir S, Bime C, Wisnivesky JP, Celedón JC. The structural and social determinants of the racial/ethnic disparities in the U.S. COVID-19 pandemic what's our role? Am J Respir Crit Care Med. American Thoracic Society. 2020;202:943–9.
11. Mitchell B, Franco J. HOLC "Redlining" Maps: the persistent structure of segregation and economic inequality [Internet]. 2018. Available from: www.ncrc.org.
12. Rothstein R. The color of law: a forgotten history of our government segregated America. New York: Liveright Publishing Corportation; 2017.
13. Bryant-Stephens TC, Strane D, Robinson EK, Bhambhani S, Kenyon CC. Housing and asthma disparities. J Allergy Clin Immunol. 2021;148(5):1121–9.
14. Nardone A, Casey JA, Morello-Frosch R, Mujahid M, Balmes JR, Thakur N. Associations between historical residential redlining and current age-adjusted rates of emergency department visits due to asthma across eight cities in California: an ecological study. Lancet Planet Health. 2020;4(1):e24–31.
15. Gaskin DJ, Dinwiddie GY, Chan KS, McCleary RR. Residential segregation and the availability of primary care physicians. Health Serv Res. 2012;47(6):2353–76.
16. Burke LM, Schwalbach J. Housing redlining and its lingering effects on education opportunity [Internet]. 2021. Center for Education Policy. Available from: http://report.heritage.org/bg3594.
17. Whittemore AH. The experience of racial and ethnic minorities with zoning in the United States. J Plan Lit. 2017;32(1):16–27.
18. Nardone A, Rudolph KE, Morello-Frosch R, Casey JA. Redlines and greenspace: the relationship between historical redlining and 2010 greenspace across the United States. Environ Health Perspect. 2021;129(1):1–9.
19. Rosenbaum E. Racial/ethnic differences in asthma prevalence: the role of housing and neighborhood environments*. J Health Soc Behav. 2008;49:131–45.
20. Brewer M, Kimbro RT, Denney JT, Osiecki KM, Moffett B, Lopez K. Does neighborhood social and environmental context impact race/ethnic disparities in childhood asthma? Health Place. 2017;44:86–93.
21. Hughes HK, Matsui EC, Tschudy MM, Pollack CE, Keet CA. Pediatric asthma health disparities: race, hardship, housing, and asthma in a national survey. Acad Pediatr. 2017;17(2):127–34.
22. Poowuttikul P, Saini S, Seth D. Inner-city asthma in children. Clin Rev Allergy Immunol. Humana Press Inc. 2019;56:248–68.
23. Matsui EC. Environmental exposures and asthma morbidity in children living in urban neighborhoods. Allergy: Eur J Allergy Clin Immunol. Blackwell Publishing Ltd. 2014;69:553–8.
24. Arshad SH. Does exposure to indoor allergens contribute to the development of asthma and allergy? Curr Allergy Asthma Rep. 2010;10:49–55.
25. Rosenstreich DL, Eggleston P, Kattan M, Baker D, Slavin RG, Gergen P, et al. The role of cockroach allergy and exposure to cockroach allergen in causing morbidity among inner-city children with asthma. N Engl J Med. 1997;336(19):1356–63.
26. Lewis SA, Weiss ST, Platts-Mills TAE, Syring M, Gold DR. Association of specific allergen sensitization with socioeconomic factors and allergic disease in a population of Boston women. J Allergy Clin Immunol. 2001;107(4):615–22.
27. Platts-Mills TAE, Rakes G, Heymann PW. The relevance of allergen exposure to the development of asthma in childhood. J Allergy Clin Immunol. 2000;105:S503–8.

28. Bacharier LB, Beigelman A, Calatroni A, Jackson DJ, Gergen PJ, O'Connor GT, et al. Longitudinal phenotypes of respiratory health in a high-risk urban birth cohort. Am J Respir Crit Care Med. 2019;199(1):71–82.
29. Morgan WJ, Crain EF, Gruchalla RS, O'connor GT, Kattan M, Evans Iii R, et al. Results of a home-based environmental intervention among urban children with asthma [Internet]. Available from: www.nejm.org.
30. Expert Panel Working Group of the National Heart L and BI (NHLBI) Administered and Coordinated NAE and PPCC (NAEPPCC), Cloutier MM, Baptist AP, Blake KV, Brooks EG, Bryant-Stephens T, et al. 2020 focused updates to the asthma management guidelines: a report from the National Asthma Education and Prevention Program Coordinating Committee Expert Panel Working Group. J Allergy Clin Immunol. 2020;146(6):1217–70.
31. Matsui EC, Peng RD. 2020 updated asthma guidelines: indoor allergen reduction. J Allergy Clin Immunol. Mosby Inc. 2020;146:1283–5.
32. Morello-Frosch R, Lopez R. The riskscape and the color line: examining the role of segregation in environmental health disparities. Environ Res. 2006;102(2):181–96.
33. Nardone A, Neophytou AM, Balmes J, Thakur N. Ambient air pollution and asthma-related outcomes in children of color of the USA: a scoping review of literature published between 2013 and 2017. Curr Allergy Asthma Rep. 2018;18(5):29.
34. Lane HM, Morello-Frosch R, Marshall JD, Apte JS. Historical redlining is associated with present-day air pollution disparities in U.S. cities. Environ Sci Technol Lett. 2022;9(4):345–50.
35. Liévanos RS. Racialized structural vulnerability: neighborhood racial composition, concentrated disadvantage, and fine particulate matter in California. Int J Environ Res Public Health. 2019;16(17):3196.
36. Kravitz-Wirtz N, Teixeira S, Hajat A, Woo B, Crowder K, Takeuchi D. Early-life air pollution exposure, neighborhood poverty, and childhood asthma in the United States, 1990–2014. Int J Environ Res Public Health. 2018;15(6):1114.
37. Delfino RJ, Wu J, Tjoa T, Gullesserian SK, Nickerson B, Gillen DL. Effect modification by residential traffic-related air pollution. Epidemiology. 2014;25(1):48–57.
38. Kim J, Natarajan S, Vaickus LJ, Bouchard JC, Beal D, Cruikshank WW, et al. Diesel exhaust particulates exacerbate asthma-like inflammation by increasing CXC chemokines. Am J Pathol. 2011;179(6):2730–9.
39. Alcala E, Brown P, Capitman JA, Gonzalez M, Cisneros R. Cumulative impact of environmental pollution and population vulnerability on pediatric asthma hospitalizations: a multilevel analysis of calenviroscreen. Int J Environ Res Public Health. 2019;16(15):2683.
40. Tuazon JA, Kilburg-Basnyat B, Oldfield LM, Wiscovitch-Russo R, Dunigan-Russell K, Fedulov AV, et al. Emerging insights into the impact of air pollution on immune-mediated asthma pathogenesis. Curr Allergy Asthma Rep. Springer. 2022;22(7):77–92.
41. Garcia E, Rice MB, Gold DR. Air pollution and lung function in children. J Allergy Clin Immunol. Mosby Inc. 2021;148:1–14.
42. Carlsten C, Salvi S, Wong GWK, Chung KF. Personal strategies to minimise effects of air pollution on respiratory health: advice for providers, patients and the public. Eur Respir J. 2020;55(6):1902056.
43. Akar-Ghibril N, Phipatanakul W. The indoor environment and childhood asthma. Curr Allergy Asthma Rep. Springer. 2020;20:43.
44. Butz AM, Matsui EC, Breysse P, Curtin-Brosnan J, Eggleston P, Diette G, et al. A randomized trial of air cleaners and a health coach to improve indoor air quality for inner-city children with asthma and secondhand smoke exposure. Arch Pediatr Adolesc Med. 2011;165(8):741–8.
45. Criteria Air Pollutants [Internet]. United States Environmental Protection Agency. 2022 [cited 2022 Sep 6]. Available from: https://www.epa.gov/criteria-air-pollutants.
46. Sullivan K, Thakur N. Structural and social determinants of health in asthma in developed economies: a scoping review of literature published between 2014 and 2019. Curr Allergy Asthma Rep. Springer. 2020;20:5.
47. Williams DR, Collins C. Racial residential segregation: a fundamental cause of racial disparities in health. Public Health Rep. 2001;116(5):404–16.

48. Williams DR, Priest N, Anderson NB. Understanding associations among race, socioeconomic status, and health: patterns and prospects. Health Psychol. American Psychological Association Inc. 2016;35:407–11.
49. Bacon SL, Bouchard A, Loucks EB, Lavoie KL. Individual-level socioeconomic status is associated with worse asthma morbidity in patients with asthma. Respir Res. 2009;17:10.
50. Sinharoy A, Mitra S, Mondal P. Socioeconomic and environmental predictors of asthma-related mortality. J Environ Public Health. 2018;2018:9389570.
51. Ilmarinen P, Stridsman C, Bashir M, Tuomisto LE, Vähätalo I, Goksör E, et al. Level of education and asthma control in adult-onset asthma. J Asthma. 2022;59(4):840–9.
52. Thakur N, Oh SS, Nguyen EA, Martin M, Roth LA, Galanter J, et al. Socioeconomic status and childhood asthma in urban minority youths: the GALA II and SAGE II studies. Am J Respir Crit Care Med. 2013;188(10):1202–9.
53. Thakur N, Martin M, Castellanos E, Oh SS, Roth LA, Eng C, et al. Socioeconomic status and asthma control in African American youth in SAGE II. J Asthma. 2014;51(7):720–8.
54. Lewis KM, Ruiz M, Goldblatt P, Morrison J, Porta D, Forastiere F, et al. Mother's education and offspring asthma risk in 10 European cohort studies. Eur J Epidemiol. 2017;32(9):797–805.
55. Chen E, Fisher EB, Bacharier LB, Strunk RC. Socioeconomic status, stress, and immune markers in adolescents with asthma. Psychosom Med. 2003;65(6):984–92.
56. Chen E, Miller GE. Stress and inflammation in exacerbations of asthma. Brain Behav Immun. 2007;21(8):993–9.
57. Miller GE, Chen E, Shalowitz MU, Story RE, Leigh AKK, Ham P, et al. Divergent transcriptional profiles in pediatric asthma patients of low and high socioeconomic status. Pediatr Pulmonol. 2018;53(6):710–9.
58. Haczku A, Panettieri RA. Social stress and asthma: the role of corticosteroid insensitivity. J Allergy Clin Immunol. 2010;125(3):550–8.
59. Hegewald MJ, Crapo RO. Socioeconomic status and lung function. Chest. 2007;132(5):1608–14.
60. Schechter SB, Lakhaney D, Peretz PJ, Matiz LA. Community health worker intervention to address social determinants of health for children hospitalized with asthma. Hosp Pediatr. 2021;11(12):1370–6.
61. Krieger J, Song L, Philby M. Community health worker home visits for adults with uncontrolled asthma: the HomeBASE trial randomized clinical trial. JAMA Intern Med. 2015;175(1):109–17.
62. Campbell JD, Brooks M, Hosokawa P, Robinson J, Song L, Krieger J. Community health worker home visits for medicaid-enrolled children with asthma: effects on asthma outcomes and costs. Am J Public Health [Internet]. 2015;105(11):2366–72. Available from: www.kingcounty.gov/healthservices/health/.
63. Kennedy S, Bailey R, Jaffee K, Markus A, Gerstein M, Stevens DM, et al. Effectiveness of evidence-based asthma interventions [Internet]. 2017;139. Article PEDIATRICS. Available from: http://publications.aap.org/pediatrics/article-pdf/139/6/e20164221/1063207/peds_20164221.pdf.
64. Elkugia N, Crocker ME, Stout JW, Bolt K, Weiner BJ, Kramer CB. Development of an asthma home-visit training program for community health workers and their supervisors in Washington State. Front Public Health. 2021;9:674843.
65. Mainardi AS, Harris D, Rosenthal A, Redlich CA, Hu B, Fenick AM. Reducing asthma exacerbations in vulnerable children through a medical–legal partnership. J Asthma. 2023;60(2):262–9.
66. Gaffney AW, Hawks L, Bor D, White AC, Woolhandler S, McCormick D, et al. National trends and disparities in health care access and coverage among adults with asthma and COPD: 1997–2018. Chest. 2021;159(6):2173–82.
67. Akenroye AT, Heyward J, Keet C, Alexander GC. Lower use of biologics for the treatment of asthma in publicly insured individuals. J Allergy Clin Immunol Pract. 2021;9(11):3969–76.
68. Mitchell SJ, Rangel A, Klein EJ, Stout JW, Lowry SJ, Wingfield E, et al. Sociodemographic differences in asthma self-management knowledge of parents seeking asthma care for

their children in pediatric emergency departments. J Health Care Poor Underserved. 2021;32(4):2191–201.
69. Cabana MD, Lara M, Shannon J. Racial and ethnic disparities in the quality of asthma care. Chest. American College of Chest Physicians. 2007;132:810S–7S.
70. Kharat AA, Borrego ME, Raisch DW, Roberts MH, Blanchette CM, Petersen H. Assessing disparities in the receipt of inhaled corticosteroid prescriptions for asthma by Hispanic and non-Hispanic white patients. Ann Am Thorac Soc. American Thoracic Society. 2015;12:174–83.
71. Foundation of America A. Ethnic disparities in the burden and treatment of asthma The Asthma and Allergy Foundation of America The National Pharmaceutical Council [Internet]. 2005. Available from: www.npcnow.org.
72. Kaiser SV, Rodean J, Bekmezian A, Hall M, Shah SS, Mahant S, et al. Effectiveness of pediatric asthma pathways for hospitalized children: a multicenter, national analysis. J Pediatr. 2018;197:165–171.e2.
73. Kaiser SV, Jennings B, Rodean J, Cabana MD, Cabana MD, Garber MD, et al. Pathways for improving inpatient Pediatric Asthma Care (PIPA): a multicenter, national study. Pediatrics. 2020;145(6):e20193026.
74. Palmer L, YES WE. CAN Children's Asthma Program [Internet]. Center for Disease Control and Prevention; 2009. Available from: https://www.cdc.gov/asthma/interventions/yes_we_can_design.htm.
75. Price D, Bjermer L, Bergin DA, Martinez R. Asthma referrals: a key component of asthma management that needs to be addressed. J Asthma Allergy. Dove Medical Press Ltd. 2017;10:209–23.
76. Diette GB, Rand C. The contributing role of health-care communication to health disparities for minority patients with asthma. Chest. American College of Chest Physicians. 2007;132:802S–9S.
77. Barnthouse M, Jones BL. The impact of environmental chronic and toxic stress on asthma. Clin Rev Allergy Immunol. Humana Press Inc. 2019;57:427–38.
78. Wright RJ. Stress and atopic disorders. J Allergy Clin Immunol. 2005;116:1301–6.
79. Felitti VJ, Anda RF, Nordenberg D, Williamson DF, Spitz AM, Edwards V, et al. Relationship of childhood abuse and household dysfunction to many of the leading causes of death in adults the adverse childhood experiences (ACE) study [Internet]. Am J Prev Med. 1998;14. Available from: www.elsevier.
80. Bhan N, Glymour MM, Kawachi I, Subramanian SV. Childhood adversity and asthma prevalence: evidence from 10 US states (2009–2011). BMJ Open Respir Res. 2014;1(1):e000016.
81. Remigio-Baker RA, Hayes DK, Reyes-Salvail F. Adverse childhood events are related to the prevalence of asthma and chronic obstructive pulmonary disorder among adult women in Hawaii. Lung. 2015;193(6):885–91.
82. Pape K, Cowell W, Sejbaek CS, Andersson NW, Svanes C, Kolstad HA, et al. Adverse childhood experiences and asthma: trajectories in a national cohort. Thorax. 2021;76(6):547–53.
83. Wing R, Gjelsvik A, Nocera M, McQuaid EL. Association between adverse childhood experiences in the home and pediatric asthma. Ann Allergy Asthma Immunol. 2015;114(5):379–84.
84. Thakur N, Hessler D, Koita K, Ye M, Benson M, Gilgoff R, et al. Pediatrics adverse childhood experiences and related life events screener (PEARLS) and health in a safety-net practice. Child Abuse Negl. 2020;108:104685.
85. Wang E, Simmons B, Holm KE, Alam R, Wamboldt FS. Intimate partner violence and adult asthma morbidity: a population-based study. J Allergy Clin Immunol Pract. 2021;9(12):4300–4309.e7.
86. Apter AJ, Garcia LA, Boyd RC, Wang X, Bogen DK, ten Have T. Exposure to community violence is associated with asthma hospitalizations and emergency department visits. J Allergy Clin Immunol. 2010;126(3):552–7.
87. Sternthal MJ, Jun HJ, Earls F, Wright RJ. Community violence and urban childhood asthma: a multilevel analysis. Eur Respir J. 2010;36(6):1400–9.
88. Chen W, Boutaoui N, Brehm JM, Han YY, Schmitz C, Cressley A, et al. ADCYAP1R1 and asthma in Puerto Rican children. Am J Respir Crit Care Med. 2013;187(6):584–8.

89. Thakur N, Barcelo NE, Borrell LN, Singh S, Eng C, Davis A, et al. Perceived discrimination associated with asthma and related outcomes in minority youth: the GALA II and SAGE II studies. Chest. 2017;151(4):804–12.
90. Thakur N, Borrell LN, Ye M, Oh SS, Eng C, Meade K, et al. Acculturation is associated with asthma burden and pulmonary function in Latino youth: the GALA II study. J Allergy Clin Immunol. 2019;143(5):1914–22.
91. Deighton S, Neville A, Pusch D, Dobson K. Biomarkers of adverse childhood experiences: a scoping review. Psychiatry Res. Elsevier Ireland Ltd. 2018;269:719–32.
92. Liberman AC, Budziñski ML, Sokn C, Gobbini RP, Steininger A, Arzt E. Regulatory and mechanistic actions of glucocorticoids on T and inflammatory cells. Front Endocrinol. Frontiers Media S.A. 2018;9:235.
93. Miller GE, Chen E. Life stress and diminished expression of genes encoding glucocorticoid receptor and 2-adrenergic receptor in children with asthma [Internet]. 2006; https://doi.org/10.1073/pnas.0506312103. Available from: www.pnas.orgcgi.
94. Ramírez F, Fowell DJ, Puklavec M, Simmonds S, Mason D. Glucocorticoids promote a TH2 cytokine response by CD4+ T cells in vitro. J Immunol. 1996;156(7):2406–12.
95. Huang YJ, Nariya S, Harris JM, Lynch SV, Choy DF, Arron JR, et al. The airway microbiome in patients with severe asthma: associations with disease features and severity. J Allergy Clin Immunol. 2015;136(4):874–84.
96. de Castro Kroner J, Knoke K, Kofler DM, Steiger J, Fabri M. Glucocorticoids promote intrinsic human TH17 differentiation. J Allergy Clin Immunol. 2018;142(5):1669–1673.e11.

Part II

Mechanistic Heterogeneity in Asthma

Phenotypes and Endotypes in Asthma

Amanda P. McIntyre and Ravi K. Viswanathan

Abstract

Asthma is a broadly encompassing diagnosis of airway inflammation with significant variability in presentation and response. Advances in molecular techniques and imaging have unraveled the delicate mechanistic tapestry responsible for the underlying inflammatory pathways in asthma. The elucidation of biomarkers and cellular components specific to these inflammatory pathways allowed for the categorization of asthma from generic phenotypes to more specific mechanistic endotypes, with two prominent subgroups emerging based on the level of Type 2 inflammation present – T2 high and T2 low (or non-T2). Sophisticated modeling and cluster analyses using a combination of clinical, physiologic, and biomarker parameters have permitted the identification of subendotypes within the broader T2 umbrella. This mechanistic-driven classification schema for asthma has dramatically altered the landscape of asthma management with the discovery and approval of targeted biologic therapies and has ushered in a new era of personalized precision medicine in asthma.

Keywords

Biomarkers · Pulmonary function · Endotypes · T helper subsets · Health care utilization · Triggers · Sputum analysis · Heterogeneity

A. P. McIntyre · R. K. Viswanathan (✉)
Division of Allergy, Pulmonary & Critical Care, Department of Medicine, University of Wisconsin School of Medicine & Public Health, Madison, WI, USA
e-mail: rviswanathan@medicine.wisc.edu

A. R. Brasier, N. N. Jarjour (eds.), *Precision Approaches to Heterogeneity in Asthma*, Advances in Experimental Medicine and Biology 1426,
https://doi.org/10.1007/978-3-031-32259-4_6

Introduction

Historically, asthma has been classified as intermittent or persistent, with the latter having additional gradations, including mild, moderate, or severe, depending upon the burden of symptoms. This schema led to the stepwise management strategy classically noted in NHLBI and GINA guidelines from as-needed bronchodilator therapy to high-dose combination inhalers or biologic therapy [1]. Given the heterogeneity noted among asthma patients, researchers and clinicians have attempted to use these characteristics to categorize patients into more distinct groups within asthma to help diagnose and manage this condition. Previous attempts at categorizing asthma led to imprecise and overlapping subgroups due to the availability of only cellular composition of airway inflammation and clinical characteristics.

The concept of asthma phenotypes (Fig. 1) was born out of the need to delineate these subgroups. An asthma phenotype is an observed combination of certain clinical, biological, and physiological features resulting from an interaction of intrinsic (genetic) and extrinsic (environmental) factors. A variety of phenotypes of asthma have been observed that are defined by distinct clinical characteristics, such as age of onset, presence of atopy, and lung function, despite having similar symptom manifestations. Identifying distinct phenotypes has led to the understanding of asthma endotypes, which furthers the asthma categorization schema by overlaying unique pathophysiological molecular mechanisms responsible for the phenotypic expression.

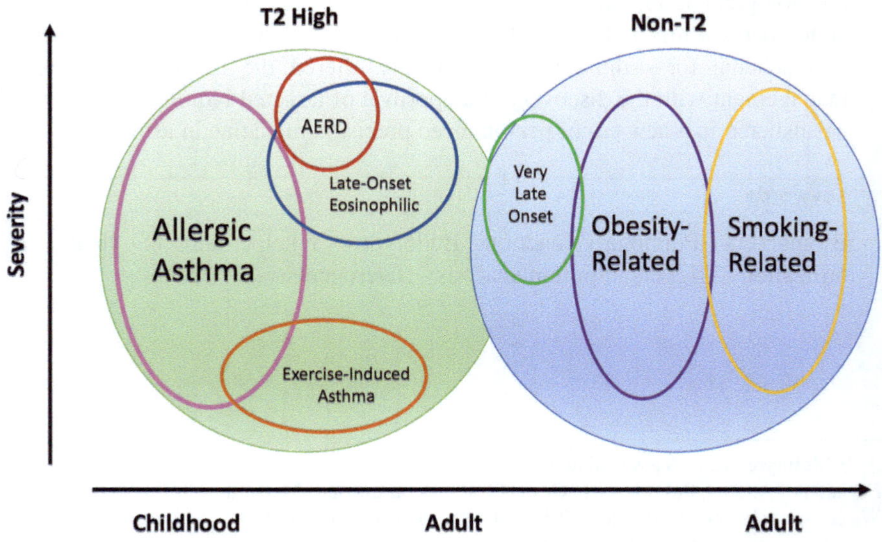

Fig. 1 Asthma phenotypes. Asthma phenotypes are classified by the age of onset and severity. This figure demonstrates overlapping features of each phenotype and the underlying type of inflammation (T2 or non-T2) as described by Wenzel [3]

While the longstanding practice of asthma management involved stepwise usage of a standard repertoire of therapies against all forms of asthma, it was recognized that a subset of patients remains refractory to management even with maximal therapy. Accounting for about 5–10% of all asthma subjects, this subset of "severe" asthma patients often have high mortality and morbidity associated with their disease and account for a disproportionately higher percentage of healthcare utilization [2]. With the advent and availability of advanced imaging, biomarkers, and pulmonary function testing, the phenotyping and endotyping of asthma have been achieved at a more nuanced level than ever before. This has allowed for a transition from a generic therapeutic approach to more precision asthma therapy, resulting in improved asthma outcomes. This will continue to be refined as our understanding of asthma endotypes and biomarkers improves and move us one step closer to "personalized" medicine. In this chapter, we will focus our discussion on the inflammatory pathways and key "players" implicated in asthma pathophysiology and delineate some approaches to the phenotyping and endotyping of asthma in general. Specific phenotyping of severe asthma will be addressed in a different section of this text.

Pathophysiology of Asthma & Characterization of Airway Inflammation

In asthma, airway inflammation is a pathognomonic feature, and mechanisms contributing to inflammation are heterogeneous. Initially, it was recognized that CD4+ T helper lymphocytes played an important role in the inflammatory cascades in certain phenotypes of asthma. Subsequently, specific T-helper cell subsets (T_H1 and T_H2) were delineated. It was recognized that T_H2 cells were critical to promoting eosinophilic airway inflammation and generation of T_H2 cytokines, namely, IL-4, 5, and 13 [4]. Consequently, an initial categorization nomenclature of Th2-high and Th2-low asthma was proposed, with an eosinophilic profile dominating the former, while a paucigranulocytic or neutrophilic or mixed profile characterized the latter. More recently, the discovery of ILC2s (type 2 innate lymphoid cells) and their ability to produce copious amounts of T2 cytokines enhanced our understanding of the inflammatory pathways implicated in asthma beyond T helper lymphocyte subsets [5]. This led to the modification of the endotypic categorization of asthma into T2-high and T2-low/non-T2 groups. The usage of additional biomarkers, such as total and specific IgE (sIgE) or skin prick testing (SPT) and fractional exhaled nitric oxide (FeNO), has allowed for further stratification of endotypes and phenotypes within the broader umbrella of T2-high and T2-low schema. This stratification has been exploited to develop targeted therapies that are directed at specific cytokines or molecules implicated in these endotypes and revolutionized asthma management, particularly difficult-to-treat asthma [6, 7].

Airway inflammation has also been described by examining induced sputum from asthma patients. Eosinophilic, neutrophilic, mixed eosinophilic/neutrophilic, and paucigranulocytic patterns have been described in both adults and children, with eosinophilic patterns generally being the more prominently observed [8]. A paucigranulocytic pattern shows no increased neutrophils or eosinophils in the sputum or blood. The paucigranulocytic pattern is most commonly seen in patients with stable asthma and seems to represent an uncoupling of bronchial obstruction from airway inflammation due to structural changes from non-inflammatory mechanisms [9]. Overall, further studies are needed to understand what these cellular patterns represent.

Type 2 Inflammation

T2 inflammation is found in 50–70% of patients with asthma and is reinforced by a number of key cellular players, namely ILC2s, eosinophils, mast cells, and T_H2 cells through the secretion of a myriad of inflammatory mediators [10]. It is characterized by specific downstream inflammatory cytokines, including IL-4, IL-5, IL-13, and GM-CSF, whose production can be triggered by aeroallergens, viruses, and pollutants. These triggers initiate various inflammatory cascades through the release of various "upstream" cytokines or "alarmins", including thymic stromal lymphopoietin (TSLP), IL-25, IL-33, derived from epithelial cells closer to the surface (Fig. 2). Aeroallergens enter the airway and are processed by dendritic cells to activate CD4+

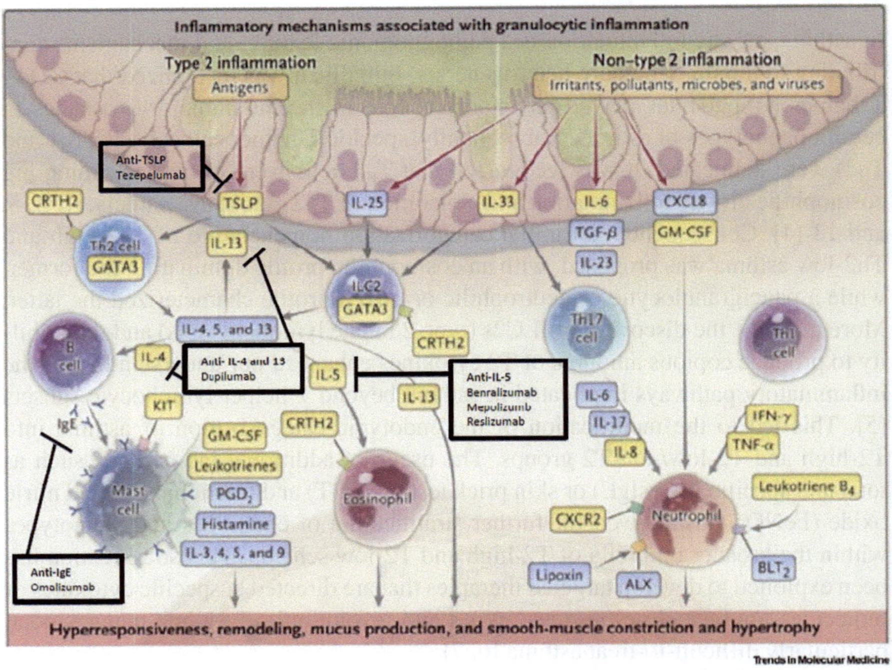

Fig. 2 Inflammatory pathways in asthma – cytokines and biologic therapies. The figure shows the T2 and non-T2 inflammatory pathways in asthma. The role of biomarkers and targets of treatment are also included. (Adapted from Figure 1 [35])

T_H2 cells using the GATA3 transcription factor. T_H2 cells are potent secretors of IL-4, IL-5, IL-13, and GM-CSF. IL-4 and IL-13 promote class-switching and IgE production, which activates mucosal mast cells, leading to the release of several broncho-constricting mediators, prostaglandins, leukotrienes, and histamines [11]. IL-5, produced by T_H2 cells and other cell types, stimulates eosinophil production. Eosinophils produce lipid mediators that stimulate smooth muscle contraction, toxic granule products that damage the airway epithelium, and cytokines such as GM-CSF and TGF-beta that play a role in airway remodeling and fibrosis. IL-4 and IL-13 also contribute to airway remodeling and mucus hypersecretion [11].

Alarmins

Environmental triggers, including respiratory viruses, aeroallergens, tobacco smoke, and pollutants, also induce the proliferation of epithelial-derived innate cytokines, called alarmins, which include IL-33, IL-25, and TSLP. These alarmins serve as the initial trigger for the inflammatory cascade by activating ILC2 cells, which further promote type 2 inflammation through the production of IL-5 and IL-13 [12–14]. IL-33 has been shown to play a central role by augmenting type 2 inflammation by acting as a positive regulator of TSLP, impacting dendritic cell signaling and activating ILC2s [15, 16]. Patients with allergic and eosinophilic asthma have higher serum levels of IL-33 than those with non-allergic and non-eosinophilic phenotypes. Levels of TSLP, IL-33, and its receptor ST2 in bronchial lavage fluid and biopsies are elevated in patients with asthma and correlate with disease severity [17]. The role of IL-25 in type 2 inflammation is less understood, but IL-25 serum levels are associated with allergic asthma and severity [18]. Loss of function mutations in IL-33 and a rare allele of IL-25 are associated with reduced incidence of asthma, while increased steroid resistance was noted in BAL fluid ILC2 cells with increased TSLP levels. While anti-TSLP treatment has been shown to reduce exacerbations exceedingly well in T2 asthma, it appears to work in non-T2 asthma. Less is known about the role of alarmins in non-T2 asthma, but this observation of efficacy suggests that alarmins may play a role in both T2 and non-T2 asthma [19]. As it stands currently, the anti-TSLP monoclonal Ab, Tezepelumab, is authorized for add-on maintenance use in adult and pediatric patients 12 years or older with severe asthma regardless of asthma phenotype or biomarker status [20]. Overall, alarmins play a central role in both direct and indirect activation of a multitude of inflammatory cascades implicated in asthma.

Cellular Components of T2 Inflammation (Eosinophils, Mast Cells/Basophils, Type 2 Innate Lymphoid Cells)

Eosinophils are critical cells associated with T2 inflammation and have a multitude of effects on the airway. Eotaxins, principally acting thru the CCR3 receptor, steer eosinophils to areas of inflammation and activate them to release a whole host of stored and newly formed mediators. These mediators include cytotoxic proteins such as eosinophil peroxidase (EPX), eosinophil cationic protein (ECP), major basic protein (MBP), and eosinophil derived neurotoxin (EDN) and result in tissue damage. T2 cytokines (IL-4, 5, 13) and cysteinyl leukotrienes are also produced with eosinophil activation and stimulation that furthers the inflammatory milieu by promoting more robust bronchoconstriction and activation of ILC2 cells that further sustains this cascade by producing additional T2 cytokines. Eosinophils also contribute to remodeling and fibrosis by stimulating bronchial fibroblasts to produce factors that cause the thickening of the reticular basement membrane and smooth muscle hypertrophy, which ultimately affects the contractility of the airway [21]. Eosinophils also create extracellular DNA traps, termed eosinophil extracellular traps (EETs). EETs are found in bronchoalveolar lavage fluid of asthmatic patients and correlate with asthma severity [22].

Mast cells and basophils also play a role in asthma pathogenesis. Mast cells are found in tissue, and basophils are blood-based but found in tissues after activation. Basophils have been shown to be increased in the bronchial walls in T2-high asthma, and tissue basophils are strongly correlated with allergic disease. Both cell types are activated by IgE cross-linking and IL-33. Upon activation, both mast cells and basophils produce histamine and lipid mediators. A subtype of mast cells has been shown to have greater potential to release prostaglandin D2 (PGD2), and the presence of PGD2-high mast cells correlates with more severe disease in patients with asthma. Mast cells also release tryptase and proteases, and basophils have been shown to secrete IL-4, which potentiates T2 inflammation [7].

Type 2 innate lymphoid cells (ILC2s) do not possess antigen-specific B or T cell receptors and are now thought to play an equally important role in fomenting T2 inflammation. Using the GATA3 transcription factor, similarly to T_H2 cells, ILC2s are activated in response to various alarmins (TSLP, IL-25, IL-33) and produce large amounts of T2 cytokines (IL-4, 5, 9, 13) to further perpetuate the T2 response. There is currently evidence to support that ILC2 cells play a role in airway remodeling and airway injury [23].

T2 Biomarkers

Absolute Eosinophil Count

Elevations in eosinophils can be measured in sputum or in peripheral blood, with the former being more invasive and less pursued. Sputum eosinophilia is generally defined as a value of 3% or greater, while there is more variability in what constitutes peripheral eosinophilia. Generally, an absolute eosinophil count (AEC) greater than 300 fits this definition, though an AEC of 150 has been utilized as a low-end cutpoint for the usage of various biologics. It is also recognized that asthma severity correlates to the level of sputum and peripheral eosinophilia [7].

FeNO

Nitric oxide (NO) is produced by airway epithelial cells, eosinophils, and macrophages by nitric oxide synthase (NOS), which is upregulated in T2-high airway inflammation [7]. Fractional excreted nitric oxide (FeNO) can be used as a marker of T2-high asthma. Per ATS guidelines, concentrations greater than 50 ppb in adults (>35 ppb in children) correlate with eosinophilic inflammation and steroid responsiveness, although a value of more than 25 ppb is considered a low-end cutpoint for T2 inflammation. This test is noninvasive, repeatable, and available at the point of care. However, various patient factors, such as other atopic diseases, age, obesity, smoking, and steroid use, can all confound the FeNO value [24]. FeNO was added to the NHLBI 2020 Focused Updates to Asthma Management guidelines when the diagnosis of asthma is uncertain or in those with persistent asthma to guide usage of more T2-directed therapies [25].

Total IgE and Aeroallergen Sensitivity (SPT or sIgE)

Serum IgE levels correlate with atopy and asthma severity in both children and adults. Aeroallergen sensitization (serum IgE or skin prick testing) serves as a marker for specific atopy and correlates with T2-high asthma. In a recent study of asthma phenotypes in inner-city children, five distinct clusters were identified with the most symptomatic cluster being characterized by the highest total serum IgE levels, specific allergic sensitization, and eosinophil counts [26]. Additionally, there is evidence to support that allergic sensitization to perennial allergens, specifically pet dander, cockroach, and molds appear to be associated with more severe forms of asthma [27]. Anti-IgE therapy, omalizumab, is effective for severe asthma but trending total serum IgE level is not useful for monitoring the response. Similarly, based on multiple studies and meta-analyses, allergen immunotherapy (AIT) has been shown to be efficacious in allergic asthma and suggested as a therapeutic option based on the most recent iteration of the joint task force practice parameter update from the AAAAI (American Academy of Allergy, Asthma, & Immunology) and ACAAI (American College of Allergy, Asthma, & Immunology) [28].

Other Potential T2 Biomarkers (Periostin, Dipeptidyl Peptidase 4, Plasma Galectin-3)

Periostin is an extracellular matrix protein that is upregulated by IL-4 and IL-13 in bronchial epithelial cells significantly more in patients with asthma than in healthy controls. It can be detected in both blood and sputum. Although initial studies suggested periostin correlated with airway eosinophilia in severe asthma patients, this was not confirmed in subsequent studies. Sputum analysis showed periostin distinguished asthma endotypes at the transcriptome level (IL-13/T_H2 and ILC-2 signatures versus a T_H17 signature), and periostin is generally considered a marker of

IL-13 pathway activation [29]. Periostin was employed as a potential biomarker to assess candidacy for IL-13 targeted biologic therapy with lebrikizumab with initial promising results in Phase 2 trials but inconsistent results in Phase 3 trials [30]. Further studies are needed to understand the utility of periostin as a biomarker in asthma, but it has the potential to distinguish T2 from non-T2 asthma.

Dipeptidyl peptidase 4 (DPP4) is a serine protease anchored by CD26 and found at high concentrations in lung connective tissue, in addition to blood and sputum. The role of DPP4 is currently unclear in asthmatic inflammation, but it is highly expressed in the bronchial epithelial cells of untreated cells. DPP4 expression is enhanced by IL-13 stimulation and is purported to be another marker for IL-13 pathway activation [31]. Another biologic directed at the IL-13 axis, tralokinumab, was employed to treat moderate to severe asthma and initially showed promise for response in a subset of patients with high DPP4 levels in Phase 2 trials but demonstrated inconsistent results in Phase 3 trials [32].

Plasma galectin-3 has been identified as another potential T2 biomarker with functional implications in airway remodeling, eosinophil recruitment, and T_H2 inflammatory skewing with increased expression of inflammatory mediators [33]. Recently, a sub-analysis of the PROXIMA study found that plasma galectin-3 levels could be used as a predictive marker to stratify and identify super responders and functional responders to omalizumab treatment in patients with severe allergic asthma [34]. Biomarkers are discussed further in Part IV.

Asthma Phenotypes by Cluster Analyses

In the past, a traditional approach to categorizing asthma using clinical and physiologic parameters was utilized. This often resulted in imprecise categorizations with multiple overlapping groups that were much harder to define. A more modern approach to this quandary involved utilizing a systems biology approach that takes into consideration multiple variables in an interactive manner to achieve more distinct and well-defined subgroups. Multiple large multi-center trials utilizing cluster analyses were conducted to delineate asthma phenotypes and endotypes. These include the Severe Asthma Research Program (SARP) in the US [36], the Unbiased Biomarkers in Prediction of Respiratory Disease Outcome (U-BIOPRED) from Europe [37], and the Leicester study in the UK [38]. There is significant variability noted in the types of patients examined, statistical/computational modeling employed, and disease characteristics utilized to formulate these clusters (Table 1). Despite this variability in cluster analyses for each study, there was a convergence in the results with at least 4 primary and distinct asthma phenotypes being described commonly among all the studies (Fig. 3). These include (1) early-onset mild allergic, (2) early-onset allergic moderate-severe asthma, (3) late-onset non-allergic eosinophilic asthma, (4) late-onset non-allergic non-eosinophilic asthma. These sophisticated analyses reinforced the importance of age of onset, lung function, and atopic status as key factors that have consistently been helpful in delineating asthma heterogeneity. The results from these cluster analyses also converge and fit into the

Table 1 Asthma phenotypes based on cluster analysis

TABLE I. Asthma heterogeneity revealed by clustering studies

	SARP[6]	Leicester, UK[7]	U-BIOPRED[8]	Yan et al[14]	Simpson et al[13]
Methodology	Ward minimum-variance unsupervised hierarchical clustering of clinical features	Unsupervised K-means clustering of clinical features and sputum eosinophil percentages	Unsupervised partition-around-medoids clustering of clinical features	K-means unsupervised clustering of sputum gene expression	Manual clustering of patients using sputum inflammatory cell cutoffs
Study size (no.)	726	187	266	100	93
Clusters	(1) Mild early-onset allergic disease (2) Moderate early-onset allergic disease (3) Late-onset eosinophilic nonallergic disease (4) Severe early-onset eosinophilic allergic disease (5) Late-onset nonallergic neutrophilic severe asthma with fixed airflow	(1) Early-onset allergic severe asthma (2) Noneosinophilic asthma with greater female predominance (3) Early-onset symptom predominant with minimal eosinophil counts (4) Late-onset eosinophilic inflammation with male predominance	(1) Moderate-to-severe asthma with allergy and low eosinophil counts (2) Overweight to obese severe late-onset asthma with the least allergy, severe airway obstruction, high eosinophil counts, and smoking history[a] (3) Severe late-onset disease similar to cluster 2 but more with more allergy and no smoking history[a] (4) Obese female patients with severe asthma and frequent exacerbations but normal lung function and allergy	(1) Severe asthma with near-fatal asthma attacks, high eosinophil counts, and lower prebronchodilator FEV_1 (2) Severe asthma with a history of more hospitalizations and least allergic (3) Mild atopic asthma with the lowest inhaled corticosteroid use	(1) Eosinophilic (2) Neutrophilic (3) Mixed granulocytic (high eosinophil and neutrophil counts) (4) Paucigranulocytic (low eosinophil and neutrophil counts) asthma
Shared clinical parameters of clustering studies			Lung function,[b] atopy[c]		
Unique clinical parameters	Age of onset, obesity	Age of onset, airway inflammation, obesity	Age of onset, airway inflammation, smoking, obesity	Age of onset, airway inflammation, obesity	Airway inflammation

[b]Pre-FEV_1 ± post-FEV_1, bronchodilator response.
[c]History of atopy or IgE, SPT, or ImmunoCAP positivity.
[a]Ten or more pack years, current smokers.

Adapted Table 1 from Kaur and Chupp [39]

This table shows the distinct phenotypes identified across different cluster analyses by large, multicenter trials

[a]Ten or more pack years, current smokers

[b]Pre-FEV_1 ± post-FEV_1, bronchodilator response

[c]History of atopy or IgE, SPT, or ImmunoCAP positivity

binary classification schema of asthma based on T2 status: T2-high vs T2-low (or non-T2).

The "-Omics" Approach to Asthma Endotyping

Recently, gene expression analysis and the availability of high-throughput molecular techniques have allowed for another intriguing way of conceptualizing asthma phenotypes/endotypes – the "omics" approach. Thus far, single-omics approaches focusing on tissue or blood transcriptomics, proteomics, epigenomics, metabolomics, and microbiomics have been attempted to characterize and segregate asthma into more informative endotypes in an unbiased manner. An initial formative study

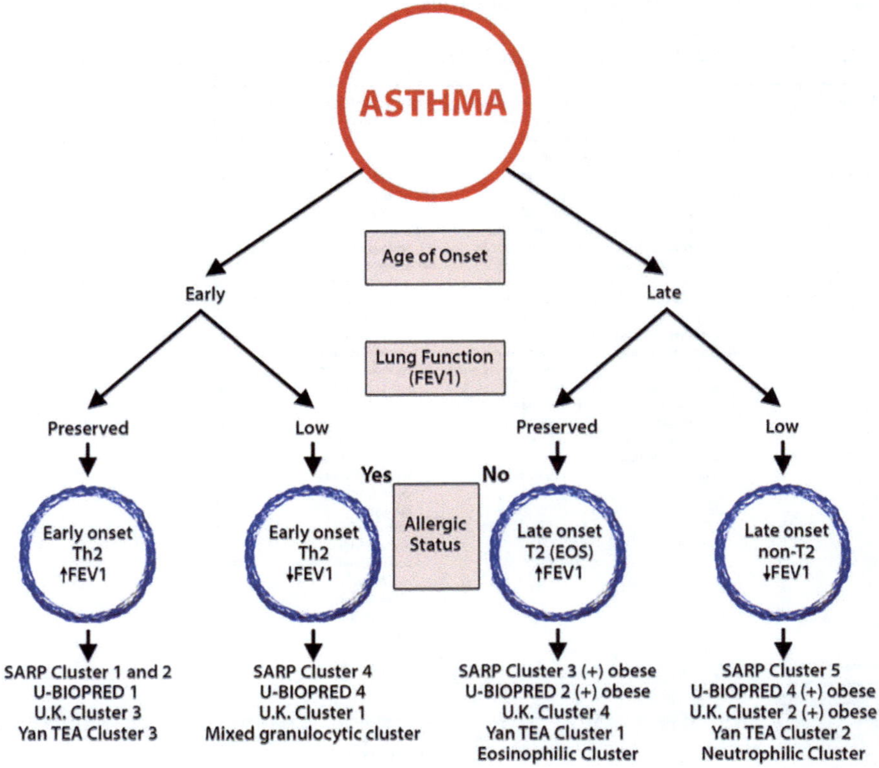

Fig. 3 Asthma phenotype classification. This figure demonstrates factors used to further delineate asthma phenotypes, including age of onset, lung function, and presence of atopy. (Adapted Figure 1 from Kaur and Chupp [39])

conducted on 42 participants with mild to moderate asthma and 28 control subjects used microarray and PCR analyses to identify two distinct subgroups of patients, T_H2-high and T_H2-low, based on differential expression of IL-5 and IL-13 inducing genes, specifically periostin, CLCA1, and serpin B2. The study paved the way for more transcriptomic analyses in asthma patients [40].

Using a similar approach, the U-BIOPRED study has expounded on molecular profiling using samples obtained in blood, sputum, bronchial brushes, bronchial biopsies, urine, and exhaled air at the transcriptomic, proteomic, metabolomics, and microbiomic levels resulting in 3 molecular phenotypes being identified – a prototypical T2 high pattern, a metabolic phenotype, and a neutrophilic pattern clustering with certain inflammasome associated genes [41]. Additionally, gene expression network analysis of sputum samples in patients with asthma has shown that expression of a T2 gene network is variable across individuals with T2 high asthma, with some subjects having a higher expression of $CD11b^+/CD103^-/IRF4^+$ who are also characterized by older age [42]. In gene expression network analysis of patients with T2 low asthma, there were increased cytotoxic CD8+ T cells and this was associated with obesity-related inflammation [43].

Genome-wide association studies (GWAS) have shown multiple genetic loci associated with asthma. Most notably, *ORMDL3/GSDMB* variants have been associated with childhood asthma, and IL-33 and *IL1RL1* SNPs have been associated with atopic asthma. *TSLP* variants have been found to be protective against T2-high asthma [44]. Multiple SNPs have also been identified that distinguish between active and inactive allergic asthma [45]. Additionally, whole genome analysis and epigenetic studies will provide further insight into the environmental influence of asthma development.

While the omics approach has certainly allowed for better profiling of asthma risk and validation of previously established phenotypes, it is still an area in relative infancy and has resulted in incongruent findings amongst different analyses. Thus far, an "omics-only" approach to asthma phenotyping has failed to be meaningfully translated into clinical practice. An integrated classification framework utilizing "multi-omics" analyses coupled with clinical features and laboratory/imaging parameters may aid in categorizing asthma into more distinct and informative subphenotypes and endotypes (Fig. 4) [42]. Further work in this area will likely contribute to our understanding of asthma pathogenesis and variable clinical expression and response to asthma therapies.

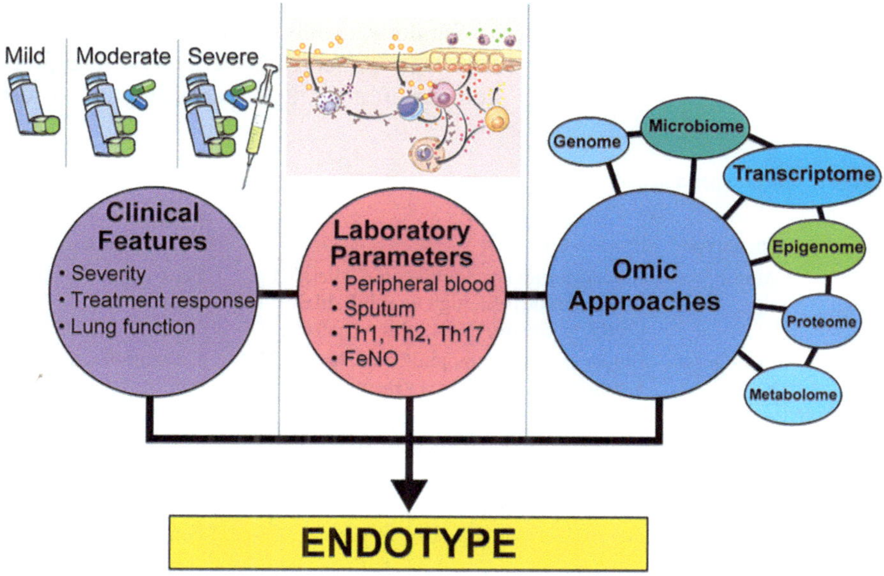

Fig. 4 Composite asthma stratification using clinical, laboratory, and omics approaches. The "-Omic" approaches are used in conjunction with clinical features and laboratory parameters to identify asthma endotypes. (Adapted Figure 1 from [46])

T2-High Phenotypes

Eosinophilia tends to correlate with T2 inflammation but does not necessarily correlate with atopy. T2 phenotypes are typically characterized by the presence or absence of atopy and age of onset based on cluster studies. In the following section, each phenotype is discussed in detail (Table 3).

Early-Onset Atopic

This phenotype is perhaps best recognized of all asthma subtypes and in the past has also been referred to as "extrinsic" asthma due to the impact of an environmental factor external to the body resulting in upper and lower airway symptoms. Allergic asthma is defined as asthma associated with sensitization to aeroallergens either by sIgE or SPT. Allergic asthma is estimated to account for 80% of asthma in children and more than 50% in adults [45]. All of the large cluster studies examining asthma phenotypes have revealed both an early and late-onset allergic phenotype. However, allergic asthma typically presents in childhood and continues to adulthood [36, 47]. Allergic asthma is often associated with a personal or family history of allergic disease and aeroallergen sensitization and occurs more in males than females [48]. In this phenotype, airway inflammation is IgE-mediated, resulting in T2-high inflammation with eosinophilic predominance and elevated FeNO levels. Lung function in allergic asthma is variable with studies showing variable differences compared to non-allergic asthma [45]. Similarly, the clinical presentation can vary from mild to severe forms and correlates with preserved or diminished lung function. Given the high concomitant presence of atopy, it is reasonable to consider

Table 3 Phenotypes and endotypes of asthma

Endotype	Phenotype	Biomarkers	Clinical Characteristics
T2 High	Atopic	Blood/sputum eosinophil count, allergic sensitization (IgE), high total IgE, high FeNO, [a]periostin, DPP4, plasma galectin-3	Early onset, steroid responsive
	Late-Onset	Blood/ sputum eosinophil count, high FeNO, [a]periostin, DPP4, plasma galectin-3	Steroid refractory, may have CRSwNP
	AERD	Blood/sputum eosinophil count, urinary LTE4, [a]periostin, DPP4, plasma galectin-3	Adult onset
Non-T2	Non-Atopic	Sputum neutrophil count	Adult onset
	Smokers	Sputum neutrophil count	Older adults
	Obesity-related	Serum IL-6	Female sex
	Very late onset	Sputum neutrophil count	>50–65 at onset

This table summarizes the endotypes and associated phenotypes of asthma along with clinical characteristics and associated biomarkers

[a]Indicates a biomarker that is still under investigation

optimal management for allergic rhinitis, including allergen immunotherapy (subcutaneous or sublingual), for this phenotype. Multiple meta-analyses have confirmed a trend toward improvement in lung function with these approaches [49].

Late-Onset Non-allergic Eosinophilic

As suggested by its terminology, this subset of asthma has an adult-onset, lacks the presence of atopy, has limited reversibility to bronchodilator, and has a markedly prominent eosinophilic preponderance often refractory to conventional therapy. This phenotype has a strong association with antecedent chronic rhinosinusitis w polyps (CRSwNP) with recurrent disease and exacerbations due to these comorbid factors. There is marked activation of ILC2 cells and exaggeration of T2-cytokines (IL-5, IL-13) in this subset, along with elevated FeNO levels. Given its propensity to be somewhat refractory to both inhaled and oral corticosteroids, this phenotype may require closer monitoring and earlier escalation to targeted biologic therapy [50, 51].

Aspirin Exacerbated Respiratory Disease (AERD)

AERD is characterized by asthma, eosinophilic chronic rhinosinusitis with nasal polyposis, and respiratory reactions upon ingesting COX-1 inhibitors. The natural history of this condition often starts in young adulthood with rhinitis and sinusitis, followed by asthma inception, nasal polyposis, and eventually an aspirin sensitivity. This condition affects about 7–8% of all patients with asthma and about 15% of patients with severe asthma. The exact pathophysiologic mechanism is unclear though it is purported to be the result of a biochemical imbalance in arachidonic acid metabolism pathways such that there is preferential shunting of mediators toward the 5-lipoxygenase (5-LO) pathway leading to the production of pro-inflammatory cysteinyl leukotrienes (CysLTs) and their receptors and simultaneous down-regulation of the cyclooxygenase (COX) pathway leading to reduced expression of prostanglandins (PGs) and its receptors. This imbalance in PGs to CysLTs results in loss of inhibition of ILC2, mast cells, and eosinophils resulting in prolific activation of T2 inflammatory cascades. A robust eosinophilia is noted in airway tissue, particularly sinuses, and in peripheral blood, leading to the clinical expression of severe recalcitrant chronic sinusitis and asthma in this phenotype/endotype.

Patients often require repeated surgical intervention to address the nasal polyposis in addition to aggressive topical therapy. Historically, aspirin desensitization with high-dose aspirin has been demonstrated to be beneficial in partially mitigating the clinical symptoms and corticosteroid dependence, even though it does not necessarily result in the reversal of many of the biochemical imbalances noted [52]. In addition to tissue and peripheral eosinophilia, this phenotype is characterized by elevated FeNO levels and elevation of urinary leukotriene E4 (uLTE4) and can serve as an important biomarker for this subset. Given the above features, this particular phenotype/endotype of asthma is quite amenable to using leukotriene modifiers (i.e., montelukast/zileuton) and various biologic therapies, particularly those that target the T2 pathway. Interestingly, while IgE and allergic sensitization are not typical or prominent players in this subset, treatment with omalizumab in a small open-label trial demonstrated a dramatic reduction in multiple biochemical metabolites (uLTE4 & PGD2) while improving baseline symptoms [53]. Aspirin desensitization was traditionally an important aspect of therapy for this phenotype until the development of biologics, including anti-IL4Ralpha, anti-IL5/IL5Ralpha, and anti-IgE, all of which have shown increased efficacy for both nasal polyposis and asthma [54–56].

Exercise Induced Bronchospasm (EIB)

EIB is used in favor of the term exercise-induced asthma as exercise has the propensity to exacerbate existing asthma (rather than cause asthma) in up to 90% of asthma patients and is defined as a transient narrowing of the lower airway after exercise in an asthmatic individual. Given this definition, whether this should be regarded as a separate phenotype may be reasonably questioned. Regardless, given the impact this entity has on asthma, we have opted to highlight this as a distinct entity. The mechanisms underlying the pathophysiology have largely focused on the production of mast cell mediators, particularly leukotrienes, in addition to histamine and prostaglandins due to a hyperosmolar state triggered by the introduction of cold air into the airways. There is concomitant eosinophil activation leading to an augmented inflammatory response causing additional mediator release, including leukotrienes, that triggers bronchoconstriction and increased neuronally stimulated mucin production [57]. This phenotype shares numerous features with AERD discussed previously and generally tends to be grouped largely within the T2 high umbrella, although there is some emerging evidence that some non-T2 mechanisms may be implicated in a subset of elite athletes with late-onset disease [58]. Given the above, the focus of therapy is directed at the bronchoconstriction by utilizing rescue and prophylactic therapy with bronchodilators as well utilizing leukotriene modifiers for longer term management.

T2-Low or Non-type 2 Inflammation

T2-low or non-T2 inflammation in asthma is less understood and typically defined as asthma without features of T2 inflammation. It is often characterized by airway neutrophilia or the absence of eosinophils or other T2 inflammatory markers [59]. Neutrophilic inflammation is thought to play a major role in non-T2 inflammation. As in T2 inflammation, environmental triggers, such as allergens, pollutants, and infection, can cause airway injury. However, in non-T2 inflammation, this leads to the release of TLRs (based on sputum studies in patients with neutrophilic asthma [60]), which induces a shift towards the T helper 1 (T_H1) and T helper 17 (T_H17) dominant inflammation.

Table 2 Comparison of T2-high vs non-T2 Features

Feature	T2-"high"	T2-"low"
Age of onset	Earlier onset	Later onset
Symptoms	May be significant	May be significant
Life-threatening exacerbations	More exacerbations	Fewer exacerbations
Obesity/metabolic dysfunction	May be present	Often present
Lung function	More obstruction	Less obstruction
Short-acting bronchodilator response	More responsive	Less responsive
Allergic sensitization	Present	Absent
Exhaled nitric oxide	Normal to elevated	Low to normal
Airway eosinophilia	Present	Absent
Airway neutrophilia	May be present	May be present
Medication requirements	More responsive to corticosteroids	Less responsive to corticosteroids

Adapted from Table 1 [6]
This table summarizes key differences between T2-high and non-T2 inflammation

In addition to the above, non-T2 asthma is often characterized by a delayed onset, reduced response to bronchodilator, higher medication burden, higher symptom burden, and relative refractoriness to corticosteroid therapy. A large proportion of patients with the non-T2 phenotypes are obese females with adult-onset asthma symptoms. Based on available data, there are a number of potential mechanisms that contribute to T2-low asthma which include (1) Type 1 (IFN-mediated – T_H1 cells and T1 ILCs) or Type 3 (IL-17 mediated – T_H17 cells) inflammation causing airway neutrophilia; (2) IL-6-mediated systemic inflammation, obesity, and metabolic dysfunction; and (3) pauci-granulocytic mechanisms. Cut-points for what constitutes airway neutrophilia vary widely between 40% and 70% neutrophils, and in some instances, it may be associated with airway eosinophilia as well [6]. The differences between T2 and non-T2 inflammation are summarized in Table 2.

Non-T2 Biomarkers (IL-6, IL-17, YKL-40)

Biomarkers and targeted therapy for non-T2 asthma have yet to be developed, although a non-T2 endotype is typically presumed in the absence of T2 biomarkers. At present, the only biologic therapy that is approved without any biomarker or phenotypic stipulation is tezepelumab (anti-TSLP). Studies with this biologic in severe asthma do demonstrate potential for its efficacy in non-T2 phenotypes, though it is far more effective when T2 markers are present and elevated [20].

IL-6 is produced by a wide variety of inflammatory stimuli. IL-6 levels correlate with systemic inflammation, metabolic dysfunction, and asthma severity. In a study of severe asthma patients, systemic IL-6 inflammation and metabolic dysfunction occurred more commonly in a subset of both obese and non-obese patients and were associated with more severe asthma [61].

Chitinase-3-like-protein 1 (CHI3L1), also known as YKL-40, has been shown to enhance the effector response of IL-13 in T2 inflammation. However, other studies have shown that YKL-40 does not correlate with T2 biomarkers in adults and children with severe asthma. When asthma patients were stratified phenotypically, YKL-40 was found to be increased in patients with non-eosinophilic asthma and asthma patients with obesity. YKL-40 can be produced by neutrophils and correlates with neutrophil-related cytokines, including IL-6 and IL-1beta. A meta-analysis showed the YKL-40 level in the serum is associated with asthma and asthma severity in adults and children. Additionally, the YKL-40 level is higher in patients with COPD and COPD-asthma overlap compared to patients with asthma [62]. Further studies are needed to confirm its utility as a non-T2 biomarker.

As previously discussed, IL-17 is produced by T_H17 cells after activation by IL-1beta, IL-6, and TGF-beta. IL-17 plays a key role in asthma pathogenesis via a variety of mechanisms. It induces neutrophil attractants from the airway, mucus hypersecretion and goblet cell hyperplasia, myofibroblast differentiation, and airway smooth muscle proliferation, which promote airway remodeling. Glucocorticoid treatment in asthma also enhances IL-17 production. The exact role of IL-17 in the airway is still being investigated [63].

T1-Mediated T2-Low Asthma

Th1 cells produce interferon-gamma and TNF-alpha that stimulates neutrophils [59]. In a subset of asthma patients, there appears to be increased interferon signaling and a higher presence of IFN-g + CD4 T_H1 cells in BAL of severe asthma patients compared to those with mild-moderate asthma, corresponding with increased expression of CXCL10 that has been proposed as an inflammatory marker for asthma exacerbations and worsening asthma [64]. These findings have been replicated in mouse models of airway inflammation and demonstrate corticosteroid resistance [65]. In addition, these studies suggest that T1-mediated inflammation fosters increased bronchial hyper-responsiveness and airway resistance and renders the airway less responsive to traditional corticosteroid therapy. IL-12-dependent T_H1 lymphocytes, which produce interferon-gamma and TNF-alpha have also been shown to contribute to pathobiology in non-T2 asthma [66].

T_H-17-Mediated (T3 Inflammation) T2-Low Asthma

T_H17 cell development depends on IL-1beta, IL-6, and TGF-beta [67, 68]. T_H17 cells produce IL-8, IL-17, and IL-22. IL-17 plays a dominant role in neutrophil recruitment by inducing bronchial epithelial cells and airway fibroblasts to produce potent neutrophil attractants (CXCL8 and CXCL-1/GRO-alpha) [69]. Additionally, the production of IL-21 by T_H17 cells further stimulates the T_H17 response [68].

Neutrophil extracellular traps (NETs), consisting of DNA–histones complexes and proteins from activated neutrophils, also promote T_H17 polarization [70]. Additionally, non-T2 asthma may be driven by structural abnormalities in airway smooth muscle and irregular neuronal activation [59]. It has been shown that IL-17-mediated, neutrophilic asthma is characterized by airway hyper-responsiveness in both mice and humans and is often refractory to corticosteroids, which prolong neutrophil survival via an anti-apoptotic mechanism [71]. Interestingly, a randomized placebo-controlled trial of brodalumab, a targeted IL-17a receptor antagonist, in moderate to severe asthma patients inadequately controlled on inhaled steroids failed to demonstrate any treatment differences in clinical measures, although a possible signal was noted in a high-reversibility subgroup (>20% post BD) for improvement in ACQ scores [72]. There has been postulation that T_H17 skewing and airway neutrophilia may be occurring in patients who receive targeted T2 blockade and that dual blockade of T2 and T_H17 pathways may be an alternate approach to consider in some patients.

Pauci-Granulocytic (Non-inflammatory) T2-Low Asthma

This subset of T2-low asthma is poorly understood. It is inclusive of those asthma patients who lack both eosinophilia and neutrophilia and are effectively devoid of any observable airway inflammation. Understandably, this mechanistic concept has been called into question, given that active treatment could conceivably be suppressing inflammatory responses. Potential mechanisms that have been put forth for this pattern of asthma inception include a neurogenic mechanism involving the parasympathetic axis and an intrinsic airway remodeling and smooth muscle dysfunction that is less understood than the traditional inflammatory mechanisms leading to airway remodeling in T2 asthma [73].

Obesity Associated or Systemic Inflammation Associated T2-Low Asthma

It is generally accepted that obesity is a risk factor for asthma, and it may represent a unique asthma phenotype. This has been confirmed by several studies, which have shown distinct metabolic fingerprints and clinical/inflammatory biomarkers in obese patients with asthma compared to obese or non-obese patients with asthma [63]. Additionally, a meta-analysis determined children whose mothers were obese during pregnancy had a higher risk of developing asthma [64].

Mechanical factors, as well as inflammation and immune response associated with the obese state, are thought to contribute to asthma pathogenesis in obese patients. The airways of obese patients with asthma have been shown to be deficient in nitric oxide, which may contribute to airway dysfunction and explain why these patients often have reduced responses to corticosteroids. It is also known that weight loss can improve asthma control in this subset of patients [63]. A subset of asthma patients suffers from a conglomerate of obesity, metabolic dysfunction (often characterized by the presence of insulin resistance), and a high degree of systemic inflammation which appears to originate in adipose tissue secreting higher levels of the pro-inflammatory cytokine IL-6 [61, 74]. This finding has been noted in some asthma patients with elevated plasma IL-6 concentrations correlating with lowered lung function and increased propensity for exacerbations. Researchers have also observed that the presence of obesity skews toward T_H1 and T_H17 pathways to some extent. There has been a recognition of a new type 3 innate lymphoid cell (ILC3) that produces IL-17 and IL-22 and has been linked with this phenotype [75]. Given these intriguing findings, IL-6 blockade is now being explored as a potential management option for a subset of severe asthma patients through an adaptive design platform in the PReCISE research trial (Clinicaltrails.gov Identifier: NCT04129931).

Smoking Associated T2-Low Asthma

Smoking is a potent and well-known trigger of asthma. This phenotype of T2-low asthma also has a neutrophilic preponderance and steroid insensitivity. Oxidative stress caused by smoking ultimately leads to neutrophil and macrophage activation while also predisposing the individual to increased allergen sensitization and increased IgE production, even though it is associated with a reduction in FeNO levels [76]. Within this phenotype, features of asthma and COPD present in an overlapping fashion, and the term asthma-copd overlap syndrome (ACOS) has been appropriately designated with strict diagnostic criteria put forth by GINA (Global Initiative for Asthma) and GOLD (Global Initiative for Chronic Obstructive Lung Disease). These include persistent airflow obstruction with several features usually associated with asthma (history of atopy, eosinophilia, and bronchodilator reversibility) and several features usually associated with COPD (i.e., smoking >10 pack years) [77]. No specific biomarker profile pattern is definitively associated with this phenotype.

A summary of T2 and non-T2 phenotypes/endotypes is depicted in Table 3 at the end of this chapter. A composite biomarker-driven diagnostic categorization schema is presented in Fig. 5 below:

Other Asthma Phenotypes

Given the overlapping features of T2 and non-T2 inflammation are noted with childhood wheezing and asthma with PAL (persistent airflow limitation), we are discussing these below as distinct entities or phenotypes that may not strictly fit within the construct of other phenotypes that were previously discussed.

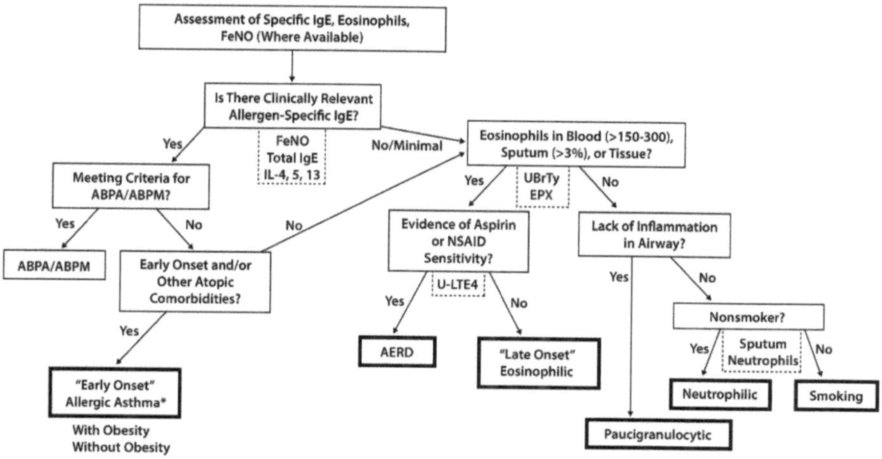

Fig. 5 Biomarker-Driven Diagnostic Categorization of Asthma into Phenotypes/Endotypes. The use of biomarkers is helpful in determining asthma phenotypes. This figure demonstrates a clinical decision tree for further asthma characterization. (Adapted from Carr et al. [78])

Asthma with Persistent Airflow Limitation

About 8–49% of patients with asthma develop PAL or fixed airway obstruction (FAO) that is not reversible [79]. PAL is present in both mild and severe asthma. In patients with mild asthma, PAL/FAO is associated with eosinophilic inflammation and a higher risk of exacerbations [80]. Among patients with asthma-onset before 25 years of age with persistent airflow limitation, the majority of the decline in pulmonary function is established before 25 years of age [81]. In adults of 55 years of age or older, advanced age, African American race, increased asthma duration, and male gender were associated with PAL/FAO [79]. This pathogenesis of PAL/FAO is thought to be the result of airway remodeling. Patients with severe asthma and PAL/FAO were found to have a greater smooth muscle surface area with features of both T2 and non-T2 inflammation. IL-12, IL-13, and IFN-gamma were the dominant cytokines present. Compared to patients with severe asthma without PAL/FAO, neither airway imaging via high-resolution computed tomography nor sputum analysis was specific to each phenotype [82]. Additionally, some patients with PAL/FAO may have coexisting COPD and asthma.

Wheezing in Childhood

Wheezing is common in early childhood, but only some children develop persistent respiratory symptoms. Early epidemiological studies classified childhood wheeze based on the age of onset and resolution and lung function changes [8, 83]. Male sex, a history of wheezing with lower respiratory tract infections, a history of parental asthma, allergic comorbidities, and allergic sensitization have been found to be predictive of the development of an asthma phenotype [84]. This is consistent with the aforementioned cluster analyses that found primarily T2 inflammation in adults with early age of asthma onset. T2 and non-T2 inflammation has been demonstrated in children with asthma [85]. Cluster analyses of childhood asthma in the Severe Asthma Research Program (SARP) and Childhood Asthma Research and Education (CARE) networks showed distinct phenotypes characterized by asthma symptoms, level of allergic inflammation, and lung function [85–87]. Overall, asthma phenotyping is more difficult in children as risk factors and disease expression may be more variable over time. Asthma across the ages is discussed in further detail in Chapter 3.

Conclusion

Over the past two decades, we have made a tremendous stride in supplanting the generic categorization of asthma with a more sophisticated stratification schema, though this schema remains incomplete. With the identification and utilization of pathway-specific biomarkers in conjunction with clinical and physiologic features and advanced computational modeling, asthma heterogeneity is being continuously and masterfully demystified into distinct phenotypes and endotypes. Most of the elucidated mechanisms focus on T2 inflammation with a relative paucity of understanding of non-T2 inflammation (T2-low) in asthma. This has led to the development of a multitude of targeted biologic therapies directed at the T2-high axis and revolutionized the management of asthma, especially those with severe disease. There remain gaps in our knowledge for a number of these endotypes, especially those characterized by T2-low inflammation, and there is promise on the horizon for employing state-of-the-art molecular and gene expression analysis tools in conjunction with clinical, laboratory/biomarker/imaging parameters to explicate these pathways further and allow for the development of additional targeted therapies and more precise employment of such therapies.

References

1. National Asthma Education and Prevention Program. Expert Panel Report 3 (EPR-3): guidelines for the diagnosis and management of asthma-summary report 2007. J Allergy Clin Immunol. 2007;120:S94–138.
2. Backman H, et al. Severe asthma-a population study perspective. Clin Exp Allergy. 2019;49:819–28.
3. Wenzel SE. Asthma phenotypes: the evolution from clinical to molecular approaches. Nat Med. 2012;18:716–25.
4. Venkayya R, et al. The Th2 lymphocyte products IL-4 and IL-13 rapidly induce airway hyperresponsiveness through direct effects on resident airway cells. Am J Respir Cell Mol Biol. 2002;26:202–8.
5. Spits H, Di Santo JP. The expanding family of innate lymphoid cells: regulators and effectors of immunity and tissue remodeling. Nat Immunol. 2011;12:21–7.

6. Fitzpatrick AM, et al. T2-“Low” asthma: overview and management strategies. J Allergy Clin Immunol Pract. 2020;8:452–63.
7. Kuruvilla ME, et al. Understanding asthma phenotypes, endotypes, and mechanisms of disease. Clin Rev Allergy Immunol. 2019;56:219–33.
8. Henderson J, et al. Associations of wheezing phenotypes in the first 6 years of life with atopy, lung function and airway responsiveness in mid-childhood. Thorax. 2008;63:974–80.
9. Tliba O, Panettieri RA Jr. Paucigranulocytic asthma: uncoupling of airway obstruction from inflammation. J Allergy Clin Immunol. 2019;143:1287–94.
10. Peters MC, et al. Refractory airway type 2 inflammation in a large subgroup of asthmatic patients treated with inhaled corticosteroids. J Allergy Clin Immunol. 2019;143:104–113.e14.
11. Bousquet J, et al. Eosinophilic inflammation in asthma. N Engl J Med. 1990;323:1033–9.
12. Kumar RK, et al. Respiratory viral infection, epithelial cytokines, and innate lymphoid cells in asthma exacerbations. J Leukoc Biol. 2014;96:391–6.
13. Karta MR, et al. Insights into group 2 innate lymphoid cells in human airway disease. Curr Allergy Asthma Rep. 2016;16:8.
14. Duerr CU, Fritz JH. Regulation of group 2 innate lymphoid cells. Cytokine. 2016;87:1–8.
15. Jackson DJ, et al. IL-33–dependent type 2 inflammation during rhinovirus-induced asthma exacerbations in vivo. Am J Respir Crit Care Med. 2014;190:1373–82.
16. Murakami-Satsutani N, et al. IL-33 promotes the induction and maintenance of Th2 immune responses by enhancing the function of OX40 ligand. Allergol Int. 2014;63:443–55.
17. Ying S, et al. Thymic stromal lymphopoietin expression is increased in asthmatic airways and correlates with expression of Th2-attracting chemokines and disease severity. J Immunol. 2005;174:8183–90.
18. Hasegawa T, et al. Increased serum IL-17A and Th2 cytokine levels in patients with severe uncontrolled asthma. Eur Cytokine Netw. 2017;28:8–18.
19. Porsbjerg CM, et al. Anti-alarmins in asthma: targeting the airway epithelium with next-generation biologics. Eur Respir J. 2020;56:2000260.
20. Menzies-Gow A, et al. Tezepelumab in adults and adolescents with severe, uncontrolled asthma. N Engl J Med. 2021;384:1800–9.
21. Halwani R, et al. Mechanisms of asthma and allergic inflammation: 127 eosinophils enhance airway smooth muscle cell proliferation via the release of cysteinyl leukotrines. World Allergy Organ J. 2012;5:S59.
22. Lu Y, et al. Eosinophil extracellular traps drive asthma progression through neuro-immune signals. Nat Cell Biol. 2021;23:1060–72.
23. Hirose K, et al. Allergic airway inflammation: key players beyond the Th2 cell pathway. Immunol Rev. 2017;278:145–61.
24. Loewenthal L, Menzies-Gow A. FeNO in asthma. Semin Respir Crit Care Med. 2022;43:635–45.
25. Expert Panel Working Group of the National Heart, Lung, and Blood Institute (NHLBI) Administered and Coordinated National Asthma Education and Prevention Program Coordinating Committee (NAEPPCC), et al. 2020 focused updates to the asthma management guidelines: a report from the National Asthma Education and Prevention Program Coordinating Committee Expert Panel Working Group. J Allergy Clin Immunol. 2020;146:1217–70.
26. Zoratti EM, et al. Asthma phenotypes in inner-city children. J Allergy Clin Immunol. 2016;138:1016–29.
27. Lombardi C, et al. Is allergic sensitization relevant in severe asthma? Which allergens may be culprit? World Allergy Organ J. 2017;10:2.
28. Cox L, et al. Allergen immunotherapy: a practice parameter third update. J Allergy Clin Immunol. 2011;127:S1–55.
29. Matsumoto H. Roles of periostin in asthma. Adv Exp Med Biol. 2019;1132:145–59.
30. Hanania NA, et al. Efficacy and safety of lebrikizumab in patients with uncontrolled asthma (LAVOLTA I and LAVOLTA II): replicate, phase 3, randomised, double-blind, placebo-controlled trials. Lancet Respir Med. 2016;4:781–96.

31. Shiobara T, et al. Dipeptidyl peptidase-4 is highly expressed in bronchial epithelial cells of untreated asthma and it increases cell proliferation along with fibronectin production in airway constitutive cells. Respir Res. 2016;17:28.
32. Panettieri RA Jr, et al. Tralokinumab for severe, uncontrolled asthma (STRATOS 1 and STRATOS 2): two randomised, double-blind, placebo-controlled, phase 3 clinical trials. Lancet Respir Med. 2018;6:511–25.
33. Gao P, et al. Galectin-3: its role in asthma and potential as an anti-inflammatory target. Respir Res. 2013;14:136.
34. Riccio AM, et al. Plasma Galectin-3 and urine proteomics predict FEV1 improvement in omalizumab-treated patients with severe allergic asthma: results from the PROXIMA sub-study. World Allergy Organ J. 2020;13:100095.
35. McIntyre A, Busse WW. Asthma exacerbations: the Achilles heel of asthma care. Trends Mol Med. 2022. https://doi.org/10.1016/j.molmed.2022.09.001.
36. Moore WC, et al. Identification of asthma phenotypes using cluster analysis in the Severe Asthma Research Program. Am J Respir Crit Care Med. 2012;181:315–23.
37. Lefaudeux D, et al. U-BIOPRED clinical adult asthma clusters linked to a subset of sputum omics. J Allergy Clin Immunol. 2017;139:1797–807.
38. Haldar P, et al. Cluster analysis and clinical asthma phenotypes. Am J Respir Crit Care Med. 2008;178:218–24.
39. Kaur R, Chupp G. Phenotypes and endotypes of adult asthma: moving toward precision medicine. J Allergy Clin Immunol. 2019;144:1–12.
40. Woodruff PG, et al. T-helper type 2-driven inflammation defines major subphenotypes of asthma. Am J Respir Crit Care Med. 2009;180:388–95.
41. Kuo C-HS, et al. A transcriptome-driven analysis of epithelial brushings and bronchial biopsies to define asthma phenotypes in U-BIOPRED. Am J Respir Crit Care Med. 2017;195:443–55.
42. Peters MC, et al. A transcriptomic method to determine airway immune dysfunction in T2-high and T2-low asthma. Am J Respir Crit Care Med. 2019;199:465–77.
43. Aneas I, et al. Asthma-associated genetic variants induce IL33 differential expression through an enhancer-blocking regulatory region. Nat Commun. 2021;12:6115.
44. Ntontsi P, et al. Genetics and epigenetics in asthma. Int J Mol Sci. 2021;22:2412.
45. Akar-Ghibril N, et al. Allergic endotypes and phenotypes of asthma. J Allergy Clin Immunol Pract. 2020;8:429–40.
46. Tyler SR, Bunyavanich S. Leveraging -omics for asthma endotyping. J Allergy Clin Immunol. 2019;144:13–23.
47. Miranda C, et al. Distinguishing severe asthma phenotypes: role of age at onset and eosinophilic inflammation. J Allergy Clin Immunol. 2004;113:101–8.
48. Romanet-Manent S, et al. Allergic vs nonallergic asthma: what makes the difference? Allergy. 2002;57:607–13.
49. Viswanathan RK, Busse WW. Allergen immunotherapy in allergic respiratory diseases: from mechanisms to meta-analyses. Chest. 2012;141:1303–14.
50. 2022 GINA Report, Global Strategy for Asthma Management and PreventionGlobal Initiative for Asthma.
51. Amelink M, et al. Three phenotypes of adult-onset asthma. Allergy. 2013;68:674–80.
52. Buchheit KM, Laidlaw TM. Update on the management of aspirin-exacerbated respiratory disease. Allergy Asthma Immunol Res. 2016;8:298–304.
53. Hayashi H, et al. Omalizumab for aspirin hypersensitivity and leukotriene overproduction in aspirin-exacerbated respiratory disease. A randomized controlled trial. Am J Respir Crit Care Med. 2020;201:1488–98.
54. Wangberg H, et al. Appraisal of the real-world effectiveness of biologic therapies in aspirin-exacerbated respiratory disease. J Allergy Clin Immunol Pract. 2022;10:478–484.e3.
55. Oykhman P, et al. Comparative efficacy and safety of monoclonal antibodies and aspirin desensitization for chronic rhinosinusitis with nasal polyposis: a systematic review and network meta-analysis. J Allergy Clin Immunol. 2022;149:1286–95.
56. Lyly A, et al. Pathomechanisms of AERD-recent advances. Front Allergy. 2021;2:734733.

57. Weiler JM, et al. Exercise-induced bronchoconstriction update-2016. J Allergy Clin Immunol. 2016;138:1292–1295.e36.
58. Rasmussen SM, et al. Asthma in elite athletes – do they have Type 2 or non-Type 2 disease? A new insight on the endotypes among elite athletes. Front Allergy. 2022;3:973004.
59. Sze E, et al. Mechanisms and therapeutic strategies for non-T2 asthma. Allergy. 2020;75:311–25.
60. Chang HS, et al. Neutrophilic inflammation in asthma: mechanisms and therapeutic considerations. Expert Rev Respir Med. 2017;11:29–40.
61. Peters MC, et al. Plasma interleukin-6 concentrations, metabolic dysfunction, and asthma severity: a cross-sectional analysis of two cohorts. Lancet Respir Med. 2016;4:574–84.
62. Jin Y, et al. Association between YKL-40 and asthma: a systematic meta-analysis. Sleep Breath. 2022;26:1011–22.
63. Ramakrishnan RK, et al. Role of IL-17 in asthma pathogenesis and its implications for the clinic. Expert Rev Respir Med. 2019;13:1057–68.
64. Raundhal M, et al. High IFN-γ and low SLPI mark severe asthma in mice and humans. J Clin Invest. 2015;125:3037–50.
65. Gauthier M, et al. Severe asthma in humans and mouse model suggests a CXCL10 signature underlies corticosteroid-resistant Th1 bias. JCI Insight. 2017;2:e94580.
66. Pelaia C, et al. Molecular targets for biological therapies of severe asthma. Front Immunol. 2020;11:603312.
67. Huang G, et al. Regulation of TH17 cell differentiation by innate immune signals. Cell Mol Immunol. 2012;9:287–95.
68. Zhou L, et al. IL-6 programs T(H)-17 cell differentiation by promoting sequential engagement of the IL-21 and IL-23 pathways. Nat Immunol. 2007;8:967–74.
69. Aujla SJ, Alcorn JF. T(H)17 cells in asthma and inflammation. Biochim Biophys Acta. 2011;1810:1066–79.
70. Krishnamoorthy N, et al. Neutrophil cytoplasts induce T_H17 differentiation and skew inflammation toward neutrophilia in severe asthma. Sci Immunol. 2018;3:eaao4747.
71. Saffar AS, et al. The molecular mechanisms of glucocorticoids-mediated neutrophil survival. Curr Drug Targets. 2011;12:556–62.
72. Busse WW, et al. Randomized, double-blind, placebo-controlled study of brodalumab, a human anti-IL-17 receptor monoclonal antibody, in moderate to severe asthma. Am J Respir Crit Care Med. 2013;188:1294–302.
73. Panettieri RA Jr. Neutrophilic and pauci-immune phenotypes in severe asthma. Immunol Allergy Clin North Am. 2016;36:569–79.
74. Peters U, et al. Obesity and asthma. J Allergy Clin Immunol. 2018;141:1169–79.
75. Kim HY, et al. Interleukin-17-producing innate lymphoid cells and the NLRP3 inflammasome facilitate obesity-associated airway hyperreactivity. Nat Med. 2014;20:54–61.
76. Takahashi K, et al. Sputum proteomics and airway cell transcripts of current and ex-smokers with severe asthma in U-BIOPRED: an exploratory analysis. Eur Respir J. 2018;51:1702173.
77. Morissette M, et al. Asthma COPD overlap: insights into cellular and molecular mechanisms. Mol Aspects Med. 2022;85:101021.
78. Carr TF, et al. Eosinophilic and noneosinophilic asthma. Am J Respir Crit Care Med. 2018;197:22–37.
79. Bennett GH, et al. Risk factors and clinical outcomes associated with fixed airflow obstruction in older adults with asthma. Ann Allergy Asthma Immunol. 2018;120:164–168.e1.
80. Kole TM, et al. Predictors and associations of the persistent airflow limitation phenotype in asthma: a post-hoc analysis of the ATLANTIS study. Lancet Respir Med. 2022. https://doi.org/10.1016/S2213-2600(22)00185-0.
81. Guerra S, et al. The course of persistent airflow limitation in subjects with and without asthma. Respir Med. 2008;102:1473–82.
82. Kaminska M, et al. Airway remodeling in subjects with severe asthma with or without chronic persistent airflow obstruction. J Allergy Clin Immunol. 2009;124:45–51.e1–4.
83. Martinez FD, et al. Diminished lung function as a predisposing factor for wheezing respiratory illness in infants. N Engl J Med. 1988;319:1112–7.

84. Guilbert TW, et al. Childhood asthma-predictive phenotype. J Allergy Clin Immunol Pract. 2014;2:664–70.
85. Conrad LA, et al. Defining pediatric asthma: phenotypes to endotypes and beyond. Pediatr Res. 2021;90:45–51.
86. Chang TS, et al. Childhood asthma clusters and response to therapy in clinical trials. J Allergy Clin Immunol. 2014;133:363–9.
87. Teague WG, et al. Baseline features of the severe asthma research program (SARP III) cohort: differences with age. J Allergy Clin Immunol Pract. 2018;6:545–554.e4.

Heterogeneity of Treatment Response to Asthma

Tetsuro Maeda and Sandhya Khurana

Abstract

The definition of asthma has evolved over the years with significant heterogeneity of the disease increasingly recognized. Complex gene and environment interactions result in different pheno-endotypes of asthma that respond differently to the same treatment. Multiple studies have revealed pharmacogenomic and endo-phenotypic factors that predict treatment response to standard therapies for asthma. Recent advances in biologic medications have enabled a more tailored approach to the care of patients with moderate to severe asthma, taking into consideration clinical traits and measurable biomarkers. This chapter will review heterogeneity in treatment response to different medication classes for asthma: inhaled and systemic corticosteroids, beta-2 agonists, leukotriene modifiers, muscarinic antagonists, macrolides, and biologics.

Keywords

Adherence · Corticosteroids · Beta-2 agonists · Leukotriene modifiers · Muscarinic antagonists · Macrolides · Biologics · Pharmacogenomics · Medication adherence · Heterogeneity

T. Maeda · S. Khurana (✉)
University of Rochester School of Medicine and Dentistry, Division of Pulmonary and Critical Care Medicine, Rochester, NY, USA
e-mail: sandhya_khurana@urmc.rochester.edu

A. R. Brasier, N. N. Jarjour (eds.), *Precision Approaches to Heterogeneity in Asthma*, Advances in Experimental Medicine and Biology 1426,
https://doi.org/10.1007/978-3-031-32259-4_7

Introduction

US-national and international guidelines recommend a stepwise approach to pharmacologic therapy for asthma [1, 2]. Inhaled corticosteroids (ICS), often in combination with long-acting beta-2 agonists (LABA), will accomplish the prevention of asthma exacerbations and restoration of lung function in most patients with asthma. However, some patients will have a suboptimal response and go on to require additional pharmacologic agents such as leukotriene modifiers (LTM), long-acting muscarinic antagonists (LAMA), macrolides, systemic corticosteroids, and/or biologic therapies such as monoclonal antibodies targeting immunoglobulin E (IgE), interleukin (IL)-5, IL-4/IL-13 and thymic stromal lymphopoietin (TSLP). There are multiple factors affecting treatment response to these medications (Fig. 1). Although genetic variation is estimated to account for as much as 60–80% of treatment response heterogeneity [3], other factors including adherence, comorbidities, environmental factors, and disease severity and subtype also play an important role [4, 5].

Pharmacogenomic studies have revealed that treatment response may be altered by heritable mechanisms that affect drug absorption, distribution, metabolism, excretion, mechanism of action on biological pathways and disease pathobiology [6]. Further advances in low-cost research technology have enabled genome-wide association studies (GWAS), including those focused on therapeutic responses to certain medications in asthma [7]. Recent advances in understanding asthma phenotypes and endotypes, both heritable and acquired, have made more tailored approaches possible for individual patients, particularly those suffering from severe asthma [8, 9].

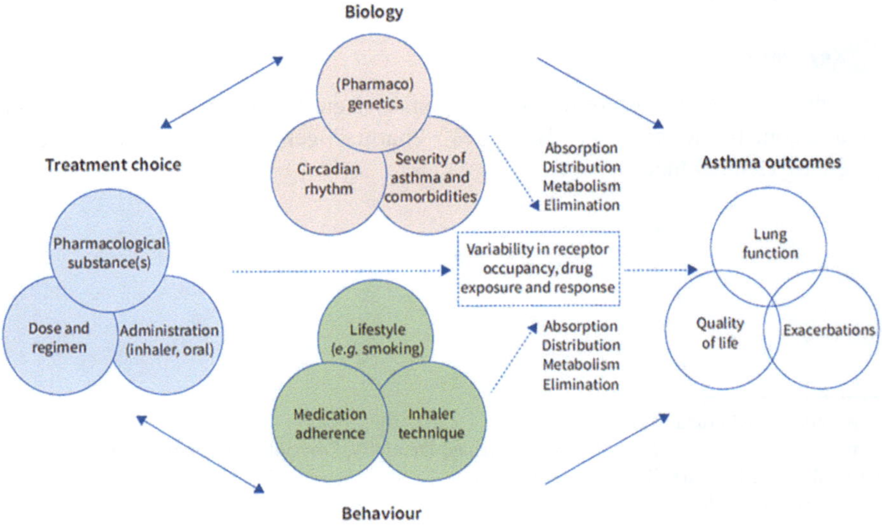

Fig. 1 Factors that explain variability in asthma drug response. (Reproduced with permission of the ERS 2022, from van Boven [5])

Heterogeneity in Response to Inhaled and Systemic Corticosteroids

ICS has been the mainstay of asthma controller therapy and considered first-line therapy in asthma. Systemic (oral and parenteral) corticosteroids are utilized for the treatment of asthma exacerbations and, reluctantly, an add-on maintenance therapy for patients with severe asthma refractory to standard treatments. Some patients respond less well to corticosteroids than others. In the study by Malmstrom et al., although the primary aim was to compare the efficacy of inhaled beclomethasone (ICS) and oral montelukast (LTM), there was a large heterogeneity for change in forced expiratory volume in 1 second (FEV1) in response to inhaled beclomethasone as initial asthma therapy. While 50% of the patients receiving inhaled beclomethasone had improvement in FEV1, 22% did not [10]. Similarly, Chan et al. in their retrospective study found that as many as 24% of severe asthma patients were "steroid-insensitive" and failed to improve their lung function after oral prednisone therapy [11].

The PRICE trial investigated predictors of long-term ICS response and demonstrated that baseline bronchodilator response, FEV1/forced vital capacity (FVC) and FEV1 (% predicted) correlated with short-term ICS response in patients with asthma. The study also showed that short-term ICS response predicted long-term ICS response [12]. Much work has been done to investigate the reason for such heterogeneity in treatment response to ICS.

Pharmacogenomics

As it became increasingly recognized that there is a large variation in response to ICS treatment, many pharmacogenomic studies were undertaken to further elucidate the mechanisms. One of the early discoveries was of sequence variants in corticotropin-releasing hormone receptor 1 (CRHR1) gene associated with greater improvement in lung function with ICS in children [13]. However, this finding was not replicated in a subsequent study looking at long-term effects of ICS in adult patients [14]. TBX21 gene, encoding for the transcription factor T-bet, was also found to be associated with the improvement of lung function in pediatric patients receiving ICS [15]. Later, Tantisira et al. conducted a GWAS, based on a family-based screening algorithm, and identified single nuclear polymorphism (SNP) rs37972, which maps to the glucocorticoid-induced transcript 1 gene (GLCCI1) and is perfectly correlated with rs37973, as a novel pharmacogenetic determinant of the response to ICS. They further found that the rs37973 variant is associated with significantly decreased luciferase reporter activity, that patients who were homozygous for the mutant rs37973 allele had poorer response to ICS compared to those who were homozygous for the wild-type allele, and that the genotype accounts for about 6.6% of overall ICS response variability [16]. However, a larger scale pooled analysis from seven clinical trials did not confirm this association [17]. The discordance between these studies is possibly due to the differences in population (pediatric vs adult) and endpoints studied.

Another GWAS identified SNPs (rs3127412 and rs6456042) tightly linked to a new pharmacogenetic locus ("*T* gene") that contributes to variation in response to ICS [18]. A more recent study by Sharma et al. identified "glucocorticoid genes" that were differentially expressed in subjects with asthma compared to healthy controls and were enriched in subjects with asthma after treatment with inhaled corticosteroids. Developmental glucocorticoid genes also influenced susceptibility to asthma and treatment response to glucocorticoid [19].

Baines et al. studied gene expression in sputum samples from an asthma patient cohort, and its association with asthma phenotypes. They discovered that the expression of genes CLC, CPA3 and DNASE1L3 are increased in eosinophilic asthma and IL1B, ALPL and CXCR2 in neutrophilic asthma. The gene expression signature predicted treatment response (change in FEV1 and fractional exhaled nitric oxide (FeNO)) to ICS [20]. SNP rs10044254, associated with decreased expression of FBXL7 gene, improved symptomatic response to ICS in pediatric cohorts [21]. VNN-1 gene, whose expression could be measured by sampling nasal epithelium mRNA, predicted response to systemic corticosteroids in pediatric patients with asthma exacerbation [22]. FAM129A gene was also shown to be a modulator of corticosteroid responsiveness [23].

Endophenotypes

It is known that eosinophilic inflammation tends to respond better to corticosteroids, while sputum neutrophilia predicts a poorer response to ICS [24]. A study comparing the efficacy of ICS vs. LTM showed that higher FeNO predicts a more favorable response to ICS compared to LTM in the pediatric population with asthma (Fig. 2) [25].

More recently, studies have suggested that certain immune-related cells and the resultant production of certain cytokines lead to corticosteroid resistance. Possible mechanisms include alteration of glucocorticoid receptors, reduced histone deacetylase 2, and increased expression of proinflammatory transcription factors (Table 1) [26, 27].

Many studies have been performed and elucidated molecular-level mechanisms of steroid resistance using a variety of specimens ranging from blood, sputum, and nasal epithelial cells. Hew et al. compared the suppression of lipopolysaccharide-induced cytokine release from peripheral blood mononuclear cells in severe and non-severe asthma patients at baseline and in response to systemic dexamethasone administration. While baseline spontaneous and stimulated release were similar between the groups, the cytokine release after dexamethasone administration was less suppressed in severe asthma patients. They also demonstrated that nuclear histone deacetylase (HDAC) and histone acetyltransferase activities were reduced in the severe asthma group compared to the non-severe asthma group, and that reduced

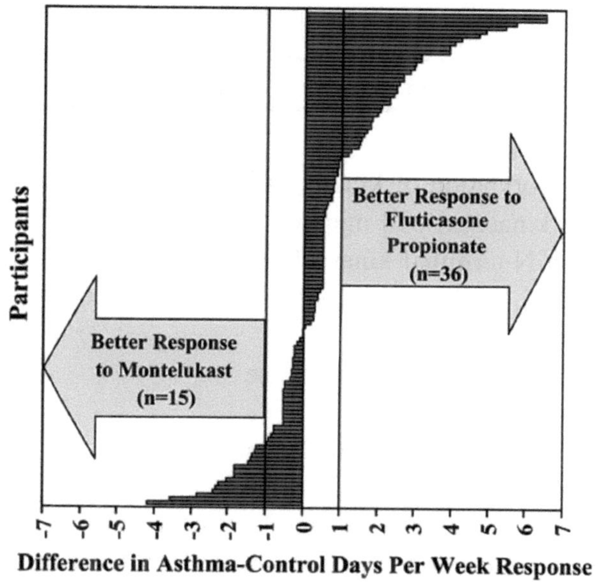

Fig. 2 Difference in asthma control days between fluticasone propionate and montelukast for individual patients. (Reprinted with permission from Elsevier, Zeiger et al. [25])

HDAC activity correlated to the steroid insensitivity of granulocyte-macrophage colony-stimulating factor and interferon (IFN)-γ release [28].

Woodruff et al. performed genome-wide profiling of airway epithelial cells and found that CLCA1, periostin, serpin B2 are upregulated in asthma patients. ICS treatment response correlated with an increase in these markers and a decrease in FK506-binding protein 51 (FKBP51) [29]. Irvin et al. demonstrated "dual-positive" Th2/Th17 cells in the bronchoalveolar lavage (BAL) fluid from asthma patients that were associated with increased BAL fluid IL-17, greater disease severity and decreased sensitivity to glucocorticoids [30]. Zhang et al. found that nasal epithelial samples from pediatric asthma patients showed different responses to dexamethasone in terms of DNA methylation in certain genes dependent or independent of SNPs, which correlated to clinical response to corticosteroids [31]. Similarly, Fayon et al. showed a differential response of FKBP51 to dexamethasone, influenced by tumor necrosis factor (TNF)-α, IL-6, and leukemia inhibitory factor (an IL-6 family neurotrophic cytokine) [32].

Clemmer et al. used principal component analysis to determine a composite corticosteroid responsiveness phenotype and found that it was more accurate than individual clinical phenotypes in predicting steroid responsiveness in asthma [33].

Table 1 Summary of possible mechanisms of steroid resistance in severe asthma

Possible mechanisms
Dysfunctional GR-α
• ↑phosphorylation of kinases (p38 mitogen-activated protein kinase α, p38 mitogen-activated protein kinase γ, c-Jun-N-terminal kinase 1)→↓ nuclear translocation
• ↑ IL-2, IL-4, IL-13→↓ phosphorylation of GR→↓ nuclear translocation
• ↑inducible nitric oxide synthase
Increased GR-β
• ↑IL-17→↑GR-β in epithelial cells
Histone acetylation
• ↓histone deacetylase 2→↓inflammatory gene repression
Increased expression of proinflammatory transcription factors
• ↑AP-1 and NF-κB

Reprinted with permission from Elsevier, from Al Heialy et al. [27]
GR glucocorticoid receptor, *IL* interleukin, *AP-1* activator protein-1, *NF-κB* Nuclear factor kappa B

Other Factors

Smoking Lazarus et al. studied the effects of smoking on treatment response in patients with mild asthma and found that smoking attenuates steroid responsiveness [34]. Chandhuri et al. compared response to oral prednisone in current, former, and never smokers with chronic stable asthma and baseline bronchodilator reversibility. While never and former smokers experienced an improvement in lung function, such response was not observed in current smokers, suggesting that active smoking impairs treatment response to systemic corticosteroids [35].

Obesity Sutherland et al. studied in vitro response of peripheral blood mononuclear cells and BAL fluid cells to dexamethasone. They demonstrated that elevated body mass index (BMI) in subjects with asthma was associated with a blunted in vitro response to dexamethasone, while this association was not observed in subjects without asthma [36]. Forno and colleagues evaluated the interaction between BMI and ICS treatment response in subjects enrolled in the childhood asthma management program trial. They found that overweight and obese children had decreased response to ICS on measures of lung function and emergency department visits/hospitalizations compared to children of normal weight [37].

Race/ethnicity The impact of race and ethnicity on the heterogeneity of treatment response in asthma is not well studied due to an overall lack of diversity in pharmacogenomic studies. Wells et al. sought to investigate the association of self-reported race-ethnicity and genetic ancestry with ICS response and found that self-reported race-ethnicity and genetic ancestry did not predict ICS response [38]. Hernandez-Pacheco and colleagues performed one of the first and largest meta-GWAS of ICS response in Hispanic/Latino and African American children and young adults with asthma and found an association of *APOBEC3B* and *APOBEC3C* genes with ICS response. They also validated the association of the L3MBTL4-ARHGAP28 genomic region in admixed populations, which was previously described in a GWAS of ICS response in subjects of European descent [39].

In summary, though more mechanisms are being unraveled, there appear to be several ways by which response to corticosteroid treatment is modified. It is likely that a combination of, rather than a single, factors account for the heterogeneity of response to corticosteroids [40, 41].

Heterogeneity in Response to Beta-2 Agonists

Short-acting beta agonist (SABA) is the most used reliever medication for asthma. LABA is the most common add-on therapy to ICS in patients with asthma. Several pharmacogenetic studies have been published on the effects of these medications.

Lung Function

Early studies on the effect of regularly scheduled SABA, compared to as-needed use only, provided conflicting data regarding the deleterious effect of regularly scheduled SABA treatment on asthma control in patients with mild asthma [42, 43]. Since more knowledge became available regarding genetic variation of treatment response, each of the study cohorts was further evaluated, dividing the subjects by the β2 adrenergic receptor (ADRB2) genotype. They still yielded conflicting results: while Hancox et al. found no association between genotype and change in asthma control, Israel et al. found that the Arg/Arg genotype at the codon 16 of the ADRB2 gene was associated with significantly lower PEFR when patients were using regularly scheduled inhaled albuterol, whereas there was no decline in PEFR in patients who used inhaled albuterol only on as-needed basis [44, 45]. Consequently, another pharmacogenetic study specifically designed to investigate this association was performed. The BARGE trial was a crossover study with regularly administered inhaled albuterol and placebo to investigate the variation of the FEV1 in association with the ADRB2 genotype. It demonstrated that patients with the Arg/Arg genotype at the 16th amino acid position of the ADRB2 had a significant reduction in the morning PEFR, as well as genotype-specific effects in FEV1, symptoms, and use of reliever medications compared to those with Gly/Gly genotype [46]. This led to the conclusion that regularly scheduled SABA treatment should be avoided in the population with the Arg/Arg genotype, which consists of as many as 15% of patients with asthma [47].

Subsequently, further studies were undertaken to see if LABA could also cause deleterious effects on lung function in patients with asthma with such genotype. Bleecker et al. published multiple pharmacogenetic studies that examined the association between the ADRB2 genotype and clinical response with scheduled inhaled salmeterol, finding no pharmacogenetic effect of ADRB2 variation on therapeutic response in patients with asthma who were on another concomitant controller therapy [47–49]. Additionally, the LARGE randomized controlled crossover trial, stratifying by ADRB2 genotypes, specifically investigated the association of ADRB2 genotypes and response to ICS (beclomethasone) plus LABA (salmeterol) therapy as compared to ICS plus placebo. The study demonstrated no differences between the Arg/Arg and Gly/Gly genotypes with respect to LABA therapy [50].

However, additional studies in the younger patient population showed yet conflicting results compared to adult population, perhaps due to differences between the age groups in response to regularly scheduled beta-2 agonist therapy. Two studies, one with an age range of 3–22 years [51] and the other 4–12 years [52], reported that Arg16 genotype was associated with increased asthma exacerbations when exposed to regularly used albuterol or salmeterol. A subsequent meta-analysis of pediatric trials also showed consistent results [53]. Using montelukast in place of LABA as a second controller in patients with Arg16 genotype led to favorable outcomes in children [54] and adolescent population [55], making tailored treatment by genotype more plausible. Further genetic study by Ortega et al. in a multiethnic cohort revealed that rare variants of the ADRB2, Ile164 and -376ins, had an association with adverse events during LABA therapy [56].

Bronchodilator Response

In addition to the effect of SABA and LABA on lung function, there have been studies on certain genetic factors with regard to bronchodilator response (BDR). Litonjua et al. identified Arginase 1 (ARG1) as another gene whose SNPs were associated with BDR [57]. Duan et al. performed a pharmacogenetic study showing that individuals with ARG1 haplotype 1, compared to haplotypes 2 and 3, were significantly more likely to have BDR [58]. A GWAS identified a new genome-wide significant locus (rs350729, rs1840321, rs1384918, and rs1319797) on Chromosome 2 near the ASB3 gene, variants of which account for BDR reduction by 20% compared to wild type [59]. Drake et al. performed another GWAS in Latino children, finding that more rare variants, related to genes such as SLC, contribute to differential drug response [60]. Further, Mak et al. performed a GWAS including children in minority groups that revealed population-specific and shared genetic variants associated with BDR (DNAH5, NFKB1, PLCB1, ADAMTS3, COX18) [61].

Heterogeneity in Response to Leukotriene Modifiers

Leukotriene modifiers, including 5-lipoxygenase (ALOX5) inhibitors (zileuton) and CysLT1 receptor antagonist (montelukast, zafirlukast, pranlukast), are another class of asthma controller therapy. The previously cited study by Malmstrom et al. showed that oral montelukast, similar to inhaled beclomethasone, had a large heterogeneity in treatment response [10]. Pharmacogenetic studies have also played an important role in the understanding of treatment response heterogeneity to leukotriene modifiers. Drazen et al. demonstrated that ABT-761 (atreleuton, not available for clinical use), a direct inhibitor of ALOX5, resulted in a greater increase in FEV1 in asthma patients with wild-type and heterozygous genotypes for ALOX5 than in those with mutant genotype [62]. Lima et al. performed another pharmacogenetic study to investigate the association of variation in leukotriene pathway candidate genes with treatment response to montelukast. They demonstrated that SNPs in the ALOX5 (rs2115819) and MRP1 (rs119774) genes were associated with changes in FEV1; SNPs in LTC4S (rs730012) and LTA4H (rs2660845) with exacerbation rates; and ALOX5 haplotypes with risk of exacerbations [63]. ALOX5 promoter genotype was also found to be associated with an increase in FEV1 and reduced need for SABA in patients treated with montelukast [64]. Such SNPs were further shown to be associated with greater response to zileuton, indicating a common pharmacogenomic mechanism for these two leukotriene modifiers [65]. Later, a GWAS by Dahlin et al. found that SNP rs6475448 is associated with increased FEV1 after treatment with montelukast [66]. Maroteau et al. compared asthma patients on montelukast with and without exacerbations for the previous 1 year and found that rs2660845 variant, presumably due to up-regulation of LTA4H activity, is associated with increased odds of exacerbations [67].

Beyond pharmacogenomic studies, montelukast has been shown to be efficacious in a unique subpopulation of asthma. The study by Lazarus et al. on the effect of smoking in patients with mild asthma also addressed the response to oral montelukast. Interestingly, montelukast significantly increased AM PEFR only in smokers, suggesting some favorable effects of LTRA on smokers [34]. Exercise-induced bronchoconstriction [68] and aspirin-exacerbated respiratory disease [69] are other asthma phenotypes in which montelukast appears to have better efficacy.

Heterogeneity in Response to Long-Acting Muscarinic Antagonists (LAMA)

LAMAs are long-acting bronchodilators that exert their effects by binding to muscarinic receptor(s), a different mechanism of action than beta-2 agonists. While they have been more commonly used in chronic obstructive lung disease, they have a role as add-on therapy in moderate-severe asthma. Tiotropium and umeclidinium (as part of single-inhaler combination therapy with fluticasone furoate and vilanterol) are the only LAMAs currently approved by the US FDA for the treatment of asthma.

Data from a double-blind, 3-way, crossover trial in patients with asthma inadequately controlled on inhaled steroids were subsequently analyzed for individual and differential treatment responses to salmeterol (LABA) and tiotropium (LAMA). An acute response to SABA, decreased FEV1/FVC, and high cholinergic tone (low resting heart rate) predicted a response to tiotropium [70, 71]. The SIENA trial, a double-blind crossover study in patients with mild persistent asthma, examined differential treatment responses to inhaled mometasone (ICS) and tiotropium compared to a placebo for each. In the subgroup with low sputum eosinophil level (<2%), inhaled mometasone did not have a significantly better treatment response than placebo but inhaled tiotropium did. On the other hand, in the subgroup with high sputum eosinophil level (≥2%), inhaled mometasone had a better treatment response than placebo but inhaled tiotropium did not [72].

Heterogeneity in Response to Macrolides

Macrolide antibiotics, particularly azithromycin, have been used in various pulmonary diseases for their anti-inflammatory effects. The efficacy of azithromycin has been studied in several clinical trials in patients with eosinophilic and non-eosinophilic asthma.

The AZISAST randomized controlled trial compared azithromycin (250 mg daily for 5 days followed by three times weekly for a total of 26 weeks) to placebo in patients with uncontrolled asthma on ICS/LABA therapy and showed that azithromycin group did not have a significant reduction in the composite primary endpoint of severe exacerbations and lower respiratory tract infection. However, it did show a significant reduction in the composite primary endpoint in the subgroup of patients with the non-eosinophilic phenotype (blood eosinophil ≤ 200/μL) [73]. Subsequently, the AMAZES randomized controlled trial, comparing azithromycin (500 mg three times weekly for 48 weeks) to placebo in patients with uncontrolled asthma on ICS/LABA therapy, showed that azithromycin group had a significant reduction in exacerbations in both subgroups of eosinophilic and non-eosinophilic phenotypes, as well as in subgroups with and without pulmonary pathogens isolated from sputum culture [74].

Heterogeneity in Response to Biologics

Over the past decade, the availability of multiple newer biologic medications has revolutionized the management of severe asthma [1, 75]. In efforts to deliver precision medicine in severe asthma, the research into pheno-endotyping of this heterogenous disease using biomarkers has also greatly expanded. Only biomarkers of type-2 inflammation in asthma are currently available for clinical use (blood eosinophil count, serum total IgE and allergen sensitization profile, and FeNO), while several other biomarkers are in development [76]. Asthma phenotypes and endotypes are reviewed in more detail in another chapter of this book.

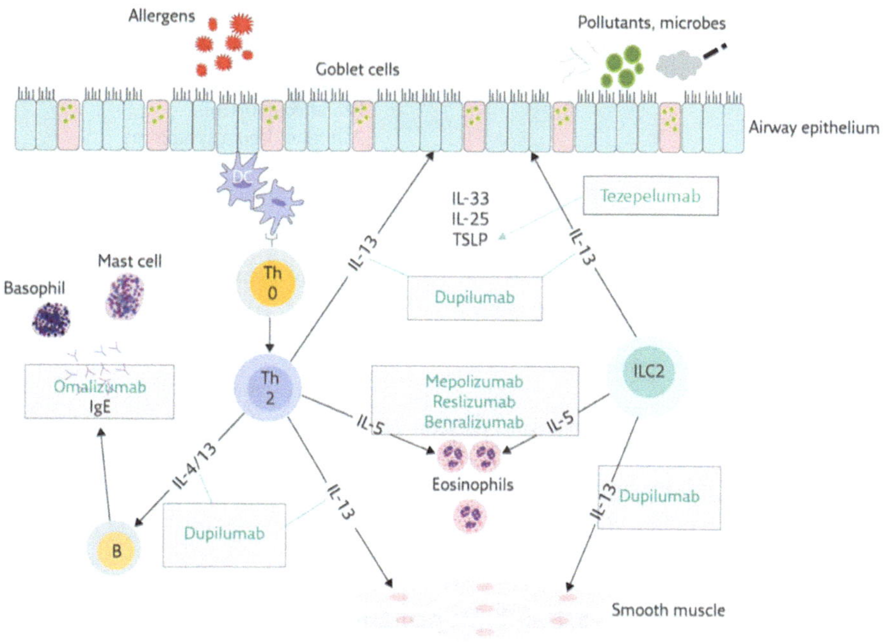

Fig. 3 Available asthma biologics and targets. DC: dendritic cell; Th0: naïve T-cell; Th2: T-helper 2 cell; B: B-cell; IL: interleukin; TSLP: thymic stromal lymphopoietin; ILC2: type 2 innate lymphoid cell. (Reproduced with permission of the ©ERS 2022: Khurana et al. [79] Published 25 September 2020)

Currently, biologic medications are targeting IgE, interleukin (IL)-5 pathway, IL-4Rα and thymic stromal lymphopoietin (TSLP) (Fig. 3) [77–79]. Clinical features and biomarkers that predict treatment response to these medications are summarized in Table 2.

Biologic Targeting IgE (Omalizumab)

Omalizumab, a monoclonal antibody against IgE, is the first biologic medication approved by the US FDA for use in asthma in 2003. It was studied in patients with moderate–severe asthma with sensitization to one or more perennial aeroallergen and total serum IgE level 30–700 IU/mL. Treatment with omalizumab was associated with significant improvement in exacerbation rates and lung function [80, 81]. Interestingly, baseline IgE level did not predict treatment response [82]. A post hoc analysis of the EXTRA study, a multicenter randomized study showed that high values of type 2 biomarkers (FeNO ≥19.5 ppb, blood eosinophil count ≥260/μL, and serum periostin ≥50 ng/mL) predicted greater reduction of asthma exacerbations in response to treatment with omalizumab compared to placebo [83]. However,

Table 2 Predictors of treatment response to biologic medications for asthma

Asthma Biologic	Mechanism of action	Predictors of treatment response
Omalizumab	Anti-IgE	FeNO ≥ 20 ppb Blood eosinophilia Allergen-driven symptoms Childhood-onset asthma
Mepolizumab	Anti-IL-5	Blood eosinophilia Baseline exacerbation frequency Adult-onset asthma Nasal polyposis
Reslizumab		
Benralizumab	Anti-IL-5 Receptor α	
Dupilumab	Anti-IL-4 Receptor α	FeNO Blood eosinophilia
Tezepelumab	Anti-TSLP	FeNO Blood eosinophilia

IL interleukin, *TSLP* thymic stromal lymphopoietin, *FeNO* fraction of exhaled nitric oxide

a real-world study (PROSPERO) failed to show such predictive values of FeNO and blood eosinophil count [84].

Biologics Targeting IL-5/IL-5R (Mepolizumab, Reslizumab, Benralizumab)

This group of biologic therapies for asthma medications targets the IL-5 pathway, an important type-2 inflammatory cytokine that mediates eosinophil activation, maturation and survival. Depletion of eosinophils via IL-5 blockade results in the improvement of eosinophilic inflammation.

Mepolizumab and reslizumab are monoclonal antibodies against IL-5, and benralizumab is a monoclonal antibody against IL-5Rα (alpha subunit of IL-5 receptor). Anti-IL5 biologics have demonstrated efficacy in improving exacerbation rates, lung function, asthma control and oral steroid-sparing effect (for mepolizumab and benralizumab) in severe eosinophilic asthma [85–92]. Analysis of the DREAM study found that baseline exacerbation frequency and blood eosinophil counts correlate with response to mepolizumab therapy [87]. Similarly, the SIROCCO randomized controlled trial demonstrated that benralizumab improved exacerbation rates and FEV1 in patients with severe asthma with blood eosinophil ≥300/μL and frequent exacerbations, but that the clinical efficacy was less significant in a similar patient population with blood eosinophil <300/μL [91]. A similarly designed CALIMA randomized controlled trial also showed that benralizumab significantly reduced asthma exacerbations and improved lung function in the patient population with blood eosinophil ≥ 300/μL, but also in that with blood eosinophil < 300/μL [93]. Subsequent pooled analysis of both studies showed that blood eosinophil count and frequency of exacerbations are both predictors of enhanced treatment response to benralizumab [94].

Biologic Targeting IL-4Rα (Dupilumab)

Dupilumab is an anti-IL4Rα (alpha subunit of IL4 receptor) that blocks the effects of both cytokines IL-4 and IL-13. Studies of biologics targeting only IL-13 yielded clinically disappointing results but found relevant differential effects by certain biomarkers [95]. For instance, subgroup analysis of a phase 2 study of lebrikizumab (an IgG4 monoclonal antibody against IL-13 which blocks its interaction with IL-4Rα) showed that patients with high serum periostin levels had a better response than those with low serum periostin level in terms of improvement in FEV1 [96]. However, subsequent replicate phase 3 randomized controlled trials failed to show a clinically significant reduction in asthma exacerbations, even in biomarker-high subgroup (serum periostin ≥ 50 ng/mL or blood eosinophil ≥ 300/μL) [97]. Dupilumab has shown efficacy in patients with severe asthma, with a reduction in exacerbation rate, improvement in lung function, and a steroid-sparing effect [98, 99]. Initial phase 2 study by Wenzel et al. suggested a beneficial effect of dupilumab in both patient groups with eosinophil count ≥300/μL as well as <300/μL [100]. Analysis of data from subsequent phase 3 study found that elevated FeNO (≥25 ppb) and/or elevated blood eosinophil count (≥150/μL) were associated with significant treatment response [98].

Biologic Targeting TSLP (Tezepelumab)

Tezepelumab targets an epithelial cytokine or alarmin, TSLP, that exerts its effects upstream of other asthma biologics. TSLP has a role in mediating both type 2 and non-type 2 inflammation. Tezepelumab is the newest biologic to receive US FDA approval and the only biologic available for the treatment of non-type 2 severe asthma. The phase 2 PATHWAY randomized controlled trial showed that tezepelumab in uncontrolled asthma achieved a reduction of exacerbation rates, regardless of eosinophil count [101]. The phase 3 NAVIGATOR randomized controlled trial showed that tezepelumab in severe asthma patients resulted in significantly fewer exacerbations and better lung function, asthma control, and health-related quality of life than those who received a placebo. This effect was seen in patients regardless of type-2 status (blood eosinophil count and FeNO). However, subgroup analysis suggested greater efficacy with increasing FeNO levels and blood eosinophil count [102]. Similarly, the SOURCE randomized controlled trial, looking at steroid-sparing effects of Tezepelumab, showed greater efficacy in patients with type 2 asthma [103].

Summary

In this chapter, we have summarized evidence from studies evaluating the significant treatment response heterogeneity for different medications across the spectrum of asthma severity. Research concerning pharmacogenomics and endophenotypes of asthma suggests that there are many mechanisms behind this treatment response heterogeneity. Patient's phenotypic characteristics (exacerbation frequency, lung function, comorbidities, etc.) combined with measurable biomarkers can help guide the selection of the asthma therapy likely to be most effective for the patient. Significant research efforts, including NIH's PrecISE (Precision Interventions for Severe and Exacerbation Prone Asthma) program, are underway to further guide optimal treatment selection for patients with severe asthma and provide guidance for precision and personalized treatment decision in these patients.

References

1. Global Initiative for Asthma. Global strategy for asthma management and prevention. 2022. Available from: www.ginasthma.org. Accessed on 30 Aug 2022.
2. 2020 focused updates to the asthma management guidelines: a report from the National Asthma Education and Prevention Program Coordinating Committee Expert Panel Working Group I NHLBI, NIH. 2021. Available from: https://www.nhlbi.nih.gov/resources/2020-focused-updates-asthma-management-guidelines. Accessed on 30 Aug 2022.
3. Drazen JM, Silverman EK, Lee TH. Heterogeneity of therapeutic responses in asthma. Br Med Bull. 2000;56(4):1054–70.
4. Hall IP. Pharmacogenetics of asthma. Eur Respir J. 2000;15(3):449–51.
5. van Boven JFM, Dierick BJH, Usmani OS. When biology meets behaviour: can medication adherence mask the contribution of pharmacogenetic effects in asthma? Eur Respir J. 2021;58(3):2100304.
6. Wechsler ME, Israel E. How pharmacogenomics will play a role in the management of asthma. Am J Respir Crit Care Med. 2005;172(1):12–8.
7. Meyers DA, Bleecker ER, Holloway JW, Holgate ST. Asthma genetics and personalised medicine. Lancet Respir Med. 2014;2(5):405–15.
8. Haldar P, Pavord ID, Shaw DE, Berry MA, Thomas M, Brightling CE, et al. Cluster analysis and clinical asthma phenotypes. Am J Respir Crit Care Med. 2008;178(3):218–24.
9. Kelly RS, Mendez KM, Huang M, Hobbs BD, Clish CB, Gerszten R, et al. Metabo-endotypes of asthma reveal differences in lung function: discovery and validation in two TOPMed cohorts. Am J Respir Crit Care Med. 2022;205(3):288–99.
10. Malmstrom K, Rodriguez-Gomez G, Guerra J, Villaran C, Pineiro A, Wei LX, et al. Oral montelukast, inhaled beclomethasone, and placebo for chronic asthma. A randomized, controlled trial. Montelukast/Beclomethasone Study Group. Ann Intern Med. 1999;130(6):487–95.
11. Chan MT, Leung DY, Szefler SJ, Spahn JD. Difficult-to-control asthma: clinical characteristics of steroid-insensitive asthma. J Allergy Clin Immunol. 1998;101(5):594–601.
12. Martin RJ, Szefler SJ, King TS, Kraft M, Boushey HA, Chinchilli VM, et al. The predicting response to inhaled corticosteroid efficacy (PRICE) trial. J Allergy Clin Immunol. 2007;119(1):73–80.
13. Tantisira KG, Lake S, Silverman ES, Palmer LJ, Lazarus R, Silverman EK, et al. Corticosteroid pharmacogenetics: association of sequence variants in CRHR1 with improved lung function in asthmatics treated with inhaled corticosteroids. Hum Mol Genet. 2004;13(13):1353–9.
14. Dijkstra A, Koppelman GH, Vonk JM, Bruinenberg M, Schouten JP, Postma DS. Pharmacogenomics and outcome of asthma: no clinical application for long-term steroid effects by CRHR1 polymorphisms. J Allergy Clin Immunol. 2008;121(6):1510–3.
15. Tantisira KG, Hwang ES, Raby BA, Silverman ES, Lake SL, Richter BG, et al. TBX21: a functional variant predicts improvement in asthma with the use of inhaled corticosteroids. Proc Natl Acad Sci U S A. 2004;101(52):18099–104.

16. Tantisira KG, Lasky-Su J, Harada M, Murphy A, Litonjua AA, Himes BE, et al. Genomewide association between GLCCI1 and response to glucocorticoid therapy in asthma. N Engl J Med. 2011;365(13):1173–83.
17. Hosking L, Bleecker E, Ghosh S, Yeo A, Jacques L, Mosteller M, et al. GLCCI1 rs37973 does not influence treatment response to inhaled corticosteroids in white subjects with asthma. J Allergy Clin Immunol. 2014;133(2):587–9.
18. Tantisira KG, Damask A, Szefler SJ, Schuemann B, Markezich A, Su J, et al. Genome-wide association identifies the T gene as a novel asthma pharmacogenetic locus. Am J Respir Crit Care Med. 2012;185(12):1286–91.
19. Sharma S, Kho AT, Chhabra D, Qiu W, Gaedigk R, Vyhlidal CA, et al. Glucocorticoid genes and the developmental origins of asthma susceptibility and treatment response. Am J Respir Cell Mol Biol. 2015;52(5):543–53.
20. Baines KJ, Simpson JL, Wood LG, Scott RJ, Fibbens NL, Powell H, et al. Sputum gene expression signature of 6 biomarkers discriminates asthma inflammatory phenotypes. J Allergy Clin Immunol. 2014;133(4):997–1007.
21. Park HW, Dahlin A, Tse S, Duan QL, Schuemann B, Martinez FD, et al. Genetic predictors associated with improvement of asthma symptoms in response to inhaled corticosteroids. J Allergy Clin Immunol. 2014;133(3):664–9.e5.
22. Xiao C, Biagini Myers JM, Ji H, Metz K, Martin LJ, Lindsey M, et al. Vanin-1 expression and methylation discriminate pediatric asthma corticosteroid treatment response. J Allergy Clin Immunol. 2015;136(4):923–31.e3.
23. McGeachie MJ, Clemmer GL, Hayete B, Xing H, Runge K, Wu AC, et al. Systems biology and in vitro validation identifies family with sequence similarity 129 member A (FAM129A) as an asthma steroid response modulator. J Allergy Clin Immunol. 2018;142(5):1479–88.e12.
24. Green RH, Brightling CE, Woltmann G, Parker D, Wardlaw AJ, Pavord ID. Analysis of induced sputum in adults with asthma: identification of subgroup with isolated sputum neutrophilia and poor response to inhaled corticosteroids. Thorax. 2002;57(10):875–9.
25. Zeiger RS, Szefler SJ, Phillips BR, Schatz M, Martinez FD, Chinchilli VM, et al. Response profiles to fluticasone and montelukast in mild-to-moderate persistent childhood asthma. J Allergy Clin Immunol. 2006;117(1):45–52.
26. Barnes PJ. Corticosteroid resistance in patients with asthma and chronic obstructive pulmonary disease. J Allergy Clin Immunol. 2013;131(3):636–45.
27. Al Heialy S, Ramakrishnan RK, Hamid Q. Recent advances in the immunopathogenesis of severe asthma. J Allergy Clin Immunol. 2022;149(2):455–65.
28. Hew M, Bhavsar P, Torrego A, Meah S, Khorasani N, Barnes PJ, et al. Relative corticosteroid insensitivity of peripheral blood mononuclear cells in severe asthma. Am J Respir Crit Care Med. 2006;174(2):134–41.
29. Woodruff PG, Boushey HA, Dolganov GM, Barker CS, Yang YH, Donnelly S, et al. Genome-wide profiling identifies epithelial cell genes associated with asthma and with treatment response to corticosteroids. Proc Natl Acad Sci U S A. 2007;104(40):15858–63.
30. Irvin C, Zafar I, Good J, Rollins D, Christianson C, Gorska MM, et al. Increased frequency of dual-positive TH2/TH17 cells in bronchoalveolar lavage fluid characterizes a population of patients with severe asthma. J Allergy Clin Immunol. 2014;134(5):1175–86.e7.
31. Zhang X, Biagini Myers JM, Yadagiri VK, Ulm A, Chen X, Weirauch MT, et al. Nasal DNA methylation differentiates corticosteroid treatment response in pediatric asthma: a pilot study. PLoS One. 2017;12(10):e0186150.
32. Fayon M, Lacoste-Rodrigues A, Barat P, Helbling JC, Nacka F, Berger P, et al. Nasal airway epithelial cell IL-6 and FKBP51 gene expression and steroid sensitivity in asthmatic children. PLoS One. 2017;12(5):e0177051.
33. Clemmer GL, Wu AC, Rosner B, McGeachie MJ, Litonjua AA, Tantisira KG, et al. Measuring the corticosteroid responsiveness endophenotype in asthmatic patients. J Allergy Clin Immunol. 2015;136(2):274–81.e8.

34. Lazarus SC, Chinchilli VM, Rollings NJ, Boushey HA, Cherniack R, Craig TJ, et al. Smoking affects response to inhaled corticosteroids or leukotriene receptor antagonists in asthma. Am J Respir Crit Care Med. 2007;175(8):783–90.
35. Chaudhuri R, Livingston E, McMahon AD, Thomson L, Borland W, Thomson NC. Cigarette smoking impairs the therapeutic response to oral corticosteroids in chronic asthma. Am J Respir Crit Care Med. 2003;168(11):1308–11.
36. Sutherland ER, Goleva E, Strand M, Beuther DA, Leung DY. Body mass and glucocorticoid response in asthma. Am J Respir Crit Care Med. 2008;178(7):682–7.
37. Forno E, Lescher R, Strunk R, Weiss S, Fuhlbrigge A, Celedon JC, et al. Decreased response to inhaled steroids in overweight and obese asthmatic children. J Allergy Clin Immunol. 2011;127:741–9.
38. Wells KE, Cajigal S, Peterson EL, Ahmedani BK, Kumar R, Lanfear DE, et al. Assessing differences in inhaled corticosteroid response by self-reported race-ethnicity and genetic ancestry among asthmatic subjects. J Allergy Clin Immunol. 2016;137(5):1364–9.e2.
39. Hernandez-Pacheco N, Farzan N, Francis B, Karimi L, Repnik K, Vijverberg SJ, et al. Genome-wide association study of inhaled corticosteroid response in admixed children with asthma. Clin Exp Allergy. 2019;49(6):789–98.
40. Mosteller M, Hosking L, Murphy K, Shen J, Song K, Nelson M, et al. No evidence of large genetic effects on steroid response in asthma patients. J Allergy Clin Immunol. 2017;139(3):797–803.e7.
41. Qiu W, Guo F, Glass K, Yuan GC, Quackenbush J, Zhou X, et al. Differential connectivity of gene regulatory networks distinguishes corticosteroid response in asthma. J Allergy Clin Immunol. 2018;141(4):1250–8.
42. Sears MR, Taylor DR, Print CG, Lake DC, Li QQ, Flannery EM, et al. Regular inhaled beta-agonist treatment in bronchial asthma. Lancet. 1990;336(8728):1391–6.
43. Drazen JM, Israel E, Boushey HA, Chinchilli VM, Fahy JV, Fish JE, et al. Comparison of regularly scheduled with as-needed use of albuterol in mild asthma. Asthma Clinical Research Network. N Engl J Med. 1996;335(12):841–7.
44. Hancox RJ, Sears MR, Taylor DR. Polymorphism of the beta2-adrenoceptor and the response to long-term beta2-agonist therapy in asthma. Eur Respir J. 1998;11(3):589–93.
45. Israel E, Drazen JM, Liggett SB, Boushey HA, Cherniack RM, Chinchilli VM, et al. The effect of polymorphisms of the beta(2)-adrenergic receptor on the response to regular use of albuterol in asthma. Am J Respir Crit Care Med. 2000;162(1):75–80.
46. Israel E, Chinchilli VM, Ford JG, Boushey HA, Cherniack R, Craig TJ, et al. Use of regularly scheduled albuterol treatment in asthma: genotype-stratified, randomised, placebo-controlled cross-over trial. Lancet. 2004;364(9444):1505–12.
47. Bleecker ER, Yancey SW, Baitinger LA, Edwards LD, Klotsman M, Anderson WH, et al. Salmeterol response is not affected by beta2-adrenergic receptor genotype in subjects with persistent asthma. J Allergy Clin Immunol. 2006;118(4):809–16.
48. Bleecker ER, Postma DS, Lawrance RM, Meyers DA, Ambrose HJ, Goldman M. Effect of ADRB2 polymorphisms on response to longacting beta2-agonist therapy: a pharmacogenetic analysis of two randomised studies. Lancet. 2007;370(9605):2118–25.
49. Bleecker ER, Nelson HS, Kraft M, Corren J, Meyers DA, Yancey SW, et al. Beta2-receptor polymorphisms in patients receiving salmeterol with or without fluticasone propionate. Am J Respir Crit Care Med. 2010;181(7):676–87.
50. Wechsler ME, Kunselman SJ, Chinchilli VM, Bleecker E, Boushey HA, Calhoun WJ, et al. Effect of beta2-adrenergic receptor polymorphism on response to longacting beta2 agonist in asthma (LARGE trial): a genotype-stratified, randomised, placebo-controlled, crossover trial. Lancet. 2009;374(9703):1754–64.
51. Basu K, Palmer CN, Tavendale R, Lipworth BJ, Mukhopadhyay S. Adrenergic beta(2)-receptor genotype predisposes to exacerbations in steroid-treated asthmatic patients taking frequent albuterol or salmeterol. J Allergy Clin Immunol. 2009;124(6):1188–94.e3.
52. Zuurhout MJ, Vijverberg SJ, Raaijmakers JA, Koenderman L, Postma DS, Koppelman GH, et al. Arg16 ADRB2 genotype increases the risk of asthma exacerbation in chil-

dren with a reported use of long-acting beta2-agonists: results of the PACMAN cohort. Pharmacogenomics. 2013;14(16):1965–71.
53. Turner S, Francis B, Vijverberg S, Pino-Yanes M, Maitland-van der Zee AH, Basu K, et al. Childhood asthma exacerbations and the Arg16 beta2-receptor polymorphism: a meta-analysis stratified by treatment. J Allergy Clin Immunol. 2016;138(1):107–13.e5.
54. Lipworth BJ, Basu K, Donald HP, Tavendale R, Macgregor DF, Ogston SA, et al. Tailored second-line therapy in asthmatic children with the Arg(16) genotype. Clin Sci (Lond). 2013;124(8):521–8.
55. Ruffles T, Jones CJ, Palmer C, Turner S, Grigg J, Tavendale R, et al. Asthma prescribing according to Arg16Gly beta-2 genotype: a randomised trial in adolescents. Eur Respir J. 2021;58(2):2004107.
56. Ortega VE, Hawkins GA, Moore WC, Hastie AT, Ampleford EJ, Busse WW, et al. Effect of rare variants in ADRB2 on risk of severe exacerbations and symptom control during longacting beta agonist treatment in a multiethnic asthma population: a genetic study. Lancet Respir Med. 2014;2(3):204–13.
57. Litonjua AA, Lasky-Su J, Schneiter K, Tantisira KG, Lazarus R, Klanderman B, et al. ARG1 is a novel bronchodilator response gene: screening and replication in four asthma cohorts. Am J Respir Crit Care Med. 2008;178(7):688–94.
58. Duan QL, Gaume BR, Hawkins GA, Himes BE, Bleecker ER, Klanderman B, et al. Regulatory haplotypes in ARG1 are associated with altered bronchodilator response. Am J Respir Crit Care Med. 2011;183(4):449–54.
59. Israel E, Lasky-Su J, Markezich A, Damask A, Szefler SJ, Schuemann B, et al. Genome-wide association study of short-acting beta2-agonists. A novel genome-wide significant locus on chromosome 2 near ASB3. Am J Respir Crit Care Med. 2015;191(5):530–7.
60. Drake KA, Torgerson DG, Gignoux CR, Galanter JM, Roth LA, Huntsman S, et al. A genome-wide association study of bronchodilator response in Latinos implicates rare variants. J Allergy Clin Immunol. 2014;133(2):370–8.
61. Mak ACY, White MJ, Eckalbar WL, Szpiech ZA, Oh SS, Pino-Yanes M, et al. Whole-genome sequencing of pharmacogenetic drug response in racially diverse children with asthma. Am J Respir Crit Care Med. 2018;197(12):1552–64.
62. Drazen JM, Yandava CN, Dube L, Szczerback N, Hippensteel R, Pillari A, et al. Pharmacogenetic association between ALOX5 promoter genotype and the response to anti-asthma treatment. Nat Genet. 1999;22(2):168–70.
63. Lima JJ, Zhang S, Grant A, Shao L, Tantisira KG, Allayee H, et al. Influence of leukotriene pathway polymorphisms on response to montelukast in asthma. Am J Respir Crit Care Med. 2006;173(4):379–85.
64. Telleria JJ, Blanco-Quiros A, Varillas D, Armentia A, Fernandez-Carvajal I, Jesus Alonso M, et al. ALOX5 promoter genotype and response to montelukast in moderate persistent asthma. Respir Med. 2008;102(6):857–61.
65. Tantisira KG, Lima J, Sylvia J, Klanderman B, Weiss ST. 5-lipoxygenase pharmacogenetics in asthma: overlap with Cys-leukotriene receptor antagonist loci. Pharmacogenet Genomics. 2009;19(3):244–7.
66. Dahlin A, Litonjua A, Lima JJ, Tamari M, Kubo M, Irvin CG, et al. Genome-wide association study identifies novel pharmacogenomic loci for therapeutic response to montelukast in asthma. PLoS One. 2015;10(6):e0129385.
67. Maroteau C, Espuela-Ortiz A, Herrera-Luis E, Srinivasan S, Carr F, Tavendale R, et al. LTA4H rs2660845 association with montelukast response in early and late-onset asthma. PLoS One. 2021;16(9):e0257396.
68. Edelman JM, Turpin JA, Bronsky EA, Grossman J, Kemp JP, Ghannam AF, et al. Oral montelukast compared with inhaled salmeterol to prevent exercise-induced bronchoconstriction. A randomized, double-blind trial. Exercise Study Group. Ann Intern Med. 2000;132(2):97–104.
69. Dahlen SE, Malmstrom K, Nizankowska E, Dahlen B, Kuna P, Kowalski M, et al. Improvement of aspirin-intolerant asthma by montelukast, a leukotriene antagonist: a randomized, double-blind, placebo-controlled trial. Am J Respir Crit Care Med. 2002;165(1):9–14.

70. Peters SP, Kunselman SJ, Icitovic N, Moore WC, Pascual R, Ameredes BT, et al. Tiotropium bromide step-up therapy for adults with uncontrolled asthma. N Engl J Med. 2010;363(18):1715–26.
71. Peters SP, Bleecker ER, Kunselman SJ, Icitovic N, Moore WC, Pascual R, et al. Predictors of response to tiotropium versus salmeterol in asthmatic adults. J Allergy Clin Immunol. 2013;132(5):1068–1074.e1.
72. Lazarus SC, Krishnan JA, King TS, Lang JE, Blake KV, Covar R, et al. Mometasone or tiotropium in mild asthma with a low sputum eosinophil level. N Engl J Med. 2019;380(21):2009–19.
73. Brusselle GG, Vanderstichele C, Jordens P, Deman R, Slabbynck H, Ringoet V, et al. Azithromycin for prevention of exacerbations in severe asthma (AZISAST): a multicentre randomised double-blind placebo-controlled trial. Thorax. 2013;68(4):322–9.
74. Gibson PG, Yang IA, Upham JW, Reynolds PN, Hodge S, James AL, et al. Effect of azithromycin on asthma exacerbations and quality of life in adults with persistent uncontrolled asthma (AMAZES): a randomised, double-blind, placebo-controlled trial. Lancet. 2017;390(10095):659–68.
75. Chung KF, Wenzel SE, Brozek JL, Bush A, Castro M, Sterk PJ, et al. International ERS/ATS guidelines on definition, evaluation and treatment of severe asthma. Eur Respir J. 2014;43(2):343–73.
76. Desai D, Newby C, Symon FA, Haldar P, Shah S, Gupta S, et al. Elevated sputum interleukin-5 and submucosal eosinophilia in obese individuals with severe asthma. Am J Respir Crit Care Med. 2013;188(6):657–63.
77. McGregor MC, Krings JG, Nair P, Castro M. Role of biologics in asthma. Am J Respir Crit Care Med. 2019;199(4):433–45.
78. Brusselle GG, Koppelman GH. Biologic therapies for severe asthma. N Engl J Med. 2022;386(2):157–71.
79. Khurana S, Bush A, Holguin F. Management of severe asthma: summary of the European Respiratory Society/American Thoracic Society task force report. Breathe (Sheff). 2020;16:200058. https://doi.org/10.1183/20734735.0058-2020.
80. Busse W, Corren J, Lanier BQ, McAlary M, Fowler-Taylor A, Cioppa GD, et al. Omalizumab, anti-IgE recombinant humanized monoclonal antibody, for the treatment of severe allergic asthma. J Allergy Clin Immunol. 2001;108(2):184–90.
81. Humbert M, Beasley R, Ayres J, Slavin R, Hebert J, Bousquet J, et al. Benefits of omalizumab as add-on therapy in patients with severe persistent asthma who are inadequately controlled despite best available therapy (GINA 2002 step 4 treatment): INNOVATE. Allergy. 2005;60(3):309–16.
82. Bousquet J, Rabe K, Humbert M, Chung KF, Berger W, Fox H, et al. Predicting and evaluating response to omalizumab in patients with severe allergic asthma. Respir Med. 2007;101(7):1483–92.
83. Hanania NA, Wenzel S, Rosen K, Hsieh HJ, Mosesova S, Choy DF, et al. Exploring the effects of omalizumab in allergic asthma: an analysis of biomarkers in the EXTRA study. Am J Respir Crit Care Med. 2013;187(8):804–11.
84. Casale TB, Luskin AT, Busse W, Zeiger RS, Trzaskoma B, Yang M, et al. Omalizumab effectiveness by biomarker status in patients with asthma: evidence from PROSPERO, a prospective real-world study. J Allergy Clin Immunol Pract. 2019;7(1):156–64.e1.
85. Haldar P, Brightling CE, Hargadon B, Gupta S, Monteiro W, Sousa A, et al. Mepolizumab and exacerbations of refractory eosinophilic asthma. N Engl J Med. 2009;360(10):973–84.
86. Nair P, Pizzichini MM, Kjarsgaard M, Inman MD, Efthimiadis A, Pizzichini E, et al. Mepolizumab for prednisone-dependent asthma with sputum eosinophilia. N Engl J Med. 2009;360(10):985–93.
87. Pavord ID, Korn S, Howarth P, Bleecker ER, Buhl R, Keene ON, et al. Mepolizumab for severe eosinophilic asthma (DREAM): a multicentre, double-blind, placebo-controlled trial. Lancet. 2012;380(9842):651–9.
88. Ortega HG, Liu MC, Pavord ID, Brusselle GG, FitzGerald JM, Chetta A, et al. Mepolizumab treatment in patients with severe eosinophilic asthma. N Engl J Med. 2014;371(13):1198–207.

89. Bel EH, Wenzel SE, Thompson PJ, Prazma CM, Keene ON, Yancey SW, et al. Oral glucocorticoid-sparing effect of mepolizumab in eosinophilic asthma. N Engl J Med. 2014;371(13):1189–97.
90. Castro M, Zangrilli J, Wechsler ME, Bateman ED, Brusselle GG, Bardin P, et al. Reslizumab for inadequately controlled asthma with elevated blood eosinophil counts: results from two multicentre, parallel, double-blind, randomised, placebo-controlled, phase 3 trials. Lancet Respir Med. 2015;3(5):355–66.
91. Bleecker ER, FitzGerald JM, Chanez P, Papi A, Weinstein SF, Barker P, et al. Efficacy and safety of benralizumab for patients with severe asthma uncontrolled with high-dosage inhaled corticosteroids and long-acting beta2-agonists (SIROCCO): a randomised, multicentre, placebo-controlled phase 3 trial. Lancet. 2016;388(10056):2115–27.
92. Nair P, Wenzel S, Rabe KF, Bourdin A, Lugogo NL, Kuna P, et al. Oral glucocorticoid-sparing effect of benralizumab in severe asthma. N Engl J Med. 2017;376(25):2448–58.
93. FitzGerald JM, Bleecker ER, Nair P, Korn S, Ohta K, Lommatzsch M, et al. Benralizumab, an anti-interleukin-5 receptor alpha monoclonal antibody, as add-on treatment for patients with severe, uncontrolled, eosinophilic asthma (CALIMA): a randomised, double-blind, placebo-controlled phase 3 trial. Lancet. 2016;388(10056):2128–41.
94. FitzGerald JM, Bleecker ER, Menzies-Gow A, Zangrilli JG, Hirsch I, Metcalfe P, et al. Predictors of enhanced response with benralizumab for patients with severe asthma: pooled analysis of the SIROCCO and CALIMA studies. Lancet Respir Med. 2018;6(1):51–64.
95. Nair P, O'Byrne PM. The interleukin-13 paradox in asthma: effective biology, ineffective biologicals. Eur Respir J. 2019;53(2):1802250.
96. Corren J, Lemanske RF, Hanania NA, Korenblat PE, Parsey MV, Arron JR, et al. Lebrikizumab treatment in adults with asthma. N Engl J Med. 2011;365(12):1088–98.
97. Hanania NA, Korenblat P, Chapman KR, Bateman ED, Kopecky P, Paggiaro P, et al. Efficacy and safety of lebrikizumab in patients with uncontrolled asthma (LAVOLTA I and LAVOLTA II): replicate, phase 3, randomised, double-blind, placebo-controlled trials. Lancet Respir Med. 2016;4(10):781–96.
98. Castro M, Corren J, Pavord ID, Maspero J, Wenzel S, Rabe KF, et al. Dupilumab efficacy and safety in moderate-to-severe uncontrolled asthma. N Engl J Med. 2018;378(26):2486–96.
99. Rabe KF, Nair P, Brusselle G, Maspero JF, Castro M, Sher L, et al. Efficacy and safety of dupilumab in glucocorticoid-dependent severe asthma. N Engl J Med. 2018;378(26):2475–85.
100. Wenzel S, Castro M, Corren J, Maspero J, Wang L, Zhang B, et al. Dupilumab efficacy and safety in adults with uncontrolled persistent asthma despite use of medium-to-high-dose inhaled corticosteroids plus a long-acting beta2 agonist: a randomised double-blind placebo-controlled pivotal phase 2b dose-ranging trial. Lancet. 2016;388(10039):31–44.
101. Corren J, Parnes JR, Wang L, Mo M, Roseti SL, Griffiths JM, et al. Tezepelumab in adults with uncontrolled asthma. N Engl J Med. 2017;377(10):936–46.
102. Menzies-Gow A, Corren J, Bourdin A, Chupp G, Israel E, Wechsler ME, et al. Tezepelumab in adults and adolescents with severe, uncontrolled asthma. N Engl J Med. 2021;384(19):1800–9.
103. Wechsler ME, Menzies-Gow A, Brightling CE, Kuna P, Korn S, Welte T, et al. Evaluation of the oral corticosteroid-sparing effect of tezepelumab in adults with oral corticosteroid-dependent asthma (SOURCE): a randomised, placebo-controlled, phase 3 study. Lancet Respir Med. 2022;10:650.

Imaging Regional Airway Involvement of Asthma: Heterogeneity in Ventilation, Mucus Plugs and Remodeling

Mark L. Schiebler, Nanae Tsuchiya, Andrew Hahn, Sean Fain, Loren Denlinger, Nizar Jarjour, and Eric A. Hoffman

Abstract

The imaging of asthma using chest computed tomography (CT) is well-established (Jarjour et al., Am J Respir Crit Care Med 185(4):356–62, 2012; Castro et al., J Allergy Clin Immunol 128:467–78, 2011). Moreover, recent advances in functional imaging of the lungs with advanced computer analysis of both CT and magnetic resonance images (MRI) of the lungs have begun to play a role in quantifying regional obstruction. Specifically, quantitative measurements of the airways for bronchial wall thickening, luminal narrowing and distortion, the amount of mucus plugging, parenchymal density, and ventilation defects that could contribute to the patient's disease course are instructive for the entire care team. In this chapter, we will review common imaging methods and findings that relate to the heterogeneity of asthma. This information can help to guide treatment decisions. We will discuss mucous plugging, quantitative assessment of bronchial wall thickening, delta lumen phenomenon, parenchymal low-density lung on CT, and ventilation defect percentage on MRI as metrics for assessing regional ventilatory dysfunction.

M. L. Schiebler (✉)
Cardiothoracic imaging, Department of Radiology, School of Medicine and Public Health, University of Wisconsin-Madison, Madison, WI, USA
e-mail: mschiebler@uwhealth.org

N. Tsuchiya
Department of Radiology, School of Medicine, University of the Ryukyus, Okinawa, Japan

A. Hahn
Department of Radiology, University of Iowa, Iowa City, IA, USA
e-mail: Andrew-hahn@uiowa.edu

A. R. Brasier, N. N. Jarjour (eds.), *Precision Approaches to Heterogeneity in Asthma*, Advances in Experimental Medicine and Biology 1426,
https://doi.org/10.1007/978-3-031-32259-4_8

Keywords

Computed tomography · Small airways disease · Hyper Polarized · Disease probability measure · Airway network flow model · Magnetic resonance imaging · Regional airflow · Mucus plugs · Remodeling

Abbreviations

COPD	chronic obstructive pulmonary disease
D Lumen	change in luminal area
DPM	disease probability measure
FAN	full-scale airway network flow model
FEV_1(%)	Forced expiratory Volume in one minute (percent predicted for age and sex)
FRC	Functional Reserve Capacity
fSAD	functional small airways disease HP- Hyper Polarized
HU	Hounsfield Units
MRI	magnetic resonance imaging
Pi10	luminal perimeter of 10 mm
PRM	parametric response mapping
TLC	Total Lung Capacity

Introduction

The imaging of asthma using chest computed tomography (CT) is well-established [1, 2]. Moreover, recent advances in functional imaging of the lungs with advanced computer analysis of both CT and magnetic resonance images (MRI) of the lungs have begun to play a role in quantifying regional obstruction. Specifically, quantitative measurements of the airways for bronchial wall thickening, luminal narrowing and distortion, the amount of mucus plugging, parenchymal density, and ventilation defects that could contribute to the patient's disease course are instructive for the

S. Fain
Department of Radiology, Biomedical Engineering, and Human Physiology, University of Iowa, Iowa City, IA, USA
e-mail: sean-fain@uiowa.edu

L. Denlinger · N. Jarjour
Department of Medicine, School of Medicine and Public Health, University of Wisconsin-Madison, Madison, WI, USA
e-mail: lcd@medicine.wisc.edu; nnj@medicine.wisc.edu

E. A. Hoffman
Departments of Radiology, Medicine and Biomedical Engineering, University of Iowa, Iowa City, IA, USA
e-mail: eric-hoffman@uiowa.edu

entire care team. In this chapter, we will review common imaging methods and findings that relate to the heterogeneity of asthma. This information can help to guide treatment decisions. We will discuss mucous plugging, quantitative assessment of bronchial wall thickening, delta lumen phenomenon, parenchymal low-density lung on CT, and ventilation defect percentage on MRI as metrics for assessing regional ventilatory dysfunction.

Imaging Methods

Pulmonary functional imaging has evolved to become a robust tool for lung evaluation that is both precise and repeatable. This is important for the testing of new asthma interventions. In a series of review articles, it has recently been shown how the many methods for imaging the lung have matured to their current state of the art [3–5].

CT and MRI Metrics for Asthma Quantification

Advanced computer analysis of chest CT [5] allows automated lung, lobe, and sublobar segmentation, as well as airway segmentation, labeling with measurement of the major airway branches out to two generations beyond the segmental bronchi (Fig. 1) [6]. In parallel, histogram analysis of the lung parenchymal density is used to measure low-density lung associated with air trapping (expiration) and hyperinflation (inspiration) [7, 8]. Key measures of airway remodeling [9] that are of interest in asthma include airway wall thickness, airway diameter and wall area, and lumen diameter and area, and the changes of these measures between expiratory and inspiratory lung inflation states due to mucus plugging and mechanical collapse (Fig. 2). Characterization of air trapping provides an index of functional small airways disease (fSAD) in areas of the lung beyond CT imaging's ability to directly measure airway structure. More recently, MRI measures of regional ventilation acquired pre-/post-bronchodilation using hyperpolarized gas MRI have shown promise to show regional patterns of ventilation heterogeneity and non-reversible airway inflammation associated with more severe Type 2 asthma [10]. To standardize these measures for subject size and airway segment, CT and MRI measures are typically normalized to reference measures and reported as percentages relative to airway size or lung volume.

Airway Measures

Measures of segmental airway branch geometry are typically performed at inspiratory lung volume, with a coached breath-hold at total lung capacity (TLC). These measures can be grouped generally into segmental measures and global measures, with segmental measures inclusive of measurements for each labeled branch of the airway tree. In low-dose studies using standardized protocols with high spatial

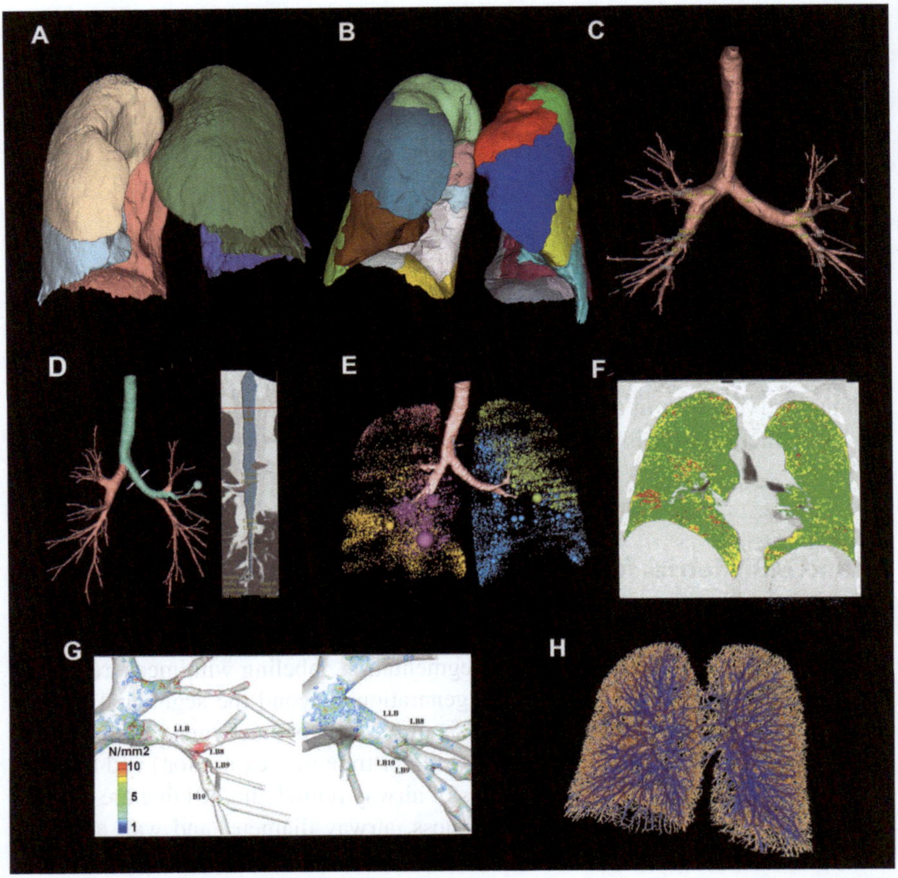

Fig. 1 Composite of methods used to assess the lungs via volumetric CT. (**a**) Identification of lungs and lobes. (**b**) Identification of sub-lobar segments. (**c**) Identification and automatic labeling of airway segments. (**d**) Measurement of airway segments perpendicular to the segment long axis. (**e**) Emphysema or Hyper-inflation assessed as voxels <−950 HU on full inspiration. (**f**) Air trapping assessed as voxels below −856 HU on full expiration, (**g**) DPM assessed functional small airways disease via image matching (PRM or DPM: see text), (**h**) Pulmonary vascular tree for assessment of vascular pruning or arterial expansion associated with downstream resistance, (**i**) Computational Fluid Dynamic modeled particle distribution based upon imaged airway tree structure

multi-slice (64 or greater) resolution chest CT scans [11], the airway tree can be labeled down to the subsegmental airways with accurately reported measures for ≥2 mm diameter airways [5]. Segmental airway measures of remodeling are typically airway wall thickness normalized to average airway diameter (%WT), and wall area percent normalized to total airway area (WA%).

A global index of airway remodeling is the empirical estimate of the square root of the wall area at a theoretical airway with an internal luminal *perimeter* of 10 mm (Pi10) – this corresponds to an internal luminal *diameter* of 1.78 mm [12]. Pi10 is estimated from a linear fit of the internal luminal perimeter vs. square root of the

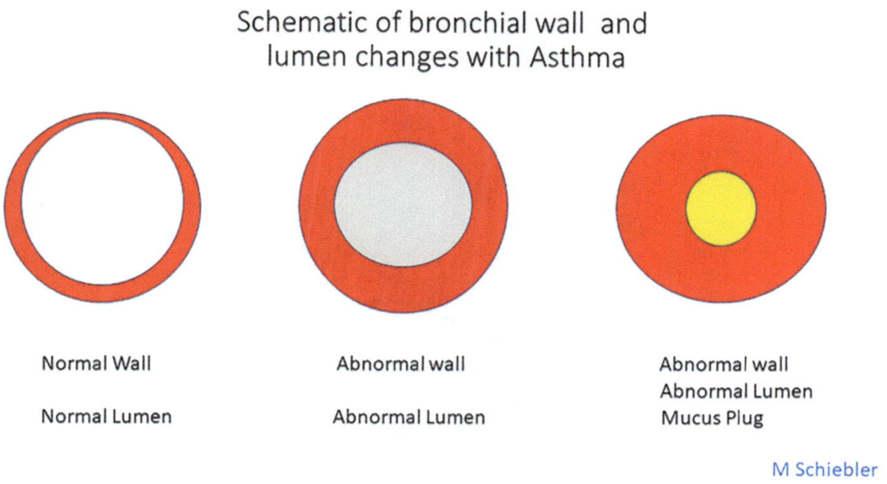

Fig. 2 Schematic of bronchial wall changes that can be seen by CT in asthma. From right to left, the first drawing shows the normal cross-section of a bronchus as seen from computed tomography with a thin wall and capacious lumen. The second drawing shows the situation of bronchial wall swelling which in turn reduces the internal diameter of the lumen. The third drawing shows the abnormally swollen bronchial wall and small lumen that is occluded by a mucus plug. Remember, these changes can be seen in many other diseases as well

wall area for airway segments throughout the lungs. The index, therefore, extrapolates to estimate this theoretical airway at the most distal branches visible on CT. More recently, another global measure using total airway count [13, 14] has been explored as a measure of overall airway narrowing and pruning in obstructive disease such as COPD and asthma and depends on the truncation of the airway segmentation mask due to lumen narrowing or mucus plugging, also as a measure of overall distal airway obstruction/loss.

Several more computationally advanced mechanical measures relevant to airflow have also been explored. These include airway branch bifurcation angle, circularity, hydraulic diameter (Dh), and change in lumen area (Delta Lumen). Importantly, these measures have enabled additional computational modeling of airway collapse and air/fluid dynamics that have identified clusters of individuals based only on imaging metrics that overlap with unique clinical presentations [15, 16]. These image-only clusters have been modeled to show different patterns of inhaled particle deposition which in turn may provide estimates regarding susceptibility to environmental pollutants or inhaled drug delivery [15].

Parenchymal Density Measures

The most common parenchymal density measures use histogram analysis to identify thresholds that depend on global lung density at expiratory lung volume, achieved via a coached breath-hold to either functional residual capacity (FRC) or

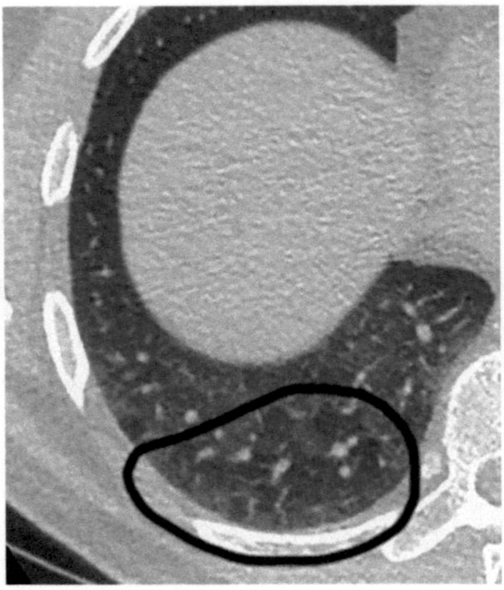

Fig. 3 Air trapping is a common feature of asthma and reflects small airways disease. Air can get in slowly to the terminal bronchi and alveolar sacs by diffusion but is not able to leave very quickly. At full expiration air trapping shows as areas of incomplete lung collapse that are of lower Hounsfield unit density than the surrounding normal lung. In general, these areas of air trapping correspond to ventilation defects seen on HP Nobel gas MRI ventilation imaging

residual volume (RV). RV is most appropriate for the assessment of air trapping (Fig. 3). This matches what is done in a pulmonary function laboratory. In the past, FRC was used when CT imaging of the full lung required a breath hold of 20 seconds or more. The current scanner design allows for imaging as fast as under one second with low radiation dose and high spatial resolution [17]. The measurements of low attenuation are highly dependent on lung inflation, so this is a key consideration for protocol design. A commonly used threshold for low-density lung indicative of air trapping is the low attenuation area below −856 Hounsfield Units (HU), while for emphysema, the threshold −950 HU at full inspiration is commonly used. The −856 HU threshold, when normalized to water density at 0 HU, translates to:

$$\rho = \frac{\rho_w\left(\left[\mathrm{HU}\right]+1000\right)}{1000} = \frac{\rho_w\left(-856+1000\right)}{1000} = 0.144\,\mathrm{g}/\mathrm{cm}^3.$$

However, a key limitation of these histogram threshold methods for lung parenchymal density quantification is their strong dependence on noise, and, thus, variations in CT dose, patient size, and CT scanner make and model [18]. Relative measures for longitudinal studies can be supported if the same, or similar, scanner make, and model are used [11, 18].

Moreover, the combination of emphysema and air trapping is an important consideration. Specifically, air trapping is partially reversible at expiration, while emphysema is a fixed area of low attenuation below −950 HU. These two types of lung disease can coexist with one another. In fact, this is more the rule than the exception. Several more advanced techniques have been introduced, in part, to overcome these limitations of histogram-based measures. The parametric response mapping (PRM) method [19] uses deformable registration between lung inspiratory and expiratory lung volumes to isolate low-density lung at inspiratory (i.e., emphysema) from low-density lung at expiratory lung volumes (i.e., air trapping renamed "functional small airways disease (fSAD)"). The fSAD measure still depends on the −856 HU threshold for defining air trapping but has the advantage of identifying air-trapped lung regions that are not associated with parenchymal damage due to emphysema. One can then follow a region with fSAD longitudinally to evaluate the consequences of small airway involvement over time. Variations on this approach use the Jacobian deformation (local lung volume changes) and changes in lung density between inspiratory and expiratory lung volumes to locally estimate ventilation and thus delineate air-trapped regions without relying on a fixed threshold [20, 21]. This is potentially more robust to variability in CT dose and scanner make and model and has been termed the *disease probability measure* (DPM).

MRI Methods for Asthma

Hyperpolarized gas MRI using either ^{3}He or ^{129}Xe [(noble gas isotopes has been shown to be of value for the measurement of regional ventilation in asthma (Figs. 4 and 5) [22]. A central advantage of hyperpolarized gas MRI is its ability to directly image gas distribution within the air spaces of the lungs [23]. The method typically uses a single inhalation of a hyperpolarized gas volume that is scaled to a fraction of the subject's predicted lung volume [24]. The method is safe, does not require ionizing radiation, and the gas rapidly washes out of the lungs (Table 1). Thus, the method can be performed before and after interventions

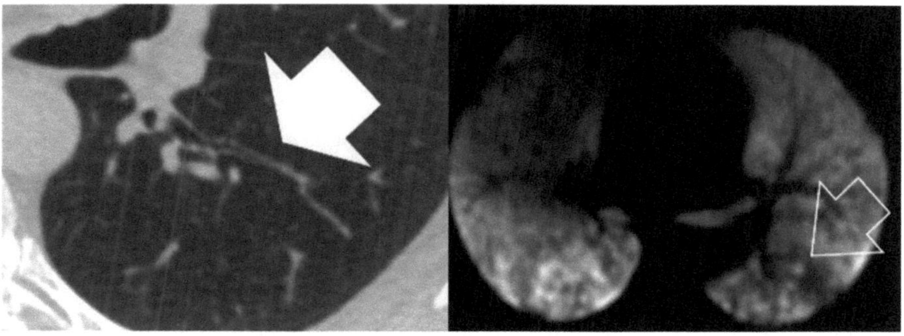

Fig. 4 Complementary imaging of the regional heterogeneity of asthmatic bronchitis with chest CT and HP Nobel Gas ventilation imaging from an asthmatic participant. From left to right, the first figure shows a long mucus plug cast/long segment bronchial lumen occlusion in the left lower lobe lateral segmental bronchus (arrow). The second figure, to the right, is the HP 3He ventilation image showing the downstream ventilation defect (open arrow). To obtain the ventilation defect percentage, the entire lung volume is used as the denominator and the total of the ventilation defect volume is summed to yield the numerator. The VDP is simply derived from the following
VDP = (total volume of defects/ total lung volume) × 100%

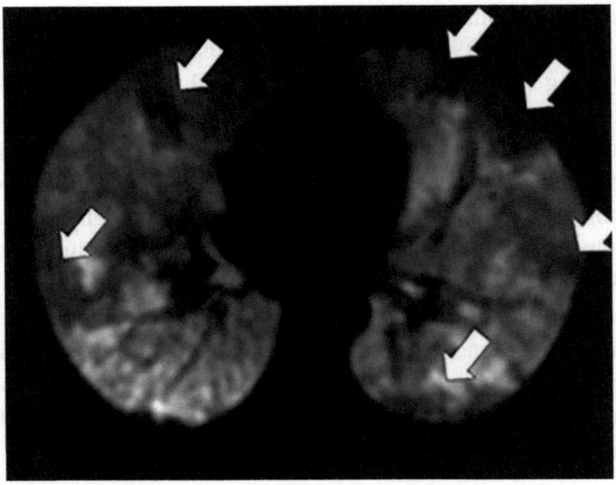

Fig. 5 Imaging the regional extent of asthmatic bronchitis with hyperpolarized Nobel gas MRI Ventilation imaging. This asthma subject shows multiple ventilation defects on HP Helium 3 ventilation imaging Gas Imaging. The bright signal intensity is from the HP gas within the alveolar sacs and bronchial structures. Note that the heart and pulmonary vasculature is black (low signal intensity) as no gas is within those

such as exercise [25] and bronchodilation [10]. Of particular interest regarding ^{129}Xe is its ability to characterize gas transfer across the alveolar walls. The signal changes as the xenon moves from the air to the alveolar wall and into the blood [26]. Thus, imaging allows one to now move from air trapping to ventilation defects down to the physiologic dysfunction at the site of oxygen delivery and carbon dioxide elimination, although this emerging method has not yet been systematically explored in asthma as of yet.

Mucous Plugging

While seemingly innocuous, mucus plugs are the primary killer in status asthmaticus. Dunican et al. showed that in autopsy studies in patients with fatal asthma, the airways are filled with small mucus plugs [31]. There are many other entities that can give rise to endobronchial debris that can mimic a mucus plug (Tables 2, 3, and 4).

Dunican et al. [33] showed in a landmark paper that mucous plugging in asthma could be enumerated on CT images in a reliable fashion. In their study of 146 participants with asthma, they found that mucus plugs were found in 58%, while for the control group of normal volunteers, only 5% showed one or more. They showed that in participants with a high mucous score, with four or more lung segments involved, there were important clinical findings as well: [1] 67% showed an FEV_1 of less than 60%. They also hypothesized that treatment of these plugs with agents that interfere

Table 1 Safety assessment study of hyperpolarized MRI for asthma

Publication	Tsuchiya[a] (2020)	Shukla[b] (2012)	Lutey[c] (2008)	Fain[d] (2007)
Hyperpolarized Gas	3He	129Xe	3He	3He
N	66 Subjects, 127 exams	33 subjects, 136 exams	100 subjects	60 exams
Age	8–18 years	20–79 years	Adult	–
Subject population	Asthma	Healthy Asthma COPD Cystic fibrosis	Healthy Mixed lung disease	Healthy Asthma COPD Sarcoidosis
Asthma population	100% (61/61)	21% (7/33)	7% (7/100)	+
MRI protocol	7–20 sec breathhold	12–14 sec breathhold	≤15 sec Breathhold	16–20 sec breathhold
Serious adverse event	None	None	None	None
Minor adverse event	2.3% (3/127) Dizziness Headache	1% (2/136) Light-headedness Headache	10% (10/100) Lightheadedness Mild headache Cough, Nausea Mild dyspnea Chest tightness Chest heaviness Sore shoulders Sweaty palms Epigastric discomfort	Less than 10% ECG ST wave depression Urinary tract infection Elevated CPK
SpO2	Nadir SpO2 ≤ 95% in 11% (14/127) of exams	No reduction from baseline in SpO2 of 5% and more	Nadir SpO2 < 90% in 26% (20/77) of subjects	–
Heart rate	Increase 3beats/min at post imaging assessment	Increase 1beats/min at post imaging assessment	Decrease 5beats/min at post imaging assessment	–

References:
[a][27]
[b][28]
[c][29]
[d][30]

with eosinophil peroxidase may be helpful in limiting crosslinking of the thiol groups of cysteine in these mucus plugs.

Figures 6, 7, and 8 help to demonstrate the imaging spectrum of mucus plugs and their mimics. Figure 9 shows typical mucous plugs. There is always a differential diagnosis for mucus plugs that need to be considered. A schematic (Fig. 10) shows the chest CT mucus plug enumeration method. The outer 2 cm of lung is excluded, and segmental involvement with at least one mucus plug is tallied. Please note this

Table 2 Developmental Endobronchial debris/lesions/masses

Etiology	Known Frequency	Bronchiectasis	Air trapping	Post obstructive consolidation	Centrilobular nodules
Cystic Fibrosis	1/2500	+++	++	++	+++
Diffuse Panbronchiolitis	1.1/10000		+++		+++
Primary Ciliary Dyskinesia (a.k.a. Kartagener's)	1/20000	+++ Lower lung	+/−	+/−	+/−
Bronchial Atresia	1.2/100000	+++	+++		

Table 3 Acquired Endobronchial debris/lesions/masses

Etiology	Relative Frequency	Bronchiectasis	Air trapping	Post obstructive consolidation	Centrilobular nodules	Tree In bud
Aspiration	++++	++	+/−	++	+++	++++
Infection	++++	+/−	+/−	++		
Mucus Plug/ Cast/mucoid pseudotumor	++++	+/−	+			
Long segment bronchial lumen occlusion	++	++	++++	+		
Poor Cough Reflex/ Debilitation	+++	+		+/−	+	
Foreign Body	+	+	+++	+		
Squamous Cell Ca	+		+	+++		
ABPA	+	++	+			
Carcinoid	+	+/−	+	++		
Plastic Bronchitis	Very rare	++	+	++		
Neutrophilic Allograft Dysfunction	Very rare	++	+	+	+++	++

Abbreviations: *ABPA* allergic bronchopulmonary aspergillosis

is not the total number of mucous plugs in all the bronchi. This method is simply a representative sample of each segment and determines whether or not at least one mucous plug is seen. If twenty plugs are seen in a segment, this will have the same score as one plug [33].

The reader should note that the current enumeration of mucus plugs is not detailed enough to show what portion of the segment will be involved, as it is simply

Table 4 Causes of Bronchial Wall thickening found on Chest CT

Etiology	Known Frequency (Relative Frequency)	Degree of Bronchial Wall Thickening	Bronchiectasis	Air Trapping	Emphysema
Particulates <2.5 microns	(++++)	+			
Viral Infections	(++++)	++		++	
COPD	(++++)	++	++	+++	++++
Pulmonary Edema (a.k.a. Cardiac Asthma)	(++++)	+	Never	Never	Never
Asthma	1/12	++	+/−	++	
ABPA	2/1000	+	+++	++	++
Cystic Fibrosis	1/2500	++++	++++	+	+
GPA	1.3/100000	++		+	
Inhalation injury					
Lymphangitic spread of malignancy	(+)	+			
Amyloidosis	(+/−)	+			

Abbreviations: Granulomatosis with Polyangiitis (a.k.a. Wegener's granulomatosis), # estimated to be 4.8 million persons globally [32]

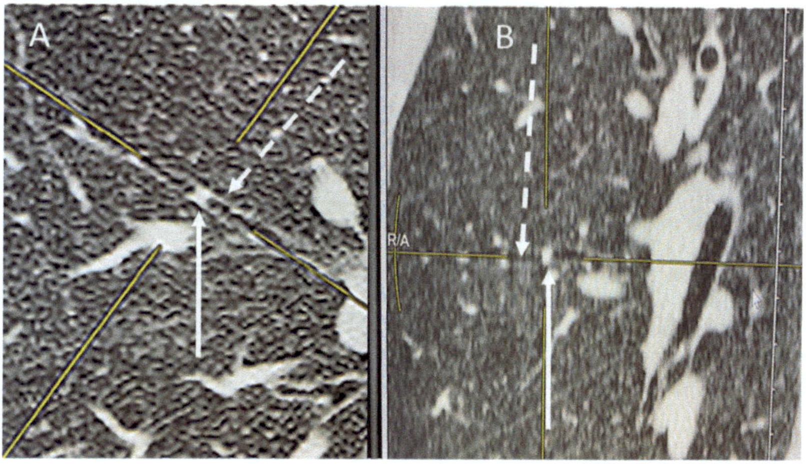

Fig. 6 Small well-defined mucus plug without associated bronchial wall thickening in an asthma participant: (**a**) Sagittal reformation with arrow showing the discrete mucus plug and the thin bronchial wall (dashed arrow), (**b**) Axial reformation showing the well-defined plug (arrow) and the thin bronchial wall (dashed arrow)

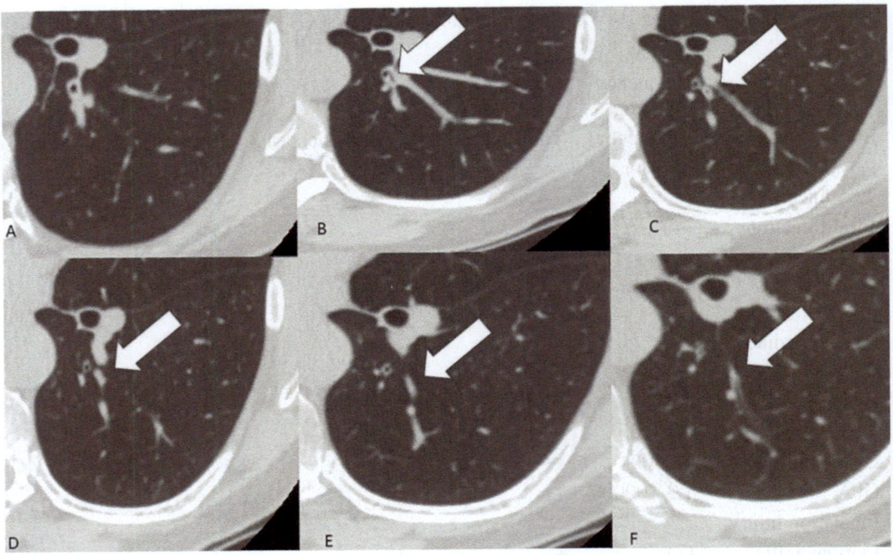

Fig. 7 Chest CT of long segment bronchial lumen occlusion from mucosal hypertrophy or possibly a long mucus plug/cast: Panel of six sequential chest computed tomography (CT) images (**a**–**f**) in the left lower lobe of an asthma participant showing the left lower lobe posterior segmental bronchus with a long mucus plug (arrow) that ends in the last image (**f**)

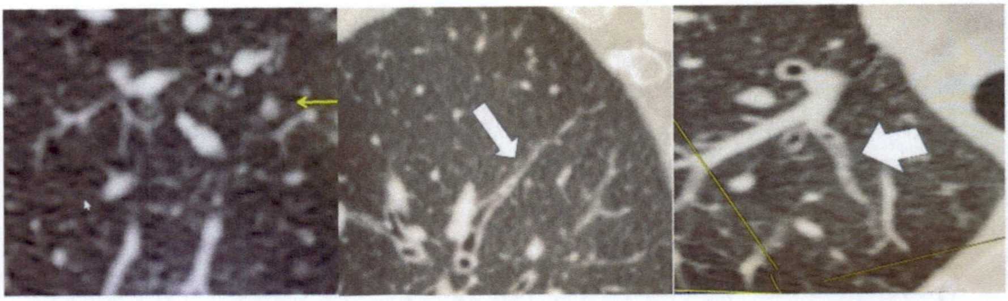

Fig. 8 Multiplanar chest CT reformations of a long segment bronchial lumen occlusion/ mucus plug cast (arrows) in the right upper lobe. From right to left, arrows show the round plug in the coronal plane (en face), sagittal plane (reformated to the centerline of the bronchus) and the axial plane (reformatted to the centerline of the bronchus). Multiplanar reformations can help to define the beginning and the end of the bronchial lumen occlusion/mucus plug cast

a yes/no determination of plug presence. This method does not take into consideration the plug size, volume, location (proximal or peripheral), or the total number of plugs within a segment (Table 5). For example, a single mucus plug in a peripheral bronchus will have the same mucus plug score as another patient with 20 plugs in all the proximal bronchi of the same lung segment. What is needed, of course, is an automated way to determine the number, volume, and location of these plugs. This will allow for a complete estimation of the number of possible ventilation defects

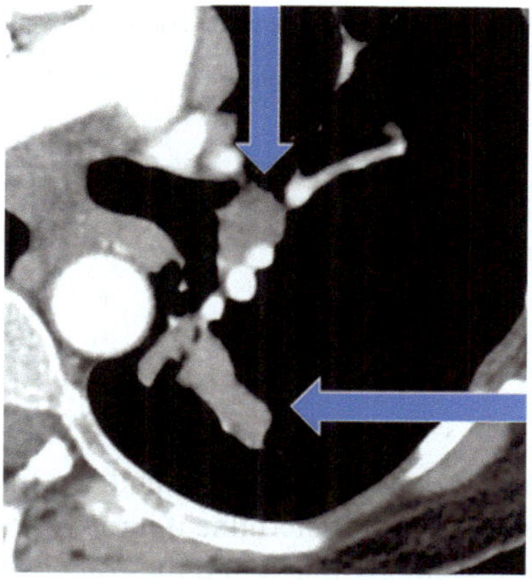

Fig. 9 Differential diagnosis of mucus plugs. Axial Chest CT angiogram with special attention to the left lower lobe showing multiple debris filled bronchi that have a relatively high attenuation value. These proved to be related to allergic bronchopulmonary aspergillosis (ABPA). The high attenuation value of the endobronchial debris is thought to be related to the iron content of this material. There are many causes of cellular bronchiolitis. Mucus plugs are not as common as simple aspiration

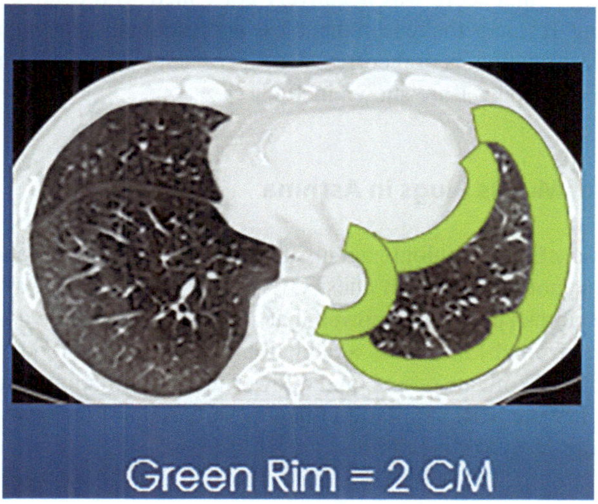

Fig. 10 Schematic drawing of how to score each lung segment for the presence of mucus plugs. The outer 2 cm (green margin) is excluded from the analysis in the axial, sagittal and coronal planes of each lung

and the associated changes in pulmonary function that are likely to be present as a result. There are congenital and acquired causes of endobronchial disease that also need to be considered in the full evaluation of any asthma patient (Tables 2 and 3). At the risk of being redundant, we must emphasize that there is a differential

Table 5 Imaging criteria for determining the presence of a mucus plug on Chest CT

CT Finding	Always present	Sometimes present	Absent
Sharp beginning	+/−	+/−	
Sharp end	+/−	+/−	
Normal bronchus before the plug		+	
Normal bronchus after the plug		+/−	+/−
Bronchial wall thickening	+/−	++/−	+/−
Debris in the airways		+/−	++/−
Bronchiectasis		+/−	++/−

diagnosis for endobronchial debris and lesions, and careful evaluation of each suspected mucus plug must be made to determine if it really fits the imaging criteria for a plug. Sometimes, it is just impossible to tell the composition of the matrix/debris in the bronchus.

Tang et al. showed that mucus plugs can be persistent over a 3-year period of observation [34]. In their analysis of 164 asthma participants, they found that the mucus plug lung segmental score was similar between the baseline exam and year 3. Those individuals with a plug at baseline were more likely to have plugs on the follow-up exam (Risk Ratio 2.8; $p < .001$). They also found that the change in mucus plug segmental score was negatively correlated with change in the $FEV_1\%$ and air trapping. It stands the reason that these persistent mucous plugs are a distinct phenotype of asthma that is oral corticosteroid-resistant. The commonly used criteria for mucus plugs on CT are shown in Table 5.

Treatment of Mucus Plugs in Asthma

With the introduction of biologic agents that limit cytokine production in the high mucus plug phenotype found on chest CT, there is now a new way to treat these asthmatics. Mcintosh et al. [35] showed that 28 days after a single dose of anti-IL-5 therapy (benraluzumab) there was a depletion of circulating blood eosinophils, an improvement in CT mucus scores, and a lowering of ventilation defect percentage found at HP (129)Xe MRI ventilation imaging in those participants with a baseline CT mucus score of five or greater.

Mucus Plugs in Chronic Obstructive Pulmonary Disease (COPD)

COPD is also associated with mucus plugs. Dunican showed that of 400 smoking participants, 57% had at least one mucus plug [36]. They also showed that COPD participants with high mucus plug scores had worse COPD self-assessment scores, more exacerbations, and a shorter 6-minute walk distance ($p < .001$).

Deep Learning Methods for Regional Ventilation

Perhaps the most important application of machine learning to chest CT imaging has been in dose reduction. Shiri et al. [37] have shown that an ultra-low-dose chest CT protocol with deep learning reconstruction showed similar image quality to full-dose exams with an 89% reduction in dose.

Segmentation of the airway tree to the 30th generation has also been achieved. Gleeson et al. 1 have shown through the full-scale airway network flow model (FAN) model that the number of small airways can be interpolated from routine high-resolution CT data, and the ensuing number of functional secondary lobules can be estimated [38]. Using the FAN model and CT Xenon ventilation imaging, Kim et al. [38] showed that the percentage of lobar ventilation computed (estimated) using the method FAN showed a high degree of correlation with (128) Xe enhanced dual-energy CT ventilation exams ($r = 0.7$, $P < .001$). This is quite exciting from a clinical perspective. When the post-processing methodology for FAN becomes routinely available, each subject with pulmonary insufficiency can be evaluated for their personal regional ventilation from a simple chest CT exam. From these data, longitudinal assessment of therapeutic response can be determined.

Capaldi et al. [39] have shown that a deep learning approach can transform the information from free breathing proton MRI exams using room air to generate ventilation maps that correlated with HP 3He MRI ventilation maps. In 114 paired examinations, proton MRI ventilation, and HP 3He ventilation MRI, their deep learning ventilation maps derived from the free-breathing exams had a mean Dice similarity coefficient of 0.9. The ventilation defect percentage for DL ventilation MRI was highly correlated with [3] He MRI ventilation defect percentage (rS = 0.83, $P < .001$). They also showed that their deep learning ventilation MRI (rS = −0.51) and [3] He MRI (rS = −0.61) ventilation defect percentages were correlated with FEV_1. With this powerful application of deep learning to a non-contrast MRI ventilation exam, it is now possible to get functional ventilation defect percentage from free breathing proton MRI [39].

Bronchial Wall Thickening

There are several causes of bronchial wall thickening. While asthma can certainly be a common cause of this observation on chest CT, the number of etiologies leading to changes in the bronchial wall and bronchiectasis are important to understand (Table 4). Bronchial wall thickening can be transient, as in the situation of an acute asthma attack, or progressive.

Interestingly, in the situation of anaphylaxis, there can be such severe acute bronchial wall swelling that respiratory arrest may quickly result unless treated aggressively with epinephrine. This situation most often occurs with drug or iodinated contrast reactions.

These sites of persistent obstruction can be identified on serial chest CT exams and are amenable to bronchial thermoplasty treatment. Chaudhuri et al. have reviewed the safety and effectiveness of bronchial thermoplasty after 10 years in subjects with severe asthma and show that this treatment is effective and has an excellent safety profile [40]. Follow-up of bronchial thermoplasty changes in asthma subjects can also be performed with CT, as shown by Zanon et al. [41].

Remodeling

This is a complicated topic and is not easily simplified into a few paragraphs of a book chapter. As the lung grows from infancy to adulthood, the degree of complexity is maximized with respect to the bronchi, terminal bronchi, and number of alveoli. A host of factors work to simplify this near-perfect set-up between the needs of the body for oxygenation and cardiopulmonary reserve.

First, there are common infections, or more recently COVID-19, which can serve to damage and permanently scar the lung. The amount of damage is, of course, variable, as is the capacity for recovery. As a rule, the lung does not grow new alveoli past young adulthood. With complete damage, the lung shrinks down to scar for the area that was involved, and the surrounding lung expands to fill the space left behind. As this occurs, the absolute density of the lung is reduced as more air is introduced into the remaining alveolar sacs and/or emphysematous areas.

Second, with aging, there is a loss of surplus lung and oxygen exchange capacity. Thus, the loss of gas exchange surface area is a known consequence of normal aging, irrespective of infections or pulmonary emboli.

Third, chronic bronchiectasis or bronchial obstruction will cause the functionality of the lung downstream to decrease. While there are pores of Kohn that allow for collateral ventilation, and the lung may seem to be expanded on CT exams performed at full inspiration, these areas of lung will demonstrate air trapping at end-expiration. They can also show ventilation defects on HP Nobel gas ventilation MRI [42, 43] and via regional Jacobians derived from Inspiration-Expiration matching of CT images [15, 44].

Some long segmental bronchial lumen occlusions have a tapered ending to them that suggests downstream remodeling/resorption of the previously non-ventilated lung. These changes that suggest permanent lung loss from chronic bronchial occlusion/mucus casts will require more longitudinal CT examinations to confirm and correlation with total airway counts and lung density measures.

The Delta Lumen mechanical phenomenon was first described by Shim et al. (Figs. 11 and 12) [16]. This refers to a measure of mechanical airway collapsibility first demonstrated in asthma. As described earlier, the Delta lumen phenomenon is defined by lumen area change at equivalent airway segments from the third through sixth generations. In severe asthma, Delta Lumen was shown to have a lesser change in the lumen area based on asthma outcomes of exacerbation leading to hospitalization and systemic corticosteroid use to control asthma, both phenomena expected to impact remodeling in the central airways. In asthma, more generally, some patients with mild/moderate asthma showed greater Delta Lumen at end-expiration, suggestive of tracheobronchomalacia. As might be expected, this is a nonobstructive cause

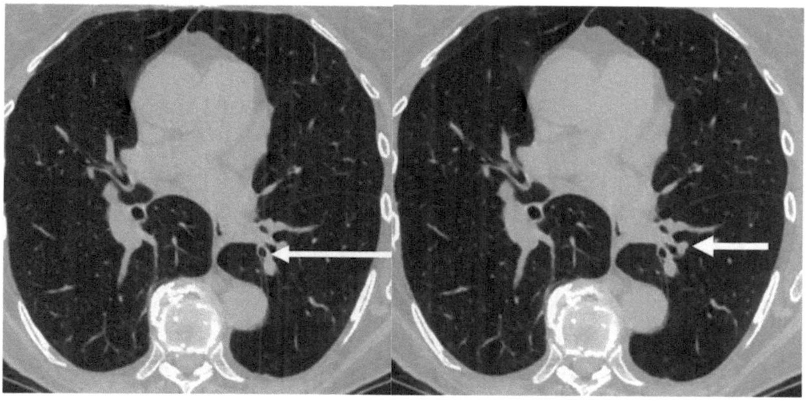

Fig. 11 The Delta Lumen Phenomenon refers to tracheobronchomalacia like changes that are seen in the 3–7th order bronchi in asthmatics. From left to right, the arrows show a normal appearing left lower lobe anterior segment, lateral segment, and posterior segmental bronchi (arrow) at full inspiration/total lung capacity (TLC) of a chest CT exam in a from an asthmatic participant. With full expiration, the lumens of these bronchi are severely narrowed (arrow). In our experience, the delta lumen phenomenon is more often seen in women than men. Body mass index also seems to be important for this phenomenon. The trachea and second order bronchi are not affected by this pathology in asthmatics. In other words, this is a different disease than tracheobronchomalacia

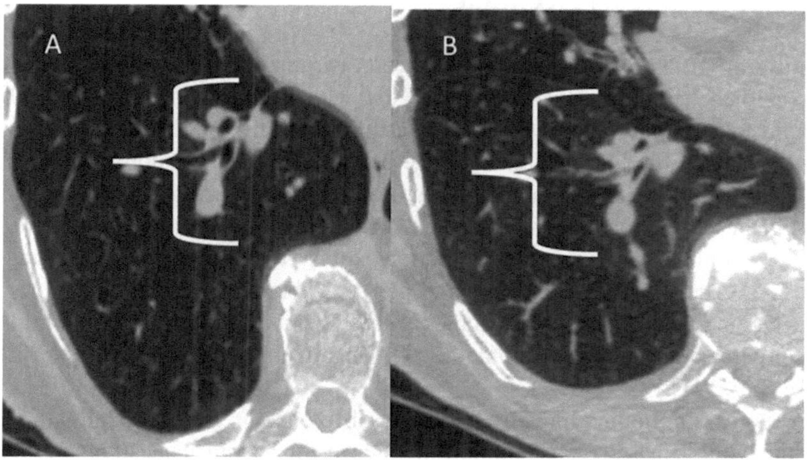

Fig. 12 Borderline changes of the delta lumen phenonomenon. From left to right, chest CT images from an asthmatic participant shows right lower lobe medial basal and posterior segmental bronchi that have mild bronchial wall thickening at TLC (bracket). On the second image to the right, there is partial, but not complete, collapse of these bronchi (bracket)

of change in the pulmonary function testing whereby FEV_1 would be reduced. These findings in general indicated a heterogeneous pattern of mechanical collapse specific to asthma suggestive of variable degrees of airway remodeling. Moreover, this change is more pronounced in individuals with high body mass index. There

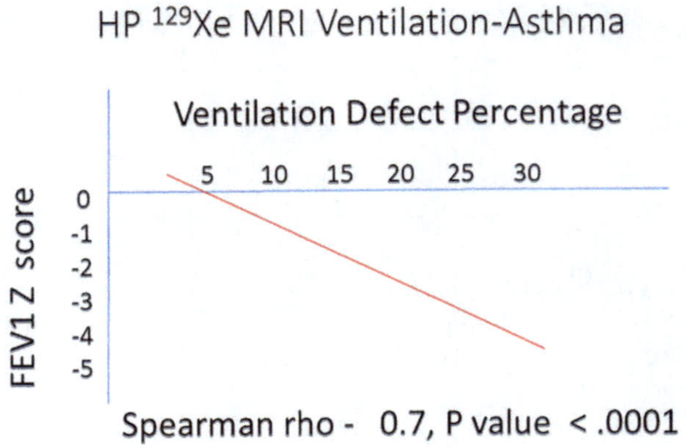

Fig. 13 Schematic graph modified from Mussel et al. [49]: Ventilation defect score found on Hyperpolarized 129 Xe ventilation MRI is negatively correlated with FEV_1 Z score [49]

may be a role for this dynamic measure of airway mechanics as an indicator of bronchial phenotypes of airway remodeling.

Ventilation Defect Percentage

The use of HP Noble gas imaging has been instrumental in helping to understand the regional heterogeneity of asthma. First with HP Helium and then with HP Xenon, many authors have shown that asthma is a diverse disease with variable involvement of the lung that can change between imaging sessions used to determine repeatability. The standard measure of ventilation using hyperpolarized gas MRI is the ventilation defect percentage (VDP), this measure is the volume of lung with poor or entirely obstructed ventilation expressed as a percent of total lung volume [45]. The VDP has been shown to be repeatable with minimal bias for single [46, 47] and multi-center [48] studies. Mussel et al. showed that the ventilation defect percentage found in asthmatics was negatively correlated with the FEV_1 Z score (Fig. 13) [49].

A typical minimum clinically important difference in VDP for asthma is 4% of the lung volume based on its relationship to the asthma control questionnaire (ACQ) [50] and prospective assessment of asthma exacerbation frequency [42]. Importantly, the VDP has been extended to a regional measure through combination with quantitative airway analysis on chest CT to evaluate bronchial thermoplasty [51] and mucus plugging [53]. Importantly, the HP 129Xe MRI method has been shown to help delineate phenotypes that may be more or less responsive to emerging biologic therapies in severe asthma [52] and to better characterize complex cases of severe asthma that are refractory to conventional and emerging therapies [49] Mummy et al. [53] have shown that the location of mucus plugs matches the ventilation defect location found on HP ventilation MRI imaging. Also, this same group found that the burden of mucus plugs increases the amount of ventilation defects.

Conclusion

Pulmonary imaging can show the regional heterogeneity of asthmatic bronchitis and the underlying pathology within the bronchi. This information can be used to sort the patient into the proper asthmatic phenotype for treatment. The high quality and reproducibility of imaging data from CT and HP MRI ventilation continue to be useful in the longitudinal assessment of treatment response for asthma.

Acknowledgments The authors wish to thank all the SARP investigators for their knowledge and expertise in helping to develop these imaging methods and the treatment paradigms this information has spawned. We also wish to extend our appreciation to the numerous participants in SARP that allowed our team to analyze their clinical and imaging data for scientific purposes.

Funding Sources NIH/NHLBI: R01 HL126771-05
NIH/NHLBI: RFA-HL-11-018

References

1. Jarjour NN, Erzurum SC, Bleecker ER, Calhoun WJ, Castro M, Comhair SA, et al. Severe asthma: lessons learned from the National Heart, Lung, and Blood Institute Severe Asthma Research Program. Am J Respir Crit Care Med. 2012;185(4):356–62.
2. Castro M, Fain SB, Hoffman EA, Gierada DS, Erzurum SC, Wenzel S, et al. Lung imaging in asthmatic patients: the picture is clearer. J Allergy Clin Immunol. 2011;128(3):467–78.
3. Gefter WB, Lee KS, Schiebler ML, Parraga G, Seo JB, Ohno Y, et al. Pulmonary functional imaging: part 2-state-of-the-art clinical applications and opportunities for improved patient care. Radiology. 2021;299(3):524–38.
4. Ohno Y, Seo JB, Parraga G, Lee KS, Gefter WB, Fain SB, et al. Pulmonary functional imaging: part 1-state-of-the-art technical and physiologic underpinnings. Radiology. 2021;299(3):508–23.
5. Hoffman EA. Origins of and lessons from quantitative functional X-ray computed tomography of the lung. Br J Radiol. 2022;95(1132):20211364.
6. Trivedi A, Hall C, Hoffman EA, Woods JC, Gierada DS, Castro M. Using imaging as a biomarker for asthma. J Allergy Clin Immunol. 2017;139(1):1–10.
7. Krings JG, Goss CW, Lew D, Samant M, McGregor MC, Boomer J, et al. Quantitative CT metrics are associated with longitudinal lung function decline and future asthma exacerbations: results from SARP-3. J Allergy Clin Immunol. 2021;148(3):752–62.
8. Busacker A, Newell JD Jr, Keefe T, Hoffman EA, Granroth JC, Castro M, et al. A multivariate analysis of risk factors for the air-trapping asthmatic phenotype as measured by quantitative CT analysis. Chest. 2009;135(1):48–56.
9. Aysola RS, Hoffman EA, Gierada D, Wenzel S, Cook-Granroth J, Tarsi J, et al. Airway remodeling measured by multidetector CT is increased in severe asthma and correlates with pathology. Chest. 2008;134(6):1183–91.
10. Svenningsen S, Eddy RL, Lim HF, Cox PG, Nair P, Parraga G. Sputum eosinophilia and magnetic resonance imaging ventilation heterogeneity in severe asthma. Am J Respir Crit Care Med. 2018;197(7):876–84.
11. Sieren JP, Newell JD Jr, Barr RG, Bleecker ER, Burnette N, Carretta EE, et al. SPIROMICS protocol for multicenter quantitative computed tomography to phenotype the lungs. Am J Respir Crit Care Med. 2016;194(7):794–806.
12. Kim V, Desai P, Newell JD, Make BJ, Washko GR, Silverman EK, et al. Airway wall thickness is increased in COPD patients with bronchodilator responsiveness. Respir Res. 2014;15:84.

13. Eddy RL, Svenningsen S, Kirby M, Knipping D, McCormack DG, Licskai C, et al. Is computed tomography airway count related to asthma severity and airway structure and function? Am J Respir Crit Care Med. 2020;201(8):923–33.
14. Kirby M, Smith BM, Tanabe N, Hogg JC, Coxson HO, Sin DD, et al. Computed tomography total airway count predicts progression to COPD in at-risk smokers. ERJ Open Res. 2021;7(4):00307.
15. Choi S, Hoffman EA, Wenzel SE, Castro M, Fain S, Jarjour N, et al. Quantitative computed tomographic imaging-based clustering differentiates asthmatic subgroups with distinctive clinical phenotypes. J Allergy Clin Immunol. 2017;140(3):690–700 e8.
16. Shim SS, Schiebler ML, Evans MD, Jarjour N, Sorkness RL, Denlinger LC, et al. Lumen area change (Delta Lumen) between inspiratory and expiratory multidetector computed tomography as a measure of severe outcomes in asthmatic patients. J Allergy Clin Immunol. 2018;142(6):1773–80 e9.
17. Hammond E, Sloan C, Newell JD Jr, Sieren JP, Saylor M, Vidal C, et al. Comparison of low- and ultralow-dose computed tomography protocols for quantitative lung and airway assessment. Med Phys. 2017;44(9):4747–57.
18. Chen-Mayer HH, Fuld MK, Hoppel B, Judy PF, Sieren JP, Guo J, et al. Standardizing CT lung density measure across scanner manufacturers. Med Phys. 2017;44(3):974–85.
19. Galban CJ, Han MK, Boes JL, Chughtai KA, Meyer CR, Johnson TD, et al. Computed tomography-based biomarker provides unique signature for diagnosis of COPD phenotypes and disease progression. Nat Med. 2012;18(11):1711–5.
20. Kirby M, Yin Y, Tschirren J, Tan WC, Leipsic J, Hague CJ, et al. A novel method of estimating small airway disease using inspiratory-to-expiratory computed tomography. Respiration. 2017;94(4):336–45.
21. Ostridge K, Gove K, Paas KHW, Burke H, Freeman A, Harden S, et al. Using novel computed tomography analysis to describe the contribution and distribution of emphysema and small airways disease in chronic obstructive pulmonary disease. Ann Am Thorac Soc. 2019;16(8):990–7.
22. Samee S, Altes T, Powers P, de Lange EE, Knight-Scott J, Rakes G, et al. Imaging the lungs in asthmatic patients by using hyperpolarized helium-3 magnetic resonance: assessment of response to methacholine and exercise challenge. J Allergy Clin Immunol. 2003;111(6):1205–11.
23. Kruger SJ, Nagle SK, Couch MJ, Ohno Y, Albert M, Fain SB. Functional imaging of the lungs with gas agents. J Magn Reson Imaging. 2016;43(2):295–315.
24. Niedbalski PJ, Hall CS, Castro M, Eddy RL, Rayment JH, Svenningsen S, et al. Protocols for multi-site trials using hyperpolarized (129) Xe MRI for imaging of ventilation, alveolar-airspace size, and gas exchange: a position paper from the (129) Xe MRI clinical trials consortium. Magn Reson Med. 2021;86(6):2966–86.
25. Kruger SJ, Niles DJ, Dardzinski B, Harman A, Jarjour NN, Ruddy M, et al. Hyperpolarized Helium-3 MRI of exercise-induced bronchoconstriction during challenge and therapy. J Magn Reson Imaging. 2014;39(5):1230–7.
26. Ebner L, Kammerman J, Driehuys B, Schiebler ML, Cadman RV, Fain SB. The role of hyperpolarized (129)xenon in MR imaging of pulmonary function. Eur J Radiol. 2017;86:343–52.
27. Tsuchiya N, Schiebler ML, Evans MD, Cadman RV, Sorkness RL, Lemanske RF Jr, Jackson DJ, Jarjour NN, Denlinger LC, Fain SB. Safety of repeated hyperpolarized helium 3 magnetic resonance imaging in pediatricasthma patients. Pediatr Radiol. 2020 May;50(5):646-655. PMID: 31980848.
28. Shukla Y, Wheatley A, Kirby M, Svenningsen S, Farag A, Santyr GE, et al. Hyperpolarized 129Xe magnetic resonance imaging: tolerability in healthy volunteers and subjects with pulmonary disease. Acad Radiol. 2012;19(8):941–51.
29. Lutey BA, Lefrak SS, Woods JC, Tanoli T, Quirk JD, Bashir A, et al. Hyperpolarized 3He MR imaging: physiologic monitoring observations and safety considerations in 100 consecutive subjects. Radiology. 2008;248(2):655–61.

30. Fain SB, Korosec FR, Holmes JH, O'Halloran R, Sorkness RL, Grist TM. Functional lung imaging using hyperpolarized gas MRI. J Magn Reson Imaging. 2007;25(5):910–23.
31. Dunican EM, Watchorn DC, Fahy JV. Autopsy and imaging studies of mucus in asthma. Lessons learned about disease mechanisms and the role of mucus in airflow obstruction. Ann Am Thorac Soc. 2018;15(Suppl 3):S184–S91.
32. Denning DW, Pleuvry A, Cole DC. Global burden of chronic pulmonary aspergillosis complicating sarcoidosis. Eur Respir J. 2013;41(3):621–6.
33. Dunican EM, Elicker BM, Gierada DS, Nagle SK, Schiebler ML, Newell JD, et al. Mucus plugs in patients with asthma linked to eosinophilia and airflow obstruction. J Clin Invest. 2018;128(3):997–1009.
34. Tang M, Elicker BM, Henry T, Gierada DS, Schiebler ML, Huang BK, et al. Mucus plugs persist in asthma, and changes in mucus plugs associate with changes in airflow over time. Am J Respir Crit Care Med. 2022;205(9):1036–45.
35. McIntosh MJ, Kooner HK, Eddy RL, Jeimy S, Licskai C, Mackenzie CA, et al. Asthma control, airway mucus, and (129)Xe MRI ventilation after a single Benralizumab dose. Chest. 2022;162(3):520–33.
36. Dunican EM, Elicker BM, Henry T, Gierada DS, Schiebler ML, Anderson W, et al. Mucus plugs and emphysema in the pathophysiology of airflow obstruction and hypoxemia in smokers. Am J Respir Crit Care Med. 2021;203(8):957–68.
37. Shiri I, Akhavanallaf A, Sanaat A, Salimi Y, Askari D, Mansouri Z, et al. Ultra-low-dose chest CT imaging of COVID-19 patients using a deep residual neural network. Eur Radiol. 2021;31(3):1420–31.
38. Kim M, Doganay O, Hwang HJ, Seo JB, Gleeson FV. Lobar ventilation in patients with COPD assessed with the full-scale airway network flow model and Xenon-enhanced dual-energy CT. Radiology. 2021;299(2):E257.
39. Capaldi DPI, Guo F, Xing L, Parraga G. Pulmonary ventilation maps generated with free-breathing proton MRI and a deep convolutional neural network. Radiology. 2021;298(2):427–38.
40. Chaudhuri R, Rubin A, Sumino K, Lapa ESJR, Niven R, Siddiqui S, et al. Safety and effectiveness of bronchial thermoplasty after 10 years in patients with persistent asthma (BT10+): a follow-up of three randomised controlled trials. Lancet Respir Med. 2021;9(5):457–66.
41. Zanon M, Strieder DL, Rubin AS, Watte G, Marchiori E, Cardoso PFG, et al. Use of MDCT to assess the results of bronchial thermoplasty. AJR Am J Roentgenol. 2017;209(4):752–6.
42. Mummy DG, Carey KJ, Evans MD, Denlinger LC, Schiebler ML, Sorkness RL, et al. Ventilation defects on hyperpolarized helium-3 MRI in asthma are predictive of 2-year exacerbation frequency. J Allergy Clin Immunol. 2020;146(4):831–9 e6.
43. Zha W, Kruger SJ, Cadman RV, Mummy DG, Evans MD, Nagle SK, et al. Regional heterogeneity of lobar ventilation in asthma using hyperpolarized Helium-3 MRI. Acad Radiol. 2018;25(2):169–78.
44. Choi S, Hoffman EA, Wenzel SE, Tawhai MH, Yin Y, Castro M, et al. Registration-based assessment of regional lung function via volumetric CT images of normal subjects vs. severe asthmatics. J Appl Physiol (1985). 2013;115(5):730–42.
45. Fain SB, Gonzalez-Fernandez G, Peterson ET, Evans MD, Sorkness RL, Jarjour NN, et al. Evaluation of structure-function relationships in asthma using multidetector CT and hyperpolarized He-3 MRI. Acad Radiol. 2008;15(6):753–62.
46. Zha W, Kruger SJ, Johnson KM, Cadman RV, Bell LC, Liu F, et al. Pulmonary ventilation imaging in asthma and cystic fibrosis using oxygen-enhanced 3D radial ultrashort echo time MRI. J Magn Reson Imaging. 2018;47(5):1287–97.
47. Svenningsen S, McIntosh M, Ouriadov A, Matheson AM, Konyer NB, Eddy RL, et al. Reproducibility of hyperpolarized (129)Xe MRI ventilation defect percent in severe asthma to evaluate clinical trial feasibility. Acad Radiol. 2021;28(6):817–26.

48. Niles DJ, Kruger SJ, Dardzinski BJ, Harman A, Jarjour NN, Ruddy M, et al. Exercise-induced bronchoconstriction: reproducibility of hyperpolarized 3He MR imaging. Radiology. 2013;266(2):618–25.
49. Mussell GT, Marshall H, Smith LJ, Biancardi AM, Hughes PJC, Capener DJ, et al. Xenon ventilation MRI in difficult asthma: initial experience in a clinical setting. ERJ Open Res. 2021;7(3):00785.
50. Eddy RL, Svenningsen S, McCormack DG, Parraga G. What is the minimal clinically important difference for helium-3 magnetic resonance imaging ventilation defects? Eur Respir J. 2018;51(6):1800324.
51. Thomen RP, Sheshadri A, Quirk JD, Kozlowski J, Ellison HD, Szczesniak RD, et al. Regional ventilation changes in severe asthma after bronchial thermoplasty with (3)He MR imaging and CT. Radiology. 2015;274(1):250–9.
52. Svenningsen S, Eddy RL, Kjarsgaard M, Parraga G, Nair P. Effects of anti-T2 biologic treatment on lung ventilation evaluated by MRI in adults with prednisone-dependent asthma. Chest. 2020;158(4):1350–60.
53. Mummy DG, Dunican EM, Carey KJ, Evans MD, Elicker BM, Newell JD Jr, et al. Mucus plugs in asthma at CT associated with regional ventilation defects at (3)He MRI. Radiology. 2022;303(1):184–90.

Brain–Airway Interactions in Asthma

William Busse, Christie Bartels, and Melissa Rosenkranz

Abstract

Asthma and brain interactions have long been appreciated and initially centered on increased anxiety and depression. Epidemiology studies have shown that early life stressors and situational disadvantages are risk factors for asthma. Conversely, the presence of asthma is a risk for mood and anxiety disorders, thus indicating a bidirectional effect between asthma and brain-related health. To substantiate asthma–brain interactions, validated instruments indicate and elucidate that communication likely exists between asthma and the brain. For example, provocation of an asthmatic response with an allergen challenge modulates how the brain responds to emotion-laden information. As detected by imaging studies, emotion-related brain activation is associated with generating airway inflammation. However, the specific mediators and processes mediating airway communication with the brain have yet to be established.

Systemic inflammation is also associated with asthma and can affect other organ systems such as the cardiovascular system and the brain. Epidemiology

W. Busse (✉)
Department of Medicine, Division of Allergy Pulmonary and Critical Care Medicine, University of Wisconsin School of Medicine and Public Health, Madison, WI, USA
e-mail: wwb@medicine.wisc.edu

C. Bartels
Department of Medicine, Division of Rheumatology, University of Wisconsin School of Medicine and Public Health, Madison, WI, USA
e-mail: cb4@medicine.wisc.edu

M. Rosenkranz
Department of Psychiatry, University of Wisconsin School of Medicine and Public Health, Madison, WI, USA
e-mail: melissa.rosenkranz@wisc.edu

A. R. Brasier, N. N. Jarjour (eds.), *Precision Approaches to Heterogeneity in Asthma*, Advances in Experimental Medicine and Biology 1426,
https://doi.org/10.1007/978-3-031-32259-4_9

studies have shown that asthma is a risk factor for dementia and Alzheimer's disease. In support of the importance of asthma as a risk factor for impaired cognitive function, imaging studies have shown changes to the white matter of the brain in asthma patients that resemble neuroinflammation changes seen in Alzheimer's disease and other neurodegenerative diseases. Therefore, bidirectional links between asthma and the brain exist with an important next research step to define asthma–brain interactions linked to neurodegeneration and dementia and explore whether treatments directed toward asthma-related inflammation can prevent the deleterious effects of asthma on brain health.

Keywords

Neurodegeneration · Alzheimers disease · Anxiety · Depression · Systemic inflammation · Asthma-brain interactions · Psychological disorders · Co-morbidity · Heterogeneity

Introduction

Asthma is a heterogeneous disease in terms of severity, age of onset, patterns of inflammation, and response to treatment [1, 2]. The heterogeneity of asthma is further complicated by the presence of comorbid diseases, which can affect disease severity, arise in relation to the presence of asthma, and/or be a consequence of the underlying pathophysiology of asthma. For example, chronic rhinosinusitis with nasal polyps (CRSwNP) is found in some patients with asthma, has reciprocal interactions with asthma severity, and the presence of aspirin sensitivity (AERD – Aspirin Exacerbated Respiratory Disease) is a risk factor for severe, life-threatening episodes of airflow obstruction [3]. The identification of abnormalities in arachidonic metabolism has provided insight into causative relationships between CRSwNP, AERD, and asthma severity [3].

Psychological disorders, i.e., anxiety and depression, and asthma, are highly prevalent and are important comorbid predictors of asthma expression, severity, and response to treatment [4]. Although anxiety and depression are recognized as important and relevant to many facets of asthma, our understanding and identification of how psychological factors contribute to the pathophysiology of asthma have not been fully elucidated, nor has the significance of their interrelationships been fully appreciated. Our chapter will begin by reviewing associations of anxiety and depression with asthma through epidemiological studies – a first step to recognize asthma–brain and brain-asthma interactions. We will then direct our discussion toward emerging insights toward bi-directional interactions between asthma and the brain, discuss new insights which confirm and more fully define the communications between the lung in asthma and the brain, and finally explore whether asthma is a risk factor for dementia, an ultimately important pathway of systemic inflammation in asthma and a potential relationship that affects the expression of cognition and unexplored target for the treatment of dementia.

What Is the Evidence for a Brain–Asthma Connection?

Interest and recognition that psychological factors contribute to asthma date back to Moses Maimondes when, in his "Treatise on Asthma," he suggested that behavioral changes may be a measure of cure [5]. Sir William Osler, in the late 1800s, concluded that many asthma cases had a strong neurotic element. These assertions reflected the concept that, at least in part, asthma was an archetypal psychosomatic disease, and, as such, asthma was felt to reflect a suppressed cry for the mother's help or attention, often referred to as "a mothered syndrome." These initial efforts focused solely on psychological factors as a cause or contributor to asthma, distracting efforts to explore interactions more carefully and thoroughly between asthma and the brain, and important bidirectional communications, which can affect function in both organ systems. Recognition and acceptance of bidirectional interactions between asthma and the brain also build on the concept of asthma as a systemic disease.

Initial efforts to link asthma and mental health disorders were based on case-controlled studies showing that these two conditions often co-exist. For example, studies found that rates of anxiety disorders were increased in asthma [6, 7]. Conversely, a greater prevalence of asthma has been found among patients hospitalized for psychiatric and post-traumatic stress disorders [8]. These observations suggest a bidirectionality in the relationship between asthma and psychological disorders. There exist a number of limitations to these observations, as pointed out by Goodwin et al. [9], including the absence or lack of precision in diagnosing both asthma and psychopathology.

To overcome these shortcomings, Goodwin et al. [9] identified a sample population of 13,222 German patients who ranged in age from 18 to 74 years of age. He used the following approach to both evaluate and validate asthma and psychopathology and then proceeded to examine the relationships between asthma and psychological disorders. After an initial self-reported questionnaire was completed, a study physician reviewed the responses to the question, "Do you have asthma?" Following this review, a standardized and specific examination was completed to confirm the presence of asthma and define its severity. In addition, a validated instrument was used to determine whether psychopathology was present. With this more precise approach to both conditions, current asthma was found to be significantly associated with more than double the odds for anxiety disorders, specifically phobia, panic disorders; severe asthma showed a four-fold odds for panic attacks. As the authors indicated, their studies were the first to investigate and substantiate the link between asthma and psychological disorders using greater precision for both underlying conditions: physician-diagnosed asthma and standardized validated measures of mental health disorders. However, although these observations paralleled confirmed earlier findings, though with less precise documentation, the use of standardized methods placed the interrelationships between asthma and the brain on firmer scientific grounds.

Although the cited results of Goodwin et al. [9] supported the relationship between asthma and anxiety disorders, there remained disparate findings and a lack of consensus in the existing literature. To address this lack of clarity and some skepticism, Ye et al. [10] conducted a meta-analysis of 19 studies that included 106,813 participants. Compared to control populations, the risks of co-existing anxiety symptoms and comorbid anxiety disorders were nearly double in patients with asthma vs. those without asthma. In a discussion of their results, the authors indicated that the mechanisms which underlie these relationships remain to be established, but their findings were foundational for research to examine the causes of these now more fully established relationships. Thus, the cited studies are examples that strongly support the concept that interactions exist between asthma and altered brain health via anxiety and depression.

What Insights Have Been Gained on the Origins of Asthma and Brain Interactions?

Pre- and post-natal maternal stress and depression have also been linked to the subsequent development of recurrent wheezing in young children, an interaction which also underscores that psychological factors are risks in the development of asthma. Using the URECA (Urban Environment and Childhood Asthma) birth cohort, it was possible to assess the relationships between maternal depression and anxiety on the later development of asthma in offspring at 10 years of age [11]. URECA represents a high-risk birth population for asthma and consists of 442 urban children from low-income neighborhoods. Furthermore, through longitudinal studies that were possible in URECA, phenotypes of asthma and allergic disease in this population could be prospectively identified and characterized based on the severity of wheezing (a surrogate for asthma) and state of atopy (reflects the level of allergic sensitization). In previous work, maternal perceived stress and depression in early life were found to be associated with wheezing at 1 year of age and recurrent wheezing at age 3 years, possibly indicating the development of asthma [12]. When an evaluation was made of the effects of maternal stress during the first 3 years of life and the presence of asthma in their offspring at 10 years of age, five well-defined phenotypes were identified. Maternal perceived stress and depression contributed to wheezing severity in children who were phenotypically characterized by moderate-wheeze and low-atopy (i.e., allergic sensitization) at age 10 years, particularly if these asthma symptoms had persisted through the first 3 years of life for the children. These findings point to the possibility that early life exposures to maternal stress and depression affect risks for wheezing early in life and have a lasting effect that manifests as asthma later in childhood.

It is not established whether these psychological effects on asthma occur pre- or postnatally. To extend observations on the potential inter-relationships between asthma and mental health disorders, Goodwin and co-workers [13, 14] evaluated data from the Western Australian Pregnancy Cohort, a cohort that consists of 2868 children who have been followed since birth. The investigators addressed two specific questions: (1) whether childhood asthma at age 5 years is associated with increased vulnerability to a range of psychological disorders, including anxiety and depression, up to age 17 years, and (2) whether asthma severity and/or its persistence influenced the relationship between asthma and psychopathology. This evaluation also allowed for the examination of the bidirectionality between psychological factors and asthma. At age 5 years, 18% of children in this Australian cohort had a diagnosis of asthma, with the majority of subjects having mild disease. The prevalence of anxiety, oppositional defiance, and behavioral problems were highest at age 5 years in children with asthma. Finally, the authors found that the greatest odds for psychological dysfunction occurred in those with more severe or persistent asthma (Fig. 1). Thus, the presence of asthma was a risk for the development of affective disorders, and conversely, psychopathology created a risk for asthma. Collectively, these two cited studies indicate that bidirectionality exists between asthma and the brain.

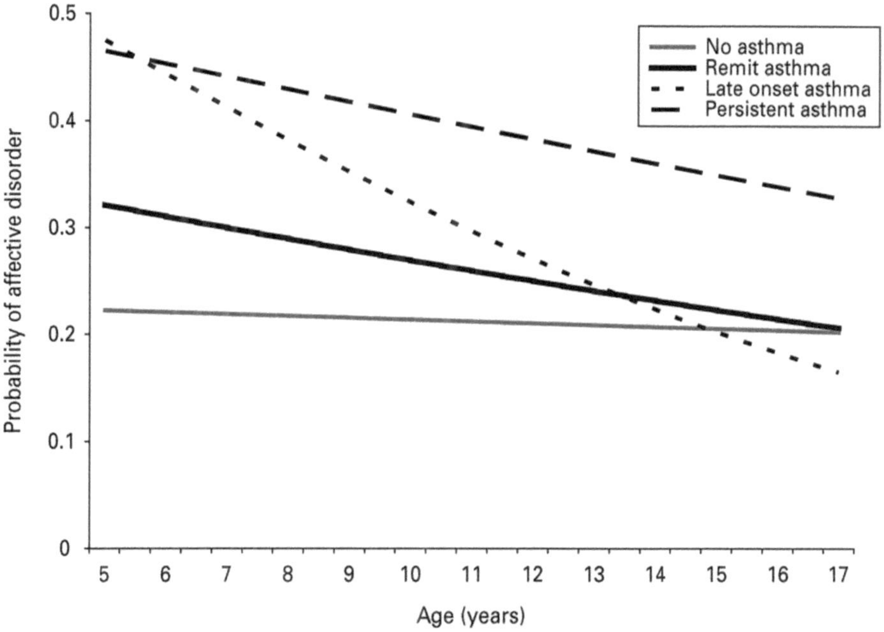

Fig. 1 The probability of affective problems at each age for asthma onset and persistence groups. (Reprinted with permission from Goodwin et al. [13])

What Evidence Has Addressed the Causal Directionality and Comorbid Association of Mental Disorders and Asthma in Adults?

To this point, our focus on understanding and identifying risk factors for disease development has largely centered on studies in children. As asthma begins around 6 years of age for most patients, this approach is reasonable, perhaps more feasible, as the development of asthma in children is better established, and, for many, asthma is a lifelong disease. There is, however, a sizable segment of the population in whom asthma develops later in life. There are distinctions between asthma which begin in later life vs. early life onset. Late-onset asthma has been defined as beginning after 18 years of age, is more prevalent in females, tends to be more severe, is less associated with allergic sensitization, and has frequent comorbid conditions, such as CRSwNP [3]. In addition, most late-onset asthma is characterized by type (T) 2 inflammation, which is characterized by elevated peripheral blood eosinophils [15]. Despite the importance of late-onset asthma for many patients, insights into etiological risk factors in that group are more limited because of the complexity of identifying at-risk subjects for study.

To explore risks for late-onset asthma, Alonso, and co-workers [16], evaluated the temporal relation of the contribution of existing psychopathology (i.e., prior to age 21 years), to asthma in patients whose disease developed after age 21 years. Their study sample was derived from 19 countries and consisted of 98,714 participants. The investigators found antecedent psychopathology, including anxiety, was linearly associated with the subsequent development of adult-onset asthma. Although there are shortcomings to a large international observational study, i.e., retrospective limitations to the confirmation of asthma by recall, these data indicate a potential risk of childhood psychopathology in the development of adult-onset asthma. This study serves to expand the risk relationship between mental health disorders on the development of asthma across the ages and suggests a more universal interaction.

Beyond temporality, another important aspect of the relationship of brain-lung communication to asthma is directionality. As reported by Chida et al. [17], these relationships are bidirectional. That is, asthma appears to raise risks for the development of psychopathology, and, conversely, psychopathology appears to affect asthma development and its resulting characteristics. Furthermore, data indicate that there is bidirectionality in this relationship: asthma-to-brain and brain-to-asthma. The next step is to consider how and under what conditions, these interactions occur and can be identified.

How Does Asthma Generate Signals to Affect Brain Activity?

A previous shortcoming of research between asthma and psychological disorders had been an absence of collaborative investigation between the asthma community and psychoneuroimmunology researchers. In 1993, the National Institute of Heart,

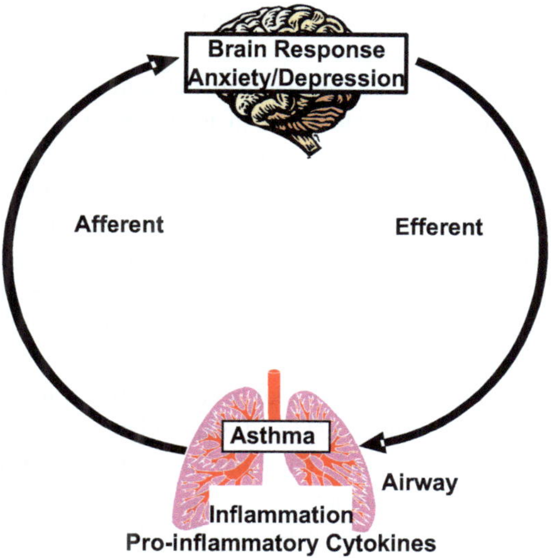

Fig. 2 The interactions between asthma and the brain are bidirectional with the brain affecting the airway and the airway in asthma affecting brain function

Lung, and Blood convened a Workshop on Stress and Asthma [18]. Although observations existed, and were generally appreciated, that stress and emotions worsen asthma, little data were available to identify the mechanisms which underlaid these relationships. A conclusion of this NIH Workshop was to re-examine the role of psychological stress in asthma, both as a contributor to acute airflow obstruction, and, perhaps more importantly, as an etiological component to asthma development, eventual severity, and characteristics, i.e. phenotype. Research in the ensuing decades has seen fruitful incorporation of studies to identify how signals from the airways in asthma interact with the brain, to identify the characteristics of these asthma-driven interactions, to detect the signals that participate in asthma–brain interactions, and finally, to begin to dissect the nature that regulates the bidirectionality (Fig. 2).

What Models of Study Have Provided Insight into Asthma-Stress Interactions?

Establishing the mechanisms of airway inflammation in response to risk factors of disease, such as allergens and respiratory viruses, has provided valuable insight to the pathophysiology of asthma and guidance for more effective treatment [19–22]. The next step was efforts to integrate stress into these airway models and responses. As noted from epidemiology studies, prolonged periods of stress or depression are risk factors for the onset of asthma [23–27]. Similarly, anxiety and depression are associated with shifts toward a Th2 profile of inflammation [28, 29] and are related to increased disease severity and diminished responsiveness to existing treatments, including corticosteroids (CS). It has yet, however, to be fully determined how emotional factors, like chronic stress, contribute to the pathophysiology of asthma, and, importantly, whether certain patients are most vulnerable to these psychological-associated factors.

Although the mechanisms through which stress and inflammation are mutually influential and interactive in asthma have yet to be fully delineated, the participation and contribution of inflammatory-promoting cytokines have been of particular interest. However, it remains to be identified which cytokines and how they are regulated in these interactions. Initial efforts to identify the primary mediators of stress-enhanced inflammation have focused on Th2 cytokines, and there is support for their role under certain circumstances [28, 29]; however, IL-1β/IL-17 pathways are also emerging as promising candidates. In both human and rodent models, psychological stress evokes increases in peripheral levels of IL-1β, and an IL-1β infusion induces anxiety-like behavior [30]. Similarly, IL-17 promotes depression-like behavior in mice [31] and is associated with depression in humans [32]. Members of the IL-17 family have also been linked to asthma severity and decreased responsiveness to treatment, including CS [33]. Further support for interactions between stress and asthma was gained with the report that the epigenetic and/or genetic variation in the gene encoding for adenylate-cyclase activating polypeptide 1 is linked to post-traumatic disorders in children and adults [34]. Extending these studies, Brehm et al. [35] examined whether the gene, ADCYAP1R1, which regulates stress, was associated with a decreased bronchodilator response. In high-stress children with the single nucleotide polymorphism (rs34548976) of ADCYAP1R1, there was a decreased bronchodilator response, likely related to the down-regulation of the beta-2 adrenergic receptor.

In the study of stress and asthma, it is important to recall that both stress and asthma are heterogeneous. Acute stress is associated with the mobilization of adrenal gland mediators, including catecholamines and corticosteroids, both of which can reduce inflammation. In contrast, chronic stress is proposed to have different effects on cell responsiveness, particularly in regard to corticosteroid generation and cellular responses to its anti-inflammatory regulation [36]. Chronic stress down-regulates the response to corticosteroids, and this down-regulation is associated with a shift toward a T2 inflammatory pattern with increases in IL-4, -5, and -13 cytokine expression. Thus, studies to evaluate stress in asthma need to consider and appreciate the differential importance of acute vs. chronic stress and realization that chronic stress is likely to have greater relevance in asthma.

Given this background, several relevant models have been used to assess the effects of stress on asthma. Final examinations in school have been a model that reflects sustained stress and have been associated with the provocation of asthma [37]. Inhaled allergen challenges can also provide insight into the development of airway inflammation and the associated development of airflow obstruction [38]. Using these two models (final examinations for stress and an inhaled allergen challenge for asthma and inflammation), Liu et al. [39] led investigations to evaluate the potential effects of stress from college final examinations on airway inflammatory responses to an allergen challenge in asthma patients.

To utilize this approach, college students were selected as subjects and studied on two separate occasions – (1) mid-semester, at a time of low stress and (2) immediately post-final examinations, a presumed time of high stress. In this study with a population of 20 college-aged students, levels of anxiety and depression significantly increased with final examinations. However, there was no increase in asthma symptoms or change in lung function in relation to the final examination, which may reflect that these subjects had mild asthma, were without symptoms at the time of enrollment, and did not have a need for maintenance medication to control their asthma.

The airway inflammatory responses, however, were influenced by the stress associated with the final examinations. During the immediate post-final examination study period, the inhaled allergen challenge led to a greater increase in sputum total white blood cells at 6- and 24-hours post-allergen challenge than had been seen in the mild stress study period. Although baseline sputum eosinophils were not increased during the post-final examination period, they were significantly elevated 6- and 24-hours post-challenge and tended to remain increased up to 7 days post-allergen challenge (Fig. 3). In addition, sputum measures of the eosinophil granule protein, eosinophil neurotoxin (EDN), were increased along with the increase in eosinophils. These relationships suggest that not only was there an increase in airway eosinophils, but that these cells recruited to the airway were activated as indicated by a release of EDN. Interestingly, blood eosinophils were also significantly increased prior to the allergen challenge during the post-final examination period. This finding suggests a heightening of risks for eosinophilic inflammation when an allergen challenge was introduced as the rise in blood eosinophils post-final examination may have provided an increased pool of available eosinophils for recruitment to the airways at the time of the allergen challenge. These findings also imply that there was a "priming" for the eosinophilic inflammation by the external or environmental event of stress from final examinations and fit with risks for increased airway inflammation as proposed by Simon Couillard et al. [40]. Finally, a significant correlation between increases in sputum eosinophils and fall in FEV_1 at 24 hours post-allergen challenge suggested that the increase in eosinophil recruitment to the airway (i.e., sputum) was the functional consequence of the heightened airway eosinophilia. That is, the more eosinophils recruited to the airway at the time of allergen challenge in association with the stress from final examinations, the greater the likelihood of airflow obstruction following the provocation of inflammation.

Stress can cause a shift toward T2 inflammation [36]. mRNA for IL-5 and interferon-γ was measured in the sputum as representatives for T2 and T1 inflammation patterns, respectively. When stress and an allergen challenge were performed during the final examination period, the increase in sputum eosinophils was associated with a shift toward a T2 pattern of inflammation (Fig. 4). Collectively, these data suggest that the stress experienced during final examinations caused a change in the regulation of the airway inflammatory response to a provocation with an allergen that either had preceded or resulted in an accentuation, or shift, towards T2-associated patterns of inflammation; this was reflected by greater eosinophil recruitment and IL-5 generation. Psychological stress did not necessarily provoke asthma but primed or increased the risk for a subject to experience a greater inflammatory response and risk for provocation of asthma. These increased risk

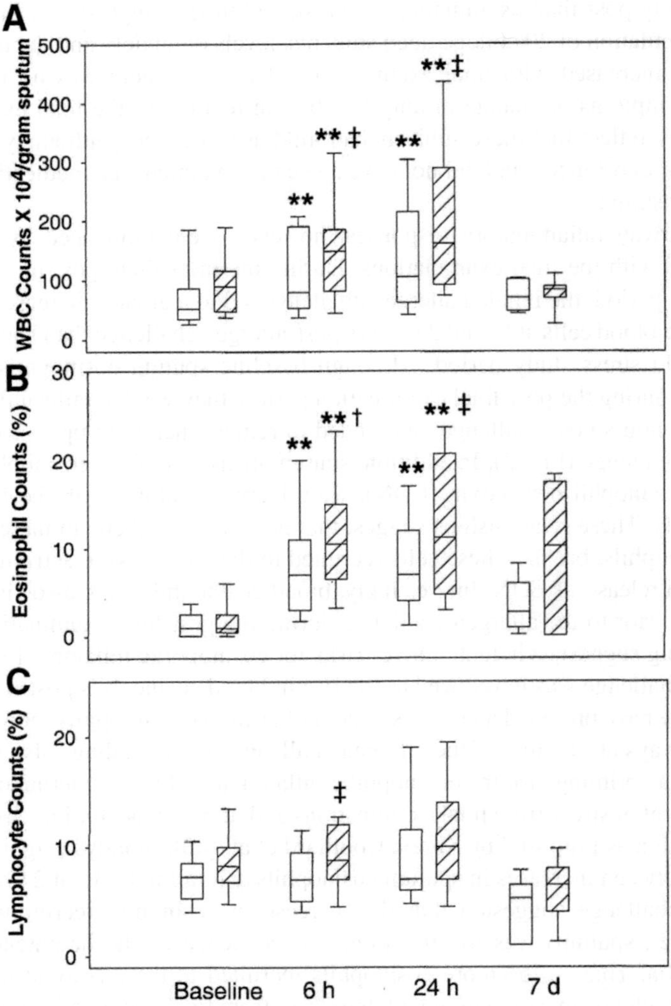

Fig. 3 Effect of final examination on sputum cellular profile following antigen challenge. The number of sputum white blood cells (**a**) and the percentages of sputum eosinophils (**b**) and lymphocytes (**c**) were determined in both stress (diagonal-hatched boxes) and low-stress phases (open boxes) at baseline and six and 24 hours and 7 days postantigen challenge. Boxes represent medians with 25 and 75% quartiles, vertical lines represent the 10th and 90th percentiles. $^{**}p < 0.01$, versus baseline; $\dagger p < 0.05$, $\ddagger p < 0.01$, stress versus low-stress. (Reprinted with permission from Liu et al. [39])

relationships could be extended to other triggers, like respiratory viruses, and explain an enhanced risk for exacerbations with stress from final examinations. Whether CNS-generated signals were also produced to affect airway inflammation was not established.

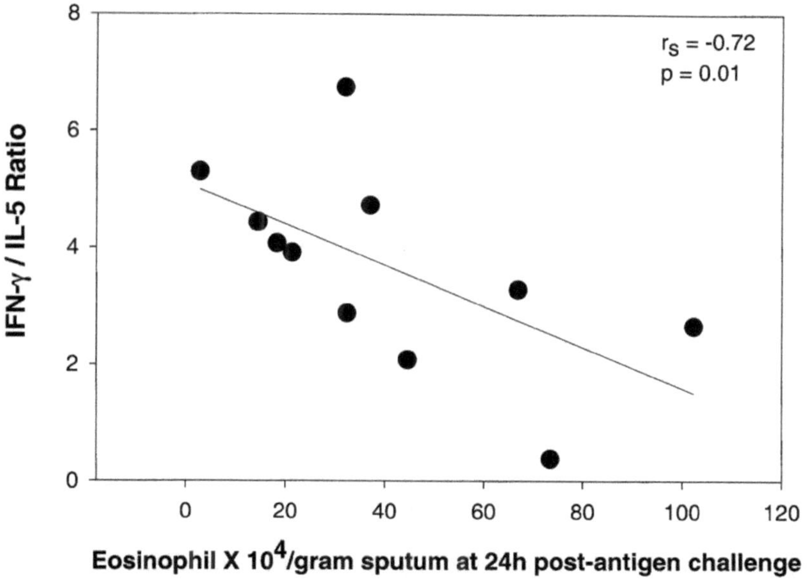

Fig. 4 Effect of final examination on the correlation of IFN-γ/IL-5 ratio and sputum eosinophils after 24 hours antigen challenge. During the stress phase, the ratio IFN-γ/IL-5 negatively correlated with the numbers of sputum eosinophil at 24 hours postantigen challenge (rs = −0.72, $p = 0.01$). (Reprinted with permission from Liu et al. [39])

These studies with allergen challenges to provoke an asthmatic inflammatory response provided pivotal data on potential links between the brain and risks for provocation of asthma by, or in association with, stress and implied that these interactions were bidirectional. Absent from the study by Liu et al. [39] was evidence that brain function was impacted by events in the airway and was related to features of asthma–eosinophilic inflammation. Neuroscientists have developed imaging methodologies to detect specific neural circuitry engagement with functional magnetic resonance imaging (fMRI). With this brain imaging technology now available, it is possible to identify specific regions and neural circuits of the brain through which emotion processing and regulation may interact with and potentially regulate physiological responses in the airway relative to asthma.

What Has Been Learned from fMRI Images of the Brain in Relationship to Asthma?

To determine the impact of airway inflammation on the neural response to emotional cues, a modified version of the Stroop Task was used, in which asthma-related, generally negative, and valence-neutral words were visually presented to participants during an fMRI scan [41]. The neural response to asthma-related words (e.g., cough, wheeze, etc.) is compared to the response to negative and neutral words to identify regions of the brain that may be differentially engaged by asthma-relevant processing. To assess relationships between the brain and inflammatory responses in airways of the lung, bronchial challenges were used with selected and different stimuli: saline (no inflammation or airflow obstruction), methacholine (airflow obstruction but no inflammation), and allergen (both inflammation and airflow obstruction), in conjunction with brain images with fMRI. Among six asthma patients who had an immediate and late phase response (LPR) to an allergen, significantly greater neural responses to asthma-related cues, relative to neutral cues, were found in the anterior cingulate cortex (ACC) and insula following allergen challenge compared to responses measured following saline or methacholine. Moreover, these potentiated fMRI responses were associated with the intensity of airflow obstruction (fall in FEV_1) at 6 hours post-allergen challenge, i.e., the late phase response, which also reflects the development of airway inflammation (Fig. 5). Furthermore, the intensity of the increased fMRI signal in these regions was significantly related to the inflammatory response as measured by the number of eosinophils recruited to the airway and measured in sputum collected 24 hours post-challenge, a time of maximal airway inflammation during allergen-provoked LPRs. As this increase in the fMRI signal contrasted to a lack of response to methacholine-provoked airflow obstruction, these results suggest that the allergen-provoked effect on brain responsivity was generated by signals related to airway inflammation. Interestingly, the increase in fMRI measurements occurred prior to the development of the late phase fall in FEV_1, and the presumed development of inflammation, as noted in the later appearance of sputum eosinophilia, suggesting that the brain response to airway provocation precedes the airway reaction and may reflect a brain-mediated response that regulates these inflammatory associated reactions [42]. Also of considerable interest were the specific regions of the brain that showed enhanced responses following the airway reaction to the allergen, the ACC and insula, both of which have been implicated in emotional and motivational aspects of sensation, affective components of pain, and coordinating the homeostatic consequences in response to physiological perturbations. Thus, for the first time, evidence was now available that detected asthma-to-brain communication and, importantly, the relationship of this activated brain response to disease-relevant characteristics of asthma-eosinophilic inflammation and airflow obstruction.

To further assess the directionality effects of stress upon airway inflammatory reactions and response of the brain, Rosenkranz et al. [43] conducted the following investigation to address the influence of stress, presumably representing brain-generated signals, on asthmatic inflammation. To accomplish this goal, there was an experimental induction of stress in the laboratory of asthma patients, with the subsequent response by the brain and airway then addressed. The objective of this study was to determine the mind-body interactions in response to stress, which brain regions are responsive to psychosocial stress, as determined by positron emission tomography (PET) scans, and to evaluate whether they may have a role in the subsequent regulation of provoked airway inflammation.

An important aspect of the methodology of this investigation was the use of the Trier Social Stress Test (TSST) and also the inclusion of subjects with low or high

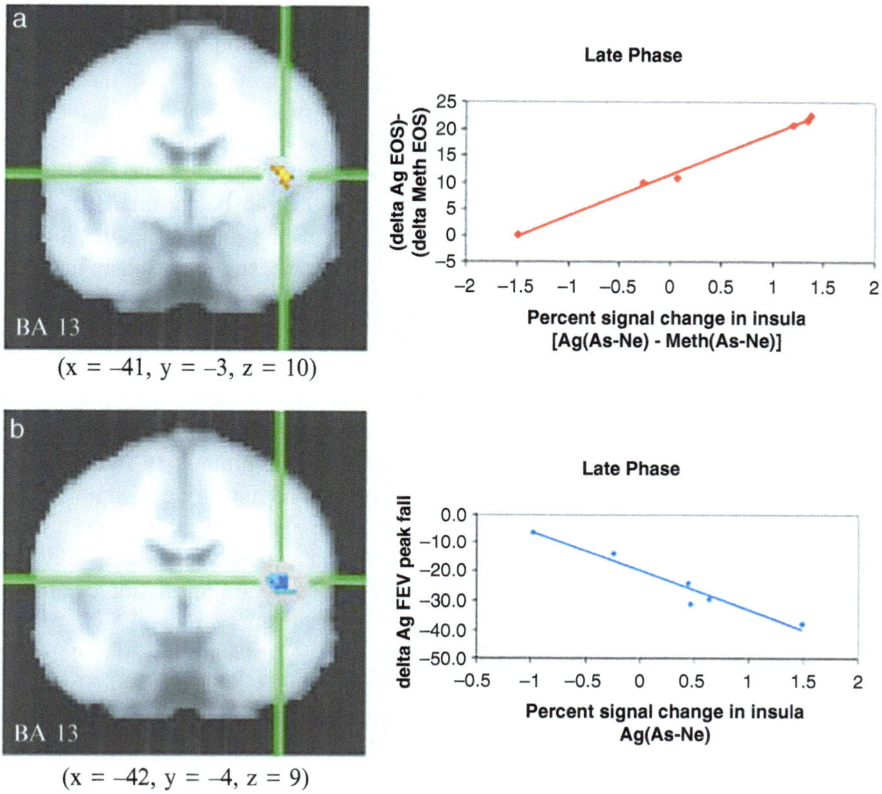

Fig. 5 Insula activity predicts peripheral measures of inflammatory potential and lung function. Percent signal change in the left insula in response to As compared with Ne words [As - Ne] and percentage of EOS during late-phase antigen relative to methacholine challenge [Ag - Meth] (**a**) ($r = 0.99$, $P < 0.001$) and peak fall FEV1 during late phase antigen [Ag] challenge (**b**) ($r = -0.97$, $P < 0.01$). Note that clusters in (**a**) and (**b**) were identified independently and are not identical, although they largely overlap. For analyses with FEV1, only values for the Ag challenge were used because one participant was missing data from the Meth challenge. (Reprinted with permission from Rosenkranz et al. [42])

chronic [44]. By selecting subjects with low and high stress, it would be possible to compare the effects of chronic stress (high stress) on the regulation of a provoked inflammatory response in the airway of asthma patients. For study, the enrolled subjects were challenged by a stressful TSST or a non-stressful control condition that otherwise matched the TSST. The background stress of the subjects was assessed by the UCLA Life Stress Interview (LSI), and based on these results, subjects were classified to high or low chronic stress in their lives in the previous 6 months. Importantly and informatively, salivary cortisol, which is released during stress, was obtained to assess the effects of high or low stress at baseline and then the cortisol response to an acute challenge, either control or TSST. The presence of airway inflammation was monitored by measurements of blood and sputum eosinophils and FeNO as well as in relationship to the TSST or control, and in both low and high-stress groups. PET scans were used to determine regions of the brain that were activated during these various s conditions.

In response to stress from the TSST, cortisol values increased in both the high and low chronic stress groups. However, despite the increase in cortisol to the stress TSST challenge, there was no suppression of blood eosinophil values at 24 hours post-allergen challenge in those asthma subjects with high background stress. In contrast, subjects with a low background stress level had the anticipated inverse relationship between eosinophil count and cortisol release, stress caused an increase in cortisol and a fall in blood eosinophils (Fig. 6). The lack of suppression of blood eosinophils by a rise in cortisol in the high-stress group suggests that chronic stress downregulated or desensitized the cortisol receptor from prolonged exposure to sustained elevation of cortisol and prevented its activation or ability from diminishing the eosinophil response to a stress challenge. Furthermore, regions of the brain that were engaged during TSST were the same regions that we previously showed were impacted by allergen challenge – including the insula and anterior cingulate cortex. Both brain regions play a role in assigning meaning and salience to stimuli from environmental, including psychological factors; consequently, it is possible that these regions of the brain may participate in responses to and regulation of airway inflammation.

Several conclusions can be drawn from these studies that are relevant to defining asthma–brain and brain–asthma interactions. First, intrinsic psychological patient characteristics contribute to the response to stress and the subsequent effect stress has on markers of inflammation. In subjects with underlying high stress, the anti-inflammatory actions of cortisol may be insufficient to normally regulate allergic inflammatory responses. Second, the regions of the brain engaged by psychosocial stress are compatible with ascribed and known functions, and may be involved in coordinating the regulation and intensity of an inflammatory response. Finally, in interpreting the effects of stress on airway inflammation in asthma, it is important to define the underlying levels of stress in at-risk subjects – low or high stress. These findings are evidence that the adverse effects of stress on asthma are more likely to manifest with ongoing chronic stress.

Over the past decade, there has been a heightened awareness that asthma is not one disease but characterized by multiple aspects of heterogeneity based, for example, upon the age of onset, disease severity, patterns of inflammation, and responsiveness to treatment, to name a few [1, 2]. To determine if asthma phenotypes also exist when evaluated in relationship to psychological characteristics and whether these heterogenous asthma characteristics may reflect neurophenotypes, the following study was undertaken [45]. Asthma subjects were identified and underwent an inhaled allergen challenge; the response to the inhaled allergen challenge was designated by (1) as an immediate and late phase response (LPR) or (2) no late phase reaction (NLPR). The LPR characterizes and reflects the development of inflammation and, in this sense, is compatible with or representative of airway events seen in

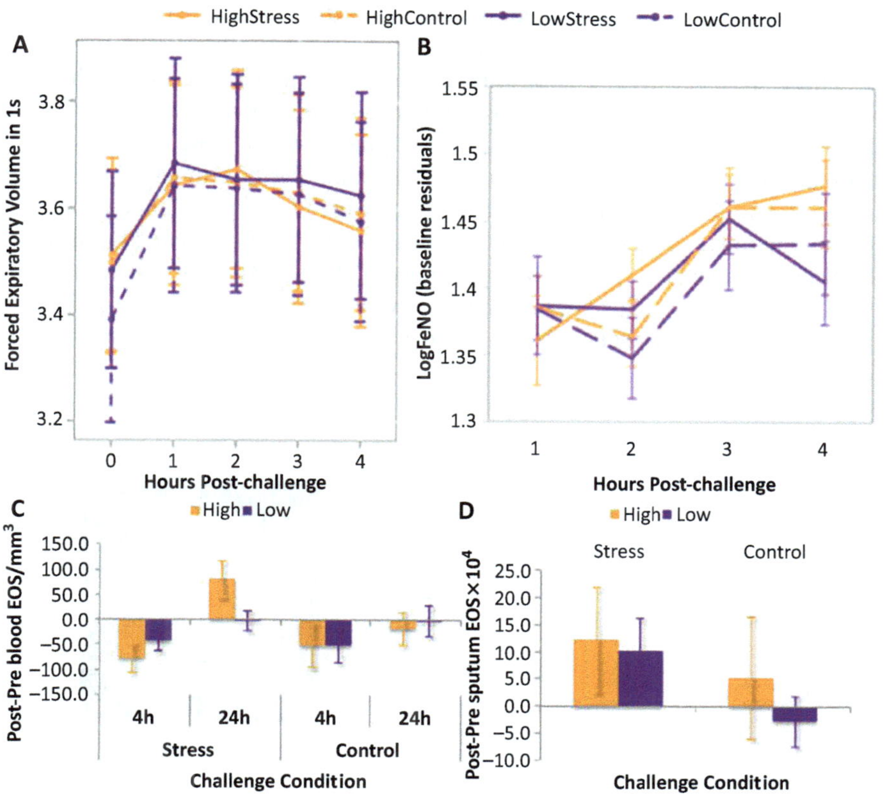

Fig. 6 Change in lung function and asthma-related inflammatory markers in response to stress and control tasks. Means displayed by group and challenge for (**a**) FEV1, (**b**) FeNO, (**c**) blood EOS, and (**d**) sputum EOS. Error bars represent standard error of the mean. (Reprinted with permission from Rosenkranz et al. [43])

clinical asthma. The LPR to the inhaled allergen is characterized by an immediate and late (4 to 6 hours post-allergen challenge) fall in FEV_1 and, during this LPR, an increase in sputum eosinophils to reflect the development of inflammation.

To investigate the possibility of an airway phenotypic response, i.e., LPR vs. NLPR, to an inhaled allergen relates to "neurophenotypes," the allergen challenges were conducted on all recruited asthma subjects. Prior to the allergen challenges, fMRI data were collected during the performance of the Stroop Task (as described above) and then repeated post-allergen challenge to determine the effects of the airway responses to the allergen on brain responses to asthma-related cues. In addition, sputum samples were obtained pre- (baseline inflammation) and post-allergen challenge to measure allergen-provoked inflammation. Therefore, the following characteristics of an allergen reaction were available to categorize the subjects: (1) asthma phenotype in terms of response to an allergen challenge – LPR vs. NLPR,

(2) airway inflammation – sputum eosinophils, and (3) brain responsivity, with a focus on the insula based on our prior work.

There were positive relationships between the increase in eosinophils and change in neural responsivity across all participants, where the increase in both brain activation and eosinophils was most prominent for those asthma subjects with an LPR (Fig. 7). These data suggest that a marker of asthma or possibly its severity, i.e., (LPR vs. NLPR) was associated with a more pronounced impact on brain function, and this differential response to inhaled allergen and brain activation may represent a specific neurophenotype. These experimental results provide evidence that the patterns of the airway response, both obstruction and inflammation, are associated with distinct changes in the neural processing of emotional information.

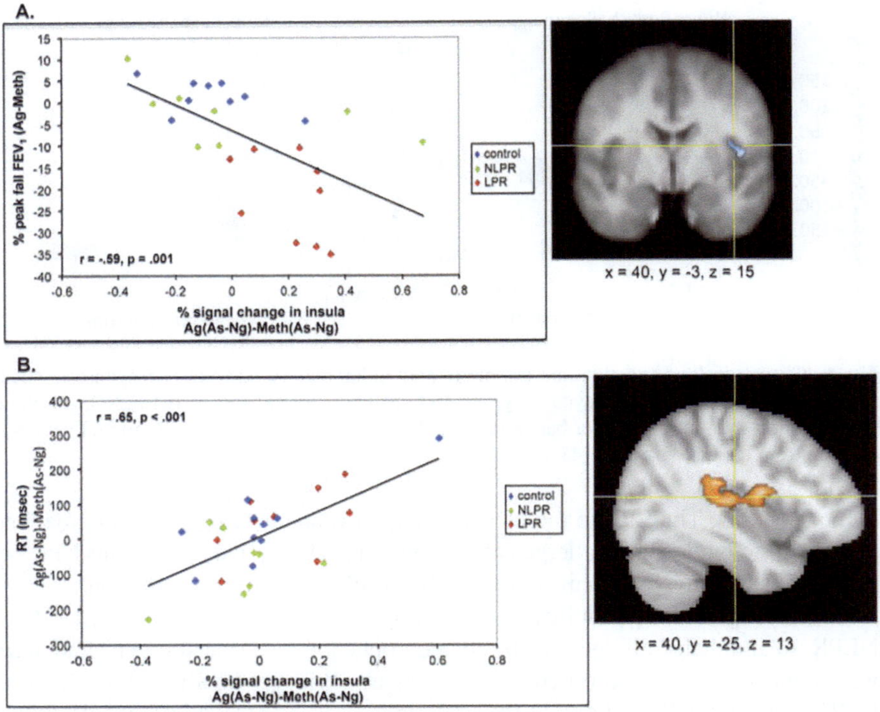

Fig. 7 Insula activity predicts magnitude of airway obstruction and cognitive interference of disease-related information during the late phase response to Ag. Percent signal change in the right mid- and posterior insula in response to asthma compared to negative words [As-Ng] and (**a**) percent peak fall in FEV1 [Ag-Meth] ($r = -0.59, p = 0.001$) and (**b**) reaction time to identify the color of As words, relative to Ng words [Ag(As-Ng)-Meth(As-Ng)] ($r = 0.66, p < 0.001$) during late phase antigen relative to methacholine challenge. Coordinates are in MNI space. (Reprinted with permission from Rosenkranz [45])

Are There Implications of Asthma–Mind–Brain Interactions for the Treatment of Asthma?

Achievement of disease control in asthma, particularly with severe disease, is assisted by an identification of the inflammatory profile and factors contributing to the pathophysiology of the disease [15]. The results of seminal studies have linked the molecular contribution or regulation of inflammation to altered lung function and risk factors for severe events such as exacerbations. These studies have initially advanced the use of biomarkers to define asthma phenotypes to guide the selection with greater specificity and efficacy of a biologic treatment with monoclonal antibodies for T2 inflammation. However, an integration of the roles of stress, anxiety, and depression to asthma pathophysiology and treatment has lagged behind our appreciation of the roles of respiratory viruses and allergens on asthma, for example. The preceding discussion indicated that psychological factors may also have a profound influence on asthma, including inflammation and severity. Therefore, integrating psychological factors for more effective treatment remains an unmet need.

To underscore this point, nearly two decades ago, Strunk and colleagues conducted a case-control study on the clinical characteristics associated with death from asthma [46]. Compared to an age-matched control population that had not died, the investigators found several discriminating characteristics in children (mean age of 13 years) with death from asthma, including a recent 50% decrease in prednisone. However, what was of interest to the investigators, and relevant to our discussion, was a striking and frequent difference in psychological factors in children who died from asthma. These psychological characteristics included a history of conflicts with staff and parents, the manipulative use of asthma, depression, emotional disturbances, and family dysfunction. The authors concluded that an emphasis should be placed on recognizing the importance of existing psychological problems in managing patients, particularly those with severe asthma. Whether such characteristics will represent a specific phenotype in which key psychological problems exist to promote risks for death will need further study.

Do Treatable Traits Link Asthma and Psychological Conditions?

Agusti and co-authors [47] proposed the concept that the clinical expression of asthma is determined or influenced by "clinical traits," which also serve as therapeutic targets to gain disease control. As noted by the authors, the expression of these traits forms phenotypes and endotypes of asthma, and their identification lends greater precision to the disease expressed in individual patients. Included in their concept of treatable traits are (A) pulmonary factors, i.e., airflow obstruction, (B) extrapulmonary, i.e., systemic inflammation, and (C) behavioral factors. Behavioral factors include, according to their concept [36], psychiatric disorders, like depression and anxiety. It is appreciated that depression and anxiety may be present in up to 25% of patients with severe asthma. However, the inclusion of these traits as targets for treatment is often absent in management plans and approaches. As Peter Sterk ably stated in an accompanying editorial to the Agusti et al. paper on "treatable traits [47]," it is perhaps more effective and patient-centric "to guide patient management by phenotypic characteristics rather than diagnostic labels," such as mild, moderate, or severe [48].

To test the hypothesis of treatment specific for identified treatable traits, McDonald and co-workers in Australia evaluated the use of treatment guided by treatable traits vs. standard guideline care [49]. Depression and/or anxiety were found in approximately 16% of the severe asthma cohort. Treatment was directed toward recognizing these symptoms, management of panic events, and the use of antidepressant medications. Asthma treatment, which included treatment of specific treatable traits, significantly improved quality of life and asthma control, an end-point of benefit. Treatment of psychological traits improved asthma outcomes but was to brain function.

Higgins and co-workers investigated the effects of Mindfulness-Based Stress Reduction in Asthma [50]. Mindfulness-Based Stress Reduction (MBSR) reduces stress in several medical conditions. In this study, patients with asthma were identified, and characteristics of T2 airway inflammation were established by measures of FeNO ≥30 ppb, blood eosinophils ≥150 cells/μL, or sputum eosinophils ≥2%. Half of the enrolled subjects underwent MBSR training, and the other portion remained "in waiting" as controls. Asthma outcomes from MBSR included changes in ACQ-6. In addition, psychological assessments were made using the Beck Depression Inventory and Beck Anxiety Inventory. Several observations were made that are relevant and applicable to our discussion. Improved asthma control (ACQ-6) occurred with MBSR and improvement was proportional to baseline measures of stress and depression. The greater the indices of stress and depression, the poorer asthma control, i.e., higher ACQ-6 scores. MBSR reduced indices of psychological distress, which then translated to improved asthma control, i.e., lower ACQ-6 scores. Interestingly, these beneficial effects continued to improve up to 4 months post-intervention, suggesting a lasting benefit from MBSR training. Finally, there was a significant reduction in FeNO over the study with MBSR. Collectively, these data suggest that a behavioral intervention, which is focused on reducing stress and emotional reactivity, improves measures of psychological health, which, in turn, led to improved asthma control and a reduction in airway inflammation as measured by FeNO. These data also imply that a brain-directed intervention can affect airway inflammatory features of the disease, i.e., reduced FeNO, to translate to improved asthma control. Finally, these effects were more apparent in those subjects with greater symptoms of depression at baseline; this association further promotes the concept of identifying neurophenotypes in asthma and including this information in formulating a more effective treatment strategy of management.

What Are the Airway Signals That Interact with the Brain?

This is an area of research that is, perhaps, least well developed in the ongoing search for how asthma–brain interactions occur; what are the cellular and molecular mediators that signal the brain, and conversely, the brain mediators or processes that communicate with asthma. To begin the search for asthma communications with the brain, we have again used an allergen challenge to elicit allergic airway inflammation. As noted earlier, allergic provocation of eosinophilic airway inflammation is associated with elicitation of a signal in the brain as detected by fMRI. It is unlikely that eosinophils migrate to the brain and elicit a response. It is, however, more likely that the generation of cytokines, which increase eosinophils, contributes to signaling the brain. Airway inflammation in asthma has been categorized as being Type (T) high inflammation or T2 low inflammation [15]. The airway inflammatory response to an allergen is a largely T2-high reaction and is associated with an increased expression of the cytokines IL-4, IL-5, or IL-13. Evidence to link an overexpression of the T2-cytokines to an activation of the brain is presently lacking. Also of interest in brain activation are products of the T2-low pathway. In contrast to the T2-high pathways and eosinophilia, the predominant inflammatory cell in the T2-low pathway is the neutrophil. The T2-low response is associated with the generation of IL-17. Increased concentrations of IL-17 have also been linked to depression and anxiety [51]. An association of IL-17 with depression makes this family of cytokines of particular interest to more fully understand and appreciate asthma and brain axis. However, it is important to recall that airway inflammation activates airway sensory fibers, which, in turn, can activate the afferent vagus nerves to communicate with the brain.

There is also evidence that eosinophils have direct effects on airway sensory nerves, as noted by Drake et al. [52]. In airway biopsies from asthma patients of varying degrees of severity, these investigators evaluated the association of markers of T2 inflammation on airway nerve elongation and branching, features that affect the function of nerves. In the presence of eosinophils, there was an increase in nerve branching which was associated with a lack of a bronchodilator response to albuterol. In addition, an increase in nerve length caused by eosinophils was associated with increased sensitivity to irritants. Collectively, eosinophils are associated with altered airway nerve structures and sensory activation, which may contribute to bronchoconstriction and airway responsiveness. It is possible that the effects of eosinophils on airway sensory nerve function may also directly influence brain activation.

Although other immune or inflammatory cytokines can influence brain function and emotions, their associations with asthma are more by inference than direct observations in asthma. For example, interferon-α has been used in the treatment of melanoma, and its administration has been associated with the development of depressive symptoms. However, the role of interferon-α in asthma has yet to be fully defined; therefore, linking this cytokine to asthma and altered brain function remains conjectural at present [53]. The proposed mechanisms by which cytokines may affect the brain, or its function, can occur through a variety of means. First, increased generation of airway cytokines leads to increases in circulating levels, which can penetrate the brain via leaks in the blood-brain barrier. Peripheral nerves, including the vagus, have cytokine receptors, which, when activated, induce an afferent response to in turn activate the brain. Finally, brain cells, including microglial cells, can be activated to release cytokines that contribute to neuroinflammation and, if persistent, neurodegeneration.

Is There Evidence of Organ Pathology Outside the Airway in Asthma?

A major advance in understanding the variability of asthma has been the recognition of disease heterogeneity [1, 2]. The emergence of asthma phenotypes has identified cohorts of patients who are at greater risk for severe disease, more likely to have an enhanced progressive decline in lung function and, based upon their pattern of inflammation, responsive to T2 biologics [15]. In addition, it is now recognized that asthma may have systemic manifestations of disease and affect other organ systems, including the brain.

Agusti et al. [54] analyzed the data from three large patient cohorts (Framingham Offspring Cohort, CARDIA and GenIII) to test the hypothesis that low lung function, presumably a surrogate marker of disease severity and persistence in asthma, was associated with a higher prevalence and earlier incidence of respiratory, cardiovascular and metabolic abnormalities, as well as premature death. If correct, this association would suggest that the consequences of asthma are not limited to dysfunction of the diseased target organ, the lung, but may also cause symptoms and disease not directly tied to the pathophysiology of asthma. In all three cohort groups [54], the presence of low lung function was associated with a higher prevalence of smoking, use of more medications, and more frequent diagnosis of respiratory diseases. Of considerable interest were the findings that low lung function was also associated with a greater frequency of ECG abnormalities and cardiovascular disease. Moreover, subjects with low lung function had a greater prevalence of diabetes. These data are strong indicators that another heterogeneous feature of asthma is the presence of systemic inflammation, which may have significant effects on other organ systems.

Peters and co-investigators [55] reported observations that supported and amplified the findings of Agusti and co-workers [54]. Peters et al. [55] hypothesized that risks for systemic inflammation are increased in severe asthma and that IL-6 is a biomarker of systemic inflammation and a potential contributor to non-lung pathology. To determine whether associations exist between concentrations of IL-6, metabolic dysfunction, and asthma severity, they identified two patient populations for study. These two populations included subjects with non-severe asthma and patients with severe asthma from the Severe Asthma Research Project (SARP). Several patient clinical characteristics were associated with high IL-6 values: increased BMI, a higher prevalence of hypertension and diabetes, more frequent asthma exacerbations, and lower FEV_1 values. These associations suggest that some patients with severe asthma and co-existing obesity have an increased generation of IL-6, which may cause metabolic dysfunction and reflect a consequence of systemic inflammation in specific phenotypes of asthma. The findings of Agusti et al. [54] and Peters and co-workers [55] also suggest that organ systems outside the lung are at risk for the consequences of asthma and that systemic inflammation is another heterogeneous component of asthma. Another important consequence of systemic inflammation in asthma is the presence of neurological dysfunction, including long-term reductions in cognition and the eventual development of dementia.

Limited but convincing early-stage data suggest that asthma is a risk factor for dementia. Peng and colleagues [56] used data from the Taiwan Longitudinal Health Insurance Database to compare HR (hazard ratio) for dementia in asthma vs. an age-matched non-asthma population. The study was longitudinal and cross-sectional and contained 12,771 newly diagnosed asthma patients who were compared to a matched control cohort of 51,084 subjects. The cumulative incidence of dementia in asthma was significantly higher as assessed by a Kaplan–Meier Analysis over 10 years ($p < 0.0001$). The investigators also found the association between asthma and dementia was amplified by the frequency of asthma-associated ER visits and hospitalizations to suggest a dose-response relationship between the frequency of exacerbations and disease severity. These data suggest that asthma phenotypic characteristics for the greatest risk for dementia include disease severity marked by frequent exacerbations. Furthermore, these data also suggest that heightened inflammation of severe disease and accentuation of inflammation with exacerbations contribute to altered brain function and the development of dementia.

Chen et al. [57] also used data from the Taiwan National Health Insurance Database to further evaluate relationships between asthma and cases of dementia and Alzheimer's disease. Subjects included were 45 years or older, enrolled between 1998 and 2008, and followed through 2011. There were 11,030 asthma participants and 44,120 age- and sex-matched controls. There were several key findings. Overall, any form of dementia was greater in asthma (HR = 2.17, [CI 1.71–4.02]). Importantly, the HR for dementia was greater in those ≤65 years of age (HR = 2.48 [1.80–3.41]) vs. 2.06 (CI 1.74–2.44) for those older than 65 years. These findings suggest that the increase in dementia in less than 65-year-old patients with asthma is not a reflection of an age-dependent loss of cognition. There was also a higher HR for those with comorbid depressive disorders, diabetes mellitus, and cardiovascular disease (Table 1). Importantly, using inhaled corticosteroids in asthma was not associated with a reduction in the HR for dementia. These findings suggest that dementia was not related to ICS use, or, importantly, treatment with ICS to control asthma and reduce airway inflammation was not sufficient to prevent dementia. Furthermore, as no measures of airway inflammation or systemic inflammation were made, it is not possible to determine whether the potential risks for dementia related to airway inflammation, systemic inflammation, or, most likely, a combination of the two inflammatory processes.

Using the United States Medicare 20% sample cohort, Bartels et al. [58] compared the incidence and prevalence of Alzheimer's disease and any dementia in asthma vs. a control population. Among 5,406,732 beneficiaries, 678,730 asthma

Table 1 Risk of developing dementia in patients with asthma vs control subjects

	<65 Years	≥65 Years	Total
	HR (95% CI)	HR (95% CI)	HR 95% (CI)
	Any Dementia*		
Asthma, presence vs absence Comorbidities, presence vs absence	**2.48 (1.80–3.41)**	**2.06 (1.74–2.44)**	**2.17 (1.87–2.52)**
Depressive disorder	**3.74 (2.84–4.92)**	**2.79 (2.36–3.30)**	**3.12 (2.71–3.60)**
Alcohol-related disorders	0.65 (0.16–2.64)	**2.03 (1.08–3.81)**	1.56 (0.88–2.77)
Diabetes mellitus	**1.31 (1.03–168)**	**1.13 (1.00–1.28)**	**1.18 (1.05–1.31)**
Cerebrovascular diseases	**2.39 (1.85–3.08)**	**1.50 (1.33–1.68)**	**1.63 (1.46–1.82)**
Use of inhaled steroid, presence vs absence	1.11 (0.79–1.57)	0.96 (0.81–1.15)	0.98 (0.84–1.15)

Reprinted with permission from Chen et al. [57]
*Adjusted by demographic data, health system utilization, medical comorbidities, and use of inhaled steroid and asthma as a binary variable. Bold type indicates statistical significance

Table 2 Observed incidence & prevalence of Alzheimers and any dementia in asthma vs controls in Medicare Database (n = 5,406,732)

	Controls n = 4,782,002 (4,359,060) n (%)	Cases/10,000 (Controls)	Asthma n = 678,730 (578,846) n (%)	Cases/100,000 (Asthma)	p
2 yr Alzheimers Incidence	48,186 (1.0)	1105	8153 (1.2)	1408	<0.001
Ever Alzheimers Prevalence	258,635 (5.4)	5409	53,175 (7.8)	7834	<0.001
2 yr Any Dementia Incidence	193,314 (4.3)	4435	37,594 (5.5)	6495	<0.001
Ever Dementia Prevalence	616,256 (12.9)	12,887	137,478 (20.3)	20,255	<0.001

Reprinted with permission from Bartels et al. [58]

cases were identified and compared to 4,359,060 controls without asthma, inflammatory bowel disease, or inflammatory rheumatic diseases (Table 2). Both the appearance of Alzheimer's disease and the incidence of any dementia over 2 years or ever prevalence were significantly greater in asthma. Per 100,000 patients with asthma over 2 years, 303 more incidents of Alzheimer's cases occurred in asthma: There were 2425 more prevalent Alzheimer's cases in asthma versus without asthma ($p < 0.001$). Multivariable models showed asthma had a greater odds of two-year Alzheimer's incidence, OR 1.21 (95% CI 1.18–1.24), and prevalence, OR 1.48 (CI 1.47–1.50). Prevalence data showed 7386 more cases of any dementia per 100,000 in asthma vs. controls – a highly significant finding given the prevalence of both conditions. These data, in a large nationally representative US population, are convincing evidence that asthma elevates the risk for dementia and Alzheimer's disease. Collectively, these epidemiological studies indicate an increase in risk for both dementia and Alzheimer's disease in asthma but do not identify the asthma-associated mechanisms that cause the deterioration in brain health.

Do Structural Changes of the Brain Occur in Asthma?

To further explore the relationships between asthma and brain microstructure, Rosenkranz et al. [59] analyzed neuroimaging data (diffusion-weighted MRI) and blood-based biomarkers associated with neuroinflammation, neurodegeneration, and Alzheimer's disease (AD) in asthma and control subjects. The 111 asthma subjects had a confirmed diagnosis of asthma of varied severity from mild to severe. Diffusion-weighted imaging metrics of brain white matter microstructure were compared between asthma and age-matched non-asthma controls ($n = 135$). Blood-based biomarkers included p-Tau 181, which is specific for AD. Neurofilament light chain (NfL) was used as a biomarker to indicate neurodegeneration; glial fibrillary acidic protein (GFAP) reflects astrocyte activation and is an index of neuroinflammation. Changes in white matter microstructure indicate greater diffusion or loss of

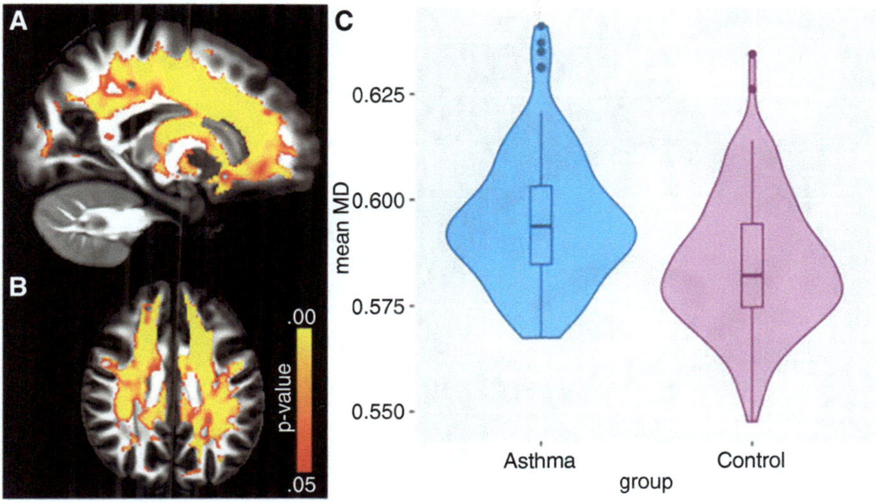

Fig. 8 Group difference in white matter microstructure between asthma ($n = 111$) and control ($n = 135$) groups displayed on a standard white matter template. Representative sagittal (**a**) and axial (**b**) slices displaying the overall test (across diffusion-weighted imaging [DWI] metrics) of the group difference. Yellow-orange areas represent regions where there is a significant group difference. Images were thresholded at a corrected $P < 0.05$. (**c**) Distribution of individual mean diffusivity (MD) means for each group, averaged over all voxels where MD was significantly greater in the asthma group, relative to the control group, in the whole brain analysis displayed in (**a**) and (**b**). MD was chosen as a representative DWI metric for plotting purposes, but this effect was observed across most of the DWI metrics examined. (Reprinted from Rosenkranz et al. [59])

white matter integrity. As noted in Fig. 8, there is greater deterioration in white matter structure in asthma compared to control subjects; these analyses controlled for age. Changes in white matter were greater with severe asthma, which was defined and based upon a collective assessment of medication use, lung function, and other disease severity parameters. Finally, the degree of white matter alteration correlated with the concentrations of the serological biomarkers for neurodegeneration (NfL) and neuroinflammation (GFAP) (Fig. 9). None of the asthma subjects, however, had evidence of clinically defined cognitive impairment.

These observations extend the scope and public health importance of the impact of asthma on the brain to now include structural changes in the brain as detected by sensitive neuroimaging methods. Importantly, these imaging findings indicate significant and deleterious alterations in white matter structure, which resemble, both

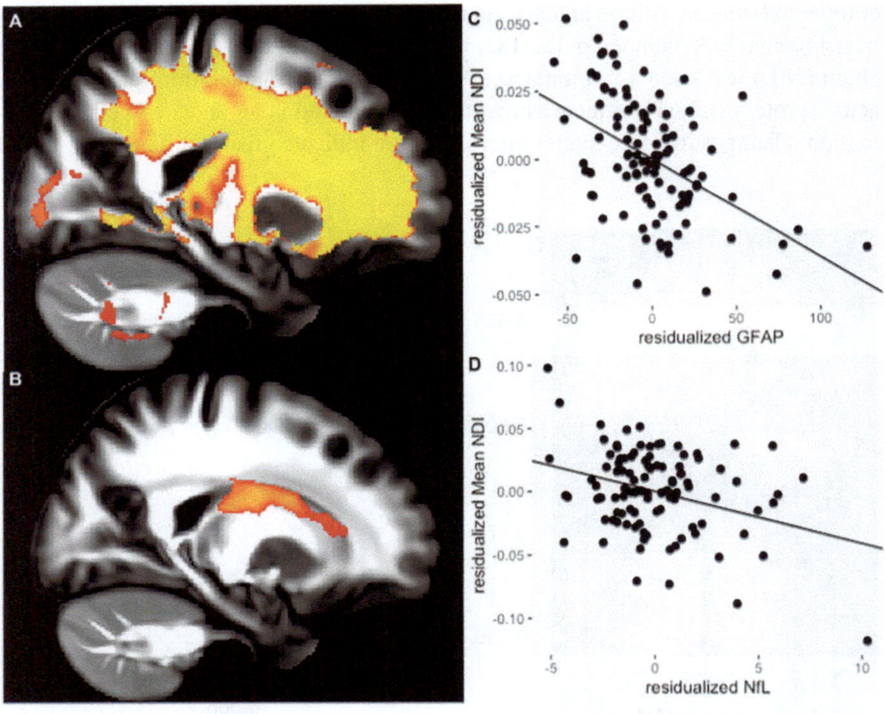

Fig. 9 Higher plasma biomarker concentrations are associated with less white matter integrity. Representative sagittal slices of a standard white matter template displaying voxels where plasma biomarker concentration is significantly associated with neurite density index (NDI). (**a**) Voxels where glial fibrillary acidic protein (GFAP) concentration showed a significant negative relationship with NDI. (**b**) Voxels where neurofilament light (NfL) concentration showed a significant negative relationship with NDI. This cluster is primarily composed of corona radiata and internal capsule fibers (including corticospinal tract and thalamic radiations). All images were thresholded at a corrected $P < 0.05$. (**c, d**) Scatterplots displaying the relationship between mean NDI, averaged across all voxels in the clusters shown in (**a**) and (**b**) and GFAP (**c**) and NfL (**d**), with variance accounted for by age and sex removed from both variables. (Reprinted from Rosenkranz et al. [59])

in extent and magnitude, findings observed in neurodegenerative diseases. A conclusion of these imaging observations in asthma suggests that these findings likely reflect neuroinflammation and neurodegeneration, as the noted associations with increased plasma concentrations of GFAP (neuroinflammation) and NfL (neurodegeneration) are independent of normal aging. A translation of these brain imaging findings to clinical dementia was not possible, though subtle changes in cognition were detected.

Although these findings are descriptive and do not yet identify causal mechanisms, the public health consequences are very significant given the high prevalence of asthma world-wide. Furthermore, asthma for many patients is a lifelong disease with ongoing airway inflammation, thus suggesting that the nearly 5 million US asthma patients may be at risk for the early or more rapid decline in cognition from accelerated neurodegeneration and are at greater risk for the development of Alzheimer's disease dementia. The influence of treatment, either protective or causative, was not assessed, but in severe disease, even when patients had good disease control, neuroimaging detected deleterious brain microstructural changes. Therefore, future studies will need to identify the mechanisms of these asthma-associated impacts on brain health and, in particular, determine what are the inflammatory pathways and products that contribute to or cause these structural changes in the brain.

What Are the Next Steps to Translate the Importance, Relevance, and Impact of Asthma–Brain Interactions to Diminish the Overall Burden of Asthma on Neurocognitive Function?

The concept of asthma as a systemic inflammatory disease is a new relationship but has parallels with other inflammatory diseases, like rheumatoid arthritis. Furthermore, the experiences from rheumatoid arthritis are of considerable interest as treatments with biologics in autoimmune diseases have been longstanding and highly effective; the experiences with biologics also provide potential insight into the therapeutic intervention of asthma–brain interactions and approaches to reduce associated risks for dementia and Alzheimer's disease in asthma, as well as other organ–disease interactions: cardiovascular and metabolic.

Chou and co-investigators [60] reviewed the associations between rheumatoid arthritis and Alzheimer's disease. As this investigative group noted, Alzheimer's disease is a disease of the elderly, and its incidence increases with age. By 85 years of age, AD is estimated to affect 32% of the US population. The pathology defining AD includes the deposition of amyloid plaques and accumulation of neurofibrillary tangles in the brain. We have already reviewed the association of surrogate biomarkers for AD and stages of neuroinflammation to neurodegeneration and eventual expression of altered cognition in asthma. As pointed out by Chou et al. [60], there has been interesting but inconsistent evidence that tumor necrosis factor (TNF)-α is a biomarker and mediator of brain inflammation. Increased concentrations of TNF-α have been found in the cerebral spinal fluid (CSF) of some AD patients, and there is also evidence that its presence relates to disease progression.

Using a large database of medical and pharmacy claims, Chou et al. [60] identified a cohort of subjects who were diagnosed with RA. From these data, investigators determined the prevalence of AD in their overall population (which included normal subjects) and, of interest to the focus of this chapter, patients with and without RA (Table 3). The prevalence of AD was almost 8-fold greater in RA vs. the non-RA subjects (0.79% vs. 0.11%, $p < 0.0001$). A second analysis was undertaken to determine the potential beneficial effects of anti-TNF-α treatment for RA on AD prevalence. Prednisone and sulfasalazine had no effect in reducing AD in RA. In contrast, anti-TNF-α agents, in general, reduced the expression of AD in RA, and, interestingly, the benefit had some specificity for etanercept, a competitive TNF receptor. A general and perhaps far-reaching conclusion from these observations is that by blocking the actions of TNF-α, the expression and consequences of RA on the development of AD pathology in the brain and/or progression to dementia were prevented. A US Medicare study found that patients on biologic DMARDs (Disease Modifying Anti-Rheumatic Drugs) had 19% lower risk for dementia than those on traditional DMARDs, HR 0.81 (95% CI 0.76–0.87), without differences between TNF and non-TNF agents suggesting that aggressive control of inflammation control matters in reducing the consequence of these processes on the brain [61].

Epidemiologic studies regarding the long-term use of biologic treatment in asthma lag the experiences in RA. However, an important question is whether it will be possible to reduce systemic manifestations of inflammation in asthma with biologics and whether this approach with biologics in asthma will have similar beneficial systemic outcomes to that observed with anti-TNF-α in RA? Several steps are necessary to address this question and expectations more clearly and fully. First, it is necessary to identify the phenotype of asthma associated with risks for dementia and AD. Second, it is essential to identify the asthma-generated inflammatory signals that elicit and lead to neuroinflammation and neurodegeneration. With this information, long-term, real-world databases of biologic use in asthma can be used to assess the effects of biologics on risks for AD and potential mechanisms of action. Given the current lack of effective therapeutics for AD, these asthma-related relationships are exciting, of considerable interest and potential benefit to a large population of asthma patients, and further evidence of the importance of an asthma–brain interaction.

Table 3 Prevalence of AD in the overall population of patients with and without RA

	AD (overall population)				
RA	+	–	Total	Prevalence of AD (%)(95% CI)	*p* value
+	325	40,784	41,109	0.79 (0.71–0.88)	<0.0001
–	8928	8,450,417	8,459,345	0.11 (0.10–0.11)	
Total	9253	8,491,201	8,500,454		

Reprinted with permission from Chou [42]
AD Alzheimer's disease, *CI* confidence interval, *RA* rheumatoid arthritis

Summary

There has been an amazing expansion of information in the last decade on brain–asthma interactions and their importance to the longtime health of millions of patients with asthma. Although the associations and interactions with psychological conditions, including anxiety and depression, remain of considerable importance and are "traits" for targeted treatment to gain disease control, efforts to identify mechanisms and the bidirectional effects of these interactions have gained considerable clinical and biological relevance. The possibility that asthma and existing airway inflammation create systemic manifestations to affect brain function is of critical interest and health-related importance. The evidence to suggest asthma is a risk factor for Alzheimer's disease has provided a pathway of future investigation to define these asthma–brain interactive mechanisms and to develop treatments to prevent neuroinflammation, neurodegeneration, and the eventual development of dementia. The development of precision in these efforts will greatly advance the science of asthma and brain interactions and provide specific and effective treatments to prevent the devastating and life-altering consequence of dementia and Alzheimer's disease in a large population of at-risk patients world-wide.

References

1. Moore WC, Meyers DA, Wenzel SE, Teague WG, Li H, Li X, et al. Identification of asthma phenotypes using cluster analysis in the Severe Asthma Research Program. Am J Respir Crit Care Med. 2010;181(4):315–23.
2. Gauthier M, Ray A, Wenzel SE. Evolving concepts of asthma. Am J Respir Crit Care Med. 2015;192(6):660–8.
3. Laidlaw TM, Boyce JA. Aspirin-exacerbated respiratory disease--new prime suspects. N Engl J Med. 2016;374(5):484–8.
4. Del Giacco SR, Cappai A, Gambula L, Cabras S, Perra S, Manconi PE, et al. The asthma-anxiety connection. Respir Med. 2016;120:44–53.
5. Rosner F. Moses Maimonides' treatise on asthma. Thorax. 1981;36(4):245–51.
6. Katon WJ, Richardson L, Lozano P, McCauley E. The relationship of asthma and anxiety disorders. Psychosom Med. 2004;66(3):349–55.
7. Calam R, Gregg L, Simpson B, Morris J, Woodcock A, Custovic A. Childhood asthma, behavior problems, and family functioning. J Allergy Clin Immunol. 2003;112(3):499–504.
8. Koltek M, Wilkes TC, Atkinson M. The prevalence of posttraumatic stress disorder in an adolescent inpatient unit. Can J Psychiatr. 1998;43(1):64–8.
9. Goodwin RD, Jacobi F, Thefeld W. Mental disorders and asthma in the community. Arch Gen Psychiatry. 2003;60(11):1125–30.
10. Ye G, Baldwin DS, Hou R. Anxiety in asthma: a systematic review and meta-analysis. Psychol Med. 2021;51(1):11–20.
11. Ramratnam SK, Lockhart A, Visness CM, Calatroni A, Jackson DJ, Gergen PJ, et al. Maternal stress and depression are associated with respiratory phenotypes in urban children. J Allergy Clin Immunol. 2021;148(1):120–7.
12. Ramratnam SK, Visness CM, Jaffee KF, Bloomberg GR, Kattan M, Sandel MT, et al. Relationships among maternal stress and depression, type 2 responses, and recurrent wheezing at age 3 years in low-income urban families. Am J Respir Crit Care Med. 2017;195(5):674–81.

13. Goodwin RD, Robinson M, Sly PD, McKeague IW, Susser ES, Zubrick SR, et al. Severity and persistence of asthma and mental health: a birth cohort study. Psychol Med. 2013;43(6):1313–22.
14. Van Lieshout RJ, Macqueen G. Psychological factors in asthma. Allergy Asthma Clin Immunol. 2008;4(1):12–28.
15. Israel E, Reddel HK. Severe and difficult-to-treat asthma in adults. N Engl J Med. 2017;377(10):965–76.
16. Alonso J, de Jonge P, Lim CC, Aguilar-Gaxiola S, Bruffaerts R, Caldas-de-Almeida JM, et al. Association between mental disorders and subsequent adult onset asthma. J Psychiatr Res. 2014;59:179–88.
17. Chida Y, Hamer M, Steptoe A. A bidirectional relationship between psychosocial factors and atopic disorders: a systematic review and meta-analysis. Psychosom Med. 2008;70(1):102–16.
18. Busse WW, Kiecolt-Glaser JK, Coe C, Martin RJ, Weiss ST, Parker SR. NHLBI Workshop summary. Stress and asthma. Am J Respir Crit Care Med. 1995;151(1):249–52.
19. Busse WW, Lemanske RF Jr, Gern JE. Role of viral respiratory infections in asthma and asthma exacerbations. Lancet. 2010;376(9743):826–34.
20. Holgate ST, Polosa R. The mechanisms, diagnosis, and management of severe asthma in adults. Lancet. 2006;368(9537):780–93.
21. Ownby D, Johnson C, Peterson E. Exposure to dogs and cats in the first year of life and risk of allergic sensitization at 6 to 7 years of age. J Am Med Assoc. 2002;288:963.
22. Sporik R, Holgate ST, Platts-Mills TA, Cogswell JJ. Exposure to house-dust mite allergen (Der p I) and the development of asthma in childhood. A prospective study. N Engl J Med. 1990;323(8):502–7.
23. Wright RJ, Visness CM, Calatroni A, Grayson MH, Gold DR, Sandel MT, et al. Prenatal maternal stress and cord blood innate and adaptive cytokine responses in an inner-city cohort. Am J Respir Crit Care Med. 2010;182(1):25–33.
24. Wright RJ, Cohen RT, Cohen S. The impact of stress on the development and expression of atopy. Curr Opin Allergy Clin Immunol. 2005;5(1):23–9.
25. Cookson H, Granell R, Joinson C, Ben-Shlomo Y, Henderson AJ. Mothers' anxiety during pregnancy is associated with asthma in their children. J Allergy Clin Immunol. 2009;123(4):847–53 e11.
26. Wright RJ, Cohen S, Carey V, Weiss ST, Gold DR. Parental stress as a predictor of wheezing in infancy: a prospective birth-cohort study. Am J Respir Crit Care Med. 2002;165(3):358–65.
27. Chiu YH, Coull BA, Cohen S, Wooley A, Wright RJ. Prenatal and postnatal maternal stress and wheeze in urban children: effect of maternal sensitization. Am J Respir Crit Care Med. 2012;186(2):147–54.
28. Chen E, Miller GE. Stress and inflammation in exacerbations of asthma. Brain Behav Immun. 2007;21(8):993–9.
29. Chen E, Hanson MD, Paterson LQ, Griffin MJ, Walker HA, Miller GE. Socioeconomic status and inflammatory processes in childhood asthma: the role of psychological stress. J Allergy Clin Immunol. 2006;117(5):1014–20.
30. Rossi S, Sacchetti L, Napolitano F, De Chiara V, Motta C, Studer V, et al. Interleukin-1beta causes anxiety by interacting with the endocannabinoid system. J Neurosci. 2012;32(40):13896–905.
31. Beurel E, Harrington LE, Jope RS. Inflammatory T helper 17 cells promote depression-like behavior in mice. Biol Psychiatry. 2013;73(7):622–30.
32. Chen Y, Yang P, Li F, Kijlstra A. The effects of Th17 cytokines on the inflammatory mediator production and barrier function of ARPE-19 cells. PLoS One. 2011;6(3):e18139.
33. Vazquez-Tello A, Halwani R, Hamid Q, Al-Muhsen S. Glucocorticoid receptor-beta upregulation and steroid resistance induction by IL-17 and IL-23 cytokine stimulation in peripheral mononuclear cells. J Clin Immunol. 2013;33(2):466–78.
34. Chen W, Boutaoui N, Brehm JM, Han YY, Schmitz C, Cressley A, et al. ADCYAP1R1 and asthma in Puerto Rican children. Am J Respir Crit Care Med. 2013;187(6):584–8.

35. Brehm JM, Ramratnam SK, Tse SM, Croteau-Chonka DC, Pino-Yanes M, Rosas-Salazar C, et al. Stress and Bronchodilator response in children with asthma. Am J Respir Crit Care Med. 2015;192(1):47–56.
36. Chrousos GP. Stress, chronic inflammation, and emotional and physical well-being: concurrent effects and chronic sequelae. J Allergy Clin Immunol. 2000;106(5 Suppl):S275–91.
37. Kang DH, Coe CL, McCarthy DO, Jarjour NN, Kelly EA, Rodriguez RR, et al. Cytokine profiles of stimulated blood lymphocytes in asthmatic and healthy adolescents across the school year. J Interf Cytokine Res. 1997;17(8):481–7.
38. Gauvreau GM, Watson RM, O'Byrne PM. Kinetics of allergen-induced airway eosinophilic cytokine production and airway inflammation. Am J Respir Crit Care Med. 1999;160(2):640–7.
39. Liu LY, Coe CL, Swenson CA, Kelly EA, Kita H, Busse WW. School examinations enhance airway inflammation to antigen challenge. Am J Respir Crit Care Med. 2002;165(8):1062–7.
40. Couillard S, Steyerberg E, Beasley R, Pavord I. Blood eosinophils, fractional exhaled nitric oxide and the risk of asthma attacks in randomised controlled trials: protocol for a systemic review and control arm patient-level meta-analysis for clinical prediction modelling. BMJ Open. 2022;12(4):e058215.
41. Stroop JR. Studies of interference in serial verbal reactions. J Exp Psychol. 1935;18:643–62.
42. Rosenkranz MA, Busse WW, Johnstone T, Swenson CA, Crisafi GM, Jackson MM, et al. Neural circuitry underlying the interaction between emotion and asthma symptom exacerbation. Proc Natl Acad Sci U S A. 2005;102(37):13319–24.
43. Rosenkranz MA, Esnault S, Christian BT, Crisafi G, Gresham LK, Higgins AT, et al. Mind-body interactions in the regulation of airway inflammation in asthma: a PET study of acute and chronic stress. Brain Behav Immun. 2016;58:18–30.
44. Narvaez Linares NF, Charron V, Ouimet AJ, Labelle PR, Plamondon H. A systematic review of the Trier Social Stress Test methodology: issues in promoting study comparison and replicable research. Neurobiol Stress. 2020;13:100235.
45. Rosenkranz MA, Busse WW, Sheridan JF, Crisafi GM, Davidson RJ. Are there neurophenotypes for asthma? Functional brain imaging of the interaction between emotion and inflammation in asthma. PLoS One. 2012;7(8):e40921.
46. Strunk RC, Mrazek DA, Fuhrmann GS, LaBrecque JF. Physiologic and psychological characteristics associated with deaths due to asthma in childhood. A case-controlled study. JAMA. 1985;254(9):1193–8.
47. Agusti A, Bel E, Thomas M, Vogelmeier C, Brusselle G, Holgate S, et al. Treatable traits: toward precision medicine of chronic airway diseases. Eur Respir J. 2016;47(2):410–9.
48. Sterk PJ. Chronic diseases like asthma and COPD: do they truly exist? Eur Respir J. 2016;47(2):359–61.
49. McDonald VM, Clark VL, Cordova-Rivera L, Wark PAB, Baines KJ, Gibson PG. Targeting treatable traits in severe asthma: a randomised controlled trial. Eur Respir J. 2020;55:1901509.
50. Higgins ET, Davidson RJ, Busse WW, Klaus DR, Bednarek GT, Goldman RI, et al. Clinically relevant effects of mindfulness-based stress reduction in individuals with asthma. Brain Behav Immun Health. 2022;25:100509.
51. Barros PO, Ferreira TB, Vieira MM, Almeida CR, Araujo-Lima CF, Silva-Filho RG, et al. Substance P enhances Th17 phenotype in individuals with generalized anxiety disorder: an event resistant to glucocorticoid inhibition. J Clin Immunol. 2011;31(1):51–9.
52. Drake MG, Scott GD, Blum ED, Lebold KM, Nie Z, Lee JJ, et al. Eosinophils increase airway sensory nerve density in mice and in human asthma. Sci Transl Med. 2018;10(457):eaar8477.
53. Miller AH, Haroon E, Raison CL, Felger JC. Cytokine targets in the brain: impact on neurotransmitters and neurocircuits. Depress Anxiety. 2013;30(4):297–306.
54. Agusti A, Noell G, Brugada J, Faner R. Lung function in early adulthood and health in later life: a transgenerational cohort analysis. Lancet Respir Med. 2017;5(12):935–45.
55. Peters MC, McGrath KW, Hawkins GA, Hastie AT, Levy BD, Israel E, et al. Plasma interleukin-6 concentrations, metabolic dysfunction, and asthma severity: a cross-sectional analysis of two cohorts. Lancet Respir Med. 2016;4(7):574–84.

56. Peng YH, Wu BR, Su CH, Liao WC, Muo CH, Hsia TC, et al. Adult asthma increases dementia risk: a nationwide cohort study. J Epidemiol Community Health. 2015;69(2):123–8.
57. Chen MH, Li CT, Tsai CF, Lin WC, Chang WH, Chen TJ, et al. Risk of dementia among patients with asthma: a nationwide longitudinal study. J Am Med Dir Assoc. 2014;15(10):763–7.
58. Bartels C, Powell R, Rosenkratz M, Chen G, Bendlin B, Kramer J, et al. Alzheimer's disease incidence and prevalence with and without asthma: a National Medicare Cohort Study. JAMA Neurol. 2022; Submitted.
59. Rosenkranz MA, Dean DC 3rd, Bendlin BB, Jarjour NN, Esnault S, Zetterberg H, et al. Neuroimaging and biomarker evidence of neurodegeneration in asthma. J Allergy Clin Immunol. 2022;149(2):589–98 e6.
60. Chou RC, Kane M, Ghimire S, Gautam S, Gui J. Treatment for rheumatoid arthritis and risk of Alzheimer's disease: a nested case-control analysis. CNS Drugs. 2016;30(11):1111–20.
61. Sattui SE, Navarro-Millan I, Xie F, Rajan M, Yun H, Curtis JR. Incidence of dementia in patients with rheumatoid arthritis and association with disease modifying anti-rheumatic drugs - analysis of a national claims database. Semin Arthritis Rheum. 2022;57:152083.

Systems Biology in Asthma

Nazanin Zounemat Kermani, Ian M. Adcock,
Ratko Djukanović, Fan Chung, and James P. R. Schofield

Abstract

The application of mathematical and computational analysis, together with the modelling of biological and physiological processes, is transforming our understanding of the pathophysiology of complex diseases. This systems biology approach incorporates large amounts of genomic, transcriptomic, proteomic, metabolomic, breathomic, metagenomic and imaging data from disease sites together with deep clinical phenotyping, including patient-reported outcomes. Integration of these datasets will provide a greater understanding of the molecular pathways associated with severe asthma in each individual patient and determine their personalised treatment regime. This chapter describes some of the

N. Z. Kermani
Data Science Institute, Imperial College London, London, UK

National Heart & Lung Institute, Imperial College London, London, UK

I. M. Adcock
National Heart & Lung Institute, Imperial College London, London, UK

R. Djukanović (✉)
NIHR Southampton Biomedical Research Centre, Clinical and Experimental Sciences, Faculty of Medicine, University of Southampton, Southampton, UK
e-mail: rd1@soton.ac.uk

F. Chung
National Heart & Lung Institute, Imperial College London, London, UK

Biomedical Research Unit, Royal Brompton & Harefield NHS Trust, London, UK

J. P. R. Schofield
Centre for Proteomic Research, Institute for Life Sciences, University of Southampton, Southampton, UK

TopMD Precision Medicine Ltd, Southampton, UK

A. R. Brasier, N. N. Jarjour (eds.), *Precision Approaches to Heterogeneity in Asthma*, Advances in Experimental Medicine and Biology 1426,
https://doi.org/10.1007/978-3-031-32259-4_10

data integration methods used to combine data sets and gives examples of the results obtained using single datasets and merging of multiple datasets (data fusion and data combination) from several consortia including the severe asthma research programme (SARP) and the Unbiased Biomarkers Predictive of Respiratory Disease Outcomes (U-BIOPRED) consortia. These results highlight the involvement of several different immune and inflammatory pathways and factors in distinct subsets of patients with severe asthma. These pathways often overlap in patients with distinct clinical features of asthma, which may explain the incomplete or no response in patients undergoing specific targeted therapy. Collaboration between groups will improve the predictions obtained using a systems medicine approach in severe asthma.

Keywords

Data integration · Genomics · Transcriptomics · Proteomics · Metabolomics · Breathomics · Metagenomics · Imaging · Clustering · Next generation sequencing · Severe asthma research programm (SARP) · Unbiased Biomarkers Predictive of Respiratory Disease Outcomes (U-BIOPRED) consortium · Heterogeneity

Abbreviations

DEG	Differentially expressed genes
DIABLO	Data Integration Analysis for Biomarker discovery using Latent Components
eNOSE	Electronic nose
GC-MS	Gas chromatography-mass spectroscopy
GSEA	Gene set enrichment analysis
GSVA	Gene set variation analysis
HBEC	Human bronchial epithelial cells
IFN	Interferon
IL	Interleukin
ILC	Innate lymphoid cells
IMI	European Union Innovative Medicines Initiative
MDS	Multi-dimensional scaling
OCS	Oral corticosteroids
PCA	Principal components analysis
RNA-seq	RNA sequencing
ROC	Receiver operator curve
SARP	Severe asthma research program
SNF	Similarity network fusion
T1	Type 1 inflammation
T2	Type 2 inflammation

TAC	Transcriptome-associated clusters
TDA	Topological data analysis
TopMD	Topological analysis of omics data
U-BIOPRED	Unbiased Biomarkers Predictive of Respiratory Disease Outcomes
VOC	Volatile organic compounds

Introduction

The past three decades have seen exponential growth in research into the pathobiology of asthma and how the mechanisms underpin its clinical features and determine responses to treatment. This was greatly enabled by the advent of techniques that could safely sample the lungs, such as research bronchoscopy in the early 90s and, soon after, the development and standardisation of sputum induction with hypertonic or normal saline to sample the epithelial lining fluid. Using these techniques, in combination with ever more sophisticated methods for the analysis of blood and urine samples, a plethora of mediators have been identified and proposed as being relevant in asthma, derived from the many inflammatory and structural cell types within the bronchial mucosa, exudates from the vasculature or systemic processes. High-throughput, large-scale methods to quantify gene expression, proteins, lipids and metabolites are on offer, either as single or multiple mediator assays [1]. Some of these methods, such as RT-PCR and single protein assays, can quantify single or limited numbers of selected mediators, while others use what is termed an 'unbiased' approach, where there is no a priori selection of variables of interest and the only limitation of this approach is the power and spectrum of detection of the laboratory method. With increasing use, even the most sophisticated of these analytical tools have become more affordable and, therefore, more readily applied in research.

Transcriptomics has led the field and continues to do so, especially with the development of RNA sequencing (RNA-seq) and its application as single-cell sequencing analysis. Thus, at the time of writing, a PubMed search identified 774 transcriptomics publications associated with asthma out of 148,000 hits. Even more publications (168,000) have some element of protein analysis (proteomics) but a relatively lower number (634) of associations with asthma. Omics methods based on mass spectrometry for the detection of volatile organic compounds (VOCs), proteins, lipids and metabolites [2–4] are rapidly growing in number. These have been applied in asthma studies on samples of the airways (nasal and endobronchial biopsies and epithelial brushings, nasal and bronchial/bronchoalveolar lavage, induced sputum) as well the peripheral circulation, either by analysis of whole blood or selected peripheral blood mononuclear cells.

Somewhat lagging behind the laboratory methods has been the data analytical aspect of mechanistic studies, at least in part because this requires scientists with both laboratory and data analysis skills or close collaboration between experts. For the integration of such analyses into a clinical context, additional clinician scientist expertise is needed to enable molecular characterisation in disease based on multi-omics approaches in order to achieve full potential. This is most effectively undertaken within consortia that have the full complement of required expertise, can recruit large numbers of patients, and are well funded. Two such consortia stand out: SARP (Severe Asthma Research Program), established in 2001 and funded by the National Heart, Lung and Blood Institute (NHLBI), and U-BIOPRED (Unbiased Biomarkers Predictive of Respiratory Disease Outcomes), founded in 2009. U-BIOPRED represented a novel collaborative approach between academics, patients, medical charities and the pharmaceutical industry, resulting in a strong public-private partnership funded by the European Union Innovative Medicines Initiative (IMI) [5]. Both consortia have had a strong collaborative and integrative basis and significant funding, which made it possible to recruit large numbers of participants to create large adult and paediatric cohorts [6].

As the name implies, U-BIOPRED was based on unbiased omics platforms which create enormous datasets. The sheer size and complexity of the generated data posed a challenge and required the development of tranSMART, a high security federated data repository with the capability to analyse data across platforms without the need to export the data [7, 8]. Various models, including principal component analysis, with subsequent generation of receiver operating characteristics (ROC) curves, have been applied to find predictive biomarkers in such samples as exhaled breath. Given that the main aim of U-BIOPRED was to offer new endotypes and phenotypes, a major effort went into clustering methods, ranging from widely used consensus clustering, through Gaussian mixture model clustering, to more recent clustering methods based on topological features of data [9, 10], first applied in 2016 in respiratory research by Hinks et al. [11]. The role of previously identified genes was tested within the U-BIOPRED dataset using Gene Set Variation Analysis (GSVA) and using biomarkers of interest as metadata superimposed on to the topological data analysis (TDA) networks. Another novelty produced by this collective project was the AsthmaMap, a machine- and human-readable representation of mechanisms underlying asthma, an open project driven by a community of experts in systems biomedicine and computational systems biologists. Its goal was to provide a means of integrating existing knowledge about asthma from molecular biology, cell biology and medicine into a molecular and cellular interaction map of the disease to make possible advanced data interpretation, hypothesis generation, and validation for many related projects [12]. This map offered a conceptual model of disease mechanisms that could be updated and transformed into executable dynamic models for computational hypothesis testing and predictions.

This chapter will review the methods used, with the aim for readers to be able to appreciate their values, complexity and limitations, so that they can consider using them in their own research.

Methods for Data Integration

The initial step in data integration is often the consolidation of data from many sources into a single centralised location, which is often a data warehouse. One example of such a data warehouse is the aforementioned U-BIOPRED data management platform, tranSMART [8]. In the context of systems biology, data integration methods are statistical and machine learning methods for the integrative analysis of two or more datasets [13]. This chapter highlights examples of the array of methods developed for integrating high-dimensional datasets.

Why do we need to integrate data in systems biology? Systems biology tends to use a holistic approach to biological research, analysing the impact of large numbers of molecules rather than studying the effect of one molecule. This holistic approach has been made possible by the advent of omics technologies [1], which identify, characterise and quantify tens of thousands of biomolecules involved in the structure, function and dynamics of a cell, tissue or whole organism. Furthermore, the method holds out the possibility of shedding light on health and disease by examining how these biomolecules are affected by internal and external perturbations [14].

Most omics technologies are reproducible and, with the continually reducing costs, it is possible to interrogate samples from a significantly greater number of subjects than previously, which results in an avalanche of data. Many computational tools have been created to analyse single omics data. At the same time, data integration methods have been developed to combine more than one dataset to create a comprehensive view of a given disease or a biological process. In the following sections, we briefly discuss some questions that can be answered using omics data and how our answers could be enhanced using data integration; next, we will describe data integration strategies. None of these lists are exhaustive, but they represent a great effort by numerous researchers across many therapeutic areas. Finally, we will explain some of the data integration methods in greater detail.

Before elaborating on the data integration methods, it is important to understand what questions are being investigated using omics data sets and how we could augment our answers using data integration. A list of frequently asked questions includes: (a) what are the genes/proteins/microbiomes that are elevated in patients (e.g., asthmatics) compared to healthy participants; this type of question can be answered using multivariate statistics, (b) how many clusters of patients are there in the one omics data set? This type of question can be answered by using clustering methods, and (c) can we predict responses to medicines for individual patients based on multiple datasets (e.g., transcriptomics, proteomics and metabolomics) derived from the study of many; this type of question can be answered, at least in part, by using classification methods. Many studies have been published investigating these questions, often using data available through public repositories, such as the gene expression omnibus (https://www.ncbi.nlm.nih.gov/geo/).

The combination of publicly available data sets, either with or without personal data, together with data integration, can generate reproducible results and greater statistical power. For example, to answer the questions mentioned above, under (a), microarray or RNA-seq meta-analysis can be considered a type of data integration that can be used to integrate publicly available datasets. For question (b), data integration methods, such as similarity network fusion (SNF) [15, 16], can be used to include different omics datasets (e.g., proteomics and microbiome) in order to group individual patients or controls based on more than one omics dataset. For question (c), deep learning methods, such as variational autoencoders and multiblock partial least squares discriminant analysis classifiers, can be used to predict the drug response based on not only one data set but on the incorporation of many datasets to improve classification performance.

Data integration methods can be categorised by their strategy to combine data sets, including early integration, intermediate integration and late integration. Early integration is done by concatenating data sets, intermediate integration requires a data integration method that can manipulate and fuse all datasets, and late integration involves analysing each data set separately and combining the results. Intermediate integration is more desirable as early integration might be restrictive when data types are different; for example, the integration of protein sequences which are variable length strings with RNA-seq data, is not possible using early integration.

SNF and methods put together under Data Integration Analysis for Biomarker Discovery using Latent cOmponents (DIABLO) [17] are computationally efficient procedures for the integration of datasets. While SNF is used for clustering after integration and answers questions such as how many clusters of patients are present in the data based on multiple datasets (i.e., unsupervised clustering), DIABLO focuses on integrating multiple datasets to improve prediction performance (i.e., supervised classification). When given two or more data sets for the same set of samples (e.g., patients), SNF initially builds a network for each datatype followed by an affinity matrix. The subsequent step involves updating each network iteratively with network matrices from all other data sets, gradually aligning them closer together. Finally, these networks are fused into a single similarity network, i.e., an SNF-combined network, and the SNF-combined network is used for clustering. In DIABLO, latent components, which are linear combinations of variables, are created and the goal is to maximise the sum of covariances between all pairs of datasets. According to the design matrix (which, in its simplest format, defines the outcome we want to predict using multi-omics), each pairwise covariance is given a weight. When using a DIABLO model to predict novel samples, one prediction is created for each dataset, aggregating these predictions in a majority vote.

In this section, we have highlighted some reasons for and approaches to data integration. However, a bioinformatics pipeline needs to be developed for a successful application of the data integration method. This multi-compartment workflow begins with quality control and the normalisation of data sets. Each dataset might have a different noise pattern, which necessitates a different pre-processing technique. After pre-processing and correction for batch and technical effects, feature selection might be necessary. After data integration, using methods such as SNF or DIABLO, functional enrichment analyses of genes or proteins are performed. Data visualisation methods are also plugged in between communication of the results or quality control.

In addition to 'classical' methods, new ones have been developed recently, such as topological analysis [18]. Data acquired by omics methods represent the 'complete picture' of the molecular landscape of a biological system that is high-dimensional, typically consisting of thousands of data points. Whilst standard statistical and bioinformatic methods require and are limited by the application of dimension reduction procedures, topological analysis does not. Topology is the measurement of geometric properties of shape which are preserved despite deformations, allowing the identification of patterns within omics data representative of biological functions, robust to the noise characteristic of each omics dataset. Topological-based data analysis (TDA) uses topology to measure the shape of data, meaning that TDA can be used to identify biological pathways discriminating molecular phenotypes.

An important consideration with omics data is that, whilst it does represent the entire molecular landscape of a biological system, standard approaches reduce this dataset by 90 to >99%. A next-generation RNA-seq transcriptomics dataset may contain one million data points, representing unique mRNAs, and a standard differentially expressed gene (DEG) analysis would reduce this to <3000 data points, which amounts to >99% reduction. The problem here is that 'omics data are very noisy due to large biological and technical noise'. An emerging method of topological analysis of omics data, TopMD, uniquely does not apply feature selection, or data reduction; instead, it measures the landscape of global omics data relative to the underlying biological network, defining the peaks or features of the dataset, which are differentially modulated biological pathways [19]. TDA has been applied to identify and understand molecular phenotypes of disease [20–22]. In the analysis of the U-BIOPRED omics datasets of asthma cohorts to date, patients have been stratified by TDA of lipidomic, proteomic and transcriptomic data. In these analyses, patients were separated by reduced feature lenses, such as principal component analysis (PCA) and multi-dimensional scaling (MDS), and then clustered by a correlation metric. The resulting TDA networks contained clusters of patients, with similar molecular profiles and could be easily interrogated for enrichment of metadata, such as measured clinical characteristics, enabling the cluster of individual clinical features to be linked to omics clusters.

Different data types from a single biological sample can be integrated by TDA, for instance, in TDA Mapper, where data types can be used as 'lenses' or axes to separate the samples in different 'data type' dimensions, and then clustered into a network with a similarity metric, thereby defining groups of biological samples characterised by similar biological mechanisms. When analysing multi 'omics data of a patient cohort, patients can be separated by transcriptomic, proteomic, lipidomic and metabolomic data lenses as different dimensions, and clustered by a similarity metric, such as a correlation matrix. Gene set enrichment analysis (GSEA) has been used to characterise biological mechanisms, by matching the lists of most differentially expressed molecules in a TDA-derived cluster of patients to library gene sets, including inflammatory pathways and transcription factor-induced genes [10].

In U-BIOPRED, TDA was applied in the analysis of asthma cohorts to go beyond stratification by eosinophil and neutrophil counts. Molecular sub-phenotypes of asthma were defined by proteomic signatures for improved stratification. Proteomic data generated by unbiased label-free quantitative mass spectrometry and topological data analysis measured the proteomes of sputum supernatants from 206 participants with asthma and 40 healthy participants as a novel means of asthma stratification [10]. Transcriptomics, for the same participants, was generated by microarray analysis of sputum cells and analysed across the TDA network to reveal underlying mechanisms. Ten proteomic clusters or molecular sub-phenotypes of asthma were defined and found to be enriched for different granulocyte counts. Of the ten clusters, two were highly atopic, with comparatively low granulocytic inflammation, three were eosinophilic, and three were highly neutrophilic. GSEA of matched transciptomics data allowed measurement of differentially activated pathway across these clusters, and logistic regression analysis identified putative protein biomarkers which could be validated for future stratification of clinical cohorts.

Having provided some core understanding of the analytical tools, the chapter now discusses clinical phenotyping and gives examples of how the tools have enhanced our understanding of asthma mechanisms.

Clinical Phenotypes of Severe Asthma

A phenotype is defined as the composite observable characteristics or traits of an organism that result from genetic as well as environmental influences. Phenotypes of asthma, described from an analysis of clinico-physiologic-inflammatory characteristics, have been developed since the middle of the last century by clinicians, including various descriptive phenotypes, such as early childhood-onset and late adult onset, severe allergic asthma, eosinophilic asthma, exacerbation-prone asthma, asthma with chronic airflow obstruction, obesity-associated asthma, cigarette smoking-associated asthma and corticosteroid-insensitive asthma (Table 1) [23]. These phenotypes were based on the association of asthma with specific circumstances, clinical features and blood measurements but have been limited in terms of delivery of precision medicine to the individual asthma patient. In this domain, one

Table 1 Clinical phenotypes of severe asthma

1.	Early onset allergic asthma
2.	Late-onset eosinophilic asthma
3.	Exacerbation-prone asthma
4.	Chronic airflow obstruction
5.	Obesity-associated asthma
6.	Cigarette smoking asthma
7.	Corticosteroid-insensitive asthma

significant and useful observation in terms of therapeutic approach has been the association of eosinophilic asthma, diagnosed by either high blood or sputum eosinophils, with good clinical response to corticosteroid treatment [24] and, more recently, biologicals that target specific molecular pathways [25, 26].

Approaches to Determining Pathways and Underlying Mechanisms

While useful to the clinician, the determination of clinical phenotypes has not been able to inform on the details of the mechanistic processes at play in determining the severity of asthma. Rather, linking the characteristics or traits of severe asthma to transcripts or proteins is more likely to lead to a description of molecular phenotypes or discovery of associated pathways with clinical traits [27] and to deliver the molecular phenotypes or endotypes [28] (Fig. 1). At its source, the U-BIOPRED cohort was initially assembled using standard, widely applied criteria of asthma severity and smoking history to group participants into the following categories: severe non-smoking asthmatics, severe smoker and ex-smoker asthmatics and mild-moderate non-smoking asthmatics, with a control group of non-smoking, non-asthmatic healthy participants [29]. On the basis of this classification [30, 31], patients with severe asthma had more symptoms and exacerbations compared to patients with mild/moderate disease, with worse quality of life and higher levels of

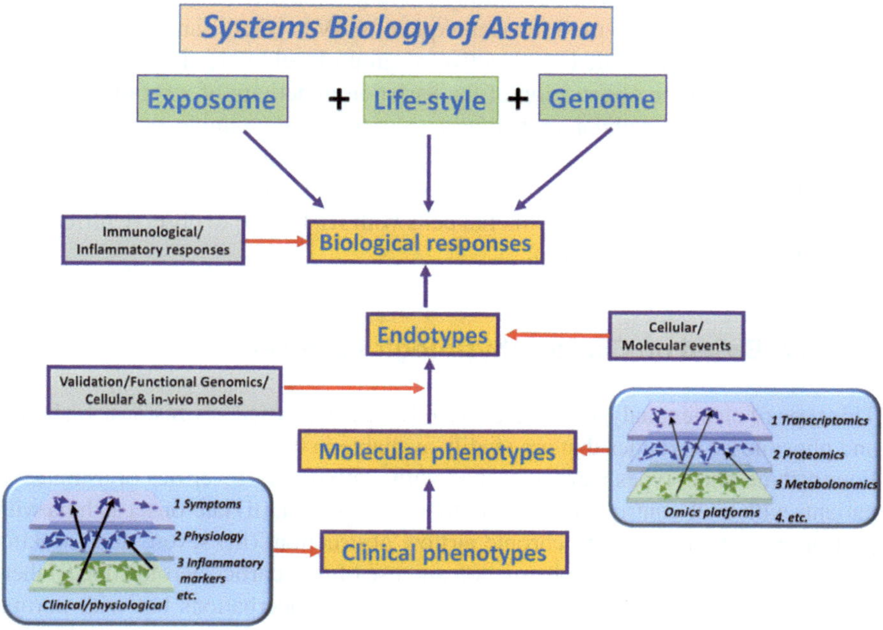

Fig. 1 The journey from clinical description to systems biology

anxiety and depression [29]. They also had a higher incidence of nasal polyps and gastro-oesophageal reflux with lower lung function. The most prevalent treatable traits were more common in severe asthma, with an increased prevalence of extra-pulmonary and behavioural or psychosocial traits that may reflect the impact of living with severe chronic respiratory conditions [32]. Sputum eosinophil counts but not blood eosinophil counts were higher in severe asthma compared to mild/moderate asthma [29]. In addition to these 'simple', readily available biomarkers, samples of blood, sputum, urine, nasal brushings and bronchoscopic specimens, bronchial brushings and bronchial biopsies were also obtained for various omics platform analyses. Bioinformatic 'large data' analytical approaches were taken in order to link the clinical, physiologic and inflammatory characteristics with the omics data so that mechanistic pathways could be derived [14].

Two approaches to identifying phenotypes were used in U-BIOPRED. The first approach was to see whether clinical characteristics or traits could provide clinical phenotypes from the application of hierarchical clustering techniques. Then, differences between clinical phenotypes in terms of the expression of transcripts or proteins could be assessed. Another approach, which could have been taken, but was not, is to undertake unsupervised analysis of the differentially expressed genes between extreme expression of specific clinical traits. The second approach used in U-BIOPRED consisted of taking individual biomarkers, e.g. sputum proteins, or gene expression data from airways-derived samples, and clustering these in an unbiased manner by hierarchical clustering and supervised learning algorithm to refine any signatures for the identified molecular clusters. Thereafter, these pathobiological clusters were characterised according to clinical features [33].

U-BIOPRED has also used the technique of GSVA, where sample-wise enrichment scores were calculated irrespective of group labels [34]. This unsupervised approach allowed the observation of the variation in activity of a set of genes that represent a specific cellular or pathway over the entire sample population. For example, to examine the expression of type-2 (T2) inflammation-related gene sets, U-BIOPRED used a T2 dataset of 34 genes that were shown to be upregulated after in vitro stimulation of airway epithelial cells with IL-13 [35] to observe differences between clinical phenotypes [36].

Clusters Based on Clinico-physiologic Features

This type of cluster analysis of the U-BIOPRED cohort used key historical, clinical and physiologic variables that are readily available in the general respiratory clinic. Four asthma phenotypes were described with a mixture of severities of asthma: (i) patients with well-controlled moderate-to-severe asthma, (ii) late-onset asthma with past or current smoking and chronic airflow obstruction, (iii) severe non-smoking asthma with chronic airflow obstruction and use of oral corticosteroids (OCS) therapy, and (iv) obese female patients with frequent exacerbations but with normal lung function [37] (Fig. 2). Two of the clusters (clusters 1 and 2) were related to chronic airflow obstruction, with cluster 1 associated with smokers and ex-smokers

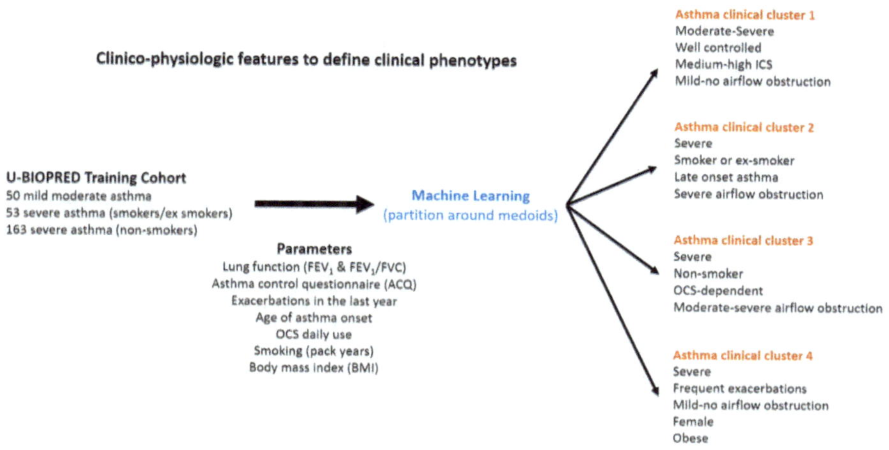

Fig. 2 Asthma clinical clusters (1–4) from the U-BIOPRED cohort are defined purely by clinical-physiologic features using partition-around-medoids clustering. (Based on Ref. [37])

with late-onset asthma with the highest blood and sputum eosinophil counts, whereas cluster 3 was associated with non-smokers receiving OCS therapy. These clusters were then characterised by molecular features. Using the focussed Somascan proteomics platform, IL-16 (lymphocyte chemoattractant factor) was detected as differentially abundant in sputum when comparing both severe asthma clusters with airflow obstruction (clusters 2 and 3) with patients with well-controlled asthma (cluster 1). Cluster 2 showed higher levels than cluster 1 of CTAP-III (CXCL7) and GM-CSF, which could reflect the effect of smoking exposure. Cluster 3 had lower sputum levels of cathepsin G compared with clusters 1 and 4.

Omics Analysis of Clinical Phenotypes Using GSVA and DEGs

In a further study aiming to give more insight into clinical phenotypes, analysis using >100 GSVA signatures was undertaken for each of the clinical phenotypes listed in Table 1. Late-onset eosinophilic asthma was analysed by applying GSVA of a signature of IL-13 regulated genes in airway epithelium; this revealed that the T2-high phenotype was more obstructed but showed no differences in exacerbations, with worse asthma control and high blood and sputum eosinophils [36]. Significant differentially enriched gene signatures were identified in nasal brushings, sputum, and endobronchial brushings in adult-onset severe asthma patients compared to childhood-onset severe asthma [38]. Signatures were associated with eosinophilic airway inflammation, mast cells, and group 3 innate lymphoid cells (ILC) were more enriched in adult-onset asthma [38]. In another study of patients with chronic airflow obstruction, differentially enriched gene signatures were associated with corticosteroid sensitivity, eosinophils, interleukin-13, interferon (IFN)α, specific CD4+ T cells and airway remodelling [39]. Analysis of frequent exacerbators (two or more severe exacerbations reported in the previous year) versus infrequent exacerbations has revealed higher type 1 (T1) and T2 inflammatory pathways with steroid insensitivity pathways. Carcinoembrionic antigen-related cell adhesion molecule 5 (CEACAM5) was the only differentially expressed single gene between frequent and infrequent exacerbators in the bronchial biopsy samples [40].

For obesity-associated asthma, an exploratory analysis of epithelial gene expression in severe asthmatics from the U-BIOPRED study identified three clusters one of which is enriched for obesity, gastro-oesophageal reflux disease and treatment with proton pump inhibitors. This represents a pauci-granulocytic group as a consequence of reduced mechanisms of cell recruitment linked to bile acid exposure and treatment with proton pump inhibitors [41].

In an analysis of bronchial biopsies and bronchial brushings for corticosteroid insensitivity, two subtypes of patients were identified as being associated with T2 and corticosteroid insensitivity [42]. Further analysis of genes that were associated with corticosteroid insensitivity and Th2 signatures showed an association between CD44 and T2-associated genes CCL26, IL1R2 and CST2. These were also associated with signatures of oxidative stress, and the patients were on the highest dose of daily oral corticosteroids [42]. Current smoking severe asthma was associated with the enrichment of oxidative stress and ER stress, while CXCL5 and MMP12 were upregulated in ex-smoking severe asthma patients [43]. The value of the GSVA or GSEA signature depends upon its selectivity for the pathway, cell or function that it represents. Removal of overlapping genes across signatures often enhances the specificity of the signal obtained, aiding in the identification of molecular phenotypes [44].

Mechanistic Pathways: Endotypes and Treatable Traits

Distinct from clinical phenotypes is the definition of endotypes, which links the clinical expression of disease to driving mechanisms. As defined by Gary Anderson in 2008, an endotype '…is a subtype of disease defined functionally and pathologically by a molecular mechanism or by treatment response…' [28]. Due to the heterogeneity of asthma and the myriad of genetic and environmental factors involved in its pathogenesis, it is likely that asthma is not composed of a single endotype, and that many distinct endotypes (co)-exist, each associated with specific clinical features but reflecting differing molecular causes and clinical responses to defined therapies.

To date, we have only one defined asthma endotype based on the presence of T2 cytokines and alarmins and high levels of blood and sputum eosinophils and FeNO [45, 46]. It is of interest that the molecular phenotype, upon which this T2 endotype is based, is derived from IL-13-stimulated human primary bronchial epithelial cells (HBECs), and it is noteworthy that anti-IL-13-directed monoclonal antibodies have failed in clinical development as a therapeutic for asthma. It would be useful to examine whether there is an even greater link between the clinical features of this asthma endotype with IL-5-stimulated HBECs [47].

Several groups and consortia have collected samples from different cells, tissues and compartments including blood, urine, breath and bronchoalveolar lavage, from patients with differing asthma severities or clinical characteristics [48]. Multi-omics analysis has been conducted on these samples with the intention of elucidating differentially expressed factors that may give insight into molecular pathways or processes that distinguish asthmatic patients with distinct clinical features and/or biochemical and physiological features [1, 49]. These omics platforms have included genetics, epigenetics, transcriptomics (gene array, bulk and single-cell RNA-seq), proteomics (unbiased and focussed multiplexing), lipidomics, metabolomics, metagenomics and breathomics [1]. A number of distinct molecular phenotypes have been described using this approach, and only a selection of these will be highlighted here.

Sputum Transcriptomics

Analysis of sputum transcriptomics and validation using focussed proteomics within the U-BIOPRED cohort identified a subset of asthmatic patients with inflammasome activation and sputum neutrophilia [50]. A concurrent semi-biased transcriptomic analysis together with unsupervised machine learning analysis of asthma sputum granulocytic samples compared with non-granulocytic asthma samples identified three transcriptome-associated clusters (TAC) [33] (Fig. 3). The first cluster, TAC1, was characterised by severe oral corticosteroid-dependent asthma, frequent exacerbations and severe airflow obstruction, highest sputum eosinophilia and FeNO, together with the expression of IL33R, CCR3 and TSLPR genes and the greatest enrichment of the IL-13/T2-inflammation gene signature. In contrast, TAC2

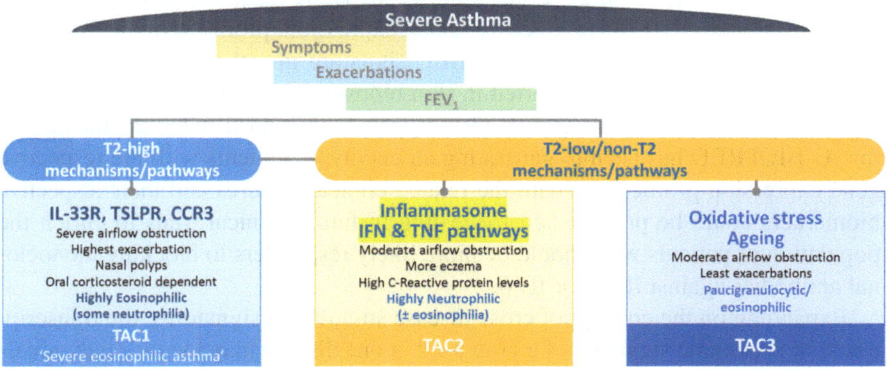

Fig. 3 Severe asthma is an umbrella term defined according to a set of clinico-physiologic features including symptoms, occurrence of exacerbations and the level of airflow obstruction according to an FEV1 measurement. These patients are subsequently classified into transcriptome-associated clusters (TAC1, TAC2 & TAC3) defining distinct molecular phenotypes as identified by hierarchical clustering of differentially expressed genes and by gene set variation analysis of specific gene sets expressed in induced sputum. T2: Type 2 inflammation. (Based on Ref. [33])

was characterised by sputum neutrophilia, inflammasome activation, IFNα and tumour necrosis factor-α–associated genes, high serum C-reactive protein levels, and a higher prevalence of eczema. However, this phenotype was also characterised by high expression of an IL-13 T2 signature, indicating that this phenotype is perhaps driven by both T1 and T2 mechanisms. TAC3 was associated with pauci-granulocytic asthma, with little airflow obstruction, and the highest expression of metabolic, mitochondrial and ubiquitination pathway genes. Of interest, in a separate study, TDA analysis of sputum proteomics data identified three potential subsets or activation states of eosinophilic and neutrophilic asthma [10] how these map to different TAC states should be studied. For endotype characterisation, future studies should investigate the potential of inflammasome inhibitors in TAC2 asthmatic subjects or of blocking the IL-1R3 co-receptor-regulated three key signalling pathways – IL-1, IL-33 and IL-36. Inflammasome inhibitors block neutrophilic asthma in mouse models [50], and a blocking mAb to human IL-1R3 (MAB-hR3) inhibits signalling via IL-1, IL-33 and IL-36 in vitro and in three distinct in vivo models of disease (crystal-induced peritonitis, allergic airway inflammation and psoriasis) [51].

Gene Signatures from Activated Airway Epithelium

In the same way that the IL-13-stimulated HBEC signature was used to identify patients with T2 asthma, stimulation of HBECs with non-T2 stimuli has been used to define patients with specific subtypes of asthma. For example, exposure of HBECs to a combination of IL-6 and sIL-6R was shown to lead to the activation of IL-6 trans-signalling (IL-6TS). This generates a gene signature that is distinct from that seen with IL-4/IL-13 HBEC stimulation and is upregulated in asthmatic subjects who do not express high levels of the T2 signature [52]. Similarly, stimulation of HBECs with IL-17A produces a gene expression profile that is expressed inversely with the IL-13-stimulated HBEC signature in asthmatic patients whose gene pattern resembles that reported in skin biopsies of patients with IL-17-driven psoriasis [53]. These two molecular phenotypes, IL-17 and IL-6TS, are the closest that U-BIOPRED has been to generating an endotype. Patients with the respective gene expression profiles and with the related clinical features and their respective biomarkers could be pre-selected for testing in future clinical trials to enrich the population of patients who should be more likely responders to blocking monoclonal antibodies against IL-17 or IL-6.

Expanding on the concept of cross-disease stimulation signatures, the transcriptomic or proteomic signature of a drug used in one disease may identify individuals who may respond to the same drug in a different disease [54]. Using transcriptomic signatures from the skin of patients with atopic dermatitis who had a good response clinically to the anti-IL-22 monoclonal antibody, Fezakinumab, it was possible to identify patients with significant enrichment of the response signature and high levels of sputum IL-22 in patients with severe neutrophilic asthma. Nasal epithelial brushings also gave a good prediction of the potential responder population. Clinical features and biomarkers identified could, therefore, be used to enrich patients in clinical trials of Fezakinumab in asthma.

Use of Gene Signatures from Activated Mast Cells

Mast cells have been widely implicated in asthma, with evidence pointing that they play an important role both in inflammation as well as in innate and adaptive immunity, but how they are involved in driving the pathways of severe asthma still remains ill-defined. To investigate this, gene signatures of mast cell activation activated not only through IgE-FceR1 interaction but also through IL-33 receptor, TLR4 and IFN receptor were used to allow detection of severe asthma endotypes. The IL-33-stimulated mast cell signature was shown to be associated with severe neutrophilic asthma, whereas the IgE-activated mast cell signature was associated with an eosinophilic phenotype [44]. These results indicated that mast cells could be induced to take on distinct transcriptional phenotypes associated with specific clinical phenotypes.

Bronchoalveolar Lavage Cells

Analysis of cells obtained from bronchoalveolar lavage by Wenzel and co-workers have identified high levels of IFNγ inversely related to secretory leukocyte peptidase inhibitor (SLPI) in patients with severe T2 low asthma treated with high dose corticosteroid therapy [55]. These results have been validated in a mouse model of severe asthma and shown to be associated with increased levels of CXCR3/CXCL10 [56] and of CCR5 [57]. Modulation of these chemokine pathways in a mouse model of Th1-high severe asthma, particularly of CCR5 using the maraviroc, the FDA-approved drug for HIV infection, blunted airway hyperresponsiveness but only partially reduced airway inflammation [57]. Trials of maraviroc, in these IFNγ-high severe asthma patients should be considered. Interestingly, CXCR3 is expressed by mast cells which may represent key target cells in all types of severe asthma [44].

Metabolomics

Metabolite levels may delineate stable severe asthma sub-phenotypes or metabotypes and also define druggable targets. Thus, urinary carnitine and related metabolites form a distinct cluster that does not alter with oral corticosteroid use and is associated with decreased fatty acid metabolism and expression of the carnitine transporter SLC22A5 in bronchial brushings and sputum. Reduced expression of SLC22A5 in U-BIOPRED was linked to specific genetic signals in its promoter [58]. Other non-invasive biomarkers including VOCs were detected using U-BIOPRED by eNOSEs or gas chromatography-mass spectroscopy (GC-MS) analysis of exhaled breath. Indeed, breathomic signatures were shown to delineate distinct TAC subsets [59].

Conclusion

Overall, several different immune and inflammatory pathways and factors have been reported to be upregulated or overactive in patients with severe asthma. Many of these pathways overlap in patients with distinct clinical features of asthma, which may explain the incomplete or no response in patients undergoing specific targeted therapy. A greater understanding of the complex interactions between these pathways and the identification of biomarkers that reflect pathways may result in better outcomes with individual patient-specific therapies [27, 60]. The advent of single-cell multi-omics and spatial transcriptomic and proteomics technologies is beginning to revolutionise our understanding of asthma mechanisms with the discovery of new cell types in the disease [61]. However, more studies in distinct asthma subtypes of severe asthma are required. These data can be used to re-analyse existing datasets using pseudo single analysis or for more detailed analysis of pathway perturbations [49]. Finally, prospective studies are needed with non-invasive markers to provide robust phenotype-based enrichment of patients to be enrolled into clinical trials, preferably using readily accessible biomarkers detected in blood, nasal, breath and urine samples.

References

1. Wheelock CE, Goss VM, Balgoma D, Nicholas B, Brandsma J, Skipp PJ, Snowden S, Burg D, D'Amico A, Horvath I, Chaiboonchoe A, Ahmed H, Ballereau S, Rossios C, Chung KF, Montuschi P, Fowler SJ, Adcock IM, Postle AD, Dahlen SE, Rowe A, Sterk PJ, Auffray C, Djukanovic R, U-BIOPRED Study Group. Application of 'omics technologies to biomarker discovery in inflammatory lung diseases. Eur Respir J. 2013;42:802–25.
2. Brinkman P, van de Pol MA, Gerritsen MG, Bos LD, Dekker T, Smids BS, Sinha A, Majoor CJ, Sneeboer MM, Knobel HH, Vink TJ, de Jongh FH, Lutter R, Sterk PJ, Fens N. Exhaled breath profiles in the monitoring of loss of control and clinical recovery in asthma. Clin Exp Allergy. 2017;47:1159–69.
3. Burg D, Schofield JPR, Brandsma J, Staykova D, Folisi C, Bansal A, Nicholas B, Xian Y, Rowe A, Corfield J, Wilson S, Ward J, Lutter R, Fleming L, Shaw DE, Bakke PS, Caruso M, Dahlen SE, Fowler SJ, Hashimoto S, Horváth I, Howarth P, Krug N, Montuschi P, Sanak M, Sandström T, Singer F, Sun K, Pandis I, Auffray C, Sousa AR, Adcock IM, Chung KF, Sterk PJ, Djukanović R, Skipp PJ, U-BIOPRED Study Group. Large-scale label-free quantitative mapping of the sputum proteome. J Proteome Res. 2018;17:2072–91.
4. Brandsma J, Goss VM, Yang X, Bakke PS, Caruso M, Chanez P, Dahlén SE, Fowler SJ, Horvath I, Krug N, Montuschi P, Sanak M, Sandström T, Shaw DE, Chung KF, Singer F, Fleming LJ, Sousa AR, Pandis I, Bansal AT, Sterk PJ, Djukanović R, Postle AD. Lipid phenotyping of lung epithelial lining fluid in healthy human volunteers. Metabolomics. 2018;14:123.
5. Riley JH, Erpenbeck VJ, Matthews JG, Holweg CTJ, Compton C, Seibold W, Higenbottam T, Wagers S, Rowe A, Myles D. U-BIOPRED: evaluation of the value of a public-private partnership to industry. Drug Discov Today. 2018;23:1622–34.
6. Kupczyk M, Wenzel S. U.S. and European severe asthma cohorts: what can they teach us about severe asthma? J Intern Med. 2012;272:121–32.
7. Auffray C, Adcock IM, Chung KF, Djukanovic R, Pison C, Sterk PJ. An integrative systems biology approach to understanding pulmonary diseases. Chest. 2010;137:1410–6.

8. Bussery J, Denis LA, Guillon B, Liu P, Marchetti G, Rahal G. eTRIKS platform: conception and operation of a highly scalable cloud-based platform for translational research and applications development. Comput Biol Med. 2018;95:99–106.
9. Bigler J, Boedigheimer M, Schofield JPR, Skipp PJ, Corfield J, Rowe A, Sousa AR, Timour M, Twehues L, Hu X, Roberts G, Welcher AA, Yu W, Lefaudeux D, Meulder B, Auffray C, Chung KF, Adcock IM, Sterk PJ, Djukanović R. A severe asthma disease signature from gene expression profiling of peripheral blood from U-BIOPRED cohorts. Am J Respir Crit Care Med. 2017;195:1311–20.
10. Schofield JPR, Burg D, Nicholas B, Strazzeri F, Brandsma J, Staykova D, Folisi C, Bansal AT, Xian Y, Guo Y, Rowe A, Corfield J, Wilson S, Ward J, Lutter R, Shaw DE, Bakke PS, Caruso M, Dahlen SE, Fowler SJ, Horváth I, Howarth P, Krug N, Montuschi P, Sanak M, Sandström T, Sun K, Pandis I, Riley J, Auffray C, De Meulder B, Lefaudeux D, Sousa AR, Adcock IM, Chung KF, Sterk PJ, Skipp PJ, Djukanović R. Stratification of asthma phenotypes by airway proteomic signatures. J Allergy Clin Immunol. 2019;144:70–82.
11. Hinks TS, Brown T, Lau LC, Rupani H, Barber C, Elliott S, Ward JA, Ono J, Ohta S, Izuhara K, Djukanovic R, Kurukulaaratchy RJ, Chauhan A, Howarth PH. Multidimensional endotyping in patients with severe asthma reveals inflammatory heterogeneity in matrix metalloproteinases and chitinase 3-like protein 1. J Allergy Clin Immunol. 2016;138:61–75.
12. Mazein A, Ivanova O, Balaur I, Ostaszewski M, Berzhitskaya V, Serebriyskaya T, Ligon T, Hasenauer J, De Meulder B, Overall RW, Roy L, Knowles RG, Wheelock CE, Dahlen SE, Chung KF, Adcock IM, Roberts G, Djukanovic R, Pellet J, Gawron P, Balling R, Maitland-van der Zee AH, Schneider R, Sterk PJ, Auffray C. AsthmaMap: an interactive knowledge repository for mechanisms of asthma. J Allergy Clin Immunol. 2021;147:853–6.
13. Gligorijević V, Pržulj N. Methods for biological data integration: perspectives and challenges. J R Soc Interface. 2015;12:20150571.
14. De Meulder B, Lefaudeux D, Bansal AT, Mazein A, Chaiboonchoe A, Ahmed H, Balaur I, Saqi M, Pellet J, Ballereau S, Lemonnier N, Sun K, Pandis I, Yang X, Batuwitage M, Kretsos K, van Eyll J, Bedding A, Davison T, Dodson P, Larminie C, Postle A, Corfield J, Djukanovic R, Chung KF, Adcock IM, Guo YK, Sterk PJ, Manta A, Rowe A, Baribaud F, Auffray C. A computational framework for complex disease stratification from multiple large-scale datasets. BMC Syst Biol. 2018;12:60.
15. Wang B, Mezlini AM, Demir F, Fiume M, Tu Z, Brudno M, Haibe-Kains B, Goldenberg A. Similarity network fusion for aggregating data types on a genomic scale. Nat Methods. 2014;11:333–7.
16. Zounemat Kermani N, Saqi M, Agapow P, Pavlidis S, Kuo C, Tan KS, Mumby S, Sun K, Loza M, Baribaud F, Sousa AR, Riley J, Wheelock AM, Wheelock CE, De Meulder B, Schofield J, Sánchez-Ovando S, Simpson JL, Baines KJ, Wark PA, Auffray C, Dahlen SE, Sterk PJ, Djukanovic R, Adcock IM, Guo YK, Chung KF. Type 2-low asthma phenotypes by integration of sputum transcriptomics and serum proteomics. Allergy. 2021;76:380–3.
17. Singh A, Shannon CP, Gautier B, Rohart F, Vacher M, Tebbutt SJ, Lê Cao KA. DIABLO: an integrative approach for identifying key molecular drivers from multi-omics assays. Bioinformatics. 2019;35:3055–62.
18. Skaf Y, Laubenbacher R. Topological data analysis in biomedicine: a review. J Biomed Inform. 2022;130:104082.
19. Penrice-Randal R, Dong X, Shapanis AG, Gardner A, Harding N, Legebeke J, Lord J, Vallejo AF, Poole S, Brendish NJ, Hartley C, Williams AP, Wheway G, Polak ME, Strazzeri F, Schofield JPR, Skipp PJ, Hiscox JA, Clark TW, Baralle D. Blood gene expression predicts intensive care unit admission in hospitalised patients with COVID-19. Front Immunol. 2022;13:988685.
20. Lum PY, Singh G, Lehman A, Ishkanov T, Vejdemo-Johansson M, Alagappan M, Carlsson J, Carlsson G. Extracting insights from the shape of complex data using topology. Sci Rep. 2013;3:1236.
21. Nielson JL, Paquette J, Liu AW, Guandique CF, Tovar CA, Inoue T, Irvine KA, Gensel JC, Kloke J, Petrossian TC, Lum PY, Carlsson GE, Manley GT, Young W, Beattie MS, Bresnahan

JC, Ferguson AR. Topological data analysis for discovery in preclinical spinal cord injury and traumatic brain injury. Nat Commun. 2015;6:8581.
22. Nicolau M, Levine AJ, Carlsson G. Topology based data analysis identifies a subgroup of breast cancers with a unique mutational profile and excellent survival. Proc Natl Acad Sci U S A. 2011;108:7265–70.
23. Chung KF. Asthma phenotyping: a necessity for improved therapeutic precision and new targeted therapies. J Intern Med. 2016;279:192–204.
24. Green RH, Brightling CE, McKenna S, Hargadon B, Parker D, Bradding P, Wardlaw AJ, Pavord ID. Asthma exacerbations and sputum eosinophil counts: a randomised controlled trial. Lancet. 2002;360:1715–21.
25. Nair P, Pizzichini MM, Kjarsgaard M, Inman MD, Efthimiadis A, Pizzichini E, Hargreave FE, O'Byrne PM. Mepolizumab for prednisone-dependent asthma with sputum eosinophilia. N Engl J Med. 2009;360:985–93.
26. Haldar P, Brightling CE, Hargadon B, Gupta S, Monteiro W, Sousa A, Marshall RP, Bradding P, Green RH, Wardlaw AJ, Pavord ID. Mepolizumab and exacerbations of refractory eosinophilic asthma. N Engl J Med. 2009;360:973–84.
27. Chung KF, Adcock IM. Precision medicine for the discovery of treatable mechanisms in severe asthma. Allergy. 2019;74:1649–59.
28. Anderson GP. Endotyping asthma: new insights into key pathogenic mechanisms in a complex, heterogeneous disease. Lancet. 2008;372:1107–19.
29. Shaw DE, Sousa AR, Fowler SJ, Fleming LJ, Roberts G, Corfield J, Pandis I, Bansal AT, Bel EH, Auffray C, Compton CH, Bisgaard H, Bucchioni E, Caruso M, Chanez P, Dahlen B, Dahlen SE, Dyson K, Frey U, Geiser T, Gerhardsson de Verdier M, Gibeon D, Guo YK, Hashimoto S, Hedlin G, Jeyasingham E, Hekking PP, Higenbottam T, Horvath I, Knox AJ, Krug N, Erpenbeck VJ, Larsson LX, Lazarinis N, Matthews JG, Middelveld R, Montuschi P, Musial J, Myles D, Pahus L, Sandstrom T, Seibold W, Singer F, Strandberg K, Vestbo J, Vissing N, von Garnier C, Adcock IM, Wagers S, Rowe A, Howarth P, Wagener AH, Djukanovic R, Sterk PJ, Chung KF, U-BIOPRED Study Group. Clinical and inflammatory characteristics of the European U-BIOPRED adult severe asthma cohort. Eur Respir J. 2015;46:1308–21.
30. Bel EH, Sousa A, Fleming L, Bush A, Chung KF, Versnel J, Wagener AH, Wagers SS, Sterk PJ, Compton CH. Diagnosis and definition of severe refractory asthma: an international consensus statement from the Innovative Medicine Initiative (IMI). Thorax. 2011;66:910–7.
31. Chung KF, Wenzel SE, Brozek JL, Bush A, Castro M, Sterk PJ, Adcock IM, Bateman ED, Bel EH, Bleecker ER, Boulet LP, Brightling C, Chanez P, Dahlen SE, Djukanovic R, Frey U, Gaga M, Gibson P, Hamid Q, Jajour NN, Mauad T, Sorkness RL, Teague WG. International ERS/ATS guidelines on definition, evaluation and treatment of severe asthma. Eur Respir J. 2014;43:343–73.
32. Simpson AJ, Hekking PP, Shaw DE, Fleming LJ, Roberts G, Riley JH, Bates S, Sousa AR, Bansal AT, Pandis I, Sun K, Bakke PS, Caruso M, Dahlén B, Dahlén SE, Horvath I, Krug N, Montuschi P, Sandstrom T, Singer F, Adcock IM, Wagers SS, Djukanovic R, Chung KF, Sterk PJ, Fowler SJ. Treatable traits in the European U-BIOPRED adult asthma cohorts. Allergy. 2019;74:406–11.
33. Kuo CS, Pavlidis S, Loza M, Baribaud F, Rowe A, Pandis I, Sousa A, Corfield J, Djukanovic R, Lutter R, Sterk PJ, Auffray C, Guo Y, Adcock IM, Chung KF. T-helper cell type 2 (Th2) and non-Th2 molecular phenotypes of asthma using sputum transcriptomics in U-BIOPRED. Eur Respir J. 2017;49:1602135.
34. Hanzelmann S, Castelo R, Guinney J. GSVA: gene set variation analysis for microarray and RNA-seq data. BMC Bioinformatics. 2013;14:7.
35. Silkoff PE, Laviolette M, Singh D, FitzGerald JM, Kelsen S, Backer V, Porsbjerg CM, Girodet PO, Berger P, Kline JN, Chupp G, Susulic VS, Barnathan ES, Baribaud F, Loza MJ. Identification of airway mucosal type 2 inflammation by using clinical biomarkers in asthmatic patients. J Allergy Clin Immunol. 2017;140:710–9.
36. Pavlidis S, Takahashi K, Kwong FNK, Xie J, Hoda U, Sun K, Elyasigomari V, Agapow P, Loza M, Baribaud F, Chanez P, Fowler SJ, Shaw DE, Fleming LJ, Howarth PH, Sousa AR, Corfield

J, Auffray C, De Meulder B, Knowles R, Sterk PJ, Guo Y, Adcock IM, Djukanovic R, Chung KF. “T2-high” in severe asthma related to blood eosinophil, exhaled nitric oxide and serum periostin. Eur Respir J. 2019;53:1800938.

37. Lefaudeux D, De Meulder B, Loza MJ, Peffer N, Rowe A, Baribaud F, Bansal AT, Lutter R, Sousa AR, Corfield J, Pandis I, Bakke PS, Caruso M, Chanez P, Dahlén SE, Fleming LJ, Fowler SJ, Horvath I, Krug N, Montuschi P, Sanak M, Sandstrom T, Shaw DE, Singer F, Sterk PJ, Roberts G, Adcock IM, Djukanovic R, Auffray C, Chung KF. U-BIOPRED clinical adult asthma clusters linked to a subset of sputum omics. J Allergy Clin Immunol. 2017;139:1797–807.
38. Hekking PP, Loza MJ, Pavlidis S, de Meulder B, Lefaudeux D, Baribaud F, Auffray C, Wagener AH, Ir Brinkman P, Ir Lutter R, Bansal AT, Sousa AR, Bates SA, Pandis Y, Fleming LJ, Shaw DE, Fowler SJ, Guo Y, Meiser A, Sun K, Corfield J, Howarth PH, Bel EH, Adcock IM, Chung KF, Djukanovic R, Sterk PJ. Pathway discovery using transcriptomic profiles in adult-onset severe asthma. J Allergy Clin Immunol. 2017a;141:1280–90.
39. Hekking PP, Loza MJ, Pavlidis S, De Meulder B, Lefaudeux D, Baribaud F, Auffray C, Wagener AH, Brinkman P, Lutter R, Bansal AT, Sousa AR, Bates SA, Pandis I, Fleming LJ, Shaw DE, Fowler SJ, Guo Y, Meiser A, Sun K, Corfield J, Howarth P, Bel EH, Adcock IM, Chung KF, Djukanovic R, Sterk PJ. Transcriptomic gene signatures associated with persistent airflow limitation in patients with severe asthma. Eur Respir J. 2017b;50:1602298.
40. Hoda U, Pavlidis S, Bansal AT, Takahashi K, Hu S, Kwong FNK, Rossios C, Sun K, Bhavsar P, Loza M, Baribaud F, Chanez P, Fowler SJ, Horvath I, Montuschi P, Singer F, Musial J, Dahlen B, Krug N, Sandstrom T, Shaw DE, Lutter R, Fleming LJ, Howarth PH, Caruso M, Sousa AR, Corfield J, Auffray C, De Meulder B, Lefaudeux D, Dahlen SE, Djukanovic R, Sterk PJ, Guo Y, Adcock IM, Chung KF. Clinical and transcriptomic features of persistent exacerbation-prone severe asthma in U-BIOPRED cohort. Clin Transl Med. 2022;12:e816.
41. Perotin JM, Schofield JPR, Wilson SJ, Ward J, Brandsma J, Strazzeri F, Bansal A, Yang X, Rowe A, Corfield J, Lutter R, Shaw DE, Bakke PS, Caruso M, Dahlén B, Fowler SJ, Horváth I, Howarth P, Krug N, Montuschi P, Sanak M, Sandström T, Sun K, Pandis I, Auffray C, De Meulder B, Lefaudeux D, Riley JH, Sousa AR, Dahlen SE, Adcock IM, Chung KF, Sterk PJ, Skipp PJ, Collins JE, Davies DE, Djukanović R. Epithelial dysregulation in obese severe asthmatics with gastro-oesophageal reflux. Eur Respir J. 2019;53:1900453.
42. Kuo CS, Pavlidis S, Loza M, Baribaud F, Rowe A, Pandis I, Hoda U, Rossios C, Sousa A, Wilson SJ, Howarth P, Dahlen B, Dahlen SE, Chanez P, Shaw D, Krug N, Sandstrm T, De Meulder B, Lefaudeux D, Fowler S, Fleming L, Corfield J, Auffray C, Sterk PJ, Djukanovic R, Guo Y, Adcock IM, Chung KF, U-Biopred Project Team. A transcriptome-driven analysis of epithelial brushings and bronchial biopsies to define asthma phenotypes in U-BIOPRED. Am J Respir Crit Care Med. 2017;195:443–55.
43. Takahashi K, Pavlidis S, Kwong FNK, Hoda U, Rossios C, Sun K, Loza M, Baribaud F, Chanez P, Fowler SJ, Horvath I, Montuschi P, Singer F, Musial J, Dahlen B, Dahlen SE, Krug N, Sandstrom T, Shaw DE, Lutter R, Bakke P, Fleming LJ, Howarth PH, Caruso M, Sousa AR, Corfield J, Auffray C, De Meulder B, Lefaudeux D, Djukanovic R, Sterk PJ, Guo Y, Adcock IM, Chung KF. Sputum proteomics and airway cell transcripts of current and ex-smokers with severe asthma in U-BIOPRED: an exploratory analysis. Eur Respir J. 2018;51:1702173.
44. Tiotiu A, Badi Y, Kermani NZ, Sanak M, Kolmert J, Wheelock CE, Hansbro PM, Dahlén SE, Sterk PJ, Djukanovic R, Guo Y, Mumby S, Adcock IM, Chung KF. Association of differential mast cell activation with granulocytic inflammation in severe asthma. Am J Respir Crit Care Med. 2022;205:397–411.
45. Woodruff PG, Modrek B, Choy DF, Jia G, Abbas AR, Ellwanger A, Koth LL, Arron JR, Fahy JV. T-helper type 2-driven inflammation defines major subphenotypes of asthma. Am J Respir Crit Care Med. 2009;180:388–95.
46. Pavord ID, Beasley R, Agusti A, Anderson GP, Bel E, Brusselle G, Cullinan P, Custovic A, Ducharme FM, Fahy JV, Frey U, Gibson P, Heaney LG, Holt PG, Humbert M, Lloyd CM, Marks G, Martinez FD, Sly PD, von Mutius E, Wenzel S, Zar HJ, Bush A. After asthma: redefining airways diseases. Lancet. 2018;391:350–400.

47. Barretto KT, Brockman-Schneider RA, Kuipers I, Basnet S, Bochkov YA, Altman MC, Jarjour NN, Gern JE, Esnault S. Human airway epithelial cells express a functional IL-5 receptor. Allergy. 2020;75:2127–30.
48. Silkoff PE, Moore WC, Sterk PJ. Three major efforts to phenotype asthma: severe asthma research program, asthma disease endotyping for personalized therapeutics, and unbiased biomarkers for the prediction of respiratory disease outcome. Clin Chest Med. 2019;40:13–28.
49. Luecken MD, Zaragosi LE, Madissoon E, Sikkema L, Firsova AB, De Domenico E, Kümmerle L, Saglam A, Berg M, Gay ACA, Schniering J, Mayr CH, Abalo XM, Larsson L, Sountoulidis A, Teichmann SA, van Eunen K, Koppelman GH, Saeb-Parsy K, Leroy S, Powell P, Sarkans U, Timens W, Lundeberg J, van den Berge M, Nilsson M, Horváth P, Denning J, Papatheodorou I, Schultze JL, Schiller HB, Barbry P, Petoukhov I, Misharin AV, Adcock IM, von Papen M, Theis FJ, Samakovlis C, Meyer KB, Nawijn MC. The discovAIR project: a roadmap towards the Human Lung Cell Atlas. Eur Respir J. 2022;60:2102057.
50. Rossios C, Pavlidis S, Hoda U, Kuo CH, Wiegman C, Russell K, Sun K, Loza MJ, Baribaud F, Durham AL, Ojo O, Lutter R, Rowe A, Bansal A, Auffray C, Sousa A, Corfield J, Djukanovic R, Guo Y, Sterk PJ, Chung KF, Adcock IM. 'Sputum transcriptomics reveal upregulation of IL-1 receptor family members in patients with severe asthma. J Allergy Clin Immunol. 2017;141:560.
51. Højen JF, Kristensen MLV, McKee AS, Wade MT, Azam T, Lunding LP, de Graaf DM, Swartzwelter BJ, Wegmann M, Tolstrup M, Beckman K, Fujita M, Fischer S, Dinarello CA. IL-1R3 blockade broadly attenuates the functions of six members of the IL-1 family, revealing their contribution to models of disease. Nat Immunol. 2019;20:1138–49.
52. Jevnikar Z, Östling J, Ax E, Calvén J, Thörn K, Israelsson E, Öberg L, Singhania A, Lau LCK, Wilson SJ, Ward JA, Chauhan A, Sousa AR, De Meulder B, Loza MJ, Baribaud F, Sterk PJ, Chung KF, Sun K, Guo Y, Adcock IM, Payne D, Dahlen B, Chanez P, Shaw DE, Krug N, Hohlfeld JM, Sandström T, Djukanovic R, James A, Hinks TSC, Howarth PH, Vaarala O, van Geest M, Olsson H. Epithelial IL-6 trans-signaling defines a new asthma phenotype with increased airway inflammation. J Allergy Clin Immunol. 2019;143:577–90.
53. Östling J, van Geest M, Schofield JPR, Jevnikar Z, Wilson S, Ward J, Lutter R, Shaw DE, Bakke PS, Caruso M, Dahlen SE, Fowler SJ, Horváth I, Krug N, Montuschi P, Sanak M, Sandström T, Sun K, Pandis I, Auffray C, Sousa AR, Guo Y, Adcock IM, Howarth P, Chung KF, Bigler J, Sterk PJ, Skipp PJ, Djukanović R, Vaarala O. IL-17-high asthma with features of a psoriasis immunophenotype. J Allergy Clin Immunol. 2019;144:1198–213.
54. Badi YE, Pavel AB, Pavlidis S, Riley JH, Bates S, Kermani NZ, Knowles R, Kolmert J, Wheelock CE, Worsley S, Uddin M, Alving K, Bakke PS, Behndig A, Caruso M, Chanez P, Fleming LJ, Fowler SJ, Frey U, Howarth P, Horváth I, Krug N, Maitland-van der Zee AH, Montuschi P, Roberts G, Sanak M, Shaw DE, Singer F, Sterk PJ, Djukanovic R, Dahlen SE, Guo YK, Chung KF, Guttman-Yassky E, Adcock IM. Mapping atopic dermatitis and anti-IL-22 response signatures to type 2-low severe neutrophilic asthma. J Allergy Clin Immunol. 2021;149:89.
55. Raundhal M, Morse C, Khare A, Oriss TB, Milosevic J, Trudeau J, Huff R, Pilewski J, Holguin F, Kolls J, Wenzel S, Ray P, Ray A. High IFN-γ and low SLPI mark severe asthma in mice and humans. J Clin Invest. 2015;125:3037–50.
56. Oriss TB, Raundhal M, Morse C, Huff RE, Das S, Hannum R, Gauthier MC, Scholl KL, Chakraborty K, Nouraie SM, Wenzel SE, Ray P, Ray A. IRF5 distinguishes severe asthma in humans and drives Th1 phenotype and airway hyperreactivity in mice. JCI Insight. 2017;2:e91019.
57. Gauthier M, Kale SL, Oriss TB, Scholl K, Das S, Yuan H, Hu S, Chen J, Camiolo M, Ray P, Wenzel S, Ray A. Dual role for CXCR3 and CCR5 in asthmatic type 1 inflammation. J Allergy Clin Immunol. 2022;149:113–124.e7.
58. Reinke SN, Naz S, Chaleckis R, Gallart-Ayala H, Kolmert J, Kermani NZ, Tiotiu A, Broadhurst DI, Lundqvist A, Olsson H, Ström M, Wheelock M, Gómez ÅC, Ericsson M, Sousa AR, Riley JH, Bates S, Scholfield J, Loza M, Baribaud F, Bakke PS, Caruso M, Chanez P, Fowler SJ, Geiser T, Howarth P, Horváth I, Krug N, Montuschi P, Behndig A, Singer F, Musial J, Shaw

DE, Dahlén B, Hu S, Lasky-Su J, Sterk PJ, Chung KF, Djukanovic R, Dahlén SE, Adcock IM, Wheelock CE. Urinary metabotype of severe asthma evidences decreased carnitine metabolism independent of oral corticosteroid treatment in the U-BIOPRED study. Eur Respir J. 2022;59:2101733.

59. Brinkman P, Wagener AH, Hekking PP, Bansal AT, Maitland-van der Zee AH, Wang Y, Weda H, Knobel HH, Vink TJ, Rattray NJ, D'Amico A, Pennazza G, Santonico M, Lefaudeux D, De Meulder B, Auffray C, Bakke PS, Caruso M, Chanez P, Chung KF, Corfield J, Dahlén SE, Djukanovic R, Geiser T, Horvath I, Krug N, Musial J, Sun K, Riley JH, Shaw DE, Sandström T, Sousa AR, Montuschi P, Fowler SJ, Sterk PJ. Identification and prospective stability of electronic nose (eNose)-derived inflammatory phenotypes in patients with severe asthma. J Allergy Clin Immunol. 2019;143:1811–20.e7.
60. Hastie AT, Steele C, Dunaway CW, Moore WC, Rector BM, Ampleford E, Li H, Denlinger LC, Jarjour N, Meyers DA, Bleecker ER. Complex association patterns for inflammatory mediators in induced sputum from subjects with asthma. Clin Exp Allergy. 2018;48:787–97.
61. Vieira Braga FA, Kar G, Berg M, Carpaij OA, Polanski K, Simon LM, Brouwer S, Gomes T, Hesse L, Jiang J, Fasouli ES, Efremova M, Vento-Tormo R, Talavera-López C, Jonker MR, Affleck K, Palit S, Strzelecka PM, Firth HV, Mahbubani KT, Cvejic A, Meyer KB, Saeb-Parsy K, Luinge M, Brandsma CA, Timens W, Angelidis I, Strunz M, Koppelman GH, van Oosterhout AJ, Schiller HB, Theis FJ, van den Berge M, Nawijn MC, Teichmann SA. A cellular census of human lungs identifies novel cell states in health and in asthma. Nat Med. 2019;25:1153–63.

Part III

Severe and Difficult to Treat Asthma Endotypes

Definition, Phenotyping of Severe Asthma, Including Cluster Analysis

Serena Casanova, Engi Ahmed, and Arnaud Bourdin

Abstract

Asthma is defined as severe when it is uncontrolled despite the high intensity of treatment, or that loses control when a therapeutic step down is tried.

These patients, for years, have been "uniformly" treated with massive doses of inhaled and oral corticosteroids regardless of their inflammatory state.

Initially, asthma was considered of genesis "exclusively allergic." Subsequently, thanks to the development of noninvasive tools and of human monoclonal antibodies targeting interleukin 5, a pathogenetic role has been given to eosinophils. Management of steroids based on sputum eosinophil counts has been suggested according to clinical phenotypes identified through cluster analysis.

The algorithms obtained from the cluster analysis have proved later to be poorly predictive of the inflammatory phenotype and difficult to apply in clinical practice.

In the new era of precision medicine, the greatest challenge is finding clinical or biological elements predictive of response to therapies such as biologics. Cluster analyses performed on omics data or on cohorts of patients treated with biologics are more promising in this sense.

In this article, starting from the current definition of severe asthma, we review the phenotypes proposed over time to date, showing the difficulty underlying the process of "phenotyping" due to the scarcity of available biomarkers.

S. Casanova
Respiratory Medicine, Department of Translational Medicine, University of Ferrara, Ferrara, Italy

E. Ahmed · A. Bourdin (✉)
Department of Respiratory Diseases, University of Montpellier, CHU Montpellier, INSERM, Montpellier, France

A. R. Brasier, N. N. Jarjour (eds.), *Precision Approaches to Heterogeneity in Asthma*, Advances in Experimental Medicine and Biology 1426,
https://doi.org/10.1007/978-3-031-32259-4_11

Keywords

Severe asthma · Phenotype · Cluster analysis · Precision medicine · Heterogeneity · Response prediction · Classification · Heterogeneity · Inflammation

Introduction

The apparent paradox of the current severe asthma definition should be considered as a reflection of the nature of the disease itself. Among chronic diseases, asthma is characterized by variability and reversibility, two pillars that will find expression clinically, functionally, and by the way also at the therapeutic level [1]. Not only will it be recorded on a day-to-day basis, but also over a longer period of time where the disease can spontaneously improve or degrade.

These seminal observations reported for centuries led to the appropriate conception that the disease should be considered as severe when these features are lost, leading to permanent impairment despite making all the efforts that can be done [2].

This certainly includes a certain acknowledgment of the benefits of currently available therapeutics and, as a consequence, the severity of the disease will be assessed only once a real period of observation of the disease activity treated by an incremental anti-inflammatory pressure will be available [1].

Historical confusion between acute severe symptoms (now referred to as episodes of exacerbation, or "attacks") and the overall severity of the disease increase the complexity of literature searches [3, 4].

Very few models of chronic diseases are fully superposable. Terms used, for example, in Chronic Obstructive Pulmonary Disease (COPD), diabetes, osteoporosis, or heart failure, are participating to the confusion around the term "severe." Then, after multiple definitions attempting to catch these subtleties [5, 6], the Global Initiative for Asthma proposed to define severe asthma by the step of treatment required, considering herein potentially two different situations: (1) the most common nowadays being a disease that can be controlled if a very high therapeutic pressure is required, or (2) the second being that the disease remains uncontrolled despite all the currently available therapeutic options [1].

Interestingly, the current approach supposed that a long period of time has been dedicated to understand why the disease would require such a high therapeutic step. Designers of the virtuous circle of asthma management (review-assess-adjust) should be applauded for so well taking into account the chronic, variable, and reversible essence of the disease; this recalls that before considering a disease as severe, multiple occasions are offered as opportunities to review the symptoms and adherence, assess the diagnosis and the potentially confounding comorbidities and keep dynamic the therapeutic adjustments (obviously not exclusively doses of medications or devices) before reaching the conclusion that the disease is severe [1].

At a glance, the term "severe" in asthma aims at depicting a disease that is naturally and intrinsically uncommonly active through a targetable or a poorly targetable mechanism.

There are challenges to come short in the field, and all the community will have to contribute to adapting the definition in a timely manner:

1. Patients treated by biologics classified as super responders or more provocatively "in remission" are still ranked in the highest step 5 of the current GINA version; this suggests that considerations beyond the disease itself such as costs are taken and that patients experiencing highly different burdens are nonetheless kept within the same step [7];
2. Promotion of the oral corticosteroids (OCS) stewardship aims at earlier interventions suggesting that step 5 is no longer desirable outside biologics;
3. Strategies dedicated to self-manage the daily dose of inhaled corticosteroids (ICS) through the maintenance and reliever therapy (MART) strategies blur the borders between steps 1 and 3, suggesting a potential for a simplified chart [8];
4. Improvement in pharmacoepidemiology may challenge the current definition as anyone may decide where the threshold of severity should be set: in that case, each payer may eventually apply its own definition based on poorly exchangeable metrics; whether this concerns 3% or 5% or 10% of the overall asthma population is highly debatable [9].

Box 1: Current Definition of Severe Asthma (GINA 2022)

Uncontrolled asthma is defined by one or both of these features: (i) poor symptoms control, (ii) >2 exacerbations/year requiring OCS use or >1 exacerbation requiring hospitalization.

Difficult to treat asthma is uncontrolled asthma despite GINA steps 4–5 treatment.

According to Track 1, medium dose of ICS-Formoterol as maintenance and reliever therapy (SMART-MART) together with the long acting muscarininc antagonist (LAMA) represents the gold standard for a patient in steps 4–5.

Severe asthma is defined as asthma that is uncontrolled despite an optimized treatment with high dose of ICS LABA and good adherence, or that loses control when a therapeutic step down from high doses is performed. Severe asthma is characterized by the persistence of symptoms and frequents exacerbations with repercussion in economical, respiratory, psychological, and social terms.

Source: Global Initiative for Asthma [1]

Two-thirds of the economic burden of asthma is attributable to severe asthma through the utilization of healthcare resources, notably due to medications, emergency department attendance, intensive care unit (ICU) admissions, hospitalizations, and unscheduled visits [9, 10].

For example, in Europe, the estimated average annual cost for a well-controlled asthmatic patient is 509 euro/year, but it grows to about five times more for those with uncontrolled disease [11].

More work is needed to assess indirect costs that may cover absenteeism, presenteeism, loss of positions for active people [12], and the poorly assessed costs of short- and long-term adverse effects of OCS [11–13]. Other poor outcomes related to severe asthma include the risk of exacerbation – the risk of dying from a fatal exacerbation, and all the healthcare-associated risks (here shared and amplified by OCS) [14, 15].

Disability associated with the course of asthma also associates decline in lung function and reduced effort capacities [16].

All the currently available unbiased epidemiological or cohort studies now allow confirming that severe asthma patients have a reduced life expectancy [14].

Overall OCS exposure has always been recognized as the main cause of concern in severe asthma, and it was one key defining criteria for years. Among severe patients belonging to GINA steps 4 and 5, 30.1% have received regular treatment with OCS (≥90 days in a year) while 51.1% of patients have received intermittent treatment [17].

Whereas the unacceptable safety profile of maintenance OCS use is now obvious, it was more recently highlighted that intermittent steroid intake is also a cause of concern. Because a significantly increased risk of severe adverse events starts at 0.5 g/year in all the epidemiological studies, this threshold should now be largely disseminated in the community not as a single criterion of severity but as a trigger for referral [18, 19].

GINA-2022 defined several objectives to reach in the management of uncontrolled asthma in order to minimize the risk of future exacerbations and the occurrence of persistent restriction airflow and to limit the adverse effects of treatments [1].

Although severe asthma is not a rare disease because of the quite high incidence and prevalence, the disease remains an orphan of a treatable cause. Because heterogeneity in the clinical expression and the underlying mechanism is a feature of this variable disease, it was suggested to recognize the different forms of the disease. "Nocturnal asthma," "effort-induced asthma," etc., were commonly reported clinical subtypes before the recognition of the merits of inflammometry [20, 21].

Accordingly, when the disease responds to the definition of severity – i.e., the disease needs massive amounts of treatment to be controlled and eventually fails to be controlled, a key step of "phenotyping" is suggested in order to provide accessible clues of the poorly accessible underlying pathogenic pathways involved, named "endotype" [1].

The final goal would be identifying the correct "characteristics of patient" for an individualized step up toward appropriate and personalized management [22].

Asthma Phenotypes and Endotypes

Because it was early recognized that one size does not fit all – please translate here as "all patients are not corticosteroid sensitive," a new role for phenotyping was found. Phenotype refers to the set of observable morphological and functional characteristics that results from a genetic and environmental interaction. In practice, it refers to observable or measurable traits [23].

According to GINA, "asthma phenotypes" are defined by clinical, demographic, and pathophysiological features [1].

Endotype describes a subtype of a disease defined by a distinct molecular mechanism potentially highlighted by treatment response [23].

In the past decades, asthma was clinically stratified into two main phenotypes: "extrinsic" and "intrinsic" asthma.

Initially, asthma was mostly considered as a respiratory manifestation of an exclusively allergic reaction. Extrinsic asthma was a synonym of allergic asthma characterized by childhood onset, serological evidence of high circulating immunoglobulin E levels (IgE), sensitization to pneumallergens evidenced by positive skin prick tests and atopic manifestations such as eczema, conjunctivitis, food allergy and rhinitis [24].

In 1948, Rackemann introduced for the first time the adjective "intrinsic" as he affirmed that "all is not allergy that wheezes."

Contrasting features of intrinsic asthma were: the late onset, more severe clinical course, no IgE evidence, no extra respiratory atopic manifestation but frequent aspirin intolerance [25].

Thanks to the development of noninvasive tools able to assess the different patterns of airway inflammation [26], in particular the standardization of the sputum induction process and the handling of the collected sample [27], the classification moved to eosinophilic and noneosinophilic asthma [28] where classical clinical features such as age of onset, severity, exacerbation, airflow impairment, etc., were collected in order to help to identify inflammatory patterns outside tertiary centers.

Because corticosteroids were the only anti-inflammatory drugs used to treat asthma, mechanistic phenotypes were limited to "Th2 high" or "Th2 low" [29].

At that time, the critical role played by the airway epithelium and ILC2 was not identified [30] (and this explains why an "h" was still accompanying the "T"), and the distinction was based on steroid responsiveness assessed by the change in FEV1 [29].

This helped at identifying new targets belonging to the T2 pathway, and periostin as a highly sensitive IL13 protein was selected as a candidate companion biomarker during the development of a new set of monoclonal antibodies (mAbs) targeting IL13 only, unfortunately not meeting the expectations [31, 32].

When clustering analysis became available and applicable to available large databases and cohorts, multiple approaches were proposed, with or without an a priori hypothesis, with the aim of finding new therapeutic candidates.

The Time of Clustering Approaches: Bedside to Bench

Cluster analysis is a multivariate statistical technique that starts from a heterogeneous group on the basis of preidentified variables for obtaining homogeneous groups through a more or less supervised method [33]. The advantages of unsupervised analysis are that the machine acts without any predefined hypothesis and, therefore, it is not affected by any external judgment in the definition of groups.

This allows to obtain groupings even hidden and not visible to the human eye. Many clustering methods actually exist not only based on the mathematical model used but also by the level of supervision the analysis can bring in. The use of cluster analysis had the ambitious goal of grouping asthmatic patients into groups with similar characteristics in order to identify future risks in terms of exacerbations and functional decline, underlying biological processes, and to the best feasible treatment.

The goal was to create an algorithm according to clinical phenotypes alternative to the homogenous traditional step-by-step approach [34].

All researchers aimed at identifying homogeneous groups of patients potentially helping to reveal common genetic/biological pathways, i.e., potentially aligning phenotypes with endotypes. At that time, strong doubts were raised regarding the role of eosinophils, given the poor benefits observed with mAbs targeting IL-5 [35]. In this trial, patients were recruited based on their clinical phenotype irrespective of their eosinophilic trait [36].

Evidence of benefits related to IL5 blockade finally came later in extremely eosinophilic patients [37].

Haldar et al. proposed in their seminal paper a cluster analysis supervised by the hypothesis that eosinophilic inflammation should be responsible for most of asthma symptoms in patients with a "coherent phenotype" [38].

Most of this hypothesis was built on the pivotal trial by Green et al. where management of asthma based on sputum eosinophil counts was shown to be superior to classical clinical management based on symptoms and lung function measurement [39]. This led to the identification of a potentially still relevant cluster of obese, highly symptomatic patients with limited eosinophilic inflammation [38].

In the US, Moore et al. developed a nearly totally unsupervised cluster analysis applied to the Severe Asthma Research cohort (SARP) [40].

The five clusters identified finally largely overlapped in terms of biomarkers.

When using three variables in an independent cohort (baseline FEV1, maximum FEV1 after six to eight puffs of albuterol, and age of onset), Moore et al. showed that the cluster analysis correctly assigned patients in 80% of the cases (Fig. 1).

The second figure shows the percentage of patients assigned for each cluster giving each one an oval shape. The thinnest part of the shape represents the misclassified subjects: cluster 3 had a higher (50%) content of poorly classified [40].

Unfortunately, it was later shown that this algorithm failed to predict different outcomes among the different clusters [41].

Moreover, it was later shown that these clusters noteworthy failed to demonstrate consistency over time. In other words, patients could change of cluster in just 1 year,

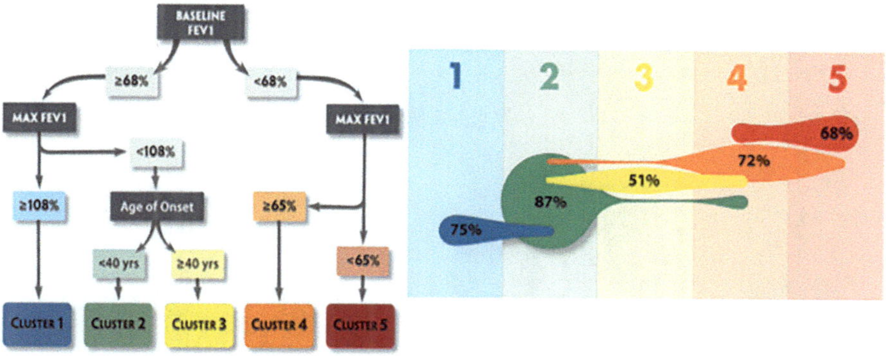

Fig. 1 The first figure shows the assignment of clusters according to the proposed flow chart using three variables (baseline FEV1, maximum FEV1 after six to eight puffs of albuterol, and age of onset). Clusters were correctly and incorrectly assigned in 80% and 20% of cases, respectively

suggesting that this kind of exercise was robust cross-sectionally but poorly informative and applicable to a chronic disease, variable and heterogenous by essence [42].

Otherwise, clustering is a mathematical model with inherent limits to the dataset available and the level of supervision set at the beginning of the analysis.

More interestingly, the response to treatment was subsequently tested and applied to an excellent set of data collected during a randomized controlled trial [43].

This approach progressively turned to introduce more mechanistic inflammatory patterns in the model.

Then, and logically, Denton et al. built clusters based on inflammatory markers such as immunoglobulin E (IgE), blood eosinophils, and fractional concentration of nitric oxide in the exhaled air (FeNo) [44].

The analysis showed that only 29% of patients had the exclusive expression of only one biomarker and that, in most cases (59%), there was a contemporary expression of two or three. Triple negativity was also reported in 12% of cases suggesting that the currently used “T2 low” asthma is by far less frequent than initially thought (Fig. 2) [44].

Useful Phenotypes in 2022

Obviously, progresses are triggered by therapeutic innovations. The dramatical response associated with currently available biologics, designed to target the key T2 cytokines, dominated the severe asthma landscape [45].

Because very few of the clinical characteristics outside biology really predict the therapeutic response, it is then currently proposed to systematically assess for a limited number of phenotypic features that will potentially impact the way in which the disease is managed (Box 2) [1].

Intriguingly, some clinical features that were nearly abandoned are now recognized as extremely meaningful characteristics in terms of mechanism and prediction of response to biologic, in particular, the presence of nasal polyposis (Fig. 3) [46].

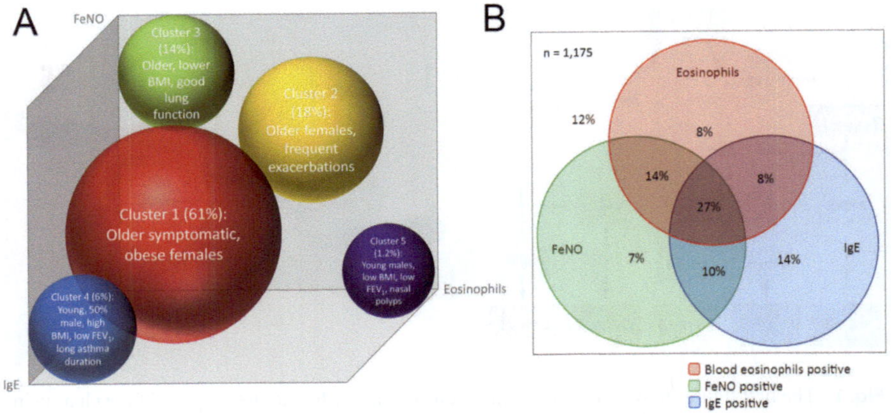

Fig. 2 (**a**) Summarizes the five identified clusters and positions them according to the inflammatory markers such as IgE, blood eosinophils, and FeNo. Although well-defined clinical groups are identified in most cases, the expression of biomarkers is not exclusive (as shown in (**b**)) and suggests the existence of multiple inflammatory mechanisms at the basis of the single phenotype [44]

Regarding biomarkers, although blood eosinophil counts (BEC) are now clearly identified as the best predictors for any of the currently approved biologics and FeNO as a potential surrogate of IL13 overexpression, the significance of total IgE counts and of SPT/specific IgE is less clear given their absence of predictive value for any of the currently available treatment, including anti-IgE [47].

Box 2: Informative Phenotypic Features Should Be Systematically Assessed in SA as It Should Modify Management

Obesity
Smoking exposure
Occupational exposure
ICU admission for asthma
Prematurity and other risk factors exposing to poor lung growth
Blood eosinophil count
Repeated FeNO
Skin prick tests
Nasal polyposis
NSAIDs intolerance
In particular circumstances
Sputum eosinophil count
Airway hyperreactivity
HRCT
ANCA1
Total IgE, precipitins, *Aspergillus fumigatus*
Others

Source: Adapted from GINA [1]

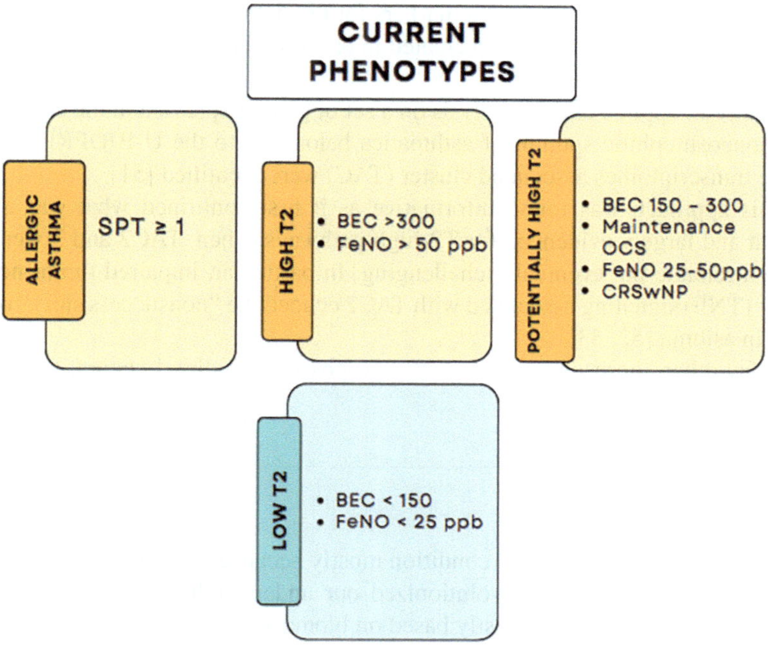

Fig. 3 Clustering based on phenotypes. SPT: skin prick test; BEC: blood eosinophil counts; CRSWNP: chronic rhinosinusitis with nasal polyposis

The currently recognized phenotypes are more or less fitting with the currently available options, even though the borders between these phenotypes are still inaccurate. The debate is interestingly reopened since the targeting of the upstream alarmin thymic stromal lymphopoietin (TSLP) showed benefits in all patients [48].

After a decade of biomarker-guided approach, shall we go back to the one size fits all? Actually, probably not, not only because all patients are not experiencing the same course and burden of the disease but also because the benefit, if present in all, is better in some.

Identifying new therapeutic candidates and new biomarker is probably highly desirable as there are still clear unmet needs in severe asthma, especially when patients are poorly responding to all the currently available options. Molecular and omics approaches are promising in this regard if the cutting-edge science behind that is informed by cutting-edge clinical data, and this is probably the greatest challenge to date [49].

Molecular Phenotypes – Omic Approach

Omics sciences are the set of disciplines that, through advanced technologies, study the molecular profile of a multi-level system of expression: genomics, transcriptomics, proteomics, and metabolomics [49].

Projects such as "Unbiased biomarkers for prediction of respiratory disease outcome" (U-BIOPRED) have been created in order to collect biological samples useful for omics studies [50].

Kuo et al. applied cluster analysis on a set of genes expressed in the eosinophilic and noneosinophilic sputum of asthmatics belonging to the U-BIOPRED cohort. Three transcriptomes associated cluster (TAC) were identified [51].

This approach was found informative as it first confirmed what was already known and largely evidenced for T2-high pathways. Then, TAC2 and 3 were considered reliable but eventually challenging. In particular, impaired tumor necrosis factor (TNF) signaling associated with TAC2 echoed the "nonsuccess story" of anti-TNF in asthma [52, 53].

Many other approaches are currently proposed, including imaging, proteomics, and metabolomics [54, 55].

Conclusions

Severe asthma is a devastating condition mostly because of the massive OCS exposure it imposes. Biologics revolutionized our understanding of the usefulness of phenotypes which are now mostly based on biomarkers, as surrogates of the underlying overactivated pathways. In an effort to find a treatment for all, the identification of new relevant biomarkers in poorly responding patients seems obvious.

References

1. Global Initiative for Asthma. Global strategy for asthma management and prevention. GINA; 2022. Available from: www.ginasthma.org.
2. Gras D, Bourdin A, Chanez P, Vachier I. Remodelage bronchique dans l'asthme – conséquences cliniques et fonctionnelles respiratoires [Airway remodeling in asthma: clinical and functional correlates]. Med Sci (Paris). 2011;27(11):959–65. French. https://doi.org/10.1051/medsci/20112711011.
3. Wenzel S. Physiologic and pathologic abnormalities in severe asthma. Clin Chest Med. 2006;27(1):29–40; v. https://doi.org/10.1016/j.ccm.2005.
4. Poon AH, Hamid Q. Severe asthma: have we made progress? Ann Am Thorac Soc. 2016;13(Suppl 1):S68–77. https://doi.org/10.1513/AnnalsATS.201508-514MG.
5. Bel EH, Sousa A, Fleming L, Bush A, Chung KF, Versnel J, Wagener AH, Wagers SS, Sterk PJ, Compton CH, Unbiased Biomarkers for the Prediction of Respiratory Disease Outcome (U-BIOPRED) Consortium, Consensus Generation. Diagnosis and definition of severe refractory asthma: an international consensus statement from the Innovative Medicine Initiative (IMI). Thorax. 2011;66(10):910–7. https://doi.org/10.1136/thx.2010.153643.
6. Chung KF, Wenzel SE, Brozek JL, Bush A, Castro M, Sterk PJ, Adcock IM, Bateman ED, Bel EH, Bleecker ER, Boulet LP, Brightling C, Chanez P, Dahlen SE, Djukanovic R, Frey U, Gaga M, Gibson P, Hamid Q, Jajour NN, Mauad T, Sorkness RL, Teague WG. International ERS/ATS guidelines on definition, evaluation and treatment of severe asthma. Eur Respir J. 2014;43(2):343–73. https://doi.org/10.1183/09031936.00202013. Epub 2013 Dec 12. Erratum in: Eur Respir J. 2014;43(4):1216.

7. Porsbjerg C, Maitland-van der Zee AH, Brusselle G, Canonica GW, Agusti A, Faner R, Vogelmeier CF, Nawijn M, van den Berge M, Rusconi F, Pilette C, Ramiconi V, Coleman C, Chaudhuri R, Chung KF, Wedzicha J, Saglani S, Van der Schee MP, Heaney L, Bourdin A, Roberts G, Djukanovic R, Kuna P, Kupczyk M, Axmann J, Staudinger H, Clarke GW, Dahlen SE, Brightling C. 3TR: a pan-European cross-disease research consortium aimed at improving personalised biological treatment of asthma and COPD. Eur Respir J. 2021;58(4):2102168. https://doi.org/10.1183/13993003.02168-2021.
8. Reddel HK, Bateman ED, Schatz M, Krishnan JA, Cloutier MM. A practical guide to implementing SMART in asthma management. J Allergy Clin Immunol Pract. 2022;10(1S):S31–8. https://doi.org/10.1016/j.jaip.2021.10.011.
9. Bourdin A, Fabry-Vendrand C, Ostinelli J, Ait-Yahia M, Darnal E, Bouee S, Laurendeau C, Bureau I, Gourmelen J, Chouaid C. The burden of severe asthma in France: a case-control study using a medical claims database. J Allergy Clin Immunol Pract. 2019;7(5):1477–87. https://doi.org/10.1016/j.jaip.2018.12.029.
10. Chen S, Golam S, Myers J, Bly C, Smolen H, Xu X. Systematic literature review of the clinical, humanistic, and economic burden associated with asthma uncontrolled by GINA steps 4 or 5 treatment. Curr Med Res Opin. 2018;34(12):2075–88. https://doi.org/10.1080/03007995.2018.1505352.
11. O'Neill S, Sweeney J, Patterson CC, Menzies-Gow A, Niven R, Mansur AH, Bucknall C, Chaudhuri R, Thomson NC, Brightling CE, O'Neill C, Heaney LG, British Thoracic Society Difficult Asthma Network. The cost of treating severe refractory asthma in the UK: an economic analysis from the British Thoracic Society Difficult Asthma Registry. Thorax. 2015;70(4):376–8. https://doi.org/10.1136/thoraxjnl-2013-204114.
12. Ong ASE, Chan AKW, Sultana R, Koh MS. Impact of psychological impairment on quality of life and work impairment in severe asthma. J Asthma. 2021;58(11):1544–53. https://doi.org/10.1080/02770903.2020.1808989. Epub 2020 Aug 24. PMID: 32777181.
13. Lefebvre P, Duh MS, Lafeuille MH, Gozalo L, Desai U, Robitaille MN, Albers F, Yancey S, Ortega H, Forshag M, Lin X, Dalal AA. Acute and chronic systemic corticosteroid-related complications in patients with severe asthma. J Allergy Clin Immunol. 2015;136(6):1488–95. https://doi.org/10.1016/j.jaci.2015.07.046.
14. Haselkorn T, Szefler SJ, Chipps BE, Bleecker ER, Harkins MS, Paknis B, Kianifard F, Ortiz B, Zeiger RS. Disease burden and long-term risk of persistent very poorly controlled asthma: TENOR II. J Allergy Clin Immunol Pract. 2020;8(7):2243–53. https://doi.org/10.1016/j.jaip.2020.02.040.
15. Bleecker ER, Menzies-Gow AN, Price DB, et al. Systematic literature review of systemic corticosteroid use for asthma management. Am J Respir Crit Care Med. 2020;201(3):276–93. https://doi.org/10.1164/rccm.201904-0903SO.
16. Bai TR, Vonk JM, Postma DS, Boezen HM. Severe exacerbations predict excess lung function decline in asthma. Eur Respir J. 2007;30(3):452–6. https://doi.org/10.1183/09031936.00165106.
17. Wang E, Wechsler ME, Tran TN, et al. Characterization of severe asthma worldwide: data from the International Severe Asthma Registry. Chest. 2020;157(4):790–804. https://doi.org/10.1016/j.chest.2019.10.053.
18. Chan JSK, Murray RB, Price D. Oral corticosteroids in asthma and beyond: moving forward. Eur Respir J. 2022;60(3):2200776. https://doi.org/10.1183/13993003.00776-2022.
19. Suehs CM, Menzies-Gow A, Price D, Bleecker ER, Canonica GW, Gurnell M, Bourdin A, Oral Corticosteroids Tapering Delphi Expert Panel. Expert consensus on the tapering of oral corticosteroids for the treatment of asthma. A Delphi study. Am J Respir Crit Care Med. 2021;203(7):871–81. https://doi.org/10.1164/rccm.202007-2721OC.
20. Sutherland ER. Nocturnal asthma. J Allergy Clin Immunol. 2005;116(6):1179–86; quiz 1187. https://doi.org/10.1016/j.jaci.2005.09.028.
21. Wenzel SE. Asthma: defining of the persistent adult phenotypes. Lancet. 2006;368(9537):804–13. https://doi.org/10.1016/S0140-6736(06)69290-8.

22. Shaw DE, Heaney LG, Thomas M, Beasley R, Gibson PG, Pavord ID. Balancing the needs of the many and the few: where next for adult asthma guidelines? Lancet Respir Med. 2021;9(7):786–94. https://doi.org/10.1016/S2213-2600(21)00021-7.
23. Agusti A, Barnes N, Cruz AA, Gibson PG, et al. Moving towards a Treatable Traits model of care for the management of obstructive airways diseases. Respir Med. 2021;187:106572. https://doi.org/10.1016/j.rmed.2021.106572.
24. Wenzel SE. Asthma phenotypes: the evolution from clinical to molecular approaches. Nat Med. 2012;18(5):716–25. https://doi.org/10.1038/nm.2678.
25. Rackemann FM. Intrinsic asthma. Bull N Y Acad Med. 1947;23(5):302–6. PMID: 19312533.
26. Zhang JY, Wenzel SE. Tissue and BAL based biomarkers in asthma. Immunol Allergy Clin N Am. 2007;27(4):623–32; vi. https://doi.org/10.1016/j.iac.2007.09.003.
27. Lemière C, Ernst P, Olivenstein R, Yamauchi Y, Govindaraju K, Ludwig MS, Martin JG, Hamid Q. Airway inflammation assessed by invasive and noninvasive means in severe asthma: eosinophilic and noneosinophilic phenotypes. J Allergy Clin Immunol. 2006;118(5):1033–9. https://doi.org/10.1016/j.jaci.2006.08.003.
28. Wenzel SE, Schwartz LB, Langmack EL, Halliday JL, Trudeau JB, Gibbs RL, Chu HW. Evidence that severe asthma can be divided pathologically into two inflammatory subtypes with distinct physiologic and clinical characteristics. Am J Respir Crit Care Med. 1999;160(3):1001–8. https://doi.org/10.1164/ajrccm.160.3.9812110.
29. Woodruff PG, Modrek B, Choy DF, Jia G, Abbas AR, Ellwanger A, Koth LL, Arron JR, Fahy JV. T-helper type 2-driven inflammation defines major subphenotypes of asthma. Am J Respir Crit Care Med. 2009;180(5):388–95. https://doi.org/10.1164/rccm.200903-0392OC. Epub 2009 May 29. Erratum in: Am J Respir Crit Care Med. 2009;180(8):796.
30. Hammad H, Lambrecht BN. The basic immunology of asthma. Cell. 2021;184(6):1469–85. https://doi.org/10.1016/j.cell.2021.02.016. Epub 2021 Mar 11. Erratum in: Cell. 2021;184(9):2521–2.
31. Busse WW, Brusselle GG, Korn S, Kuna P, Magnan A, Cohen D, Bowen K, Piechowiak T, Wang MM, Colice G. Tralokinumab did not demonstrate oral corticosteroid-sparing effects in severe asthma. Eur Respir J. 2019;53(2):1800948. https://doi.org/10.1183/13993003.00948-2018.
32. Hanania NA, Korenblat P, Chapman KR, Bateman ED, Kopecky P, Paggiaro P, Yokoyama A, Olsson J, Gray S, Holweg CT, Eisner M, Asare C, Fischer SK, Peng K, Putnam WS, Matthews JG. Efficacy and safety of lebrikizumab in patients with uncontrolled asthma (LAVOLTA I and LAVOLTA II): replicate, phase 3, randomised, double-blind, placebo-controlled trials. Lancet Respir Med. 2016;4(10):781–96. https://doi.org/10.1016/S2213-2600(16)30265-X.
33. Bourdin A, Chanez P. Clustering in asthma: why, how and for how long? Eur Respir J. 2013;41(6):1247–8. https://doi.org/10.1183/09031936.00003313.
34. Kaur R, Chupp G. Phenotypes and endotypes of adult asthma: moving toward precision medicine. J Allergy Clin Immunol. 2019;144(1):1–12. https://doi.org/10.1016/j.jaci.2019.05.031.
35. Flood-Page P, Swenson C, Faiferman I, Matthews J, Williams M, Brannick L, Robinson D, Wenzel S, Busse W, Hansel TT, Barnes NC, International Mepolizumab Study Group. A study to evaluate safety and efficacy of mepolizumab in patients with moderate persistent asthma. Am J Respir Crit Care Med. 2007;176(11):1062–71. https://doi.org/10.1164/rccm.200701-085OC.
36. O'Byrne PM. The demise of anti IL-5 for asthma, or not. Am J Respir Crit Care Med. 2007;176(11):1059–60. https://doi.org/10.1164/rccm.200708-1264ED.
37. Haldar P, Brightling CE, Hargadon B, Gupta S, Monteiro W, Sousa A, Marshall RP, Bradding P, Green RH, Wardlaw AJ, Pavord ID. Mepolizumab and exacerbations of refractory eosinophilic asthma. N Engl J Med. 2009;360(10):973–84. https://doi.org/10.1056/NEJMoa0808991. Erratum in: N Engl J Med. 2011;364(6):588.
38. Haldar P, Pavord ID, Shaw DE, Berry MA, Thomas M, Brightling CE, Wardlaw AJ, Green RH. Cluster analysis and clinical asthma phenotypes. Am J Respir Crit Care Med. 2008;178(3):218–24. https://doi.org/10.1164/rccm.200711-1754OC.

39. Green RH, Brightling CE, McKenna S, Hargadon B, Parker D, Bradding P, Wardlaw AJ, Pavord ID. Asthma exacerbations and sputum eosinophil counts: a randomised controlled trial. Lancet. 2002;360(9347):1715–21. https://doi.org/10.1016/S0140-6736(02)11679-5.
40. Moore WC, Meyers DA, Wenzel SE, Teague WG, Li H, Li X, D'Agostino R Jr, Castro M, Curran-Everett D, Fitzpatrick AM, Gaston B, Jarjour NN, Sorkness R, Calhoun WJ, Chung KF, Comhair SA, Dweik RA, Israel E, Peters SP, Busse WW, Erzurum SC, Bleecker ER, National Heart, Lung, and Blood Institute's Severe Asthma Research Program. Identification of asthma phenotypes using cluster analysis in the Severe Asthma Research Program. Am J Respir Crit Care Med. 2010;181(4):315–23. https://doi.org/10.1164/rccm.200906-0896OC.
41. Bourdin A, Molinari N, Vachier I, Varrin M, Marin G, Gamez AS, Paganin F, Chanez P. Prognostic value of cluster analysis of severe asthma phenotypes. J Allergy Clin Immunol. 2014;134(5):1043–50. https://doi.org/10.1016/j.jaci.2014.04.038.
42. Boudier A, Curjuric I, Basagaña X, Hazgui H, Anto JM, Bousquet J, Bridevaux PO, Dupuis-Lozeron E, Garcia-Aymerich J, Heinrich J, Janson C, Künzli N, Leynaert B, de Marco R, Rochat T, Schindler C, Varraso R, Pin I, Probst-Hensch N, Sunyer J, Kauffmann F, Siroux V. Ten-year follow-up of cluster-based asthma phenotypes in adults. A pooled analysis of three cohorts. Am J Respir Crit Care Med. 2013;188(5):550–60. https://doi.org/10.1164/rccm.201301-0156OC.
43. Ortega H, Li H, Suruki R, Albers F, Gordon D, Yancey S. Cluster analysis and characterization of response to mepolizumab. A step closer to personalized medicine for patients with severe asthma. Ann Am Thorac Soc. 2014;11(7):1011–7. https://doi.org/10.1513/AnnalsATS.201312-454OC.
44. Denton E, Price DB, Tran TN, Canonica GW, Menzies-Gow A, FitzGerald JM, Sadatsafavi M, Perez de Llano L, Christoff G, Quinton A, Rhee CK, Brusselle G, Ulrik C, Lugogo N, Hore-Lacy F, Chaudhry I, Bulathsinhala L, Murray RB, Carter VA, Hew M. Cluster analysis of inflammatory biomarker expression in the International Severe Asthma Registry. J Allergy Clin Immunol Pract. 2021;9(7):2680–8.e7. https://doi.org/10.1016/j.jaip.2021.02.059. Epub 2021 Mar 18. Erratum in: J Allergy Clin Immunol Pract. 2021;9(11):4182.
45. Diamant Z, Vijverberg S, Alving K, Bakirtas A, Bjermer L, Custovic A, Dahlen SE, Gaga M, Gerth van Wijk R, Giacco SD, Hamelmann E, Heaney LG, Heffler E, Kalayci Ö, Kostikas K, Lutter R, Olin AC, Sergejeva S, Simpson A, Sterk PJ, Tufvesson E, Agache I, Seys SF. Toward clinically applicable biomarkers for asthma: an EAACI position paper. Allergy. 2019;74(10):1835–51. https://doi.org/10.1111/all.13806.
46. Heaney LG, Perez de Llano L, Al-Ahmad M, Backer V, Busby J, Canonica GW, Christoff GC, Cosio BG, FitzGerald JM, Heffler E, Iwanaga T, Jackson DJ, Menzies-Gow AN, Papadopoulos NG, Papaioannou AI, Pfeffer PE, Popov TA, Porsbjerg CM, Rhee CK, Sadatsafavi M, Tohda Y, Wang E, Wechsler ME, Alacqua M, Altraja A, Bjermer L, Björnsdóttir US, Bourdin A, Brusselle GG, Buhl R, Costello RW, Hew M, Koh MS, Lehmann S, Lehtimäki L, Peters M, Taillé C, Taube C, Tran TN, Zangrilli J, Bulathsinhala L, Carter VA, Chaudhry I, Eleangovan N, Hosseini N, Kerkhof M, Murray RB, Price CA, Price DB. Eosinophilic and noneosinophilic asthma: an expert consensus framework to characterize phenotypes in a global real-life severe asthma cohort. Chest. 2021;160(3):814–30. https://doi.org/10.1016/j.chest.2021.04.013.
47. Hanania NA, Wenzel S, Rosén K, Hsieh HJ, Mosesova S, Choy DF, Lal P, Arron JR, Harris JM, Busse W. Exploring the effects of omalizumab in allergic asthma: an analysis of biomarkers in the EXTRA study. Am J Respir Crit Care Med. 2013;187(8):804–11. https://doi.org/10.1164/rccm.
48. Menzies-Gow A, Corren J, Bourdin A, Chupp G, Israel E, Wechsler ME, Brightling CE, Griffiths JM, Hellqvist Å, Bowen K, Kaur P, Almqvist G, Ponnarambil S, Colice G. Tezepelumab in adults and adolescents with severe, uncontrolled asthma. N Engl J Med. 2021;384(19):1800–9. https://doi.org/10.1056/NEJMoa2034975.
49. Galeone C, Scelfo C, Bertolini F, Caminati M, Ruggiero P, Facciolongo N, Menzella F. Precision medicine in targeted therapies for severe asthma: is there any place for "omics" technology? Biomed Res Int. 2018;2018:4617565. https://doi.org/10.1155/2018/4617565.

50. Shaw DE, Sousa AR, Fowler SJ, Fleming LJ, Roberts G, Corfield J, Pandis I, Bansal AT, Bel EH, Auffray C, Compton CH, Bisgaard H, Bucchioni E, Caruso M, Chanez P, Dahlén B, Dahlen SE, Dyson K, Frey U, Geiser T, Gerhardsson de Verdier M, Gibeon D, Guo YK, Hashimoto S, Hedlin G, Jeyasingham E, Hekking PP, Higenbottam T, Horváth I, Knox AJ, Krug N, Erpenbeck VJ, Larsson LX, Lazarinis N, Matthews JG, Middelveld R, Montuschi P, Musial J, Myles D, Pahus L, Sandström T, Seibold W, Singer F, Strandberg K, Vestbo J, Vissing N, von Garnier C, Adcock IM, Wagers S, Rowe A, Howarth P, Wagener AH, Djukanovic R, Sterk PJ, Chung KF, U-BIOPRED Study Group. Clinical and inflammatory characteristics of the European U-BIOPRED adult severe asthma cohort. Eur Respir J. 2015;46(5):1308–21. https://doi.org/10.1183/13993003.00779-2015. Epub 2015 Sep 10. Erratum in: Eur Respir J. 2017;49(6): PMID: 26357963.
51. Kuo CS, Pavlidis S, Loza M, Baribaud F, Rowe A, Pandis I, Sousa A, Corfield J, Djukanovic R, Lutter R, Sterk PJ, Auffray C, Guo Y, Adcock IM, Chung KF, U-BIOPRED Study Group. T-helper cell type 2 (Th2) and non-Th2 molecular phenotypes of asthma using sputum transcriptomics in U-BIOPRED. Eur Respir J. 2017;49(2):1602135. https://doi.org/10.1183/13993003.02135-2016.
52. Berry MA, Hargadon B, Shelley M, Parker D, Shaw DE, Green RH, Bradding P, Brightling CE, Wardlaw AJ, Pavord ID. Evidence of a role of tumor necrosis factor alpha in refractory asthma. N Engl J Med. 2006;354(7):697–708. https://doi.org/10.1056/NEJMoa050580.
53. Wenzel SE, Barnes PJ, Bleecker ER, Bousquet J, Busse W, Dahlén SE, Holgate ST, Meyers DA, Rabe KF, Antczak A, Baker J, Horvath I, Mark Z, Bernstein D, Kerwin E, Schlenker-Herceg R, Lo KH, Watt R, Barnathan ES, Chanez P, T03 Asthma Investigators. A randomized, double-blind, placebo-controlled study of tumor necrosis factor-alpha blockade in severe persistent asthma. Am J Respir Crit Care Med. 2009;179(7):549–58. https://doi.org/10.1164/rccm.200809-1512OC.
54. Tyler SR, Bunyavanich S. Leveraging -omics for asthma endotyping. J Allergy Clin Immunol. 2019;144(1):13–23. https://doi.org/10.1016/j.jaci.2019.05.015.
55. Yang Z, Qin L, Qiao J, Cheng C, Liu Y, Zhang S, Fang X, Li Z, Renz H, Liu X, Xia L, Hu Q, Xie M. Novel imaging phenotypes of naïve asthma patients with distinctive clinical characteristics and T2 inflammation traits. Ther Adv Chronic Dis. 2022;13:20406223221084831. https://doi.org/10.1177/20406223221084831.

Asthma Exacerbations: Patient Features and Potential Long-Term Implications

Praveen Akuthota

Abstract

Asthma exacerbations occur in the context of a complex interplay between external exposures and host factors. Respiratory tract viral infections, in particular rhinovirus, are dominant initiators of exacerbations, with allergens and other inhalation exposures as additional key contributors. The presence of underlying type II inflammation, with associated biomarker elevations, is a major driver of exacerbation risk and mechanism, as evidenced by the consistent reduction of exacerbations seen with biologics targeting these pathways. Several genetic polymorphisms are associated with exacerbations, and while they may individually have small effects, they are cumulatively important and magnified by environmental exposures. A history of exacerbations predicts future exacerbations with potentially negative implications on long-term lung health.

Keywords

Exacerbations · Respiratory virus · Infection · Exposures · Rhinovirus · COVID · Allergens · Anti-viral immunity · Remodeling · Heterogeneity · Allergens · Environmental exposures

P. Akuthota (✉)
Division of Pulmonary, Critical Care, Sleep Medicine, & Physiology, University of California San Diego, La Jolla, CA, USA
e-mail: pakuthota@health.ucsd.edu

A. R. Brasier, N. N. Jarjour (eds.), *Precision Approaches to Heterogeneity in Asthma*, Advances in Experimental Medicine and Biology 1426,
https://doi.org/10.1007/978-3-031-32259-4_12

Introduction

Asthma, by definition, is variable, and periods of poor control can intermittently rise to the level of an exacerbation, regardless of the baseline level of asthma severity. Asthma exacerbations are an acute worsening of respiratory symptoms, including dyspnea and wheezing, leading to a drop in lung function [1]. When severe, asthma exacerbations can result in substantial adverse outcomes, including emergency department visits, hospitalizations, respiratory failure, and even death. Even patients with asthma that had previously been considered mild are at risk for severe events [2]. Dusser et al demonstrated that 16% of near-fatal asthma events and 15–20% of fatal asthma events occurred in patients who had less than weekly symptoms in the preceding 3 months [3].

As with asthma in general, there is heterogeneity in patient characteristics and external exposures that confer an increased risk of exacerbation (Table 1). Similarly, some patients are more exacerbation-prone than others and are therefore at increased risk of morbidity, not only from the exacerbation event itself but from the potential for increased exposure to systemic corticosteroid therapy [4].

External Exposures That Contribute to Exacerbation

Viruses

Viral respiratory tract infections are a major cause of asthma exacerbations, with approximately 80–85% of children and 80% of adults having detectable respiratory viruses in association with exacerbation [5]. Rhinovirus is easily the most common culprit in both children and adults [6–9]. Hospitalization and exacerbation rates correlate with the seasonality of rhinovirus infection rates [6]. Amongst the subtypes of rhinovirus, RV-A and RV-C are more likely to be associated with exacerbation, and RV-C is linked to more severe attacks and respiratory failure [10, 11]. In addition to rhinovirus, other respiratory viruses including respiratory syncytial virus (RSV) and influenza are associated with flares of asthma [9]. RSV is often a cause of wheezing in infants but can also lead to exacerbation in older and high-risk adults [12]. Data

Table 1 Contributors to asthma exacerbations

External exposures
Respiratory viruses (e.g., rhinovirus)
Allergens
Air pollutants
Cigarette smoke
Host factors
Prior exacerbation history
Type II inflammation
Demographic factors
Comorbidities
Bronchodilator responsiveness

for associations between COVID-19 infection and asthma are still emerging, though one recent meta-analysis suggested that patients with asthma had a higher risk of hospitalization from COVID-19 compared to individuals without asthma [13]. Of particular note was the reduction of exacerbations during the initial phase of the COVID pandemic, which was likely largely driven by a decrease in other viral respiratory infections (i.e., rhinovirus) resulting from protective social measures designed to slow the spread of COVID-19 [13].

Antiviral immunity may be negatively impacted by the activation of allergic pathways in the airways, which would magnify the impact of viral exposure on exacerbation [8, 14]. Specifically, reduction in antiviral type I interferon responses is diminished in ex vivo data generated from immune cells from patients with asthma [15].

Allergens

Allergic sensitization and associated type II inflammation are major contributors to asthma pathogenesis, and exposure to allergens potentiates the risk of asthma exacerbation due to respiratory viruses. In study of children with asthma in Costa Rica, the presence of high titers of specific IgE to dust mite allergen was associated with an increased risk of wheezing provoked by rhinovirus infection [16]. *Alternaria* sensitization has been linked to the respiratory arrest in children and young adults with asthma [17]. The phenomenon of "thunderstorm asthma," described most prominently in a 2016 event in Melbourne, Australia, is a cluster of asthma exacerbations related to large increases in airborne allergens due to thunderstorm events and is illustrative of the potential for acute exposure to allergens in sensitized individuals to cause severe exacerbation [18].

Other Exposures

Worsening of asthma control, including exacerbation, can occur with a number of other inhalation exposures including cigarette smoke and environmental pollutants. Air pollution is linked to an increased risk of asthma in children and to increased risk of exacerbation in adults [19]. Similarly, secondhand smoke is a risk factor for asthma development in children, and cigarette smoking in adults is associated with increased emergency department visits and hospitalizations [20].

Patient Factors That Contribute to Exacerbation

Prior Exacerbations

A major predictor of future exacerbation is the occurrence of a prior exacerbation [21, 22]. In a study of electronic health records in England, Bloom and colleagues found that a past exacerbation increased the risk of future exacerbation more than all other factors studied [21]. However, only a small percentage of patients (0.7%) with asthma were frequent exacerbators who had at least one exacerbation per year. Over half of the patients with asthma with 7 years of records available did not have any exacerbations over the time period studied. The Severe Asthma Research Program (SARP) in the United States described an exacerbation-prone phenotype in their cohort [22]. Over a 3-year follow-up period, the exacerbation rate in year 3 was 82% in patients who had an exacerbation in years 1 and 2 and only 13% in patients who did not have an exacerbation in years 1 and 2. SARP patients with this exacerbation-prone phenotype had characteristics of metabolic syndrome, with higher rates of obesity, diabetes, and hypertension. They also had higher plasma IL-6 levels, a key inflammatory biomarker for metabolic syndrome.

Eosinophils

Blood eosinophil counts are a risk factor for asthma exacerbations. Denlinger et al. found in the SARP cohort that the rate ratio for exacerbation was 1.6 for every log unit of blood eosinophils [23]. In a pooled observational cohort study from the Netherlands, Westerhof and colleagues observed that elevated blood eosinophils were an independent risk factor for exacerbation in asthma patients with no history of smoking [24]. In these patients, a blood eosinophil cut-off of greater than or equal to 250 cells/μl resulted in an odds ratio of 2.94 for frequent exacerbations. Sputum eosinophils have also in some studies correlated with frequent exacerbations, though associations have not been as robust on multivariate regression analyses [8, 23]. Regardless of the performance of sputum eosinophil counts as a predictive biomarker for exacerbations, airways eosinophils have been demonstrated to be a key cellular mediator of asthma pathology, including that of exacerbations [25]. Indeed, several clinical trials which demonstrated the efficacy of specific eosinophil-depleting strategies with monoclonal antibodies against IL-5 or its receptor clearly illuminate the strong link between eosinophils and exacerbations [26–28]. The anti-IL-5 therapy mepolizumab was initially found to not show efficacy in clinical trials of unselected asthma patients designed around spirometric endpoints [29]. However, in the DREAM study, Pavord et al. studied selected patients with severe asthma who had an eosinophilic phenotype, as measured by sputum eosinophils, in a Phase II trial with severe exacerbations as the primary endpoint [28]. Depletion or reduction of eosinophils with mepolizumab resulted in an approximately 50% reduction in exacerbations, a finding that has been remarkably stable in subsequent Phase III studies of mepolizumab and in clinical development programs for reslizumab (anti-IL-5) and benralizumab (anti-IL-5 receptor) [26, 27, 30–32]. These data, in aggregate, have convincingly demonstrated that eosinophils play a key role in exacerbations, though there are clearly other important factors given that exacerbations are not completely eliminated by these therapies. There are also differential responses amongst patients, suggesting differential contributions of eosinophils to exacerbations even in biomarker-positive patients. Ramakrishnan et al recently reported the successful use of benralizumab in a case of near-fatal asthma exacerbation in which the use of systemic corticosteroids was felt to be contraindicated [33].

Other Type II Biomarkers

Exhaled nitric oxide is a marker of type II inflammation that is produced by the airway epithelium largely in response to stimulation by IL-13. As is the case with blood eosinophils, exhaled nitric oxide levels have been observed to be a marker for exacerbation risk [34]. In a European asthma cohort, an exhaled nitric oxide level of >45 parts per billion was associated with increased risk of asthma exacerbation [34]. It is not entirely surprising that exhaled nitric oxide and blood eosinophils provide overlapping signals with respect to exacerbation risk, given the high degree of overlap observed in type II biomarkers. Approximately half of the patients in the International Severe Asthma Registry are biomarker positive for both exhaled nitric oxide and blood eosinophils [35]. IL-13 signaling and its downstream effects, including the release of nitric oxide, occur through IL-4-receptor-alpha, and blocking this receptor with the monoclonal antibody dupilumab is approved therapy for moderate-to-severe asthma in multiple regions around the world [36]. Similar to eosinophil depletion strategies providing compelling evidence of the link between eosinophils and exacerbations, targeting IL-13 signaling and nitric oxide release with dupilumab resulted in reduced exacerbations in its phase II and phase III clinical trials [37, 38]. Those with an exhaled nitric oxide of greater than or equal to 25 parts per billion had a reduction in exacerbations, while those with lower levels did not [37].

IgE levels are also used as biomarker of type II inflammation clinically. From a canonical mechanistic perspective, IgE is an important mediator of exacerbation, binding to mast cells via the high-affinity IgE receptor, leading to the release of histamine and factors that promote bronchoconstriction and airway inflammation [36]. While elevated IgE has not been found to be associated with exacerbation in adults with severe asthma, allergic sensitization is a significant risk factor for exacerbation is children [39]. Environmental exposure to cockroach allergen in children with sensitivity to it resulted in increased hospitalization for asthma [40]. Connecting type II inflammatory factors and viral triggers of exacerbation, Teach and colleagues demonstrated that seasonal use of the monoclonal antibody against IgE, omalizumab, on return-to-school in the fall resulted in decreased asthma exacerbations and improved antiviral interferon-alpha responses [41].

Biomarkers of non-type II pathways in asthma are lacking, which extends in general to the prediction of exacerbation. However, in former smokers with asthma, elevated blood neutrophils do predict exacerbation, suggesting that these patients have a distinct inflammatory milieu promoting exacerbation [24].

Genetic Factors

Individual genetic variants have been found to confer small but detectable increased risks for asthma exacerbation [42]. A single nucleotide polymorphism (SNP) associated with loss-of-function for the P2X(7) nucleotide receptor predicted prednisone use in the prior year in a study by Denlinger et al. [43] genome-wide association studies (GWAS) in two large pediatric asthma cohorts led to the identification of SNPs in the genes *CTNNA3* (catenin (cadherin-associated protein), alpha 3) and *SEMA3D* (sema domain, immunoglobulin domain (Ig), short basic domain, secreted, (semaphorin) 3D) that were associated with exacerbation risk [44]. A Danish pediatric GWAS study identified the gene *CDHR3* (cadherin-related family member 3) as a susceptibility locus [45]. In childhood asthma, multiple studies have demonstrated interactions of genetic loci, such as those for IL-13, TGFB1, and IL-9, with environmental exposures in influencing risk of exacerbation [46–48]. Gene-by-environment interactions also likely modulate the risk of exacerbation through epigenetic changes [49].

Demographics

In a recent large registry study of patients with severe asthma in the United States (CHRONICLE), Lugogo and colleagues found several demographic factors that were associated with increased exacerbation risk, with Black and Hispanic patients experiencing increased rates of asthma exacerbation compared to White patients, female patients with an increased rate compared to male patients, and younger patients with an increased rate compared to older patients [50]. These findings are representative of complex interactions between biologic, environmental, and socioeconomic factors in asthma [51].

Physiologic Factors and Comorbidities

The degree of bronchodilator responsiveness was shown in the SARP cohort to be associated with exacerbation rate, with a rate ratio of 1.2 for a 10% increase in bronchodilator response on spirometry [23]. In addition to the association with obesity, diabetes, and hypertension noted above, gastroesophageal reflux disease, chronic rhinosinusitis, anxiety/depression, and psychosocial stressors all may contribute to exacerbation risk [23, 52–55]. Obesity not only influences airway mechanics in asthma but exists in the context of the metabolic syndrome, with systemic inflammation that is increasing recognized to be associated with adverse outcomes in asthma [22, 23, 56]. Chronic rhinosinusitis is a particularly important consideration given the often-overlapping type II inflammation that occurs in the upper and lower airways [57]. Treatment of rhinosinusitis and gastroesophageal reflux can be key components of maintaining asthma control. Also important for clinicians in improving asthma control are the recognition and treatment of anxiety and depression.

Long-Term Implications

As detailed above, a history of exacerbations is a predictor of future exacerbations. However, it is not entirely clear what the implications of asthma exacerbations are on long-term outcomes. In a case–control study of patients with moderate-to-severe asthma in British Columbia, frequent exacerbations were associated with a greater annual decline in FEV1 [58]. Each severe exacerbation correlated with a 30.2 ml greater annual decline in FEV1 over the course of the study. In another study from British Columbia using administrative records, both adult and pediatric patients identified for follow-up due to a history of an asthma exacerbation had an increased rate of subsequent exacerbation after the first follow-up event [59]. However, there was substantial heterogeneity in rates amongst individuals, suggesting variability in host and environmental factors as important considerations in exacerbation outcomes. Future studies will be needed to determine how exacerbation prevention in the short-term influences long-term outcomes, including lung function.

Conclusions

Asthma exacerbations occur in the context of a complex interplay between external exposures and host factors. Respiratory tract viral infections, in particular rhinovirus, are dominant initiators of exacerbations, with allergens and other inhalation exposures as additional key contributors. The presence of underlying type II inflammation, with associated biomarker elevations, is a major driver of exacerbation risk and mechanism, as evidenced by the consistent reduction of exacerbations seen with biologics targeting these pathways. Several genetic polymorphisms are associated with exacerbations, and while they may individually have small effects, they are cumulatively important and magnified by environmental exposures. A history of exacerbations predicts future exacerbations with potentially negative implications on long-term lung health.

References

1. Papi A, Brightling C, Pedersen SE, Reddel HK. Asthma. Lancet. 2018;391:783–800. https://doi.org/10.1016/S0140-6736(17)33311-1.
2. Golam SM, Janson C, Beasley R, FitzGerald JM, Harrison T, Chipps B, et al. The burden of mild asthma: clinical burden and healthcare resource utilisation in the NOVELTY study. Respir Med. 2022;200:106863. https://doi.org/10.1016/J.RMED.2022.106863.
3. Dusser D, Montani D, Chanez P, de Blic J, Delacourt C, Deschildre A, et al. Mild asthma: an expert review on epidemiology, clinical characteristics and treatment recommendations. Allergy. 2007;62:591–604. https://doi.org/10.1111/J.1398-9995.2007.01394.X.
4. Dougherty RH, Fahy JV. Acute exacerbations of asthma: epidemiology, biology and the exacerbation-prone phenotype. Clin Exp Allergy. 2009;39:193–202. https://doi.org/10.1111/J.1365-2222.2008.03157.X.

5. Ritchie AI, Farne HA, Singanayagam A, Jackson DJ, Mallia P, Johnston SL. Pathogenesis of viral infection in exacerbations of airway disease. Ann Am Thorac Soc. 2015;12(Suppl 2):S115–32. https://doi.org/10.1513/ANNALSATS.201503-151AW.
6. Ortega H, Nickle D, Carter L. Rhinovirus and asthma: challenges and opportunities. Rev Med Virol. 2021;31:e2193. https://doi.org/10.1002/RMV.2193.
7. Castillo JR, Peters SP, Busse WW. Asthma exacerbations: pathogenesis, prevention, and treatment. J Allergy Clin Immunol Pract. 2017;5:918–27. https://doi.org/10.1016/J.JAIP.2017.05.001.
8. Denlinger LC, Heymann P, Lutter R, Gern JE. Exacerbation-prone asthma. J Allergy Clin Immunol Pract. 2020;8:474–82. https://doi.org/10.1016/J.JAIP.2019.11.009.
9. McIntyre A, Busse WW. Asthma exacerbations: the Achilles heel of asthma care. Trends Mol Med. 2022;28:1112. https://doi.org/10.1016/J.MOLMED.2022.09.001.
10. Cox DW, Bizzintino J, Ferrari G, Khoo SK, Zhang G, Whelan S, et al. Human rhinovirus species C infection in young children with acute wheeze is associated with increased acute respiratory hospital admissions. Am J Respir Crit Care Med. 2013;188:1358–64. https://doi.org/10.1164/RCCM.201303-0498OC.
11. Bashir H, Grindle K, Vrtis R, Vang F, Kang T, Salazar L, et al. Association of rhinovirus species with common cold and asthma symptoms and bacterial pathogens. J Allergy Clin Immunol. 2018;141:822–824.e9. https://doi.org/10.1016/J.JACI.2017.09.027.
12. Falsey A, Hennessey P, Formica M, Cox C, Walsh E. Respiratory syncytial virus infection in elderly and high-risk adults. N Engl J Med. 2005;352:77. https://doi.org/10.1056/NEJMOA043951.
13. Lee B, Lewis G, Agyei-Manu E, Atkins N, Bhattacharyya U, Dozier M, et al. Risk of serious COVID-19 outcomes among adults and children with moderate-to-severe asthma: a systematic review and meta-analysis. Eur Respir Rev. 2022;31:220066. https://doi.org/10.1183/16000617.0066-2022.
14. Edwards MR, Regamey N, Vareille M, Kieninger E, Gupta A, Shoemark A, et al. Impaired innate interferon induction in severe therapy resistant atopic asthmatic children. Mucosal Immunol. 2013;6:797–806. https://doi.org/10.1038/MI.2012.118.
15. Wark PAB, Johnston SL, Bucchieri F, Powell R, Puddicombe S, Laza-Stanca V, et al. Asthmatic bronchial epithelial cells have a deficient innate immune response to infection with rhinovirus. J Exp Med. 2005;201:937–47. https://doi.org/10.1084/JEM.20041901.
16. Soto-Quiros M, Avila L, Platts-Mills TAE, Hunt JF, Erdman DD, Carper H, et al. High titers of IgE antibody to dust mite allergen and risk for wheezing among asthmatic children infected with rhinovirus. J Allergy Clin Immunol. 2012;129:1499–1505.e5. https://doi.org/10.1016/J.JACI.2012.03.040.
17. O'Hollaren M, Yunginger J, Offord K, Somers M, O'Connell E, Ballard D, et al. Exposure to an aeroallergen as a possible precipitating factor in respiratory arrest in young patients with asthma. N Engl J Med. 1991;324:390–1. https://doi.org/10.1056/NEJM199102073240602.
18. Price D, Hughes KM, Thien F, Suphioglu C. Epidemic thunderstorm asthma: lessons learned from the storm down-under. J Allergy Clin Immunol Pract. 2021;9:1510–5. https://doi.org/10.1016/J.JAIP.2020.10.022.
19. Tiotiu AI, Novakova P, Nedeva D, Chong-Neto HJ, Novakova S, Steiropoulos P, et al. Impact of air pollution on asthma outcomes. Int J Environ Res Public Health. 2020;17:1–29. https://doi.org/10.3390/IJERPH17176212.
20. Thomson NC, Chaudhuri R, Livingston E. Asthma and cigarette smoking. Eur Respir J. 2004;24:822–33. https://doi.org/10.1183/09031936.04.00039004.
21. Bloom CI, Palmer T, Feary J, Quint JK, Cullinan P. Exacerbation patterns in adults with asthma in England. A population-based study. Am J Respir Crit Care Med. 2019;199:446–53. https://doi.org/10.1164/rccm.201808-1516OC.
22. Peters MC, Mauger D, Ross KR, Phillips B, Gaston B, Cardet JC, et al. Evidence for exacerbation-prone asthma and predictive biomarkers of exacerbation frequency. Am J Respir Crit Care Med. 2020;202:973–82. https://doi.org/10.1164/RCCM.201909-1813OC.

23. Denlinger LC, Phillips BR, Ramratnam S, Ross K, Bhakta NR, Cardet JC, et al. Inflammatory and comorbid features of patients with severe asthma and frequent exacerbations. Am J Respir Crit Care Med. 2017;195:302–13. https://doi.org/10.1164/RCCM.201602-0419OC.
24. Westerhof GA, de Groot JC, Amelink M, de Nijs SB, ten Brinke A, Weersink EJ, et al. Predictors of frequent exacerbations in (ex)smoking and never smoking adults with severe asthma. Respir Med. 2016;118:122–7. https://doi.org/10.1016/j.rmed.2016.08.006.
25. Nelson RK, Bush A, Stokes J, Nair P, Akuthota P. Eosinophilic asthma. J Allergy Clin Immunol Pract. 2020;8:465–73. https://doi.org/10.1016/J.JAIP.2019.11.024.
26. Bel EH, Wenzel SE, Thompson PJ, Prazma CM, Keene ON, Yancey SW, et al. Oral glucocorticoid-sparing effect of mepolizumab in eosinophilic asthma. N Engl J Med. 2014;371:1189–97. https://doi.org/10.1056/NEJMOA1403291.
27. Ortega HG, Liu MC, Pavord ID, Brusselle GG, FitzGerald JM, Chetta A, et al. Mepolizumab treatment in patients with severe eosinophilic asthma. N Engl J Med. 2014;371:1198–207. https://doi.org/10.1056/NEJMOA1403290.
28. Pavord ID, Korn S, Howarth P, Bleecker ER, Buhl R, Keene ON, et al. Mepolizumab for severe eosinophilic asthma (DREAM): a multicentre, double-blind, placebo-controlled trial. Lancet. 2012;380:651–9. https://doi.org/10.1016/S0140-6736(12)60988-X.
29. Flood-Page P, Swenson C, Faiferman I, Matthews J, Williams M, Brannick L, et al. A study to evaluate safety and efficacy of mepolizumab in patients with moderate persistent asthma. Am J Respir Crit Care Med. 2007;176:1062–71. https://doi.org/10.1164/RCCM.200701-085OC.
30. Bleecker ER, FitzGerald JM, Chanez P, Papi A, Weinstein SF, Barker P, et al. Efficacy and safety of benralizumab for patients with severe asthma uncontrolled with high-dosage inhaled corticosteroids and long-acting β2-agonists (SIROCCO): a randomised, multicentre, placebo-controlled phase 3 trial. Lancet. 2016;388:2115–27. https://doi.org/10.1016/S0140-6736(16)31324-1.
31. FitzGerald JM, Bleecker ER, Nair P, Korn S, Ohta K, Lommatzsch M, et al. Benralizumab, an anti-interleukin-5 receptor α monoclonal antibody, as add-on treatment for patients with severe, uncontrolled, eosinophilic asthma (CALIMA): a randomised, double-blind, placebo-controlled phase 3 trial. Lancet. 2016;388:2128–41. https://doi.org/10.1016/S0140-6736(16)31322-8.
32. Castro M, Zangrilli J, Wechsler ME, Bateman ED, Brusselle GG, Bardin P, et al. Reslizumab for inadequately controlled asthma with elevated blood eosinophil counts: results from two multicentre, parallel, double-blind, randomised, placebo-controlled, phase 3 trials. Lancet Respir Med. 2015;3:355–66. https://doi.org/10.1016/S2213-2600(15)00042-9.
33. Ramakrishnan S, Camp JR, Vijayakumar B, Hardinge FM, Downs ML, Russell REK, et al. The use of benralizumab in the treatment of near-fatal asthma: a new approach. Am J Respir Crit Care Med. 2020;201:1441–3. https://doi.org/10.1164/RCCM.202001-0093LE.
34. Kupczyk M, ten Brinke A, Sterk PJ, Bel EH, Papi A, Chanez P, et al. Frequent exacerbators – a distinct phenotype of severe asthma. Clin Exp Allergy. 2014;44:212–21. https://doi.org/10.1111/cea.12179.
35. Denton E, Price DB, Tran TN, Canonica GW, Menzies-Gow A, FitzGerald JM, et al. Cluster analysis of inflammatory biomarker expression in the International Severe Asthma Registry. J Allergy Clin Immunol Pract. 2021;9:2680–2688.e7. https://doi.org/10.1016/J.JAIP.2021.02.059.
36. Busse WW, Kraft M, Rabe KF, Deniz Y, Rowe PJ, Ruddy M, et al. Understanding the key issues in the treatment of uncontrolled persistent asthma with type 2 inflammation. Eur Respir J. 2021;58:2003393. https://doi.org/10.1183/13993003.03393-2020.
37. Castro M, Corren J, Pavord ID, Maspero J, Wenzel S, Rabe KF, et al. Dupilumab efficacy and safety in moderate-to-severe uncontrolled asthma. N Engl J Med. 2018;378:2486–96. https://doi.org/10.1056/NEJMOA1804092.
38. Wenzel S, Castro M, Corren J, Maspero J, Wang L, Zhang B, et al. Dupilumab efficacy and safety in adults with uncontrolled persistent asthma despite use of medium-to-high-dose inhaled corticosteroids plus a long-acting β2 agonist: a randomised double-blind placebo-controlled pivotal phase 2b dose-ranging trial. Lancet. 2016;388:31–44. https://doi.org/10.1016/S0140-6736(16)30307-5.

39. Simpson A, Tan VYF, Winn J, Svensén M, Bishop CM, Heckerman DE, et al. Beyond atopy: multiple patterns of sensitization in relation to asthma in a birth cohort study. Am J Respir Crit Care Med. 2010;181:1200–6. https://doi.org/10.1164/RCCM.200907-1101OC.
40. Rosenstreich DL, Eggleston P, Kattan M, Baker D, Slavin RG, Gergen P, et al. The role of cockroach allergy and exposure to cockroach allergen in causing morbidity among inner-city children with asthma. N Engl J Med. 1997;336:1356–63. https://doi.org/10.1056/NEJM199705083361904.
41. Teach SJ, Gill MA, Togias A, Sorkness CA, Arbes SJ, Calatroni A, et al. Preseasonal treatment with either omalizumab or an inhaled corticosteroid boost to prevent fall asthma exacerbations. J Allergy Clin Immunol. 2015;136:1476–85. https://doi.org/10.1016/J.JACI.2015.09.008.
42. Park HW, Tantisira KG. Genetic signatures of asthma exacerbation. Allergy Asthma Immunol Res. 2017;9:191–9. https://doi.org/10.4168/AAIR.2017.9.3.191.
43. Denlinger LC, Manthei DM, Seibold MA, Ahn K, Bleecker E, Boushey HA, et al. P2X7-regulated protection from exacerbations and loss of control is independent of asthma maintenance therapy. Am J Respir Crit Care Med. 2013;187:28–33. https://doi.org/10.1164/RCCM.201204-0750OC.
44. Mcgeachie MJ, Wu AC, Tse SM, Clemmer GL, Sordillo J, Himes BE, et al. CTNNA3 and SEMA3D: promising loci for asthma exacerbation identified through multiple genome-wide association studies. J Allergy Clin Immunol. 2015;136:1503–10. https://doi.org/10.1016/J.JACI.2015.04.039.
45. Bønnelykke K, Sleiman P, Nielsen K, Kreiner-Møller E, Mercader JM, Belgrave D, et al. A genome-wide association study identifies CDHR3 as a susceptibility locus for early childhood asthma with severe exacerbations. Nat Genet. 2014;46:51–5. https://doi.org/10.1038/NG.2830.
46. Sordillo JE, Kelly R, Bunyavanich S, McGeachie M, Qiu W, Croteau-Chonka DC, et al. Genome-wide expression profiles identify potential targets for gene-environment interactions in asthma severity. J Allergy Clin Immunol. 2015;136:885–892.e2. https://doi.org/10.1016/J.JACI.2015.02.035.
47. Hunninghake GM, Soto-Quirós ME, Avila L, Su J, Murphy A, Demeo DL, et al. Polymorphisms in IL13, total IgE, eosinophilia, and asthma exacerbations in childhood. J Allergy Clin Immunol. 2007;120:84–90. https://doi.org/10.1016/J.JACI.2007.04.032.
48. Sharma S, Raby BA, Hunninghake GM, Soto-Quirós M, Avila L, Murphy AJ, et al. Variants in TGFB1, dust mite exposure, and disease severity in children with asthma. Am J Respir Crit Care Med. 2009;179:356–62. https://doi.org/10.1164/RCCM.200808-1268OC.
49. Chen YC, Tsai YH, Wang CC, Liu SF, Chen TW, Fang WF, et al. Epigenome-wide association study on asthma and chronic obstructive pulmonary disease overlap reveals aberrant DNA methylations related to clinical phenotypes. Sci Rep. 2021;11:5022. https://doi.org/10.1038/S41598-021-83185-1.
50. Lugogo N, Judson E, Haight E, Trudo F, Chipps BE, Trevor J, et al. Severe asthma exacerbation rates are increased among female, Black, Hispanic, and younger adult patients: results from the US CHRONICLE study. J Asthma. 2022;59:2495. https://doi.org/10.1080/02770903.2021.2018701.
51. Forno E, Celedón JC. Health disparities in asthma. Am J Respir Crit Care Med. 2012;185:1033–5. https://doi.org/10.1164/RCCM.201202-0350ED.
52. Price D, Wilson AM, Chisholm A, Rigazio A, Burden A, Thomas M, et al. Predicting frequent asthma exacerbations using blood eosinophil count and other patient data routinely available in clinical practice. J Asthma Allergy. 2016;9:1–12. https://doi.org/10.2147/JAA.S97973.
53. Tay TR, Radhakrishna N, Hore-Lacy F, Smith C, Hoy R, Dabscheck E, et al. Comorbidities in difficult asthma are independent risk factors for frequent exacerbations, poor control and diminished quality of life. Respirology. 2016;21:1384–90. https://doi.org/10.1111/RESP.12838.
54. ten Brinke A, Sterk PJ, Masclee AAM, Spinhoven P, Schmidt JT, Zwinderman AH, et al. Risk factors of frequent exacerbations in difficult-to-treat asthma. Eur Respir J. 2005;26:812–8. https://doi.org/10.1183/09031936.05.00037905.

55. Sandberg S, Paton JY, Ahola S, McCann DC, McGuinness D, Hillary CR, et al. The role of acute and chronic stress in asthma attacks in children. Lancet. 2000;356:982–7. https://doi.org/10.1016/S0140-6736(00)02715-X.
56. Peters MC, McGrath KW, Hawkins GA, Hastie AT, Levy BD, Israel E, et al. Plasma interleukin-6 concentrations, metabolic dysfunction, and asthma severity: a cross-sectional analysis of two cohorts. Lancet Respir Med. 2016;4:574–84. https://doi.org/10.1016/S2213-2600(16)30048-0.
57. de Groot JC, Amelink M, de Nijs SB, Plaat R, Reitsma BH, Storm H, et al. Risk factors for frequent severe exacerbations in late-onset eosinophilic asthma. Am J Respir Crit Care Med. 2015;192:899–902. https://doi.org/10.1164/rccm.201505-1003LE.
58. Bai TR, Vonk JM, Postma DS, Boezen HM. Severe exacerbations predict excess lung function decline in asthma. Eur Respir J. 2007;30:452–6. https://doi.org/10.1183/09031936.00165106.
59. Lee TY, Petkau J, Sadatsafavi M. Long-term natural history of severe asthma exacerbations and their impact on the disease course. Ann Am Thorac Soc. 2022;19:907–15. https://doi.org/10.1513/ANNALSATS.202012-1562OC.

Innate Immunity, Epithelial Plasticity, and Remodeling in Asthma

Allan R. Brasier

Abstract

Innate immune responses (IIR) of the epithelium play a critical role in the initiation and progression of asthma. The core of the IIR is an intracellular signaling pathway activated by pattern recognition receptors (PRRs) to limit the spread of infectious organisms. This chapter will focus on the epithelium as the major innate sentinel cell and its role in acute exacerbations (AEs). Although the pathways of how the IIR activates the NFκB transcription factor, triggering cytokine secretion, dendritic cell activation, and Th2 polarization are well-described, recent exciting work has developed mechanistic insights into how chronic activation of the IIR is linked to mucosal adaptive responses. These adaptations include changes in cell state, now called epithelial–mesenchymal plasticity (EMP). EMP is a coordinated, genomic response to airway injury disrupting epithelial barrier function, expanding the basal lamina, and producing airway remodeling. EMP is driven by activation of the unfolded protein response (UPR), a transcriptional response producing metabolic shunting of glucose through the hexosamine biosynthetic pathway (HBP) to protein N-glycosylation. NFκB signaling and UPR activation pathways potentiate each other in remodeling the basement membrane. Understanding of injury-repair process of epithelium provides new therapeutic targets for precision approaches to the treatment of asthma exacerbations and their sequelae.

A. R. Brasier (✉)
Department of Medicine and Institute for Clinical and Translational Research (ICTR), School of Medicine and Public Health (SMPH), University of Wisconsin-Madison, Madison, WI, USA
e-mail: abrasier@wisc.edu

A. R. Brasier, N. N. Jarjour (eds.), *Precision Approaches to Heterogeneity in Asthma*, Advances in Experimental Medicine and Biology 1426,
https://doi.org/10.1007/978-3-031-32259-4_13

Keywords

Acute exacerbations · Interferon · Epithelial mesenchymal transition · Plasticity · Chromatin remodeling · Club cells · Secretoglobin · Bromodomain containing protein 4 (BRD4) · Extracellular matrix (ECM) · Epithelial mesenchymal trophic unit (EMTU) · Unfolded protein response (UPR) · Inositol-requiring enzyme 1 (IRE1) · Endoplasmic reticulum stress · X-box binding protein · Sentinel cells · Mucosa

Abbreviations

AA	Allergic asthma
AE	Acute exacerbation;
CDH1	Epithelial cadherin
COL	Collagen
EMP	Epithelial-mesenchymal plasticity
EMTU	Epithelial-mesenchymal trophic unit
ER	Endoplasmic reticulum
FN1	Fibronectin
HBP	Hexosamine biosynthetic pathway
IFN	Interferon
IIR	Innate immune response
IRE	Inositol-requiring enzyme
IRE1α	Inositol-requiring protein 1α
MMP	Metalloproteinase
ORMDL3	ORM1 (yeast)–like protein 3/Sphingolipid Biosynthesis Regulator 3
PAMP	Pathogen-associated molecular pattern
PRR	Pattern recognition receptors
SNAI	Snail family repressor
UPR	Unfolded protein response
XBP1	X box-binding protein 1
ZEB1	Zinc finger E-box binding homeobox
αSMA	Alpha smooth muscle actin

Introduction

Asthma is a chronic disease characterized by airway hyperreactivity, episodic reversible airway obstruction and, in the majority of cases, a T-helper (Th) cell-predominant inflammation [1]. One of the most common subtypes of asthma is allergic asthma (AA), a disease that often begins early in childhood and is characterized by (i) infiltrating Th2 cells and eosinophils in the airway mucosa, (ii) thickening of the *lamina reticularis*, (iii) secretion of Th2 mediators, and (iv) production of IgE [2]. Although AA is conventionally thought to be caused by an imbalance of Th2/1-type lymphocytes, known as the "Th2 hypothesis," this hypothesis is challenged to explain clinical evidence that respiratory viral infections are linked to the initiation and exacerbations of the disease, that Th2-directed therapies are not uniformly effective for treating AA, and that innate signatures are observed in subtypes of asthmatics [3]. A broader view of innate immunity in asthma is warranted, and more mechanistic understanding is needed.

Acute Exacerbations in Asthma

Patients with asthma exhibit episodic clinical decompensations, known as acute exacerbations (AEs), responsible for the major morbidity, mortality, and health care costs associated with the management of this disease [4]. Observational studies have shown that upper respiratory tract RNA viral infections [rhinovirus, respiratory syncytial virus (RSV), and others] and/or allergen exposures are the most common precipitating factors of AEs [4]. Frequent AEs in a subtype of poorly controlled asthmatics, known as "frequent exacerbators" [5], are associated with a more rapidly declining pulmonary function [6, 7]. Declining pulmonary function is the result of structural airway remodeling.

Airway remodeling is a poorly understood, multicellular process that produces disruption of normal structural components of the airway, resulting in airway hyperresponsiveness (AHR) and fixed obstruction [8]. Specifically, remodeling is characterized by changes in epithelial cell state (mucous metaplasia and/or mesenchymal transition), disruption of the epithelial barrier, thickening of the airway smooth muscle layer, enhanced deposition of the *lamina reticularis* and angiogenesis within the airway wall [8–10]. Enhanced mucus production from the expansion of submucosal goblet cell population and hypertrophy of airway smooth muscle layers enhances small airway obstruction, reducing lung compliance and producing hyperreactivity to methacholine [8]. In this way, airway remodeling contributes to stiffer and less compliant airways in asthma and may lead to persistent, irreversible/fixed airflow obstruction in frequent exacerbators [8].

Innate Sentinel Cells of the Airway Provide First Line of Defense to Viral Infection

Innate "sentinel" cells of the airway are programmed to sense the presence of replicating microbial pathogens and trigger an innate immune response (IIR). Multiple cell types can play these roles; to date, the major innate cells of the airway that have been identified are epithelial cells, dendritic cells, and macrophages. These cell types play different roles depending on the pathogen and its primary site of replication and are reviewed extensively by Esnault et al. in this volume (see Part III, chapter "Development of Adaptive Immunity and Its Role in Lung Remodeling"). As the primary target of RNA virus infection, responsible for the majority of AEs, we will focus on the airway epithelium because this cell type plays a central role in innate antiviral responses [11, 12]. Airway epithelial cells express a spectrum of pattern recognition receptors (PRRs), including cell-surface and endosomal localized PRRs, cytosolic retinoic acid-inducible gene I (RIG-I)-like helicases (RLHs) and NOD-like receptors [13] that bind distinct viral pathogen-associated molecular patterns (PAMPs). Double-stranded RNA and 5-phosphorylated RNA produced in the airway lumen by active viral replication are PAMPs bound by membrane-associated Toll-like receptor 3 (TLR3) present in airway epithelial cells. In contrast, these same PAMPs produced by intracellular viral replication are detected by the RLHs.

PRRs recognize specific PAMP molecules. Upon binding, PRRs recruit signaling adapters into a "signalsome" complex that triggers intracellular innate signaling. PRR-associated signalsomes activate an interconnected network of intracellular signaling pathways including the MyD88-IκB kinase (IKK)-NFκB, tank binding kinase (TBK)-interferon regulatory factor (IRF), mitogen-activated protein kinase pathways, and others. These cascades trigger a genomic response resulting in the secretion of danger signals, protective IFNs, and cytokines [14] as well as determining apoptotic cell fate decisions [11, 13, 15–17].

Role of Lower Airway Epithelial Cells in Inflammation-Remodeling

Prospective observational studies show that children with severe lower respiratory tract infections with RSV exhibit long-term decreased pulmonary function [18, 19]. In these lower tract infections, RSV replication has been documented in lining epithelial cells located in the bronchioles and alveoli. This phenomenon of lower respiratory tract replication has been observed in children with naturally acquired (fatal) severe infections [20] and in normal immune volunteers experimentally infected intranasally [21].

In the human lung, the small airway forms a functional "transition zone," a region between the gas-conducting bronchioles and the gas-exchanging alveoli. The small airway (<2 mm ID without cartilage) is of particular interest because small airways are the major site of airflow limitation in patients with asthma [22]. This zone is composed of a pseudostratified columnar epithelium composed of at least five distinct epithelial phenotypes. These cell types have been elucidated by morphological criteria and using gene signatures from single-cell sequencing studies as: multi-ciliated cells, secretory/"Club" cells, mucin-producing goblet cells, ionocytes, and basal cells [11, 23–25]. These cells differ from one another in secretory functions, self-renewal, and mucociliary clearance properties. Underlying this population of epithelial cells is a basement membrane composed of collagen (COL I/III), fibronectin (FN1) and population of subepithelial fibroblasts. This epithelial-fibroblast unit is referred to as the "epithelial-mesenchymal trophic unit" (EMTU); extensive intercellular communication occurs between these cell types in the EMTU, affecting cellular function and behavior [26].

Focusing on the transition zone of the airway epithelium, emerging evidence generated from animal models points to specialized epithelial cells in this zone also as a major sensor in the initiation and maintenance of inflammation and airway remodeling [12, 27, 28]. Lower airway epithelial cells are programmed to produce distinct patterns of inflammatory and remodeling gene networks than epithelia derived from other sites. Quantitative gene-profiling experiments have shown that lower airway cells produce greater amounts of Th2-activating CCL-type chemokines than conduct airway-derived epithelia [29, 30]. These genomic studies were extended by systems levels proteomics studies that found small airway epithelial cells derived from the transition zone also produce greater amounts of myofibroblast growth factors (IL6 and TGFβ), Th2 polarizing cytokines [thymic stromal lymphopoietin (TSLP)] and mucogenic cytokines (CCL-20) than do cells from the conducting airways [27]. These findings provide mechanistic insight into why lower airway infections are associated with atopy and Th2 polarization (Fig. 1, Ref. [31])].

The important role of the innate immune response of lower airway epithelial cells in inflammation and remodeling has been extended by cell-type depletion experiments in mouse models. Selective depletion of a key arm of the innate response pathway, the NFκB subunit, RelA, by tissue-specific expression of the Cre recombinase has provided definitive proof that cells derived from Club cell lineages are the major functionally important innate sensors of viral infection. Specifically, mice depleted of RelA in the secretoglobin (Scgb1a1)+/"Club cell" population have significantly reduced chemokine response, leukocytic inflammation, and airway

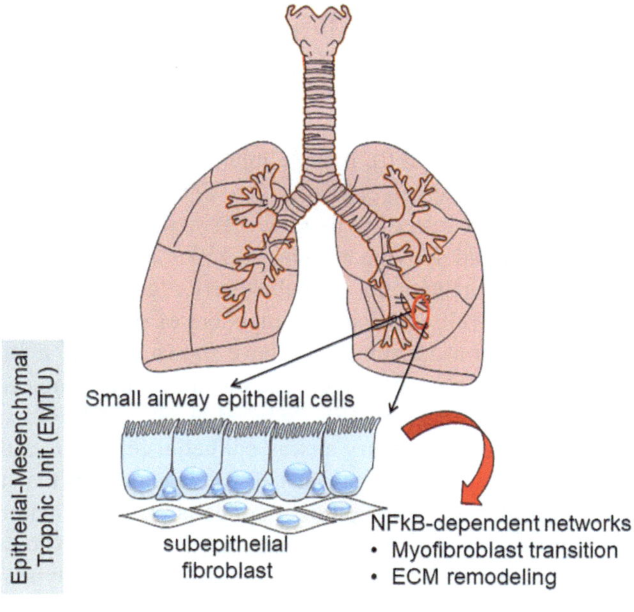

Fig. 1 Role of the transition zone in innate airway remodeling. Shown is a schematic view of the human lungs from the larynx to the distal alveoli. The lungs can be functionally separated into the conducting zone (larynx, trachea) and the respiratory zone (small bronchioles and alveoli). Of focus for this work, the distinct epithelial cell types in the transition zone play key roles in inflammation and remodeling through NF-κB regulated inducible gene expression program

obstruction in response to experimental RSV infection [12] and TLR3 stimulation [32]. This same transition zone, Scgb1a1-derived bronchiolar cells, mediate remodeling in response to *Aspergillosis* [33], cat dander allergens [28], and in response to house dust mite allergen [34]. Collectively, this evidence strongly supports the conclusion that the IIR in Scgb1a1+/Club cell progenitors in the transition zone of the lower airway is critical to the initiation of airway inflammation and remodeling from a variety of environmental sources.

Epithelial-Mesenchymal Plasticity (EMP)

In response to environmental toxicants, epithelial cells rapidly undergo necrosis and shed, releasing damage-associated molecular patterns. Surviving adjacent epithelial cells respond by de-differentiation, enabling them to migrate to repopulate the injured area [35]. The recognition that stimulated or injured epithelial cells can de-differentiate to become motile mesenchymal-like cells is now referred to as EMP [36–38]. EMP is a metastable state where cells can advance toward mesenchymal features, originally termed epithelial-mesenchymal transition (EMT), as well as reverse this state, known as mesenchymal–epithelial transition (MET). EMP is an injury-repair mechanism that has been observed in the airways of chronic asthma and in response to viral infection, smoke injury, exposure to aeroallergen, or even repeated methacholine-induced bronchoconstriction [9, 28, 39–41]. It is also important to emphasize that cells entering EMP do not become myofibroblasts per se but instead are motile and proliferate to repopulate injured alveolar type 1 and 2 cell populations [42].

At the cell biology level, EMP is a multi-step response triggered by the release of epithelial growth factors (TGFβ), matricellular components, Th2-derived cytokines (IL13), or actions of allergen proteases. One of the earliest steps in the injury response pathway involves loss of cell-surface expression of epithelial cadherin (CDH1) via proteolysis and subsequent internalization, disrupting adherens junctions [43] responsible for maintaining epithelial barrier function [44]. Disruption of the epithelial barrier is a well-established component of AA (Fig. 2). In particular, immunostaining studies of asthmatic tissues have shown reduced numbers of tight junctions, lower trans-epithelial resistance, and enhanced macromolecular permeability than seen in normal controls [45]. Reduced levels of CDH1 expression are seen [46], and isolated human primary airway epithelial cells from asthmatics lose zona occludin-1 (ZO-1) expression at lower TGFβ concentrations than that required for normal epithelial cells [47]. Disruption of epithelial barrier function, associated with fragility, epithelial denudation/shedding enhances allergen exposures, promoting allergic sensitization [48] and correlates with airway hyperreactivity in asthma [49].

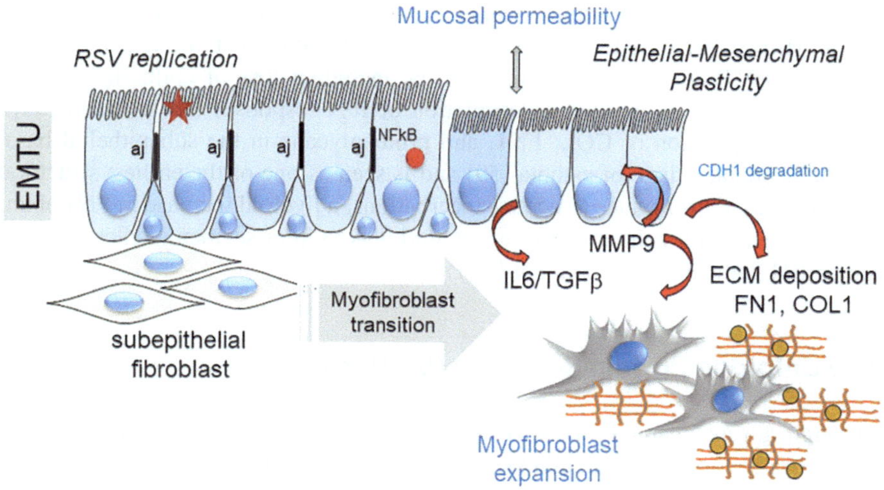

Fig. 2 Epithelial–mesenchymal plasticity (EMP) in the lower airway. Shown is a segment of transition zone airway epithelium and homeostatic properties affected by EMP. The induction of EMP is activated by viral replication and/or repetitive aeroallergen exposure. EMP results in loss of adherens junctions (aj), loss of apical-basal polarity and disruption of mucosal barrier function, and enhanced ECM deposition. TGFβ, transforming growth factor beta

Innate Signaling Drives EMP Progression Toward the Mesenchymal State

The NFκB transcription factor is a central component of PRR signaling, mediating the IIR controlling networks of cytokines and IFN genes [50]. Recent systems level genomics and proteomics studies have also found that NFκB is a master regulator of EMP activated by PRR or TGFβ signaling [51–53]. With the persistence of inflammation or injury, NFκB drives enhanced expression of the core Snail family repressor (SNAI) and Zinc Finger E-Box Binding Homeobox (ZEB1) transcription factor module. SNAI1 and ZEB1 are core mesenchymal transcription factors that drive gene activation and repression in EMP. In differentiated epithelial cells, SNA1 and ZEB1 are expressed at low levels. Here, the expression of miR-34 blocks the translation of SNAI1 mRNA [54]. In parallel, miR-200 expression inhibits the translation of ZEB1 mRNA [55]. PRR signaling or TGFβ stimulation activates NFκB, which perturbs this negative feedback loop by enhancing SNAI1 expression [52]. SNAI1 inhibits miR-34 expression, allowing SNAI1 to be translated. With SNAI1 expression, SNAI1 directly activates ZEB1 abundance, silencing miR-200 and promoting EMP transition [56]. In addition to its actions on ZEB, SNAI1 directly binds to regulatory promoter regions of CDH1 and ZO-1, leading to their repression and subsequent cellular loss of apical-basal polarity [57].

Epithelial cells in EMP express mesenchymal signatures and functionally change their differentiated cell state. Alpha smooth muscle actin (αSMA) expression results in enhanced cytokinesis, enhanced production of the intermediate filament, vimentin (VIM), produce motility, and secretion of COL, FN1, and metalloproteinase-9 (MMP9) promote ECM deposition into the basal lamina. Many cellular hallmarks of EMP are seen in chronic asthma, including loss of epithelial polarity, changes in permeability, and dysregulation of IFN responses (reviewed in [58]). Enhanced deposition of COL, FN1, and proteoglycans in the subepithelial basal lamina is characteristic of asthma [59] and is seen as one of the earliest structural changes [60]. Of relevance to acute exacerbations, a fivefold increase in FN1 abundance is seen in bronchoalveolar lavage fluid in AAs after segmental allergen challenge [61], indicating activation of EMP.

The Unfolded Protein Response (UPR) in Asthma

Through abrupt changes in ECM production, the epithelial cell entering EMP undergoes endoplasmic reticulum (ER) stress. The UPR is activated by changes in protein load within the endoplasmic reticulum and is known to be activated in asthma through a number of animal and human studies [62]. In a mouse model of severe asthma induced by ovalbumin (OVA) and TLR4 stimulation, Kim et al. implicated UPR activation to be associated with NFκB and inflammation; treatment with a chemical chaperone reduced the level of NFκB activation and inflammation, suggesting that the UPR participated in potentiating innate inflammation [63]. A separate study of humans with mild AA treated with inhaled corticosteroids using unbiased RNA sequencing studies of bronchial brushings (a sample enriched in epithelial cells) discovered increased XBP1 expression, a signature of UPR activation [64]. This finding was independently validated by demonstrating the presence of the spliced form of XBP1s, whose abundance correlated with Th2 inflammation and interferon-stimulated gene expression, and reduced lung function. A separate study examining signatures in neutrophilic and eosinophilic asthma also discovered UPR signatures in this severe asthma model [65]. Finally, genetic susceptibility studies have shown that Orosomucoid like-3/ORDML3 is the strongest replicated genetic risk factor for early-onset asthma [66]. Interestingly, single polynucleotide polymorphisms of the *ORDML3* gene are associated with increased expression. ORDML3 is involved in sphingolipid biosynthesis and reduces ER calcium content, leading to increased ER stress [67]. Collectively, these data support the findings of an intricate link between the UPR and NFκB activation in asthma.

The Unfolded Protein Response (UPR) Is a Core Pathway Driving EMP

We have recently discovered that the UPR is a driving cellular adaptation in EMP; this finding now provides a unifying mechanism relating innate inflammation with EMP and remodeling the subepithelial basement membrane (Fig. 3; Ref. [68]). The UPR is a highly conserved cellular pathway enabling cells to respond to sudden changes in secretory protein loads and has been recently implicated in promoting

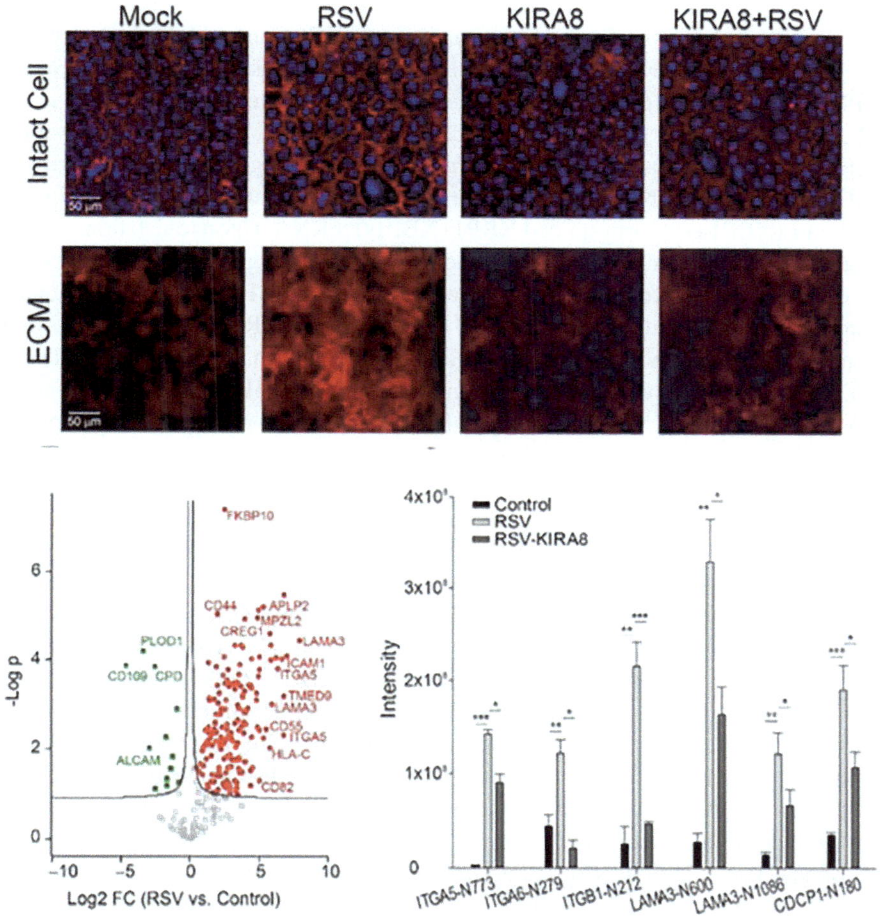

Fig. 3 RSV induces ECM remodeling via the IRE1α–XBP1 arm of the UPR. (**a**) Scgb1a1-derived hSAECs treated with solvent (DMSO) or IRE1a inhibitor (KIRA8). Cells were then mock or RSV-infected (MOI = 1, 24 h). The cells were fixed and stained for extracellular FN1 without permeabilization. Nuclei were then stained with DAPI. Red, FN1. Blue, DAPI. Scale bar 50 µm shown. Identically treated plates were decellularized and stained for FN1, and imaged. (**b**) Proteomics analysis of N-glycosylation in hSAECs. N-glycosylated peptides were enriched with lectins and then analyzed with label-free LC-MS/MS. (**a**) Volcano plot of N-glycosylated peptides (RSV vs. Control). Red circle, N-glycoproteins upregulated by RSV; green square, N-glycoproteins down-regulated by RSV infection. N-glycosylated peptides strongly induced by RSV infection and regulated by the IRE1α–XBP1 arm of UPR are shown (Student's t-test with permutation FDR < 0.05). (Data from Ref. [72])

the inflammation in asthma, as noted above. Virtually all stimuli that induce the UPR activate NFκB, trigger EMP transition and cause acute asthma exacerbations. This relationship has been documented for paramyxovirus viral infections, smoking and allergen exposure [62, 63, 69].

Two major UPR sensors/effectors are activated by changes in protein load in the ER; these are the inositol-requiring protein 1α (IRE1α) and protein kinase RNA-like ER kinase (PERK) [70]. Of these, IRE1α functions as the primary arm of the UPR linked to EMP [69]. In the presence of unfolded proteins accumulating in the ER, the HSPA5/BiP chaperone dissociates from IRE1α, resulting in a coupled dimerization-autotransphosphorylation reaction, triggering the RNAse activity of IRE1α. The IRE1α RNAse processes the mRNA encoding unspliced X box-binding protein 1 (XBP1u) to form spliced XBP1 (XBP1s) mRNA. Upon translation, XBP1s is a transcription factor that controls the transcription of genes encoding proteins involved in protein folding, ER-associated degradation (ERAD), protein quality control, and ER membrane biogenesis. Phosphorylated IRE1α also induces JUN N-terminal kinase (JNK) and IKK-NFκB pathways through interactions with TRAF adapters, interfacing with the innate signaling pathway [71]. Through these arms, the UPR restores ER homeostasis by either increasing the protein folding capacity of the ER, reducing the influx of nascent proteins into the ER, and/or degradation through ERAD.

Protein N-glycosylation enhances protein folding and efflux from the ER, relieving proteotoxic stress. We recently observed that protein N-glycosylation is enhanced by EMP (Fig. 4; Refs. [68, 69, 72]; Reviewed in [73]). To further understand the mechanisms controlling this process, we observed that EMP activates rate-limiting enzymes of the hexosamine biosynthetic pathway (HBP). The HBP is a major glucose metabolism pathway producing uridine 5′-diphosphate-N-acetyl-d-glucosamine (UDP-GlcNAc), an obligate substrate for posttranslational modification (O-GlcNAcylation) and protein N-glycosylation. The enzymes of the HBP pathway include Glutamine-fructose-6-phosphate transaminase (GFPT)-1, -2, Glucosamine-phosphate N-acetyltransferase (GNPNAT) and UDP-N-acetylglucosamine pyrophosphorylase 1 (UAP1). These enzymes shunt fructose-6 phosphate into UDP-GlcNAc, a rate-limiting substrate of the O-GlcNAc transferase, constituting the first committed step in glycoprotein formation [74].

Through the HBP, activation of EMP produces substantial intracytoplasmic formation of N-glycosylated proteins and their secretion [68, 72]. Using systems-level glycomics studies, we showed that EMP induces secretion of N-glycosylated ECM components, providing a mechanistic pathway for how the UPR triggers ECM remodeling [72]. We found that the IRE1α-XBP1s pathway of the UPR increases N-glycosylation of integrin subunits, laminins, COL, FN1, ECM-cross-linking enzymes, and MMPs. These proteins are essential for ECM organization, ECM–cell signaling, and neutrophil degranulation. Together, our data suggest that HBP is an adaptive response activated in EMP to improve the folding and secretion of ECM proteins to relieve ER stress.

Despite prevailing views, the lung ECM undergoes high levels of turnover, where COL turnover rates have been estimated to be up to 10–15% per day [75]. Increased COL-containing ECM in the subepithelial basement membrane may have important implications for respiratory mechanics [76]. In addition to mechanical effects, enhanced deposition of N-glycosylated COL and FN1s is an important signal for pulmonary mesenchymal cell populations in the EMTU to acquire myofibroblast

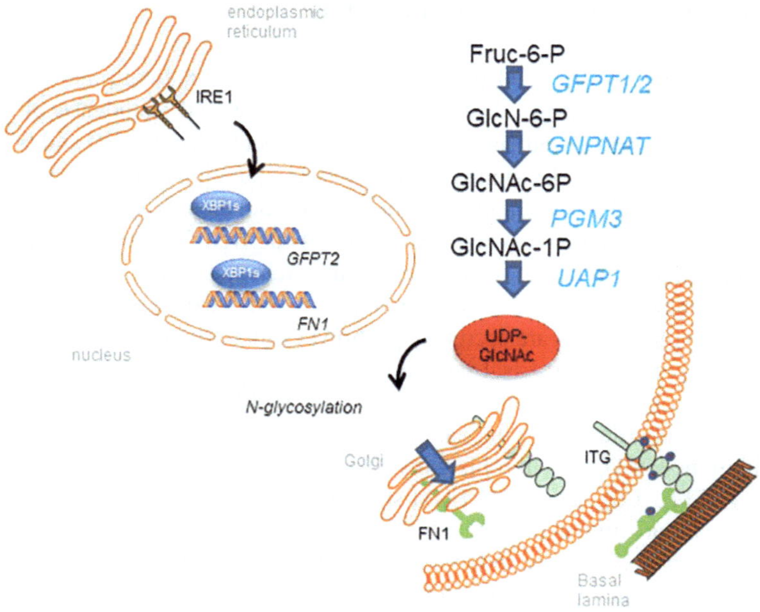

Fig. 4 Role of HBP in N-glycosylation. A schematic view of the relationship between the IRE1α–XBP1 pathway of the unfolded protein response and activation of the HBP. IRE1 activated in the ER induces alternative splicing and produces the formation of activated XBP1s, which is a transcription factor controlling the expression of the hexosamine biosynthetic pathway (HBP). With HBP activation, UDP-GlcNAc formation enhances protein N-glycosylation of fibronectin (FN1), collagen (COL), matrix metalloproteinase (MMP), and integrin (ITG), promoting remodeling of the basal lamina. After processing through the Golgi, glycosylated ECM components are presented on the cell surface and contribute to the remodeling of the basal lamina

properties [77]. The presence of high ECM stress and aberrantly processed FN1 splice products induce phosphorylation and activation of focal adhesion kinases in subepithelial myofibroblasts [78]. Myofibroblasts are thought to be major producers of ECM in the airway in many chronic obstructive lung diseases.

Crosstalk of the UPR and the IIR Inducing the HBP

A recent exciting finding found crosstalk between the IRE1α-XBP1s arm of the UPR and the IKK-NFκB pathway [79]. Here, IKK activation downstream of TGFβ stimulation directly phosphorylates – XBP1s, which activates HBP and upregulates protein N-glycosylation, preventing ER stress-induced apoptosis. These data suggest that the IKKβ-XBP1s-HBP crosstalk pathway couples inflammation and glucose metabolic reprogramming supporting EMP. Once activated, the UPR is sustained by an autoregulatory loop where XBP1s enhances Pol II binding to its own promoter [80]. This autoregulation ensures a continuous supply of *XBP1* mRNA to maintain ER proteostasis and support EMP.

Myofibroblast Activation in the EMTU

The composition of the ECM plays a critical role in maintaining the EMTU and airway physiology. Changes in the structural stiffness and abundance of matrix-associated factors produced during injury/repair affect both epithelial and fibroblast cells within the EMTU. Expanded myofibroblast populations are observed in acute asthma, severe fatal asthma, and refractory asthma and their exacerbations [81, 82]. Although these are a heterogenous group of cells derived from subepithelial fibroblasts, circulating fibrocytes, or interstitial mesenchymal cell populations, subepithelial fibroblasts contribute to the αSMA and COL1-expressing myofibroblasts [83, 84]. Myofibroblasts are secretory phenotypes that produce ECM proteins and MMPs that contribute to basal lamina expansion in chronic airway disease [76]. With myofibroblast persistence, deposition of ECM stiffens the lungs, reducing normal elastic properties and pulmonary function.

Within the EMTU, the fibroblast-myofibroblast transition is mediated by growth factors (TGFβ, IL6), changes in ECM stiffness and MMP secretion [52, 85]; all of these signals are initiated from EMP produced by epithelial injury/repair [86]. Epithelial barrier disruption induces the secretion of epithelial growth factors (IL6, TGFβ, and EGF) and fibrogenic cytokines (periostin, IL-17, and IL-11) [87]. These factors promote fibroblast motility, anti-apoptosis, and expression of ECM proteins, FN1 and COL1 (reviewed in [58]). The effects of TGFβ and fibrogenic cytokines further stimulate metabolic adaptations in the myofibroblast population, leading to epigenetic changes resulting in enhanced production of ECM, resistance to apoptosis, and migratory invasiveness [88, 89]. Myofibroblasts themselves also release inflammatory/profibrotic mediators that perpetuate epithelial injury and further promote ECM deposition [90].

MMPs released by EMP can also modify myofibroblast populations independently of changes in ECM composition. MMP9 has emerged as an important paracrine regulator of EMP in lung disease by its property to cleave CDH1 [91], reducing its cell-surface abundance, destabilizing adherens junctions, and disrupting epithelial barrier function. Intriguing studies have shown that MMP9 is directly recruited to the fibroblast membrane by lysyl hydroxylase three to activate αSMA expression and myofibroblast transition [92]. Using reconstituted epithelial-fibroblast co-cultures, we showed this mechanism plays an important role in RSV-induced remodeling and expansion of subepithelial myofibroblasts [85, 86]. These intriguing data inform the hypothesis that innate inflammation in small airway epithelial cells is linked to EMP, and EMP triggers airway remodeling by myofibroblasts, in part through basement membrane remodeling and paracrine MMP9 activity promoting myofibroblast expansion.

Mechanisms of How Allergens Trigger EMP

In contrast to the well-developed studies of viral infections, less is known how individual aeroallergens trigger EMP. Aeroallergens are plant- and animal-derived products that modify the epithelial barrier function and activate innate signaling cascades that trigger sensitization and AA [93]. A number of common and important aeroallergens have been identified and studied.

The house dust mite (HDM), *Dermatophagtoides pteronyssius* produces a complex aeroallergen containing bacterial cell wall products [lipopolysaccharide (LPS) and ß-glucan] [94] as well as mite-produced proteases [95]. The Der p1 is a cysteine protease that disrupts tight epithelial junctions by cleavage of zona occludens [39] and also activates the protease-activated receptor (PAR)-2 cleaving its NH2 terminus, irreversibly activating signaling [96]. The cockroach allergen, Per a 10, also induces innate signaling by protease activity directed to PAR-2 [97]. In contrast, Ragweed pollen contains an endogenous NADPH-driven oxidase that disrupts the epithelial barrier by forming reactive oxygen species [98], producing CXCL2 release and neutrophil recruitment [99]. The *Aspergillus*-derived alkaline protease 1 (Alp1) is an aeroallergen that disrupts epithelial tight junctions by cleaving CDH1, producing IL33/CCL2 secretion and eosinophilia [33]. Cat dander extract (CDE) activates the MD2 co-receptor and TLR4, upstream of the Myd88-NFκB pathway [100, 101], producing CXCL2 secretion and neutrophil recruitment [100]. CDE triggers a coordinated time-dependent increase of TGFβ-1, -2, and -3 production, local SMAD3 and NFκB signaling, and expression of the mesenchymal core regulatory proteins ZEB1/SNAI1 in a manner dependent on the critical Scgb1a1+/"Club cell" progenitor of the lower airway [28]. In human AA, TGFβ in the bronchoalveolar lavage fluid of chronic asthma patients as well as in sensitized experimental animals challenged with OVA, suggests a critical role of TGFβ in airway remodeling and lung EMT [102].

Repetitive (Chronic) Innate Activation Produces Airway Remodeling

In the real world, aeroallergen exposures are frequent and repetitive. The consequences of these repetitive exposures on EMP and remodeling are not fully understood. We recently demonstrated that repetitive cat dander exposure triggered ECM production, NFκB-dependent EMP, myofibroblast transition, and mucous metaplasia in a mouse model [40]. These characteristic features of airway remodeling require NFκB/RelA signaling and association with a chromatin remodeling factor known as bromodomain-containing protein 4 (BRD4).

BRD4 Mediates Inflammation and Remodeling in Acute Exacerbations

BRD4 is a member of the bromodomain and extra-terminal domain (BET) family of chromatin modifying proteins that play essential roles in epigenetic control of inflammation-inducible gene expression [40, 103], maintenance of cellular identity [104] and chromatin compaction/conformation [105]. Because BRD4 functions as a coactivator of NFκB signaling, its role in mediating acute exacerbations of airway disease is beginning to be investigated. We have found that the activated NFκB-BRD4 complex is formed in lower airway epithelial cells in response to RSV infection. Disruption of this pathway, either by antagonizing the action of NFκB, or the use of a small molecule BRD4 inhibitor treatment, reduced RSV-induced neutrophilia and airway obstruction in a mouse mode [12]. Similarly, BRD4 mediates EMP and acute inflammation in response to repetitive stimulation with TLR3 agonists [32]. Finally, the NFκB-BRD4 pathway is activated a mouse model of aeroallergen challenge, where BRD4 antagonists interfere with cat dander-induced EMP, ECM remodeling and allergic sensitization [40]. These exciting data provide mechanistic linkage between the activation of innate immunity, epithelial plasticity and ECM remodeling by both repetitive aeroallergen exposure and viral infections.

Epigenetic Control of ECM Remodeling in EMP

Epigenetic reprogramming is central for how transcription factor cascades control gene expression important in EMP [106, 107]. EMP regulatory epigenetics are mediated by posttranslational modifications of histone side chains; these are produced by chromatin regulatory proteins, including the histone acetyltransferase and reader, BRD4. A body of work shows that BRD4 plays essential roles both in the maintenance of epithelial identity and in the genomic reprogramming underlying EMP through its ability to bind acetylated histones, functioning as an RNA polymerase II COOH terminal kinase and as an enzymatically active histone acetyltransferase [108–110]. These roles include its direct role as a coactivator of master transcription factors in the EMP and, through its actions controlling cell-type identity genes via super-enhancers. Interestingly, *XBP1* is one such gene under BRD4-dependent super-enhancer control [104].

After complex formation, the sequence-specific NFκB DNA binding activity recruits BRD4 to the key EMP regulators, *SNA1I, ZEB1, Twist, IL6, FN1,* and others [111]. Here, the BRD4 complex activates EMP programs through a process of regulated transcriptional elongation [112, 113] (Fig. 5). In addition, BRD4 –SMAD3 complex also plays a key role in TGFβ-induced myofibroblast formation [114], and we have recently found that BRD4 forms complexes with members of the AP-1 complex, known to be important in lung fibrosis [115]. More work will be required to understand how these transcription factors function in a synergistic manner to reposition BRD4 to gene regulatory networks controlling EMP.

BRD4 Regulates Growth Factors and ECM Remodeling Through Chromatin Accessibility

Despite understanding that BRD4 is required for EMP, the BRD4-dependent gene regulatory network of nontransformed epithelial cells is incompletely understood. To help advance this topic, we recently investigated the effect of a specific BRD4 inhibitor on RSV-induced EMP using next-generation sequencing technology. We

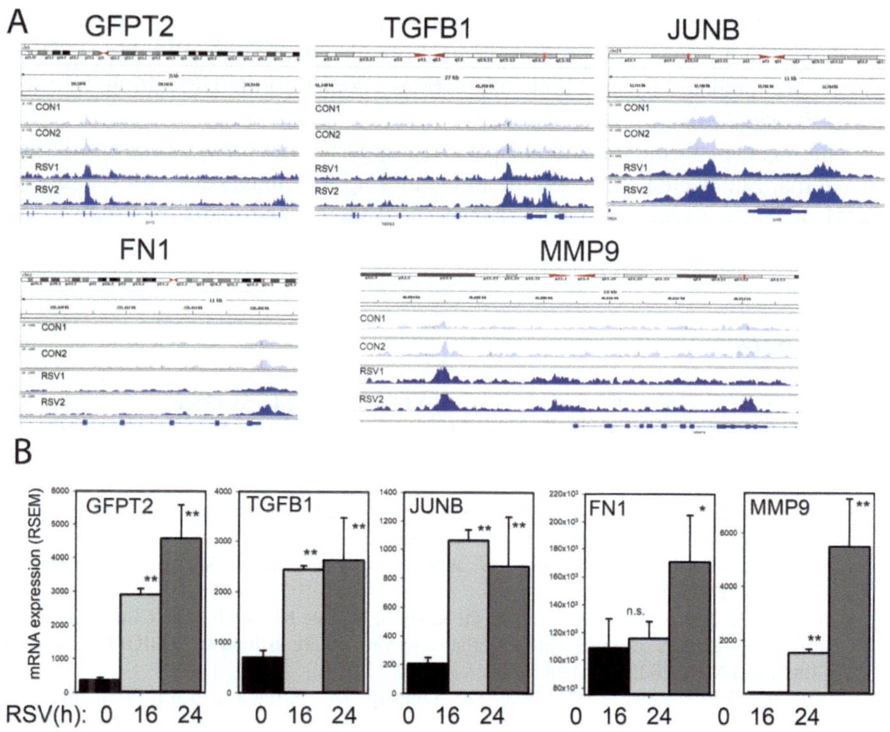

Fig. 5 RSV infection produces chromatin changes in the TGFβ growth factor-ECM pathway. (**a**) Integrated genomic viewer (IGV) of ATAC-Seq cleavage fragments mapped to the GFPT2, TGFB1, JUNB, and FN1 genes in control and RSV-infected state. Note the increased transposase digestion of promoter elements after RSV infection (**b**) mRNA changes of each gene. Shown is mean ± SD ($n = 4$ independent RNA-seq reads quantified by RSEM). *, $p < 0.05$, **, $p < 0.01$ post hoc Tukey's test. (Reproduced with permission from [116])

found BRD4 mediates RSV and produces nucleosome-free regions on the *TGFB1, JUNB, FN1, MMP9*, and GPT2 genes, associated with their increased expression [116] (Fig. 5). These studies indicate that BRD4 may play a critical role in mediating the expression of this HBP-plasticity program through enhanced chromatin opening.

Implications for Precision Approaches

From this work, it is inescapable that the innate immune response driven by epithelium plays a contributory role in the expansion of the ECM and myofibroblast population in animal and human models of AA. I have reviewed in this chapter our deep, mechanistic, understanding of the cellular adaptations produced by viral and aeroallergen-driven exacerbations producing NFκB signaling coupled to activation of Th2 cytokines, protective IFNs, and cell state changes associated with EMP. The transition of the epithelium into EMP has diverse effects on ECM synthesis, ECM mechanics and impacts on profibrotic state of subepithelial myofibroblasts. These effects are mediated by metabolic adaptations coupled to the UPR through hexosamine biosynthesis.

This mechanistic understanding opens new opportunities for precision approaches in asthma. Major challenges will be in identifying new diagnostics for monitoring ECM in AA, assaying cell state changes, and identifying endotypes with innate components.

References

1. Busse WW, Lemanske RF Jr. Asthma. N Engl J Med. 2001;344(5):350–62.
2. Martinez FD, Wright AL, Taussig LM, Holberg CJ, Halonen M, Morgan WJ. Asthma and wheezing in the first six years of life. The Group Health Medical Associates. N Engl J Med. 1995;332(3):133–8.
3. Holtzman MJ. Asthma as a chronic disease of the innate and adaptive immune systems responding to viruses and allergens. J Clin Invest. 2012;122(8):2741–8.
4. Sykes A, Johnston SL. Etiology of asthma exacerbations. J Allergy Clin Immunol. 2008;122(4):685–8.
5. Hoda U, Pavlidis S, Bansal AT, Takahashi K, Hu S, Ng Kee Kwong F, et al. Clinical and transcriptomic features of persistent exacerbation-prone severe asthma in U-BIOPRED cohort. Clin Transl Med. 2022;12(4):e816.
6. Calhoun WJ, Haselkorn T, Miller DP, Omachi TA. Asthma exacerbations and lung function in patients with severe or difficult-to-treat asthma. J Allergy Clin Immunol. 2015;136(4):1125–7.e4.
7. O'Byrne PM, Pedersen S, Lamm CJ, Tan WC, Busse WW. Severe exacerbations and decline in lung function in asthma. Am J Respir Crit Care Med. 2009;179(1):19–24.
8. Prakash YS, Halayko AJ, Gosens R, Panettieri RA Jr, Camoretti-Mercado B, Penn RB, et al. An official American thoracic society research statement: current challenges facing research and therapeutic advances in airway remodeling. Am J Respir Crit Care Med. 2017;195(2):e4–e19.
9. Hackett TL. Epithelial-mesenchymal transition in the pathophysiology of airway remodelling in asthma. Curr Opin Allergy Clin Immunol. 2012;12(1):53–9.
10. Bergeron C, Al-Ramli W, Hamid Q. Remodeling in asthma. Proc Am Thorac Soc. 2009;6(3):301–5.
11. Whitsett JA, Alenghat T. Respiratory epithelial cells orchestrate pulmonary innate immunity. Nat Immunol. 2015;16(1):27–35.
12. Tian B, Yang J, Zhao Y, Ivanciuc T, Sun H, Wakamiya M, et al. Central role of the NF-kappaB pathway in the Scgb1a1-expressing epithelium in mediating respiratory syncytial virus-induced airway inflammation. J Virol. 2018;92(11):e00441–18.
13. Alexander-Brett J, Holtzman MJ. Virus infection of airway epithelial cells. Mucosal Immunol, Chapter 53, Elsevier. 2015:1013–21. https://doi.org/10.1016/B978-0-12-415847-4.00053-7
14. Bertolusso R, Tian B, Zhao Y, Vergara L, Sabree A, Iwanaszko M, et al. Dynamic cross talk model of the epithelial innate immune response to double-stranded RNA stimulation: coordinated dynamics emerging from cell-level noise. PLoS One. 2014;9(4):e93396.
15. Czerkies M, Korwek Z, Prus W, Kochanczyk M, Jaruszewicz-Blonska J, Tudelska K, et al. Cell fate in antiviral response arises in the crosstalk of IRF, NF-kappaB and JAK/STAT pathways. Nat Commun. 2018;9(1):493.
16. Choudhary S, Boldogh I, Brasier AR. Inside-out signaling pathways from nuclear reactive oxygen species control pulmonary innate immunity. J Innate Immun. 2016;8(2):143–55.

17. Holt PG, Strickland DH, Wikstrom ME, Jahnsen FL. Regulation of immunological homeostasis in the respiratory tract. Nat Rev Immunol. 2008;8(2):142–52.
18. Martinez FD. The origins of asthma and chronic obstructive pulmonary disease in early life. Proc Am Thorac Soc. 2009;6(3):272–7.
19. Fauroux B, Simoes EAF, Checchia PA, Paes B, Figueras-Aloy J, Manzoni P, et al. The burden and long-term respiratory morbidity associated with respiratory syncytial virus infection in early childhood. Infect Dis Ther. 2017;6(2):173–97.
20. Johnson JE, Gonzales RA, Olson SJ, Wright PF, Graham BS. The histopathology of fatal untreated human respiratory syncytial virus infection. Mod Pathol. 2007;20(1):108–19.
21. Jozwik A, Habibi MS, Paras A, Zhu J, Guvenel A, Dhariwal J, et al. RSV-specific airway resident memory CD8+ T cells and differential disease severity after experimental human infection. Nat Commun. 2015;6:10224.
22. Yanai M, Sekizawa K, Ohrui T, Sasaki H, Takishima T. Site of airway obstruction in pulmonary disease: direct measurement of intrabronchial pressure. J Appl Physiol. 1992;72(3):1016–23.
23. Adams TS, Schupp JC, Poli S, Ayaub EA, Neumark N, Ahangari F, et al. Single-cell RNA-seq reveals ectopic and aberrant lung-resident cell populations in idiopathic pulmonary fibrosis. Sci Adv. 2020;6(28):eaba1983.
24. Deprez M, Zaragosi L-E, Truchi M, Becavin C, Ruiz García S, Arguel M-J, et al. A single-cell atlas of the human healthy airways. Am J Respir Crit Care Med. 2020;202(12):1636–45.
25. Zaragosi L, Deprez M, Barbry P. Using single-cell RNA sequencing to unravel cell lineage relationships in the respiratory tract. Biochem Soc Trans. 2020;48(1):327–36.
26. Evans MJ, Van Winkle LS, Fanucchi MV, Plopper CG. The attenuated fibroblast sheath of the respiratory tract epithelial–mesenchymal trophic unit. Am J Respir Cell Mol Biol. 1999;21(6):655–7.
27. Zhao Y, Jamaluddin M, Zhang Y, Sun H, Ivanciuc T, Garofalo RP, et al. Systematic analysis of cell-type differences in the epithelial secretome reveals insights into the pathogenesis of respiratory syncytial virus-induced lower respiratory tract infections. J Immunol. 2017;198(8):3345–64.
28. Skibba ME, Xu X, Weiss K, Huisken J, Brasier AR. Role of Secretoglobin(+) (club cell) NFκB/RelA-TGFβ signaling in aero-allergen-induced epithelial plasticity and subepithelial myofibroblast transdifferentiation. Respir Res. 2021;22(1):315.
29. Olszewska-Pazdrak B, Casola A, Saito T, Alam R, Crowe SE, Mei F, et al. Cell-specific expression of RANTES, MCP-1, and MIP-1alpha by lower airway epithelial cells and eosinophils infected with respiratory syncytial virus. J Virol. 1998;72(6):4756–64.
30. Zhang Y, Luxon BA, Casola A, Garofalo RP, Jamaluddin M, Brasier AR. Expression of respiratory syncytial virus-induced chemokine gene networks in lower airway epithelial cells revealed by cDNA microarrays. J Virol. 2001;75(19):9044–58.
31. Sigurs N, Aljassim F, Kjellman B, Robinson PD, Sigurbergsson F, Bjarnason R, et al. Asthma and allergy patterns over 18 years after severe RSV bronchiolitis in the first year of life. Thorax. 2010;65(12):1045–52.
32. Tian B, Liu Z, Yang J, Sun H, Zhao Y, Wakamiya M, et al. Selective antagonists of the bronchiolar epithelial NF-kappaB-bromodomain-containing protein 4 pathway in viral-induced airway inflammation. Cell Rep. 2018;23(4):1138–51.
33. Wiesner DL, Merkhofer RM, Ober C, Kujoth GC, Niu M, Keller NP, et al. Club cell TRPV4 serves as a damage sensor driving lung allergic inflammation. Cell Host Microbe. 2020;27(4):614–28.e6.
34. Tully JE, Hoffman SM, Lahue KG, Nolin JD, Anathy V, Lundblad LK, et al. Epithelial NF-κB orchestrates house dust mite–induced airway inflammation, hyperresponsiveness, and fibrotic remodeling. J Immunol. 2013;191(12):5811–21.
35. Erjefält JS, Erjefält I, Sundler F, Persson CG. In vivo restitution of airway epithelium. Cell Tissue Res. 1995;281(2):305–16.
36. Nieto MA, Huang RY-J, Jackson RA, Thiery JP. EMT: 2016. Cell. 2016;166(1):21–45.
37. Jolly MK, Tripathi SC, Jia D, Mooney SM, Celiktas M, Hanash SM, et al. Stability of the hybrid epithelial/mesenchymal phenotype. Oncotarget. 2016;7(19):27067.

38. Yang J, Antin P, Berx G, Blanpain C, Brabletz T, Bronner M, et al. Guidelines and definitions for research on epithelial–mesenchymal transition. Nat Rev Mol Cell Biol. 2020;21(6):341–52.
39. Heijink IH, Postma DS, Noordhoek JA, Broekema M, Kapus A. House dust mite-promoted epithelial-to-mesenchymal transition in human bronchial epithelium. Am J Respir Cell Mol Biol. 2010;42(1):69–79.
40. Tian B, Hosoki K, Liu Z, Yang J, Zhao Y, Sun H, et al. Mucosal bromodomain-containing protein 4 mediates aeroallergen-induced inflammation and remodeling. J Allergy Clin Immunol. 2019;143(4):1380–94.e9.
41. Grainge CL, Lau LC, Ward JA, Dulay V, Lahiff G, Wilson S, et al. Effect of bronchoconstriction on airway remodeling in asthma. N Engl J Med. 2011;364(21):2006–15.
42. Rock JR, Barkauskas CE, Cronce MJ, Xue Y, Harris JR, Liang J, et al. Multiple stromal populations contribute to pulmonary fibrosis without evidence for epithelial to mesenchymal transition. Proc Natl Acad Sci U S A. 2011;108(52):E1475–83.
43. Aiello NM, Maddipati R, Norgard RJ, Balli D, Li J, Yuan S, et al. EMT subtype influences epithelial plasticity and mode of cell migration. Dev Cell. 2018;45(6):681–95.e4.
44. Noureddine N, Chalubinski M, Wawrzyniak P. The role of defective epithelial barriers in allergic lung disease and asthma development. J Asthma Allergy. 2022;15:487–504.
45. Xiao C, Puddicombe SM, Field S, Haywood J, Broughton-Head V, Puxeddu I, et al. Defective epithelial barrier function in asthma. J Allergy Clin Immunol. 2011;128(3):549–56.e1–12.
46. De Boer W, Sharma HS, Baelemans S, Hoogsteden H, Lambrecht B, Braunstahl G-J. Altered expression of epithelial junctional proteins in atopic asthma: possible role in inflammation. Can J Physiol Pharmacol. 2008;86(3):105–12.
47. Hackett T-L, Warner SM, Stefanowicz D, Shaheen F, Pechkovsky DV, Murray LA, et al. Induction of epithelial–mesenchymal transition in primary airway epithelial cells from patients with asthma by transforming growth factor-β1. Am J Respir Crit Care Med. 2009;180(2):122–33.
48. Knight DA, Holgate ST. The airway epithelium: structural and functional properties in health and disease. Respirology. 2003;8(4):432–46.
49. Laitinen LA, Heino M, Laitinen A, Kava T, Haahtela T. Damage of the airway epithelium and bronchial reactivity in patients with asthma. Am Rev Respir Dis. 1985;131(4):599–606.
50. Tian B, Zhang Y, Luxon BA, Garofalo RP, Casola A, Sinha M, et al. Identification of NF-kappaB-dependent gene networks in respiratory syncytial virus-infected cells. J Virol. 2002;76(13):6800–14.
51. Tian B, Li X, Kalita M, Widen SG, Yang J, Bhavnani SK, et al. Analysis of the TGFbeta-induced program in primary airway epithelial cells shows essential role of NF-kappaB/RelA signaling network in type II epithelial mesenchymal transition. BMC Genomics. 2015;16:529.
52. Tian B, Patrikeev I, Ochoa L, Vargas G, Belanger KK, Litvinov J, et al. NF-kappaB mediates mesenchymal transition, remodeling, and pulmonary fibrosis in response to chronic inflammation by viral RNA patterns. Am J Respir Cell Mol Biol. 2017;56(4):506–20.
53. Tian B, Widen SG, Yang J, Wood TG, Kudlicki A, Zhao Y, et al. The NFkappaB subunit RELA is a master transcriptional regulator of the committed epithelial-mesenchymal transition in airway epithelial cells. J Biol Chem. 2018;293(42):16528–45.
54. Siemens H, Jackstadt R, Hünten S, Kaller M, Menssen A, Götz U, et al. miR-34 and SNAIL form a double-negative feedback loop to regulate epithelial-mesenchymal transitions. Cell Cycle. 2011;10(24):4256–71.
55. Ahn Y-H, Gibbons DL, Chakravarti D, Creighton CJ, Rizvi ZH, Adams HP, et al. ZEB1 drives prometastatic actin cytoskeletal remodeling by downregulating miR-34a expression. J Clin Invest. 2012;122(9):3170–83.
56. Lu M, Jolly MK, Levine H, Onuchic JN, Ben-Jacob E. MicroRNA-based regulation of epithelial–hybrid–mesenchymal fate determination. Proc Natl Acad Sci. 2013;110(45):18144–9.
57. Vincent T, Neve E, Johnson JR, Kukalev A, Rojo F, Albanell J, et al. A SNAIL1–SMAD3/4 transcriptional repressor complex promotes TGF-β mediated epithelial–mesenchymal transition. Nat Cell Biol. 2009;11(8):943–50.

58. Ijaz T, Pazdrak K, Kalita M, Konig R, Choudhary S, Tian B, et al. Systems biology approaches to understanding Epithelial Mesenchymal Transition (EMT) in mucosal remodeling and signaling in asthma. World Allergy Organ J. 2014;7(1):13.
59. Roberts CR. Is asthma a fibrotic disease? Chest. 1995;107(3):111S–7S.
60. Fedorov IA, Wilson SJ, Davies DE, Holgate ST. Epithelial stress and structural remodelling in childhood asthma. Thorax. 2005;60(5):389–94.
61. Meerschaert J, Kelly EAB, Mosher DF, Busse WW, Jarjour NN. Segmental antigen challenge increases fibronectin in bronchoalveolar lavage fluid. Am J Respir Crit Care Med. 1999;159(2):619–25.
62. Pathinayake PS, Hsu AC, Waters DW, Hansbro PM, Wood LG, Wark PAB. Understanding the unfolded protein response in the pathogenesis of asthma. Front Immunol. 2018;9:175.
63. Kim SR, Im Kim D, Kang MR, Lee KS, Park SY, Jeong JS, et al. Endoplasmic reticulum stress influences bronchial asthma pathogenesis by modulating nuclear factor κB activation. J Allergy Clin Immunol. 2013;132(6):1397–408.e11.
64. Bhakta NR, Christenson SA, Nerella S, Solberg OD, Nguyen CP, Choy DF, et al. IFN-stimulated gene expression, type 2 inflammation, and endoplasmic reticulum stress in asthma. Am J Respir Crit Care Med. 2018;197(3):313–24.
65. Pathinayake PS, Waters DW, Nichol KS, Brown AC, Reid AT, Hsu AC, et al. Endoplasmic reticulum-unfolded protein response signalling is altered in severe eosinophilic and neutrophilic asthma. Thorax. 2021;77:443–51.
66. James B, Milstien S, Spiegel S. ORMDL3 and allergic asthma: from physiology to pathology. J Allergy Clin Immunol. 2019;144(3):634–40.
67. Carreras-Sureda A, Cantero-Recasens G, Rubio-Moscardo F, Kiefer K, Peinelt C, Niemeyer BA, et al. ORMDL3 modulates store-operated calcium entry and lymphocyte activation. Hum Mol Genet. 2013;22(3):519–30.
68. Zhang J, Jamaluddin M, Zhang Y, Widen SG, Sun H, Brasier AR, et al. Type II epithelial-mesenchymal transition upregulates protein N-glycosylation to maintain proteostasis and extracellular matrix production. J Proteome Res. 2019;18(9):3447–60.
69. Qiao D, Skibba M, Xu X, Garofalo RP, Zhao Y, Brasier AR. Paramyxovirus replication induces the hexosamine biosynthetic pathway and mesenchymal transition via the IRE1alpha-XBP1s arm of the unfolded protein response. Am J Physiol Lung Cell Mol Physiol. 2021;321(3):L576–L94.
70. Kopp MC, Larburu N, Durairaj V, Adams CJ, Ali MM. UPR proteins IRE1 and PERK switch BiP from chaperone to ER stress sensor. Nat Struct Mol Biol. 2019;26(11):1053–62.
71. Hetz C. The unfolded protein response: controlling cell fate decisions under ER stress and beyond. Nat Rev Mol Cell Biol. 2012;13(2):89–102.
72. Zhao Y, Qiao D, Skibba M, Brasier AR. The IRE1α-XBP1s arm of the unfolded protein response activates N-glycosylation to remodel the subepithelial basement membrane in paramyxovirus infection. Int J Mol Sci. 2022;23(16):9000.
73. Brasier AR, Qiao D, Zhao Y. The hexosamine biosynthetic pathway links innate inflammation with epithelial-mesenchymal plasticity in airway remodeling. Front Pharmacol. 2021;12:808735.
74. Akella NM, Ciraku L, Reginato MJ. Fueling the fire: emerging role of the hexosamine biosynthetic pathway in cancer. BMC Biol. 2019;17(1):1–14.
75. Laurent G. Dynamic state of collagen: pathways of collagen degradation in vivo and their possible role in regulation of collagen mass. Am J Phys Cell Phys. 1987;252(1):C1–9.
76. Brewster CE, Howarth PH, Djukanovic R, Wilson J, Holgate ST, Roche WR. Myofibroblasts and subepithelial fibrosis in bronchial asthma. Am J Respir Cell Mol Biol. 1990;3(5):507–11.
77. Tomasek JJ, Gabbiani G, Hinz B, Chaponnier C, Brown RA. Myofibroblasts and mechano-regulation of connective tissue remodelling. Nat Rev Mol Cell Biol. 2002;3(5):349–63.
78. Hinz B, Phan SH, Thannickal VJ, Galli A, Bochaton-Piallat M-L, Gabbiani G. The myofibroblast: one function, multiple origins. Am J Pathol. 2007;170(6):1807–16.

79. Zhao Y, Zhang J, Sun H, Brasier AR. Crosstalk of the IkappaB kinase with spliced X-box binding protein 1 couples inflammation with glucose metabolic reprogramming in epithelial-mesenchymal transition. J Proteome Res. 2021;20(7):3475–88.
80. Qiao D, Skibba M, Xu X, Garofalo RP, Zhao Y, Brasier AR. Paramyxovirus replication induces the hexosamine biosynthetic pathway and mesenchymal transition via the IRE1α-XBP1s arm of the unfolded protein response. Am J Physiol Lung Cell Mol Physiol. 2021;321(3):L576–l94.
81. Boser SR, Mauad T, Araújo-Paulino BB, Mitchell I, Shrestha G, Chiu A, et al. Myofibroblasts are increased in the lung parenchyma in asthma. PLoS One. 2017;12(8):e0182378.
82. Weitoft M, Andersson C, Andersson-Sjöland A, Tufvesson E, Bjermer L, Erjefält J, et al. Controlled and uncontrolled asthma display distinct alveolar tissue matrix compositions. Respir Res. 2014;15(1):1–13.
83. Gizycki MJ, Adelroth E, Rogers AV, O'Byrne PM, Jeffery PK. Myofibroblast involvement in the allergen-induced late response in mild atopic asthma. Am J Respir Cell Mol Biol. 1997;16(6):664–73.
84. Knight D. Epithelium–fibroblast interactions in response to airway inflammation. Immunol Cell Biol. 2001;79(2):160–4.
85. Xu X, Qiao D, Dong C, Mann M, Garofalo RP, Keles S, et al. The SWI/SNF-related, matrix associated, actin-dependent regulator of chromatin A4 core complex represses respiratory syncytial virus-induced syncytia formation and subepithelial myofibroblast transition. Front Immunol. 2021;12:633654.
86. Brasier AR. RSV reprograms the CDK9*BRD4 chromatin remodeling complex to couple innate inflammation to airway remodeling. Viruses. 2020;12(4):472.
87. Holgate ST, Holloway J, Wilson S, Bucchieri F, Puddicombe S, Davies DE. Epithelial-mesenchymal communication in the pathogenesis of chronic asthma. Proc Am Thorac Soc. 2004;1(2):93–8.
88. Horowitz JC, Cui Z, Moore TA, Meier TR, Reddy RC, Toews GB, et al. Constitutive activation of prosurvival signaling in alveolar mesenchymal cells isolated from patients with nonresolving acute respiratory distress syndrome. Am J Phys Lung Cell Mol Phys. 2006;290(3):L415–L25.
89. Henderson J, O'Reilly S. The emerging role of metabolism in fibrosis. Trends Endocrinol Metab. 2021;32(8):639–53.
90. Powell D, Mifflin R, Valentich J, Crowe S, Saada J, West A. Myofibroblasts. I. Paracrine cells important in health and disease. Am J Phys Cell Phys. 1999;277(1):C1–C19.
91. Cowden Dahl KD, Symowicz J, Ning Y, Gutierrez E, Fishman DA, Adley BP, et al. Matrix metalloproteinase 9 is a mediator of epidermal growth factor–dependent E-cadherin loss in ovarian carcinoma cells. Cancer Res. 2008;68(12):4606–13.
92. Dayer C, Stamenkovic I. Recruitment of matrix metalloproteinase-9 (MMP-9) to the fibroblast cell surface by lysyl hydroxylase 3 (LH3) triggers transforming growth factor-β (TGF-β) activation and fibroblast differentiation. J Biol Chem. 2015;290(22):13763–78.
93. Lambrecht BN, Hammad H. Allergens and the airway epithelium response: gateway to allergic sensitization. J Allergy Clin Immunol. 2014;134(3):499–507.
94. Douwes J, Zuidhof A, Doekes G, van der ZEE S, Wouters I, Marike Boezen H, et al. (1→3)-β-D-glucan and endotoxin in house dust and peak flow variability in children. Am J Respir Crit Care Med. 2000;162(4):1348–54.
95. Jacquet A. The role of the house dust mite-induced innate immunity in development of allergic response. Int Arch Allergy Immunol. 2011;155(2):95–105.
96. Asokananthan N, Graham PT, Stewart DJ, Bakker AJ, Eidne KA, Thompson PJ, et al. House dust mite allergens induce proinflammatory cytokines from respiratory epithelial cells: the cysteine protease allergen, Der p 1, activates protease-activated receptor (PAR)-2 and inactivates PAR-1. J Immunol. 2002;169(8):4572–8.
97. Arizmendi NG, Abel M, Mihara K, Davidson C, Polley D, Nadeem A, et al. Mucosal allergic sensitization to cockroach allergens is dependent on proteinase activity and proteinase-activated receptor-2 activation. J Immunol. 2011;186(5):3164–72.

98. Bacsi A, Dharajiya N, Choudhury BK, Sur S, Boldogh I. Effect of pollen-mediated oxidative stress on immediate hypersensitivity reactions and late-phase inflammation in allergic conjunctivitis. J Allergy Clin Immunol. 2005;116(4):836–43.
99. Hosoki K, Aguilera-Aguirre L, Brasier AR, Kurosky A, Boldogh I, Sur S. Facilitation of allergic sensitization and allergic airway inflammation by pollen-induced innate neutrophil recruitment. Am J Respir Cell Mol Biol. 2016;54(1):81–90.
100. Hosoki K, Boldogh I, Aguilera-Aguirre L, Sun Q, Itazawa T, Hazra T, et al. Myeloid differentiation protein 2 facilitates pollen- and cat dander-induced innate and allergic airway inflammation. J Allergy Clin Immunol. 2016;137(5):1506–13.e2.
101. Hosoki K, Redding D, Itazawa T, Chakraborty A, Tapryal N, Qian S, et al. Innate mechanism of pollen- and cat dander-induced oxidative stress and DNA damage in the airways. J Allergy Clin Immunol. 2017;140(5):1436–9.e5.
102. Yamaguchi M, Niimi A, Matsumoto H, Ueda T, Takemura M, Matsuoka H, et al. Sputum levels of transforming growth factor-ß. J Investig Allergol Clin Immunol. 2008;18(3):202–6.
103. Nowak DE, Tian B, Jamaluddin M, Boldogh I, Vergara LA, Choudhary S, et al. RelA Ser276 phosphorylation is required for activation of a subset of NF-{kappa}B-dependent genes by recruiting cyclin-dependent kinase 9/cyclin T1 complexes. Mol Cell Biol. 2008;28(11):3623–38.
104. Loven J, Hoke HA, Lin CY, Lau A, Orlando DA, Vakoc CR, et al. Selective inhibition of tumor oncogenes by disruption of super-enhancers. Cell. 2013;153(2):320–34.
105. Devaiah BN, Gegonne A, Singer DS. Bromodomain 4: a cellular Swiss army knife. J Leukoc Biol. 2016;100(4):679–86.
106. Kalluri R, Weinberg RA. The basics of epithelial-mesenchymal transition. J Clin Invest. 2009;119(6):1420–8.
107. Peixoto P, Etcheverry A, Aubry M, Missey A, Lachat C, Perrard J, et al. EMT is associated with an epigenetic signature of ECM remodeling genes. Cell Death Dis. 2019;10(3):205.
108. Devaiah BN, Lewis BA, Cherman N, Hewitt MC, Albrecht BK, Robey PG, et al. BRD4 is an atypical kinase that phosphorylates serine2 of the RNA polymerase II carboxy-terminal domain. Proc Natl Acad Sci U S A. 2012;109(18):6927–32.
109. Devaiah BN, Singer DS. Cross-talk among RNA polymerase II kinases modulates C-terminal domain phosphorylation. J Biol Chem. 2012;287(46):38755–66.
110. Devaiah BN, Case-Borden C, Gegonne A, Hsu CH, Chen Q, Meerzaman D, et al. BRD4 is a histone acetyltransferase that evicts nucleosomes from chromatin. Nat Struct Mol Biol. 2016;23(6):540–8.
111. Tian B, Zhao Y, Sun H, Zhang Y, Yang J, Brasier AR. BRD4 mediates NF-κB-dependent epithelial-mesenchymal transition and pulmonary fibrosis via transcriptional elongation. Am J Physiol Lung Cell Mol Physiol. 2016;311(6):L1183–l201.
112. Tian B, Zhao Y, Sun H, Zhang Y, Yang J, Brasier AR. BRD4 mediates NFkB-dependent epithelial-mesenchymal transition and pulmonary fibrosis via transcriptional elongation. Am J Physiol Lung Cell Mol Physiol. 2016;311(6):L1183–L201.
113. Yang J, Tian B, Brasier AR. Targeting chromatin remodeling in inflammation and fibrosis. In: Donev R, editor. Advances in protein chemistry and structural biology. Cambridge: Elsevier; 2017.
114. Ijaz T, Jamaluddin M, Zhao Y, Zhang Y, Finnerty CC, Jay J, et al. Coordinate activities of BRD4 and CDK9 in the transcriptional elongation complex are required for TGFb-induced Nox4 expression and myofibroblast transdifferentiation. Cell Death Dis. 2017;8(2):e2606.
115. Mann M, Roberts DS, Zhu Y, Li Y, Zhou J, Ge Y, et al. Discovery of RSV-induced BRD4 protein interactions using native immunoprecipitation and parallel accumulation—serial fragmentation (PASEF) mass spectrometry. Viruses. 2021;13(3):454.
116. Xu X, Qiao D, Mann M, Garofalo RP, Brasier AR. Respiratory syncytial virus infection induces chromatin remodeling to activate growth factor and extracellular matrix secretion pathways. Viruses. 2020;12(8):804.

Development of Adaptive Immunity and Its Role in Lung Remodeling

Stephane Esnault and Nizar N. Jarjour

Abstract

Asthma is characterized by airflow limitations resulting from bronchial closure, which can be either reversible or fixed due to changes in airway tissue composition and structure, also known as remodeling. Airway remodeling is defined as increased presence of mucins-producing epithelial cells, increased thickness of airway smooth muscle cells, angiogenesis, increased number and activation state of fibroblasts, and extracellular matrix (ECM) deposition. Airway inflammation is believed to be the main cause of the development of airway remodeling in asthma. In this chapter, we will review the development of the adaptive immune response and the impact of its mediators and cells on the elements defining airway remodeling in asthma.

Keywords

Asthma · Mucins · Airway smooth muscle · Lung remodeling · Extracellular matrix (ECM) · Cytokine · Adaptive immunity · Fibroblast · Innate lymphoid cell · Matrix metalloproteinase · Epithelial cell · Eosinophil · Neutrophil · Mast cell · Dendritic cell · Allergen · Lymphocyte

---

S. Esnault (✉) · N. N. Jarjour
University of Wisconsin-Madison, Madison, WI, USA
e-mail: sesnault@medicine.wisc.edu

A. R. Brasier, N. N. Jarjour (eds.), *Precision Approaches to Heterogeneity in Asthma*, Advances in Experimental Medicine and Biology 1426,
https://doi.org/10.1007/978-3-031-32259-4_14

Abbreviations

AEC	Airway epithelial cells
AHR	Airway hyperreactivity
APC	Antigen-presenting cell
ASM	Airway smooth muscle
BALF	Bronchoalveolar lavage
BALF	Bronchoalveolar lavage fluid
BEC	Bronchial epithelial cells
CysLT	Cysteinyl leukotrienes
DC	Dendritic cell
ECM	Extracellular matrix
EGFR	Epidermal growth factor receptor
ELANE	Neutrophil elastase
ET	Endothelin
FEV_1	Forced expiratory volume in 1 second
FGF	Fibroblast growth factor
IFN	Interferon
IGF	Insulin-like growth factor
IL	Interleukin
ILC	Innate lymphoid cell
MAIT	Mucosal-associated invariant T
MMP	Matrix metalloproteinase
NK	Natural killer
PDGF	Platelet derived growth factor
PG	Prostaglandin
SBP-Ag	Segmental bronchoprovocation with an allergen
T2	Type-2
TGF	Transforming growth factor
Th1	Type 1 T helper lymphocyte
VEGFA	Vascular endothelial growth factor
γ/δ	Gamma-delta
αSMA	Alpha-smooth muscle cell actin

Introduction

Pulmonary insults by aero-allergens, microbes, or pollutants disrupt the bronchial epithelial barrier leading to vascular leakage in the airways and bringing inflammatory cells, growth and coagulation factors, and matrix proteins to stop the leakages and start tissue repair. This is followed by accumulation and activation of contractile and extracellular matrix protein-producing fibroblasts to enhance the repair process. While these processes favor lung protection and healing, exaggerated or prolonged response can lead to scarring and fibrosis, which are detrimental to proper lung function. Concomitantly, these events are partially orchestrated by the innate immune response that involves inflammatory cells, such as macrophages, innate lymphoid cells (ILC), and neutrophils. This primary innate response may also participate in the development of an adaptive immune response that involves antigen-presenting cells (e.g., dendritic cells) to ultimately produce antigen-specific memory B and T lymphocytes that will remember the specific initiator of the insult and more efficiently react to the next exposure from the same aggressor. This adaptive response that generates large amounts of numerous cytokines and other factors, further enhances the innate immune response, and, in the case of allergies, leads to eosinophilia and specific IgE/IgE receptors binding on mast cells.

Asthma is characterized by airflow limitations resulting from bronchial closure, which can be either reversible (e.g., airway smooth muscle contraction) or fixed due to a change in airway tissue composition and structure, also known as remodeling. In asthma, airway remodeling is defined as increased presence of mucins-producing epithelial cells (goblet cell hyperplasia), increased thickness of airway smooth muscle cells (ASMC) layer (ASMC hyperplasia and hypertrophy), angiogenesis, increased number and activation state of fibroblasts (subepithelial fibrosis) and extracellular matrix (ECM) deposition. Airway remodeling has been implicated in increased airway hyperresponsiveness (AHR) [1], decline in lung function [2], and decreased responsiveness to asthma therapies. Importantly, except for bronchial thermoplasty which is a treatment option to reduce the airway smooth muscle (ASM) layer [3], current therapeutics for asthma including the corticosteroids, do not directly target airway remodeling [4, 5]. However, because therapeutics targeting products from the immune response (e.g., biologics) exist or are being developed, in this chapter, we will describe how the adaptive immune response affects lung remodeling, particularly fibroblast, ASMC activation, and mucus production by bronchial epithelial cells (BEC).

Evidence of Remodeling in Asthma

Fibroblast Activation and Accumulation of Extracellular Matrix Proteins in the Airways

Fibroblasts are the main cell type that produces ECM proteins, and they are the direct culprit for subepithelial thickness or fibrosis. Subepithelial fibrosis is a prominent feature of airway remodeling in asthma [6, 7], particularly in severe asthma [8], and is a direct product of fibroblasts. Fibroblasts are mesenchymal cells present in the connective tissue at the base of the airway epithelial layer and have an important role in tissue repair. Following airway insults and injuries, activated local and newly recruited fibroblasts divide and differentiate into contractile (alpha-smooth muscle producers) [9] and ECM-producing myofibroblasts due to activation by transforming growth factor-β1 (TGF-β1) and other mediators [10–13]. Activated fibroblasts and myofibroblasts produce large amounts of ECM proteins (e.g., collagens, fibronectin, etc.), leading to accumulation and changes in the biochemical properties of the matrix. Fibroplasia and ECM proteins stiffnesses are important elements of lung fibrosis and the irreversible loss of lung function in asthma [14]. In addition, fibroblasts and fibrocytes differentiation into myofibroblasts participate in hyperplasia of ASM [15].

Increased deposition of collagen under the basement membrane was observed in lung biopsies from both patients who died from status asthmaticus and asymptomatic subjects with asthma, suggesting irreversibility of ECM deposition after asthmatic attacks [16]. Later, Davidson et al. noticed marked fibrosis in lung biopsies from living patients with asthma [17]. Electron microscopy and immunohistochemical analyses of subepithelial fibrosis in asthma revealed noticeable presence of collagens (I, II and V) and fibronectin in the thickened lamina reticularis [18]. Furthermore, Brewster et al. [19] found a strong correlation and co-localization of bronchial subepithelial collagen thickness and the number of α-smooth muscle actin (αSMA)-producing myofibroblasts in asthma. Later, in 1997, histopathologic features of bronchial biopsies in allergic asthma showed that subepithelial fibrosis (basement membrane thickening/collagen deposition) was associated with metacholine airway responsiveness, asthma severity, and decline in forced expiratory volume in 1 second (FEV_1) [20, 21]. Furthermore, increased peribronchial fibrosis in asthma is associated with the number of eosinophils, the thickness of basement membrane, and goblet cell area [22]. In rodents, prolonged allergen exposure leads to goblet cell hyperplasia, ECM deposition, and thickening of the airway walls [23, 24]. In bronchoalveolar lavage fluids (BALF) from subjects with mild asthma elongated, highly mobile, and ECM- and αSMA-producing fibroblasts have been found along elevated numbers of eosinophils [25]. These fibroblasts were not found in the nonasthma control group [25]. Besides fibroblasts, a circulating cell called fibrocytes possesses fibroblast-like features, including the generation of ECM and αSMA [26–28]. Fibrocytes retain antigen-presenting capabilities [29] and can further differentiate into fibroblasts and myofibroblasts in the airways in asthma [15], where their number correlates with the basement membrane thickness [30]. In asthma also, the count of vascular endothelial growth factor (VEGFA) positive cells and the vascular area is higher than in healthy individuals [31]. In that study, VEGFA level was associated with basement membrane thickness, indicating that VEGF may play a role in subepithelial fibrosis [31], and indicating that promotion of new blood vessel growth is likely part of lung remodeling.

In allergic asthma, segmental bronchoprovocation with an allergen (SBP-Ag) in subjects with mild asthma induced fivefold the amount of a locally produced ECM protein, fibronectin in BALF, which correlated with early release of histamine and accumulation of inflammatory cells [32]. In that same model, SBP-Ag increased BALF basic fibroblasts growth factor-2 (bFGF, aka FGF2), an activator of fibroblasts, and ASMC, as well as matrix metalloproteinase (MMP)-9 [33, 34], which cleaves ECM proteins and allows migration of inflammatory cells and the release of VEGFA, fibrotic factors such FGF2, active TGF-β1, platelet-derived growth factor (PDGF) and insulin-like growth factor (IGF) [35–40].

Smooth Muscle Cells

Smooth muscle cells are present from the central airway to the most peripheral parts of the lung. Increased airway smooth muscle mass is a pathological feature of asthma, with further increase airway smooth muscle area in severe versus moderate asthma [41–43]. Both hyperplasia and hypertrophy of ASMC are strongly increased in asthma versus control individuals [44–47], particularly in the more severe cases [48–50]. In vitro, ASMC from asthma subjects display a higher ability to proliferate than ASMC from nonasthmatic subjects, indicating a different and stable phenotype in asthma [51]. ASMC contraction is opposed to elastic recoil of the airways that limits bronchoconstriction, and ASM hypertrophy correlates to AHR to histamine in asthma [52]. Thus, changes of ASMC in asthma are likely a major cause of increased AHR, excessive bronchoconstriction, and airway narrowing [53, 54]. Peribronchial smooth muscle hypertrophy is often concomitant to the deposition of collagen and subepithelial fibrosis in asthma [16]. In vitro, ASMC activated with serum from asthmatic subjects produce an increased amount of ECM, compared to ASMC activated with serum from nonasthmatic individuals [55]. ASMC, particularly injured ASMC produce TGF-β and TGF-β receptors, which in an autocrine way enhances their production of glycosaminoglycans and ECM, such as collagens, and the pro-neutrophilic cytokine, IL-8 [56–59]. Furthermore, mechanical stress that mimics force generated by mechanical bronchoconstriction in asthma leads to release of endothelin-1 (ET-1) and endothelin-2 (ET-2) by epithelial cells, which can stimulate ASMC to further enhance bronchoconstriction [60–62]. Therefore, there are tight interconnections between fibrosis and ASMC activity in asthma.

Mucus Accumulation

The role of mucus plugging in asthma is addressed by Schiebler et al. (Part II, Chap. 8) of this volume. Here, we provide a brief description of mucus production and potential effects on asthma.

In the 1960s, Dunnill et al. described the presence of mucus plugs associated with the loss of ciliated mucosal cells are features of subjects dying in *status asthmaticus* [63, 64]. Later, it was proposed that the presence of bronchial mucus plugs may be due to reduced mucociliary clearance in asthma compared to control individuals [65]; and bronchial epithelial goblet cells hyperplasia with mucus accumulation was observed after death by asthma attack [66]. More recently, state-of-the-art imaging was implemented to quantify mucus plugs in asthma and examine their associations with its clinical features such as predisposition for exacerbation and presence of airway obstruction [67]. The formation of mucus plugs involves the release of gel-forming mucin proteins (e.g., MUC5AC and MUC5B). MUC5AC is produced by BEC with the implication of IL-13, epidermal growth factor receptor (EGFR) signaling pathways and neutrophil elastase (ELANE) [68–71]. In addition to gel-forming mucin proteins, mucus plugs in asthma contain fibrin clots, which have also been observed in fatal asthma [72, 73], suggesting tight imbrication between mucin proteins and fibrogenesis for the formation of mucus plugs and airway obstruction in asthma. While mucus plugging is now well-established as a major contributor to asthma pathophysiology, including the risk of exacerbation and disease severity, there are no specific therapies to date that directly address mucus plugging in asthma, making this one of the important areas for future investigation.

Development of Adaptive Immunity in Asthma

Mucosal surface in the airways is a barrier to the development of adaptive immune responses to external agents such as aero-allergens, microbes, and pollutants. The epithelial barrier may, however, suffer some damage and allow the development of an innate immune response that involves macrophages, neutrophils, basophils, dendritic cells (DC), natural killer (NK) cells, NKT cells, gamma-delta (γ/δ) T cells, mast cells, and eosinophils as well as proteins such as proteins part of the complement. Following the innate immune response, an adaptive immune response takes place, but the response can be limited by mechanisms of lymphocyte unresponsiveness such as anergy, and ultimately adaptive tolerance. However, the adaptive immune response can fully grow, producing memory T and B lymphocytes, which can amplify a damaging succeeding response to the same original external agents.

The Role of the Epithelium and the Innate Immune Cells

The development and role of the innate immune response in lung tissue remodeling are addressed in the chapter by Brasier (Part III, Chap. 13) of this volume. Yet, to put the adaptive immune response into perspective in the complex global events occurring after an airway challenge in asthma, some of the cells of the innate immune response and their products will be briefly described here vis-à-vis their role in lung remodeling and the development of an adaptive immune response.

Airway Epithelial Cells (AEC)

AEC form the main first barrier between the host and inhaled factors involved in asthma development, and they have a critical early role in the immune response against these agents. Airway inflammation is believed to be the main cause for the development of airway remodeling in asthma [74, 75]; however, AEC, particularly those from the small airways, can directly participate in promoting remodeling [see Brasier (Part III, Chap. 13) of this volume]. Allergen challenges in rats increase the number of AEC [76]. During asthma exacerbation, there is both recruitment of inflammatory cells and AEC damage, which leads to the release of proteolytic enzymes, oxygen radicals, and profibrotic factors, including TGF-β and epidermal growth factor (EGF) [77–80]. Injured epithelial cells can release TGF-β1 from the

ECM to activate myofibroblast differentiation and survival [81–83]. Also, AEC produces MMP-2 which induces the proliferation of subepithelial fibroblasts [84]. Furthermore, mechanical stress of AEC that can occur during bronchoconstriction of the airways leads to the production and release of ET-1 and ET-2 [60]. Notably, among the AEC types, enhanced differentiation of mucus-producing goblet cells is one of the main characteristics of airway remodeling, and hyperplasia and hypertrophy of the goblet cells are associated with the severity of the disease [66, 85]. In asthma, AEC display increased protease-activated receptor-2 (PAR-2) and constitutively produce the enhanced amount of cytokines (IL-8 and GM-CSF), which increases their release of MMP-9 and GM-CSF [86, 87] that stimulate neutrophil recruitment and prolong eosinophil survival, respectively [88–92]. Finally, asthma subjects display increased VEGF and VEGF receptor expression on BEC that correlates with airway remodeling (mucus-producing cells, subepithelial fibrosis, and airway smooth muscle hyperplasia), airflow obstruction, and AHR [93].

AEC also has an important role in developing the adaptive immune response. Following interactions with environmental factors, including allergens, microbes, and other aero-particles, AEC release cytokines, such as the innate cytokines thymic stromal lymphopoietin (TSLP), IL-33 and IL-25 (IL17B), which are important activators if the mucosal surface-enriched innate lymphoid cells-2 (ILC2) that produce IL-13, IL-4, IL-5, and IL-9 [94]. Through the release of these type 2 (T2) cytokines, ILC2 not only acts directly on IgE and mucus production, fibroblasts and ASMC activation, and vascular permeability [95] but also contributes to the development of the type 2 adaptive immune response [96–98]. In addition, in an ILC2-independent manner, the AEC-derived cytokines, TSLP and IL-33 also have a direct critical role in promoting a T2 adaptive immune response by activating antigen-presenting cells (APC) to promote a T2 immune response when presenting antigen to antigen-specific CD4+ T lymphocytes [99–101]. Furthermore, besides releasing TSLP and IL-33, AEC produces many chemokines including CCL2 (MCP-1), which is elevated in the airways in asthma [102]. CCL2 attracts T lymphocytes [103], favors T2 development [104], and activates the release of T2 cytokines by mast cells [105].

Eosinophils and TGF-β

The numerous effects of the eosinophil on tissue remodeling and the regulation of the adaptive immune response are well-discussed in a review by Lee et al. [106]. Eosinophils are a hallmark of asthma [107, 108], and their number in the airways correlates with asthma severity [108, 109]. Postmortem observations demonstrate lung tissue eosinophilia in patient who died from severe asthma compared to those without asthma [110]. Many more peribronchial eosinophils are observed by electron microcopy in severe symptomatic asthma versus asymptomatic asthma, indicating the role of eosinophils in asthma exacerbation [16]. Eosinophil granule toxic proteins are present in mucus plugs on damaged epithelial surfaces in patients who died from asthma, suggesting tissue damaging functions from the eosinophil products [111]. In the airways, eosinophil extracellular traps augment goblet cell hyperplasia and mucus production and activate pulmonary neuroendocrine cells to amplify allergic immune response via neuropeptides and neurotransmitters [112].

In mice, IL-5 and eotaxin are critical for eosinophil differentiation and recruitment into the lung to induce AHR [113, 114]. Conversely, the elimination of eosinophils leads to a significant reduction in IL-13 production during an allergic response [115], indicating that while the T2 immune response induces eosinophilia, eosinophils have in turn a role in T2 cytokines production. Evidence for eosinophils to affect the adaptive immune response come from their ability to interact with T lymphocytes and to release numerous cytokines and growth factors that control the T1, T17 and T2 immune responses [116–118]. Eosinophils themselves can release IL-4 and IL-13 and thus perpetuate both the T2 immune response and directly lung remodeling [118, 119]. Eosinophils release multiple other factors including IL-1β [117], a pro-T17 cytokine, and they activate fibroblasts in an IL-1α-dependent manner [120]. The sum of the factors released by eosinophils changes fibroblasts into both a pro-inflammatory and profibrotic phenotype with the significant secretion of pro-neutrophilic factors, IL-6, IL-8, and CXCL1 [120–123]. Notably, both IL-1β and IL-6 may induce hyperplasia and hypertrophy of ASMC [124].

Eosinophils are a principal source of TGF-β1 in the airways in asthma, and their granule toxic proteins may activate fibroblasts to produce more TGF-β1 [21, 125, 126]. TGF-β1 stimulates fibroblast proliferation, differentiation into myofibroblasts, and their production of collagens (types I and III), fibronectin, glycoaminoglycans, and elastin [127–130]. TGF-β induces ASMC hypertrophy [131] and increases the level of smooth muscle α-actin in SMC, enhancing their capacity to contract and migrate [132–134]. In mouse, treatment with an anti-TGF-β neutralizing antibody reduced airway ECM deposition, ASMC proliferation, and mucus production after chronic allergen challenges [135]. BAL TGF-β1 protein is high in the airways in atopic asthma, is enhanced by SBP-Ag [136], and its expression is associated with eosinophilia and asthma severity (i.e., fibrosis and lung function decline) [21]. Hoshino et al. found that subepithelial thickness correlated with the number of fibroblasts in the submucosa, which was correlated with TGF-β1 expression in asthma [137]. TGF-β1 can also convert human BEC into an elongated fibroblasts-like shape, producing α-SMA, F-actin stress fibers, collagen I and losing the epithelial marker E-cadherin, an indication of epithelial-to-mesenchymal transition (EMT) [138].

Macrophages/Monocytes

Macrophages in the lung are a first defense against pathogens due to their high phagocytosis capacities and the production of cytokines such as type I interferons. Local lung tissue macrophages may present phagocytized and processed antigens to lung DC [139] and thus amplify specific adaptive immune responses. However, depending on the number of macrophages, they can also suppress the antigen presentation from DC to T lymphocytes [139]. Macrophages also promote adaptive immunity by presenting antigens to naïve CD8+ and CD4+ T lymphocytes via major histocompatibility complex (MHC) class I and MHC class II surface molecules and the co-stimulatory molecules, CD40, CD80, and CD86 [140]. As for T lymphocytes, macrophages have different functions depending on the cytokine environment. IL-10-producing regulatory macrophages have anti-inflammatory function, and macrophages in a T2 environment (M2) have wound healing and fibrotic functions via their production of TIMP, MMP, PDGF, TGF-β1, and chitinase-like proteins [141–147]. These airway M2 macrophages also produce CCL11 (eotaxin 1), CCL17 (TARC) and CCL22, which recruits eosinophils [148] and T2 lymphocytes [149]. Alveolar macrophages can produce IL-13 [150] and may be critical to maintain IL-13-dependent lung fibrosis [151]. As major producers of IL-1β and IL-23, macrophages likely act on IL-17A and T17 immune response generation [152, 153], which are known to induce pulmonary fibrosis [154]. It is however important to note that the M2 macrophages may be more predisposed to resolve inflammation and fibrosis via the production of IL-10 and pro-resolving lipids (reviewed in Ref. [155]) [156].

As part of the myeloid cells along with macrophages and DC, monocytes have similar properties, and they can mature into macrophages to replenish tissue macrophages, present antigens to CD8+ and CD4+ T lymphocytes via MHC class I and MHC class II surface molecules, as well as be pro- or anti-inflammatory (reviewed in Ref. [157]). Monocytes accumulate in the airways during an asthma attack [158]. In addition, circulating CD14+ monocytes can transform into fibroblast-like cells, fibrocytes, which migrate to injured tissues where they are matured by TGF-β1, produce ECM (Fibronectin, collagens), and promote wound contraction [15, 27, 159]. Circulating fibrocytes number and activation state are elevated in asthmatic patients following an exacerbation [160]. Monocytes also present antigens to T lymphocytes to induce differentiation into T1, T2, and T17 lymphocytes [157]. In an airway inflammation mouse model, newly airway recruited monocytes primed the antigen-specific CD4+ T lymphocytes to become T2 type [161]. Lastly, in an autoimmune mouse model, granulocyte-macrophage colony-stimulating factor (GM-CSF)-activated monocytes produce IL-6 and IL-1β to differentiate T lymphocytes into a T17 phenotype [162].

Neutrophils and MMP-9

In a cohort enriched in patients with severe asthma, 16% of the population displayed a neutrophilic phenotype as defined by ≥76% of neutrophils in sputum samples [163], while typically, only ~40% of immune cells in sputum samples are neutrophils in healthy individuals [164]. Mixed granulocytic inflammation (>2% eosinophils plus >40% neutrophils in sputum samples) in severe asthma is associated with lower lung function, worse asthma control, increased symptoms, and health care requirements compared to patients with only increased eosinophils, neutrophils, or none [165, 166]. In addition, the number of neutrophils in BALF increases by >20-fold 48 h after SBP-Ag in subjects with mild asthma [167], suggesting that neutrophils have an active role in allergic asthma. There is evidence showing that neutrophils can enhance lung remodeling. For instance, a specific product from neutrophils, neutrophil elastase (ELANE), increases mucin gene expression, mucus-producing BEC hypertrophy and hyperplasia, and disrupts the epithelial barrier [168–171]. Also, ELANE increases the migration of fibroblasts toward epithelial cells in vitro [172]. Neutrophil extracellular DNA traps (NET) level associated with asthma severity likely via AEC damage and eosinophil degranulation [173, 174].

NET are also associated with myofibroblast differentiation [175], and they activate macrophages to produce IL-1β [176], which directly activate fibroblasts and may increase the production of IL-17A by memory CD4+ T lymphocytes [117, 152]. Furthermore, through the release of exosomes, activated neutrophils enhance ASMC proliferation [177]. Activated neutrophils release many proteases that can degrade the matrix and contribute to lung remodeling, including MMP-9 (gelatinase) [178, 179]. MMP-9 activity is critical for antigen-uptake by dendritic cells (DC) in the airways [180]. In the lung, MMP-9 may be produced by most of stimulated cell types, and it is present in endobronchial biopsies in all asthmatic subjects, while absent in healthy individuals [107]. Notably, MMP-9 can mature IL-1β release by eosinophils [181], and thus participates in the activation of memory Th-17 [117, 152]. Although MMP-9 is predominant in asthma, all MMP-2 (collagenase A), MMP-3 (stromeolysin-1), and MMP-12 (metalloelastase) are elevated in the asthmatic airways [182]. MMP is associated with enhanced ECM deposition and inflammatory cells in the epithelium and subepithelium region. ECM-degrading collagenases and stromelysins affect the function and migration of inflammatory cells and matrix deposition in asthma [183, 184]. Although MMP could be considered as anti-fibrotic by degrading ECM, by cleaving ECM, they can also facilitate the migration of inflammatory cells and the release of fibrotic factors such as FGF2, active TGF-β1 and insulin-like growth factor (IGF) thus promoting fibrosis [185–188]. The expression of MMP genes is upregulated by cytokines such as IL-1, TNF-α, and TGF-α but is inhibited by TGF-β1 or IL-4. Their latent form requires maturation by cleavage to exert their proteolytic activity.

Besides, neutrophils can directly impact the development of the adaptive immune response by secreting cytokines such as IL-1, IL-23, IL-12, and IFN-γ [189–191], and thus favor T1 and T17 adaptive immune responses. Neutrophils also transport antigens and interact with myeloid APC to activate CD8+ T cells [192] and NET prime CD4+ T lymphocytes as they reduce the lymphocytes activation threshold [193]. Finally, neutrophils produce B-cell-stimulating factors, such as B-cell-activating factor (BAFF), APRIL, and a B-cell-activating molecule, CD40L [194].

Mast Cells and Cysteinyl Leukotrienes (CysLT)

Mast cells are tissue-resident cells. The number of mast cells in bronchial biopsies is higher in asthma than in healthy individuals [31], and it correlates with airway ECM thickness [195]. Mast cell number is increased in fibrotic disease and thus may play an important role in the development of fibrosis [196–199]. In asthma, mast cells are present in the ASM layer, which associates with the severity of the disease [200–202]. Mast cells function in allergic asthma is particularly important following the development of the adaptive T2 immune response that culminates in the production of antigen-specific IgE. Mast cells produce surface high-affinity receptors for IgE (FcεRI) and degranulate by cross-linking of the antigen-IgE complexes attached to FcεRI (reviewed in Ref. [203]) [204]. In mucosal and connective tissues, degranulating mast cells release a long list of preformed and newly produced mediators, including histamine, tryptase, chymase, LTC4, LTB4, PAF, PGD2, TNF-α, IL-1β, IL-4, IL-5, IL-6, IL-13, IL-8, IL-16, GM-CSF, MCP, RANTES, eotaxin, TGF-β1, FGF2, nerve growth factor (NGF), VEGF, and PDGFA, the latter of which being a chemoattractant and mitogenic for fibroblasts (reviewed in Refs. [203, 205]) [206]. Tryptase stimulates fibroblast proliferation and the release of type I collagen production by fibroblasts via TGF-β1 [207–209]. Mast cell and ASMC interaction lead to tryptase-induced ASMC proliferation, activation, and ASMC production of TGF-β1 that enhances their contractibility [210, 211]. Besides tryptase, the reason for the tight association between mast cells and fibrosis is their release of multiple other potential profibrotic factors, such as TNF-α, FGF2, and IL-1β, as well as their interaction with fibroblasts [212], which can lead to increase fibroblast contractility [213]. In addition, mast cells activated via FcεRI also release amphiregulin (AREG), a ligand for the EGF receptor, which increases mucin gene expression in epithelial cells and fibroblast proliferation [214, 215]. Mast cells also promote Th2 cell development due to their secretion of the T2 cytokines, IL-4 and IL-13 [216], and mast cell-derived histamine enhances IL-13 expression in murine T helper-2 lymphocytes via JAK-STAT dependent pathways [217].

Along with eosinophils, neutrophils, and basophils, mast cells are an important source of arachidonic acid products, including cysteinyl leukotrienes (CysLT) [218, 219]. CysLT are powerful bronchoconstrictor agents and promote mucus production and leukocyte inflammation [220]. CysLT are issued from the activation of phospholipase A2 (PLA2), which hydrolyses arachidonic acid (AA) and activates 5-lipoxygenase (5-LO); CysLT are chemoattractants for immune cells and promote eosinophil survival [221–225]. CysLT also favor T2 versus the T1 immune response [226–228]. Furthermore, CysLT activate alveolar macrophages to release profibrotic factors, such as IL-6, FGF, and MMP [229–232]. CystLT receptors are expressed on ASMC and CysLT trigger ASMC migration, contraction, and proliferation alone or associated with EGF [233–237]. Finally, CysLT increase fibroblast chemotaxis, proliferation, and myofibroblast differentiation and activation [238–243]. Therapeutics targeting the CystLT pathways reduce eosinophilia and tissue remodeling and have shown benefits in asthma and airway inflammation [244–248].

Development of Memory T Cells and B Cells by Dendritic Cells and Lymphocytes

Memory T Cells in Asthma

The three different kinds of adaptive immune responses according to the T lymphocyte-generated profile of cytokines are T1, T2, and T17, which are all three dampened by the T-regulatory cells. The main producers of cytokines and drivers of the different immune responses are the polarized memory CD45RO+CD4+ helper T lymphocytes, which are called, type 1 T helper (Th1), type 2 T helper (Th2) and type 17 T helper (Th17) [249–251]. Th1, Th2, and Th17 are the T lymphocytes that mainly define the T1, T2, and T17 immune responses, respectively. Th1 is characterized by the T-box transcription factor (T-bet) and produces the cytokines, IFN-γ, and IL-2; Th2 displays the transcription factor, GATA3 and produce IL-13, IL-4, IL-5, and IL-9; while Th17 produce RORC2, IL-17A and IL17C. The other forth main CD4+ cell type, composed of the T-regulatory (Treg) cells express Forkhead box P3 (FOXP3) and the anti-inflammatory cytokines, IL-10 and TGF-β. It is, however, important to know that these T-cell populations are more heterogeneous and display plasticity, changing from one type to another type. In addition, these definitions hide more complex situations since, for instance, IL-17A can be produced by FOXP3+ cells [252], and differentiated memory CD4+ T lymphocytes can produce both IL-17A and IL-4 [253]. Other CD4+ T cells called type 0 T helper (Th0) can produce both IL-4 and IFN-γ while the naïve CD45RA+CD4+ T lymphocytes have not encountered a cognate antigen, and thus, they produce low amounts of cytokines. Usually, the T1 response protects against infections from bacteria and viruses, the T2 response protects from parasitic infections and enhances the humoral response (including IgE production) [254], while the T17 response has some anti-infection properties, and it is present in autoimmune diseases and tumors.

The development of a certain type of adaptive immune response is strongly regulated by the persistent cytokine environment. For instance, inhalation of an antigen in mice displaying active and persistent airway T2 inflammation leads to T lymphocyte activation, while in naïve mice, it induces tolerance rather than activation [255]. In agreement, T2 cells differentiate following antigen presentation by DC and the presence of IL-4, while T1 cells develop during antigen presentation and the presence of IL-12 and IFN-γ. The exact cocktail of cytokines required to generate Th17 during antigen presentation remains uncertain but often include IL-1β or α, IL-23, IL-6, and IL21. The increased production of IL-17A from human memory CD4+ T cells usually requires IL-1β or α plus IL-23 [256].

CD8+ T cells or cytotoxic lymphocytes are critical for the defense against intracellular pathogens and the elimination of tumor cells. The cytotoxic activity of CD8+ T cells is particularly important to achieve viral clearance in the lung. Like CD4+ T cells, CD8+ T cells produce cytokines (e.g., IFN-γ and TNF), but they are mostly known for their production of T1 cytokines and toxic proteins (granzyme B and perforin) that kill infected cells and cancer cells. While APC (DC and macrophages) use their major histocompatibility complex class II (MHC-II) to present antigens to CD4+ T cells, CD8+ T cells recognize an antigen via the MHC-I. Co-stimulation during antigen presentation is critical for the development of effector and memory CD8+ T cells (i.e., OX40/OX40L), as well as the presence of cytokines such as common γ-chain cytokines (IL-2, IL-7, and IL-15). Importantly, while CD8+ T cells are predisposed to produce T1 cytokines (i.e., IFN-γ), they can also polarize into T2-producing lymphocytes (IL-13, IL-5, and IL-4) when in a T2 environment [257–261]. These T2 CD8+ T cells have been identified in human individuals including in patients with atopic asthma [262], yet it is uncertain whether the T2 CD8+ T cells can remain long-life memory lymphocytes [263].

Although the role of T1 IFN-γ+ CD8 and CD4 T lymphocytes in asthma remains under investigation, the T1 immune response has been observed particularly in severe asthma [264–267]. IFN-γ and IFN-γ-producing Th1 cells are present in the airways in asthma [268–270]. Whereas IFN-γ is known to antagonize the T2 immune response [271], it has also been shown that IFN-γ-producing Th1 cells caused severe airway inflammation, principally via activation of macrophages/monocytes, and did not reduce AHR during a T2 immune response [272]. In another study, the transfer of both Th1 and Th2 cells before an allergen challenge resulted in the recruitment of both Th1 and Th2 cells and rapid eosinophilic inflammation in the airways [273]. In any case, the presence of T1 IFN-γ+ CD8 and CD4 T memory lymphocytes in asthma is likely reminiscent of repeated viral and bacterial infections that are a major cause for asthma development and exacerbation [274, 275].

Memory B Cells in Asthma

B cells are the central generators of humoral immunity to protect against pathogens and eliminate unwanted elements that can bind to an antibody. While T cells distinguish between self-antigen versus nonself-antigen with its T-cell receptor (TCR), a B-cell does so by using its immunoglobulin (Ig) B-cell receptor (BCR). B cells are antibody factories that produce IgM and can further class switch into IgG- or IgE-producing cells. Antigen-specific T-cell-dependent IgG+ B cells develop in germinal centers due to isotype-switched recombination, somatic hypermutation, and affinity maturation and migrate to bone marrow as high-affinity memory B cells or long-lived antibody-secreting plasma cells. B cells recognize either small soluble antigens directly via their BCR or larger antigens via a presentation by a professional APC. In addition, during the interaction with T cells, B-cell activation implicates CD40/CD40L and ICOS/ICSOSL connections, MHC class II presentation of the antigen by B cells to T cells, and the presence of cytokines, such as IL-4, IL-21, and IFN-γ. T-cell-generated IL-9 participates in the development of memory B cells [276, 277], while terminally differentiated plasma cells need pro-survival factors such as IL-6 and hyaluronic acid [278]. The strength of the interaction during cell-cell contact is a factor that determines the production of either memory B cells or plasma cells [279].

Allergy is characterized by the production of allergic-specific IgE+ B cells. The recombination switch from IgM to IgE is mainly initiated by B-cell interactions via CD40/CD40L and CD28/CD80 (or CD86) with IL-4- and IL-13-producing T cells during the presentation of allergen by DC. Notably, other factors produced by APC and neutrophils such as BAFF and APRIL also facilitate the IgE isotype switching [280]. The cross-linking of the allergen/IgE complex to FcεRI on mast cells (and basophils) and DC lead to the rapid release of inflammatory factors and to internalization of the allergen followed by its presentation to T lymphocytes, respectively. IgE also binds to B cells and DC via the low-affinity IgE receptor (CD23), which permits the uptake of the allergen and its presentation to T lymphocytes to augment the immune response. Memory B cells are long-lived cells present in circulating blood for decades. Under new exposure to the same antigen, memory B cells can quickly proliferate and differentiate into antibodies-secreting plasma cells. While the existence of memory IgE+ B cells remains controversial, it is however proposed that specific IgE-secreting B cells may be quickly generated by T2 cytokines-activated long-lived memory IgG+ B cells (reviewed in Refs. [281, 282]).

Role of Dendritic Cells

Antigen-presenting cells, such as DC, are part of a heterogeneous population [140, 283] located at mucosal surface, across the epithelial barrier, and in the subepithelial layer, with a key role in connecting the innate with the adaptive immune response. DC are significantly better than other antigen-presenting cells to differentiate naïve T cells into effector T cells [284]. After uptaking inhaled antigens, mucosal matured DC migrate to lymph nodes to present antigens to naïve T cells, which differentiate into polarized CD4+ T lymphocytes [285–287]. DC interactions with naïve lymphocytes via Class II MHC and co-stimulatory surface accessory molecules will activate CD4+ T lymphocytes toward a more or less polarized Th-1, Th-2, Th-17, and T-regulatory lymphocyte phenotype, depending on the cytokine environment, the route, the form and the antigen dose [287–292]. Generally, the stronger is the co-stimulation during DC/T lymphocyte interaction, the most likely the response will skew toward T1 versus T2, which can also be achieved by favoring the interaction of the T-cell CD28 marker with B7.1 (CD80) versus B7.2 (CD86) [293, 294]. However, in asthma, DC isolated from the lungs seem to be potent inducers of a T2 immune response [295, 296], and myeloid DC preferentially skew the immune response toward Th2 [295]. In a mouse model, a DC subset can present antigen to activate Th2 cells up to 5 weeks after antigen challenge, explaining the occurrence of chronic Th2 airway inflammation [297]. The numerous other important co-stimulatory surface molecules interacting during the antigen presentation of DC to T lymphocytes are CD2/LFA-3, ICAM-1/LFA-1, OX-40/OX40L, ICOS/ICOSL, CD27/CD70, CD30/CD30L, 4-1BB/4-1BBL, HVEM/TNFSF14, and GITR/GITRL [298, 299]. Allergen challenges trigger the production of progenitors for DC [76], and in asthma, airway DC are increased in number, and their activation status, as measured by cell surface class II MHC molecules and Fc epsilon RI-alpha (FcεRI), is higher [300–302]. DC activation is partially induced by locally produced CSF-1, GM-CSF, and the interaction of DC with T cells via CD40-CD40 ligands [303, 304]. Studies have shown the link between DC and tissue fibrosis. For instance, the IL-10-producing DC is part of the tolerance process following exposure to antigen, and they reduce cardiac inflammation and fibrosis [305, 306]. CD209+ DC are associated with fibrosis in the heart [307], and accumulation of DC is found in the lung fibrotic area and BALF in idiopathic pulmonary fibrosis (IPF) [308, 309].

The Role of T Cells, B Cells, γ/δ T, NK, and NKT Cells in the Development of Memory Lymphocytes

The development of the adaptive immune response is strongly regulated by the cytokine environment, with T2 and T1 cytokines favoring the development of Th2 and Th1, respectively. With generally increased expression of T2 cytokines, most of the subjects with asthma are classified as T2 high (i.e., high IL-13 signaling) [310]. Of note, human B cells produce the receptor for IL-13 while T cells do not, indicating that IL-13 can play a role in the development of memory B cells but do not participate in the development of memory T lymphocytes [311]. Distinctively, the

IL-4 receptor is produced by both T and B cells. In asthma, CD4+ T helper cells are increased in the airways and expressed activation markers [109, 270]. In the airways, the production of T2 cytokines (IL-4, IL-5, and IL-13) co-localizes with T lymphocytes [312], adding to compelling evidence that CD4+ T cells are the main constitutive source of T2 cytokines in the airways and thus are critical regulators of the adaptive immune response. Yet, via the secretion of IL-4 and IL-13, subsets of Class II MHC-expressing accessory cells, CD8+ T cells, γ/δ T cells and NKT cells can also promote the development of Th2 and IgE+ B cells [312–318]. T lymphocytes also produce IL-9 that participates in the development and activation of T17 lymphocytes and B cells, enhancing IL-4-mediated IgE and IgG by B cells [319, 320]. In addition to the production of T2 cytokines and IL-9, antigen-activated CD4, CD8 and γ/δ T cells can all become IL10- and TGF-β-producing CD25+ T-regulatory cells that downregulate the immune response [321–327]. Additionally, many of the unconventional and innate-like lymphocytes present in mucosal surfaces including γ/δ T cells and NKT produce IL-17A and IL-22 [328–331]. Although only representing <2% of T lymphocytes in BAL fluids in asthma, NKT cells respond to both lipids presented by APC via CD1d and cytokines to produce T2 and T17 cytokines [332, 333]. Indeed, as for the CD4+ T lymphocytes, NKT cells are a source of IL-17A and IL-22 via stimulation of DC-derived IL-1β and IL-23 [330, 334]. NKT cells also produce IFN-γ [335], and they respond to IL-25 and IL-33 to produce T2 (IL-4, IL-13) and T1 (IFN-γ) cytokines [336–339]. Besides NKT cells, NK cells can secrete high amount of IFN-γ [340], and CD161+CD3− NK cells produce IL-17A [341]. Another unconventional T lymphocyte type, the *mucosal-associated invariant T* (MAIT) cell recognizes microbial metabolites via the MHC class I-related MR1 and produces T1 (IFN-γ, IL-12), T17 (IL-17A, IL-23) and T2 (IL-4, IL-13) cytokines [342, 343]. Like NKT cells, MAIT cells can be activated by IL-1β and IL-23 to produce IL-17A, but unlike NKT cells, MAIT cells are in fact abundant in human tissues [344]. Antigen-activated MAIT cells also promote B cells and DC activation via CD40L [345]. Notably, it has been reported that the number of airway MAIT cells appears associated with asthma severity [346], and the number of IL-17-producing MAIT cells are positively correlated with severe asthma exacerbation [347]. Therefore, although the primary role of the unconventional T lymphocytes is to defend the host against pathogens, they also actively participate in the development of the adaptive immune response.

The development of the adaptive immune response described above is summarized in Fig. 1.

Role of Adaptive Immunity on Remodeling

In models of lung injury, recruitment and activation of inflammatory cells precede fibrotic changes [348, 349], suggesting a cause-effect relationship between inflammation and tissue remodeling. In this part of this chapter, we will review how the cells and the products of the adaptive immune response impact tissue remodeling in asthma.

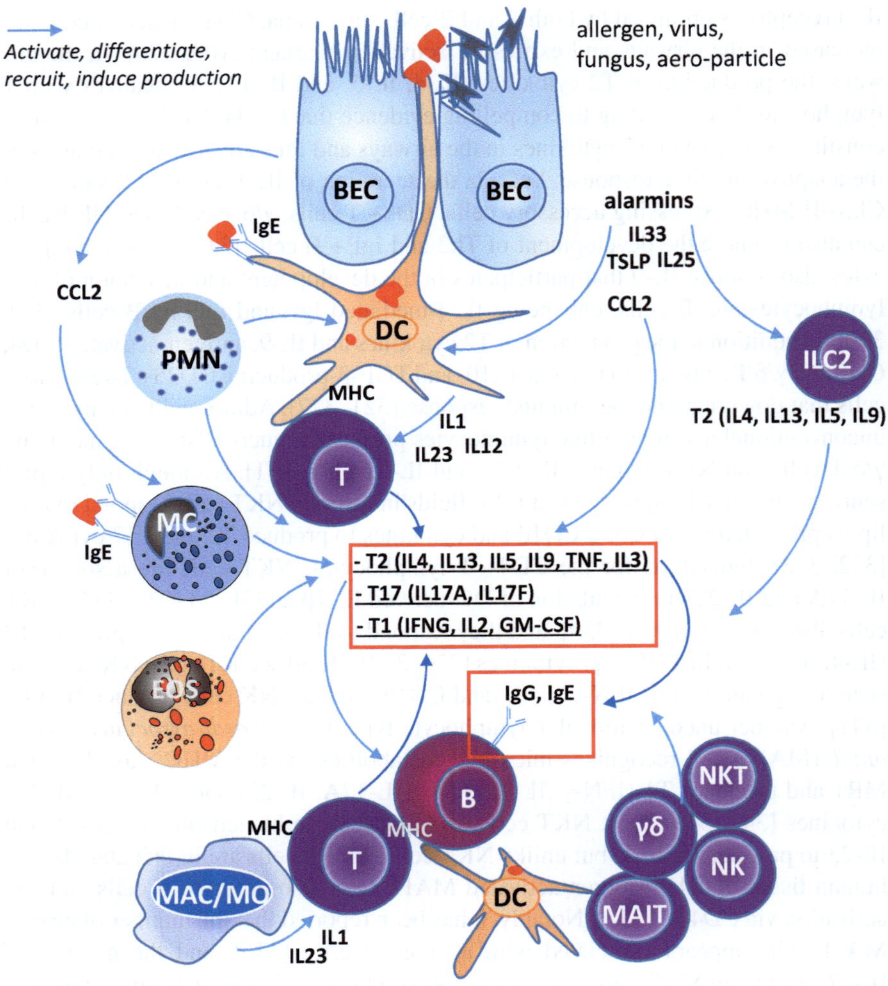

Fig. 1 Development of adaptive immunity in asthma. Adaptive immunity is composed of memory B and T lymphocytes. The different types of adaptive immune responses (T2, T17, and T1) are defined by the profile of cytokines generated by the T lymphocytes (T) and by the type of specific antibodies produced by B lymphocytes (B) as shown in the red rectangles. The fourth subset of T lymphocytes, the T-regulatory cells, produce IL-10 and TGF-β, and inhibit the development of the other three subsets, T1, T2, and T17. The profile of cytokines present in the environment will support the development of its own kind. For instance, the T2 cytokine, IL-4, will favor the development of T2 lymphocytes while the T1 cytokine, IFN-γ, will support the development of T1 lymphocytes. Following interaction with environmental factors, including allergens, microbes, and other aero-particles, the bronchial epithelial cells (BEC) release cytokines, such as the innate cytokines thymic stromal lymphopoietin (TSLP), IL-33 and IL-25 (IL17B), which are important activators of the mucosal surface-enriched innate lymphoid cells-2 (ILC2) to produce IL-13, IL-4, IL-5, and IL-9. Through the release of these type-2 (T2) cytokines, ILC2 act directly on IgE production and the development of the type-2 adaptive immune response. In addition, in an ILC2-independent manner, the BEC-derived innate cytokines also have a direct role in promoting a T2 adaptive immune response by activating dendritic cells (DC) when presenting antigen to antigen-specific T lymphocytes via the major histocompatibility complex (MHC). Furthermore, BEC also

Adaptive Immunity and Fibrosis

T2 Adaptive Immunity

It is well-known that subjects with allergic asthma express high levels of type-2 cytokines in their airways compared to nonasthmatic subjects [350]. It has been proposed by Holgate et al. that impaired epithelial repair in a T2 environment leads to myofibroblast activation, and excessive ECM deposition amplifying airway remodeling in asthma [351]. The T2 transcription factor, GATA3, enhances pulmonary fibrosis in a bleomycin mouse model [352]. Along with TGF-β, among the T2 cytokines, IL-13 predominates in its contribution to the pathophysiology of fibrosis [353]. In allergic asthma, IL-13 level is elevated in the airways compared to control [354–356], and the capacity of cord blood CD4+ T cells to produce IL-13 is a predictor for the development of atopic diseases [357]. IL-13 receptor is composed of the two subunits IL-13Rα1 and IL-4Rα, which is also a subunit of the IL-4 receptor. Presaging of the critical role of IL-13 in tissue remodeling, the IL-13 receptor is expressed on fibroblasts, ASMC, and BEC [358, 359]. The fibrotic response to IL-13 is concomitant with increases MMP and cathepsins that regulate matrix deposition [360]. Unlike HBF from healthy controls, HBF from patients with asthma responds to IL-13 to produce α-SMA and collagen type-1 (COL1A2) downstream of MMP-2 and TGF-β1 [361]. IL-13 augments TGF-β1-induced TIMP-1 expression in primary human airway fibroblasts via SMAD phosphorylation [362]. In an inducible transgenic mouse model, overexpression of lung IL-13 induced eosinophil accumulation in the airway and airway fibrosis, as examined by histology [353]. In the same model, IL-13 leads to prolonged airway inflammation, including eosinophils, neutrophils, lymphocytes, and macrophages, and persistent collagen deposition, weeks after induction of IL-13 [363]. Conversely, a neutralizing anti-IL-13 reduced AHR, subepithelial collagen deposition, and mucus production in an OVA

Fig. 1 (continued) produce CCL2, which attracts T lymphocytes, favors T2 development, and activate the release of T2 cytokines by mast cells. The antigen-presenting cells (APC), DC, macrophages (MAC) and monocytes process and present antigens to T lymphocytes via MHC-II and MHC-I to differentiate the lymphocytes into T1, T2, or T17 lymphocytes. APC secretion of IL-1 and IL-23 favors T17 differentiation, while their secretion of IL-12 favors T1 development. Similar to conventional CD4+ and CD8+ T lymphocytes, activated γδ T lymphocytes, natural killer (NK) cells, NK-T lymphocytes, and mucosal-associated invariant T (MAIT) cells also produce pro-T1, T2, and T17 cytokines and thus influence the development of the adaptive immune response, including the production of antibodies. The granulocytes, neutrophils (PMN), eosinophils (EOS), and mast cells (MC) release numerous cytokines and growth factors that directly or indirectly via APC control the T1, T17, and T2 immune responses. The release of these cytokines and other factors by tissue-resident MC is induced by the cross-linking of the complex IgE/antigen with MC surface FcεRI. B cells recognize either small soluble antigens directly or larger antigens via a presentation by a professional APC. During the interaction with T cells, B cells present antigens to T cells, and they are activated to produce antibodies in the presence of cytokines, such as IL-4 and IFN-γ. T2 cytokines enhance the humoral response leading to the production of IgE by B lymphocytes, and cross-linking of the complex IgE/antigen with DC surface FcεRI leads to the antigen internalization and presentation to T lymphocytes mouse model [364]. IL-13 stimulation also leads to reduced production of prostaglandin-E2 (PGE2), a prostaglandin that blocks TGF-β-induced effects on fibroblasts [365] as well as to induced collagen contraction by activated fibroblasts [366]. IL-13 upregulates and activates TGF-β1 [367, 368], and it stimulates the proliferation of myofibroblasts via signal transducer and activator of transcription 6 (STAT6) [369]. Subepithelial fibrosis observed in IL-13 transgenic mice is due in part to MMP-9-dependent activation of TGF-β [370]. The innate cytokine, IL-33 amplifies IL-13-induced airway macrophages differentiation toward M2 [371, 372], which produce the profibrotic TGF-β and increase eosinophil recruitment via CCL11 production [148]. Finally, adenosine is the product of the dephosphorylation of adenine nucleotides released from damaged cells and is elevated in blood and airways in asthma [373, 374]. In vivo, IL-13 induces high levels of adenosine and decreases adenosine deaminase (ADA) activity, an adenosine inhibitor which reduces IL-13-induced subepithelial fibrosis in mouse [375].

The cellular sources of IL-13 and IL-4 are mostly identical, and due to their common receptor subunit, the two T2 cytokines have often been examined simultaneously. As for IL-13 and conversely to IFN-γ, IL-4 is typically characterized as a profibrotic cytokine due to IL-4 ability to increase collagen production by fibroblasts [376, 377]. A study by Liu et al. showed that IL-4 and IL-13 increase collagen contraction by human fibroblasts but do not directly activate fibroblasts to produce TGF-β1 and fibronectin [366]. Instead, IL-4 and IL-13-induced profibrotic molecules, such as TGF-β, may occur as a result of epithelial cell activation, which was inhibited by IFN-γ [378, 379]. Besides, IL-4 and IL-13 have additive effects with oncostatin M (OSM) to increase both STAT3 phosphorylation in fibroblasts and fibroblast-mediated collagen gel contraction [380]. IL-4 and IL-13 induce human lung fibroblast-myofibroblast transition as evaluated by induction of α-SMA production through c-jun NH(2)-terminal kinase (JNK) signaling pathway [381]. In another study, both IL-4 and IL-13 increase the expression of α-SMA and collagen-III in human primary lung fibroblasts [382]. Doucet C et al. and others showed evidence that activation of fibroblast with IL-4 and IL-13 led to enhance proliferation, upregulation of cell surface adhesion molecules (integrin and vascular cell adhesion molecule 1 [VCAM-1]) as well as increase production of inflammatory cytokines and chemokines, such as IL-6, GM-CSF and monocyte chemoattractant protein 1 [MCP-1 aka CCL2] [383–385]. Finally, as indicated above, for IL-13, IL-4-activated fibroblasts also release CCL11 [379], a key chemokine for eosinophil recruitment into lung tissue. Thus, both IL-13 and IL-4 trigger intracellular signaling in fibroblasts, yet the effects of these T2 cytokines on fibroblast activation and fibrosis appear to occur partially downstream of MMP, TGF-β, adenosine, and increased granulocytes and macrophages recruitment and activation.

IL-5 is another T2 cytokine produced by memory CD4+ T lymphocytes. However, unlike for IL-13 and IL-4, the IL-5 receptor (IL5Rα) is almost exclusively produced by eosinophils, mast cells, and basophils [386, 387], and thus IL-5 cannot have a direct effect on fibroblasts. In murine models, IL-5 from CD4+ T lymphocytes is essential to develop eosinophilia [388]. In asthmatic subjects, a segmental bronchoprovocation with an allergen (SBP-Ag) leads to the recruitment of blood CD4+ T cells to the airways [389], which after ex vivo activation, produces an increase among IL-5 compared to a challenge with saline [390]. In the airways, allergic asthmatic subjects have increased levels of both IL-4 and IL-5 associated with elevated IgE and eosinophilia, while nonallergic asthmatic subjects with elevated IL-5 also displayed enhanced eosinophilia [391]. Anti-IL-5 treatments in mild atopic asthma reduce ECM protein deposition (e.g., tenascin, lumican, and procollagen-III) in bronchial subepithelial basement membrane, probably due to decrease presence of TGF-β1 [392]. In addition, IL-5 enhances the presence of surface semaphorin-7A on eosinophils, which has a role in pulmonary and liver fibrosis and can increase the production of α-SMA in HBF [393–395]. Therefore, while IL-5 does not have a direct role on fibroblasts, it may have an indirect impact on fibrosis via eosinophil-derived toxic proteins, TGF-β, profibrotic other cytokines/chemokines as well as the capability for eosinophils to enhance the T2 response including IL-13 production in vivo.

T9 Adaptive Immunity

Although IL-9 is often described as a T2 cytokine, more differentially polarized IL-9-producer cells are named Th9 cells. As for Th2 cells, Th9 cells differentiate under T2 conditions, but they also require TGF-β and IL-2 [396]. Regarding the development of IL-9–producing CD4+ T cells, various cytokine signaling pathways, such as IL-2/STAT5, IL-4/STAT6, type I IFN/STAT1, and TGF-β/SMAD, have been reported to promote IL-9 production in CD4+ T cells [397]. Yet, the two types of cells (Th2 and Th9) remain closely related to each other [398]. In fact, many kinds of cells can produce IL-9, including Th17 cells, regulatory T cells, and other subsets of immune cells, such as ILC2, mast cells, and NKT cells [399, 400]. IL-9 is a pleiotropic cytokine that promotes the development of allergic diseases [399, 401]. IL-9 receptor is present on effector T cells, mast cells, and granulocytes [402–404]. IL-9 potentiates the effect of IL-4 on B cells to produce IgE [320], enhances IL-3-induced proliferation of mast cells, and increases mast cell production of IL-6 [405]. Inducible lung-producing IL-9 mice display a strong accumulation of airway eosinophils, mast cells, and lymphocytes [406], some of these functions being likely indirect effects [407]. The strong correlation between IL-9 mRNA expression level and number of eosinophils suggests a role in eosinophil recruitment and survival [408], but it may also be due to correlations between the expression of IL-9 and the T2 cytokines. IL-9 increases eosinophil differentiation in the presence of IL-3 and IL-5; and enhances the expression of IL-5Rα [409, 410]. In mouse, neutralization of IL-9 in a house dust mite allergen challenge led to reduced airway mature mast cells, TGF-β1, VEGF, and FGF2, and to enhanced lung function [404]. In an Alternaria alternata-induced airway remodeling mouse model, IL-9 overexpression led to further accumulation of collagen and fibronectin with increase eosinophilia, RANTES (eosinophil chemoattractant), and the profibrotic CTGF in the lung [411]. In sum, as for IL-5, it is likely that IL-9 augments lung fibrosis mostly indirectly via granulocyte and mast cell activation.

T17 Adaptive Immunity

Another important mediator involved in airway remodeling is IL-17A. IL-17A is produced by CD4+ T cells, particularly CD4+CD4RO+ and CD8+CD4RO+ memory T cells [412–414]. IL-17A signals via the multimeric IL-17RA and IL-17RC receptor, which is present on fibroblasts and T lymphocytes [412, 415, 416]. IL-17 has a major role in defense against infections, notably by inducing chemoattractant factors for neutrophils and macrophages [417, 418]. Excess IL-17 is associated with lung dysfunction, including excessive airway neutrophilia [419, 420]. For instance, in vivo, neutralization of IL-17 attenuates bleomycin-induced pulmonary fibrosis including collagen deposition, α-SMA expression, and activation of MMP-2, as well as T2 and T17 inflammation [421]. IL-17A is typically associated with a more severe, neutrophilic, and corticosteroid-resistant asthma phenotype [422, 423]. IL-17A is significantly increased in sputum samples and BALF from subjects with asthma compared to healthy individuals [424] and BAL cells after SBP-Ag [117]. In asthma, airway IL-17A and the T17 immune response expression levels correlate with the number of airway neutrophils [425, 426], but it also correlates with the presence of eosinophils [427, 428] and IL-5 expression level [425]. Both IL-17A and T2 cytokines are elevated in sera in severe uncontrolled asthma [429]. As for IL-5 and IL-9, a possible mechanism for IL-17A to increase fibrosis is through the activation of eosinophils to produce TGF-β1 [430]. However, IL-17A also acts directly on fibroblasts to produce TGF-β and collagens [431, 432]. In a mouse model, intranasal instillation of LPS plus ATP-activated DC led to a mixed T2 and T17 immune response with correlation between IL-17A production and airway remodeling [431]. In that same study, IL-17A stimulated mouse fibroblasts to release TGF-β and express collagen in vitro [431]. IL-17A also enhances the effects of IL-13 on gene expression in fibroblasts likely via heightened STAT-6 activation [433]. Additionally, IL-17A increases the production of IL-6, IL-11, and IL-8 by HBF [424], and both IL17A and IL17F enhance CD40L-induced collagen production by blood monocyte-derived fibroblasts [434]. In that latter study, IL17A augmented CD40L-mediated IL-6 production while IL17F increased VEGF and angiogenin [434]. IL-17A induces the production of the chemokine IL-8 from epithelial cells, endothelial cells, fibroblasts, and macrophages, leading to the recruitment of neutrophil granulocytes [435]. Therefore, there is evidence that the T17 cytokines have direct effects on fibroblasts to trigger fibrosis.

T-Regulatory Cells

In humans, Treg cells are typically defined as FOXP3-expressing CD4+CD25^{hi}CD127^{-} T lymphocytes. Naturally occurring Treg cells are generated in the thymus and play a crucial role in the maintenance of self-tolerance by cell-cell interaction [436–439]. Conversely, inducible or adaptive Treg cells are differentiated outside the thymus, probably due to low accessory molecule implication and the presence of TGF-β [440]. Inducible Treg cells express relatively more modest levels of FOXP3 and CD25 compared to the naturally occurring Treg, and they mediate immune suppression [440]. It is well-established that Treg cells produce large amount of IL-10 and TGF-β [441–444]. Treg cells suppress the activity of many different immune cells, including T cells, B cells, NK cells, NKT cells and APC, by consuming IL2 and producing IL-10 and TGF-β. Both resting (CD45RA+) and activated (CD45RA−) Treg have immunosuppressive activity in vitro [445]. In bronchial biopsies from subjects with asthma, TGF-β1 is present in CD3+ lymphocytes [446], and TGF-β1 is produced by most immune cells [447]. TGF-β1 is an anti-inflammatory cytokine that reduces the activation of macrophages [448], the proliferation and generation of T lymphocytes [449], and the production of type-2 cytokines in T lymphocytes [450]. IL-10, mainly produced by macrophages and Treg cells, inhibits APC activity, T-cell proliferation and activation, and immune cell cytokine expression [451–457]. Treg cells limit lung collagen deposition and likely control tissue fibrosis [458–460]. In accordance with these studies, Treg depletion results in enhanced profibrotic gene expression level in skin tissue [461]. In a TGF-β-induced lung fibrosis mouse model, Treg cells inhibit fibrosis via suppression of FGF9 [462]. Treg cells can also reduce fibrosis by limiting CXCL12 production and, thus, fibrocytes recruitment to lung injury [463]. In asthma, the function of Treg cells seems reduced compared to normal [464–466], and successful immunotherapy (i.e., tolerance to allergen and reduction of symptoms) has been associated with increased Treg cell function [467, 468]. On the other hand, autocrine TGF-β-induced PDGF release by Treg directly activates lung fibroblasts; and the transfer of these Treg in vivo in mouse induced collagen deposition in the lung tissue in a noninflammatory condition [469]. All together, these studies suggest that the benefits of the functions of Treg cells on the reduction of inflammation in vivo dominates their potential induction of tissue fibrosis via TGF-β. Therefore, it seems fair to say that in an inflammatory disease such as asthma, Treg cells do not enhance fibrosis, but conversely, it inhibits fibrosis by controlling inflammation.

T1 Adaptive Immunity (IFN-γ)

Because T2 cytokines are profibrotic factors and IFN-γ enhances the T1 response at the expense of T2 development, it is logical to classify IFN-γ as an anti-fibrotic cytokine. Cell surfaced IFN-γ-enhanced IL-13Rα2, a decoy receptor for IL-13Rα1, reduces IL-13 activity [470, 471]. IFN-γ also strongly reduces T2 cytokine production by ILC2 [472]. In vivo, IFN-γ controls fibrosis in a bleomycin-induced lung fibrosis mouse model, likely by downregulating the expression of TGF-β and the procollagens [473, 474]. However, although IFN-γ decreases the synthesis of collagen by fibroblasts, in the same condition, collagen deposition in the matrix remained high and IFN-γ-activated fibroblasts displayed higher collagen receptor activity [475]. In addition, in another bleomycin mouse model, IFN-γ promotes lung inflammation and collagen accumulation [476]. IFN-γ enhances fibronectin expression level by fibroblasts [477]. In alveolar macrophages, IFN-γ increases the expression of PDGF-B [478], which disulfide-linked homodimer plays an important role in the wound healing process by recruiting fibroblasts, pericytes, and endothelial cells [479–481]. To make the dogma of IFN-γ anti-fibrotic activity more complex, CD4+ T cells that produce both IL-13 and IFN-γ have been identified in patients with pulmonary fibrosis [482]. In asthma, the expression of T2 markers in the airway remains positively correlated with IFN-γ expression, indicating the concomitant presence of both T2 and T1 immune responses in a subset of patients [426]. Moreover, in subjects with uncontrolled asthma or severe asthma, high amount of IFN-γ in exhaled breath condensate and the elevated number of IFN-γ+CD4+ T cells in BAL fluid have been reported [266, 483]. Adoptive transfer of a Th1 clone in a mouse after allergen challenge, reduced the T2 response but increased the noneosinophilic inflammation, including the accumulation of mononuclear cells [484]. Therefore, the role of IFN-γ in fibrosis in asthma remains unknown. In a disease such as idiopathic pulmonary fibrosis (IPF), IFN-γ therapy was tested on end-stage patients who subsequently suffered from acute respiratory failure following IFN-γ treatment [485]. The lack of efficacy of IFN-γ to treat IPF was further confirmed in other studies [486, 487].

B Cells

Deficiency in CD19 results in diminished B-cell responses and reduce fibrosis in bleomycin-induced lung fibrosis, and in the same model, CD19 overexpression enhances fibrosis [488]. In another study, using that same mouse model, IL-6 deficiency in B cells and blockade of BAFF led to reduced lung fibrosis [489]. Confirming these data, in vitro, IL-6-producing B cells promote collagen secretion by fibroblasts [489]. In co-culture experiments, B cells induce α-SMA and collagen expression in human dermal fibroblasts [490], while B-cell depletion using a therapeutic anti-CD20 antibody ameliorates patients with systemic sclerosis [491]. As with Treg cells, regulatory B cells (Breg) produce TGF-β to potentially enhance fibroblast activation and tissue remodeling [489, 492–494]. As mentioned above, the primary function of B cells is to produce antibodies such as IgG and IgE. Allergen-induced cross-linking of FcεRI bound IgE on the surface of mast cells has a major role in allergic asthma and airway remodeling via mast cell degranulation of numerous factors, including pro-granulocytic cytokines, TGF-β1, FGF2, VEGF, and PDGFA. There is, however, little to no evidence of an effect of B cells and immunoglobulins on fibroblasts in asthma.

The role of the adaptive immune response in fibrosis, as described above, is summarized in Fig. 2.

Adaptive Immunity and ASM Activation

In a group of subjects with asthma and chronic persistent and intermittent airway obstruction (FEV_1 <70% and ≥70%, respectively), the concentration of sputum IL-13, IL-5, IL-12, and IFN-γ is correlated with the smooth muscle area determined

Fig. 2 (continued) recruitment of neutrophils (PMN) that are a major source of MMP, which in turn triggers the release of active TGF-β. Notably, T-regulatory lymphocytes (Treg) inhibit most of the immune responses upstream of fibrosis. B cells co-cultured with fibroblasts induce expression of αSMA and collagen by fibroblasts, possibly via B-cell release of IL6. Finally, the production of specific IgE by B cells leads indirectly to increase fibroblast activation by MC degranulation of profibrotic factors following cross-linking of surface MC FcεRI with the complex IgE/antigen

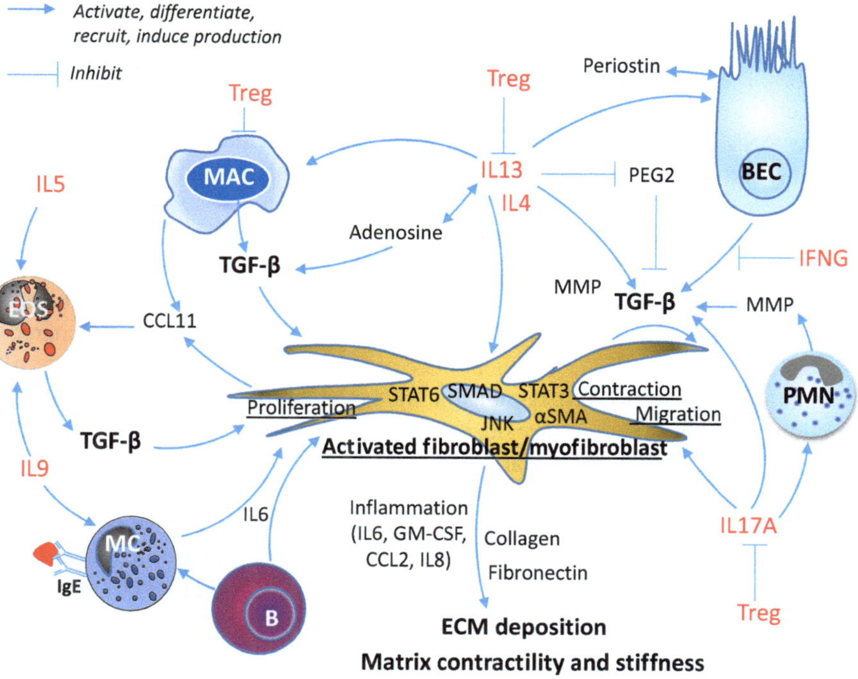

Fig. 2 Adaptive immunity in fibrosis. The fibroblast and its TGF-β-induced differentiated and α-smooth muscle actin (αSMA)-producing form, the myofibroblast, are responsible for lung subepithelial fibrosis, characterized by fibroblast hyperplasia, and increased contractility, the accumulation of extracellular matrix (ECM) proteins (e.g., collagens and fibronectin) and changes in the extracellular matrix biochemical properties (contractility, stiffness). Among the T2 cytokines, IL13 predominates in its contribution to the pathophysiology of fibrosis. Fibroblasts display functional IL13 receptors on their surfaces, and IL13 activates directly or indirectly through TGF-β, fibroblast proliferation, the production of alpha-smooth muscle actin (αSMA), and collagens via intracellular activation of STAT6 and SMAD signaling pathways. In addition, IL13 indirectly induces the production and the activation of the profibrotic factor, TGF-β through macrophages (MAC), MMP activities, and by inhibiting PGE2. IL13-activated macrophages also produce CCL11 that recruits TGF-β-producing eosinophils (EOS). Due to their common receptor subunit, IL4 possesses many similar functions as IL-13 on fibroblasts, and thus, studies have examined IL4 and IL13 functions concomitantly. IL4 and IL13 activate c-jun NH(2)-terminal kinase (JNK) and STAT3 signaling pathways in fibroblasts leading to αSMA production and fibroblast-mediated collagen contraction. Furthermore, IL4 and IL13 increase not only fibroblast proliferation but also their production of pro-inflammatory cytokines and chemokines. Finally, IL-4 and IL-13 activate epithelial cells (BEC) to produce both TGF-β, which in turn activates fibroblasts to produce collagens and fibronectin, and periostin that is part of the ECM. Notably, the induction of TGF-β in BEC by the T2 cytokines is tempered by the T1 cytokine, IFNG. The other main T2 cytokine, IL5, indirectly stimulates fibroblasts via eosinophils. Similarly, IL9 acts on fibroblasts via granulocytes such as EOS and mast cells (MC), and it enhances the T2-induced production of IgE by B lymphocytes, IL3-induced MC proliferation as well as the production of IL6 by MC. The receptor for the major T17 cytokine, IL17A, is present in fibroblasts. IL17A stimulates fibroblasts to produce TGF-β and collagen. IL17A enhances the effects of IL13 via STAT6 activation, and it increases the release of pro-inflammatory cytokines and chemokines (IL6, IL8) by fibroblasts. In vivo, IL17A has an indirect effect on fibrosis through its function on IL8 release by many airway cell types leading to the

in bronchial biopsies [495]. Interestingly, in that same study, the principal component analysis revealed that components that include IL-13, IL-12, and IFN-γ or IL-9, IL-17, and RANTES, were associated with chronic obstruction or intermittent obstruction, respectively. In vitro, contact of activated CD4+ T cells with ASMC induces ASMC proliferation [496]. Therefore, there is evidence on the role of lymphocytes and their products on the ASMC compartment in asthma.

T2 Adaptive Immunity and IL-9

T2-high asthma, as defined by the increased expression of IL-13 and IL-5 in the airway and the level of IL-13-responsive genes in BEC, is associated with increased AHR to methacholine and reticular basement membrane thickness, compared to T2-low asthma [497]. IL-13-deficient mice develop eosinophilia but fail to develop allergen-induced AHR [498], demonstrating the direct role of IL-13 on airway constriction. IL-13 is sufficient to induce AHR and enhance BAL eosinophil numbers and total IgE blood levels in naïve mice [353, 499]. Lack of IL-4 or IL-13 also reduced AHR and the presence of airway alpha-smooth muscle actin in a chronic allergen-induced mouse model [500]. IL-13 can induce AHR by direct effects on ASMC [501–503] or indirectly via STAT6 in BEC [504]. In vitro, IL-4 and IL-13-activated ASMC increase collagen contraction that is inhibited by PGE2 [366], and IL-13 increases human ASMC proliferation [505]. A possible mechanism for IL-13-enhanced ASMC proliferation is via the augmentation of surface CystLT receptors on ASMC and their interaction with LTD_4 [506]. Both CysLT and non-CysLT amounts are higher in BALF in subjects with asthma symptoms compared to controls [507], and both IL-13 and IL4 induce LT and their receptors [508], which are powerful bronchochonstrictor agents and further promote leukocyte inflammation [220]. Furthermore, IL-13 increases the production of the complement component C3 (C3) by BEC, which cleaved product, C3a causes ASMC contraction in asthma [509, 510]. Both IL-13 and IL-4 enhance histamine- and LTD_4-induced human ASMC contraction and Ca^{2+} mobilization, in vitro [511], while IL-13, but not IL-4, reduces β-adrenoceptor-induced relaxation of ASMC via MAP kinase signaling [503, 512]. Finally, in vivo, IL-13 likely increases bronchoconstriction and CCL2 production in asthmatic subjects due to ASMC activation by IL-13-induced adenosine [375, 513].

In mice, lack of IL-5 leads to reduced peribronchial smooth muscle thickness after chronic administration of allergen [514]. In passively sensitized human airways, anti-IL5 therapies reduced histamine-induced AHR [515]. However, these effects of IL-5 on the smooth muscle are likely mostly due to the deletion or reduced IL-5-induced activation of eosinophils, mast cells, and basophils.

The receptor for IL-9 is expressed on ASMC, and it activates ASMC to release pro-eosinophilic and neutrophilic chemokine, CCL11, and IL-8 via ERK and STAT3 signaling [516–518]. To our knowledge, besides IL-9 effects on ASMC to release pro-inflammatory cytokines/chemokines, there is no evidence of IL-9 affecting ASMC migration, proliferation, or contractility, and thus having direct consequences on airway smooth muscle thickening or AHR.

T17 Adaptive Immunity

In vivo and in vitro, IL-17A enhances the effects of IL-13 on AHR [433], and Th17 transfer or IL-17A overexpression in an airway allergen challenge mouse model, increased methacholine-induced AHR [519]. Unlike IL-13 and IL-4, neither IL-17A nor IL-5 enhances histamine- and LTD_4-induced human ASMC contraction and Ca^{2+} mobilization, in vitro [511]. However, there is evidence that IL-17A can directly activate ASMC. ASMC express the receptor for IL-17A [520], IL-17A stimulates the production of IL-8 by human ASMC [521], and IL-17A synergizes with OSM to induce human ASMC-release of IL-6 and CCL2 [522]. In addition, T17 cytokines increase both ASMC proliferation [523] and the production of IL-17-induced growth-related oncogene (GRO) by ASMC, which promotes ASMC migration [524]. Kudo et al. have shown that in vivo, lack of Th17, but not IL-17A-producing γδ T cells, is associated with lack of AHR after allergen challenge in mouse [525]. In that same study, in vitro, IL-17A acted directly on ASMC to enhance methacholine-induced contraction of mouse tracheal rings and human bronchi through nuclear factor κ light-chain enhancer of activated B cells (NF-κB), ras homolog gene family, member A (RhoA) and Rho-associated coiled-coil containing protein kinase 2 (ROCK2) signaling cascade [525]. IL-17A-induced ASMC contractility was later confirmed by other studies [526, 527].

T1 Adaptive Immunity (IFN-γ)

In a mouse model, IFN-γ-producing CD4+ T cells are responsible for increased AHR following a chronic low-level allergen challenge [528]. Yet, in vitro, IFN-γ inhibits SMC activities, including ASMC proliferation and chemotaxis [529–531]. IFN-γ blocks spontaneous release of VEGF by ASMC and reduces T2- and TGF-β-augmented VEGF production [532]. While the role of IFN-γ on ASMC contractility has not been reported, it has been described that IFN-γ reduced normal human intestinal SMC contractility, motility and proliferation [533]. Therefore, because most of the in vitro studies identify an inhibitory function for IFN-γ on SMC, it is possible that the role observed on AHR in vivo, in the study by Kumar et al., did not occur via ASMC.

B Cells

In asthma, serum IgE level is associated with AHR [534]. ASMC express both a functional high-affinity receptor for IgE (FcεRI) [535] and the low-affinity IgE receptor (FcεRII; CD23) that are upregulated by IL-4, TNF-α, IL-1β, and GM-CSF [536, 537]. In ASMC, IgE triggers the production of IL-8 and eotaxin, two chemokines that recruit neutrophils and eosinophils, respectively; and following FcεRI upregulation on ASMC, IgE induces the release of T1 and T2 chemokines via MAPK, Akt, and STAT3, and intracellular calcium mobilization [535, 537]. Furthermore, IgE increases ASMC proliferation and their deposition of fibronectin and collagen through MAPK and STAT3 [538, 539]. It has also been reported that IgE-sensitized ASMC produces IL-13, which can activate ASMC in an autocrine way to augment AHR [512].

Human ASMC may also express FcγRI (CD64) and FcγRIIb (CD32), the latter of which possesses an immunoreceptor tyrosine-based inhibitory motif (ITIM) and its interaction with heat aggregated-IgG (ligand for CD32) leads to inhibition of IL-1-induced production of IL-6 and IL-8 [540]. Interestingly, activation in vitro of a SMC issued from a guinea pig, with the complex IgG/antigen leads to SMC contraction [541]. However, to our knowledge, studies on the potential role of a complex IgG/antigen on ASMC in asthma remain inexistent.

Interestingly, B cells, particularly regulatory IL-10+ B cells, appear to have a beneficial regulatory effect on AHR and airway collagen deposition as examined in allergen-challenged mouse models [542].

The role of the adaptive immune response on airway smooth muscle activation, as described above, is summarized in Fig. 3.

Adaptive Immunity and Mucus Production

Airway mucus is produced by BEC and submucosal glands and forms a gel layer on the epithelial surface to protect against external agents. Mucous gel is cleared from the lower airways as a result of its transportation by epithelial ciliated cells toward the oral cavity. Impaired mucus clearance causes mucus plug formation and airway obstruction [543]. Mucus is mostly composed of water, lipids, and proteins such as mucins [544]. MUC5AC and MUC5B are the main secreted mucin proteins in the airways, and they define the biophysical characteristics of the mucous layer [544]. MUC5AC is mostly secreted from airway epithelial goblet cells, while MUC5B is produced by mucous cells in the submucosal glands [545]. In asthma, the ratio MUC5AC:MUC5B is increased relative to that in healthy individuals, notably because of the significant increase in the number of goblet cells and MUC5AC production [546–548].

T2 Adaptive Immunity and IL-9

Compared to T2-low asthma, T2-high asthma is associated with an enhanced MUC5AC:MUC5B expression ratio [497]. IL-13, IL-4, and IL-9 can induce mucus production by human BEC in vitro [549–551], and murine studies have shown that IL-9, IL-4, and IL-13 have a role in goblet cell hyperplasia and mucin gene expression, including MUC5AC and MUC2 [552]. IL-9 induces mucus production in epithelial cells during injury repair [551]. However, while transgenic mice expressing IL-4 or IL-9 in the lung display increased expression of MUC5AC and levels of mucus glycoconjugates [553, 554], IL-4 or IL-9 deficient mice still produce mucus hyperplasia in vivo [407, 555]. In fact, although IL-4 and IL-9 can directly activate BEC to produce mucus, these cytokines may also do so indirectly via IL-13 in vivo [498]. For instance, the production of IL-9 in the lung can induce T2 cytokine production, including IL-13 which thereafter stimulates mucin expression in BEC [407, 549]. Therefore, IL-13 is usually seen as the major T2 cytokine to induce mucus production by BEC in asthma. IL-13 is sufficient to induce mucus

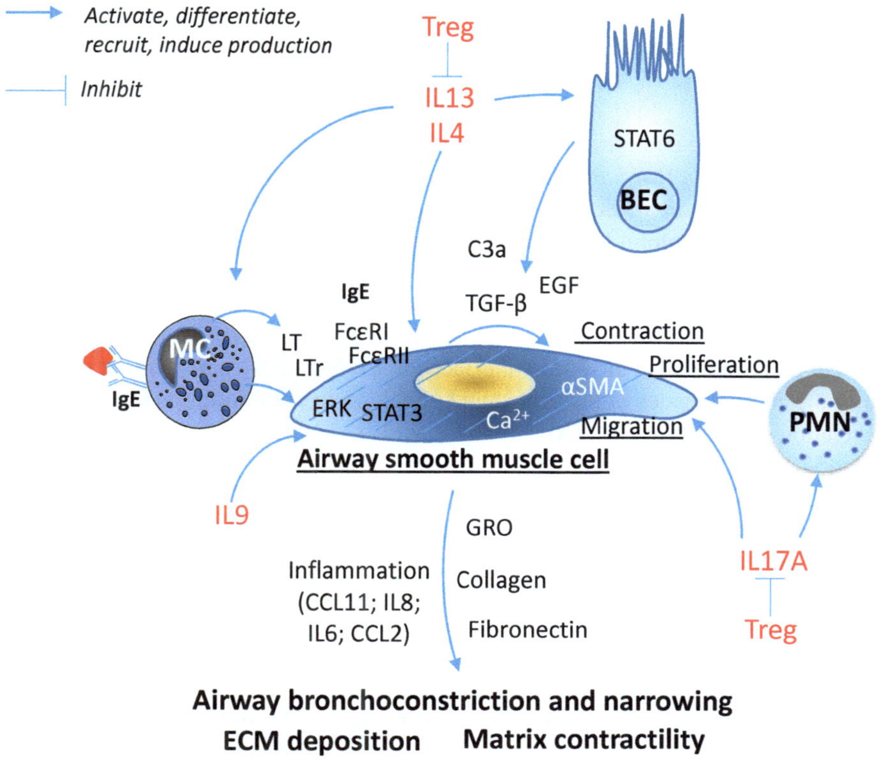

Fig. 3 Adaptive immunity in airway smooth muscle cell activation. Increased airway smooth muscle mass is a pathological feature of asthma and is a major cause for airway hyperresponsiveness, bronchoconstriction, and narrowing. In vivo, IL13 induces airway hyperresponsiveness (AHR) by direct effects on the airway smooth muscle cells (ASMC) or indirectly via activation of the STAT6 pathway in bronchial epithelial cells (BEC). In vitro, IL-13 increases ASMC proliferation directly or indirectly via the upregulation of the leukotriene receptors (LTr) on the ASMC surface. IL4 and IL13 enhance ASMC-induced collagen contraction, an event inhibited by PGE2. In addition, IL13-activated BEC augment the presence of TGF-β, EGF and C3a that lead to ASMC hypertrophy, production of alpha-smooth muscle actin (αSMA), contraction, and migration. In vitro, also, both IL13 and IL4 enhance human ASMC contraction and Ca^{2+} mobilization by the mast cell (MC) products, histamine, and LT, while cross-linking of surface MC high-affinity receptor for IgE (FcεRI) with the complex IgE/antigen induces the release of numerous activators of ASMC, including histamine, LT, and TGF-β. The receptor for IL9 is expressed on ASMC and IL9 activates ASMC to release pro-eosinophilic and neutrophilic chemokine, CCL11, and IL-8 via ERK and STAT3 signaling. ASMC also express the receptor for IL-17A. In vitro, IL17A promotes ASMC migration, proliferation, contraction, and it stimulates ASMC production of IL8, IL6, CCL2 and growth-related oncogene (GRO). In addition, downstream IL17A and the presence of the chemokine IL8, activated neutrophils enhance ASMC proliferation through the release of exosomes. ASMC express both FcεRI and the low-affinity IgE receptor (FcεRII) that are upregulated by IL-4. Via these receptors, IgE triggers intracellular calcium mobilization, and the production of IL-8, CCL11, T1, and T2 chemokines via MAPK, Akt, and STAT3. Finally, IgE increases ASMC proliferation and their deposition of fibronectin and collagen through MAPK and STAT3. T-regulatory lymphocytes (Treg) inhibit the adaptive T2 and T17 immune responses

production in naïve mice [353, 499], and in an inducible transgenic mouse model, overexpression of lung IL-13 induced mucus cell metaplasia [353]. Conversely, a neutralizing anti-IL-13 significantly reduces mucus production in an allergen challenge mouse model [364], and in many other studies using mouse models, blockage of IL-13 signaling demonstrates that IL-13 is essential for mucus production [499, 556–560]. In vitro, IL-13 increases the ratio of goblet cells:ciliated BEC [561] and BEC proliferation in an EGFR/EGFR ligand (TGFα or epigen)-dependent manner [562, 563]. Furthermore, IL-13 induces increased mucus secretion by BEC through enhanced expression of a Ca^{2+} activating chloride channel, CLCA1 [564], and STAT6 signaling [504]. Importantly, demonstration of direct induction of mucus production in BEC and goblet cell hyperplasia by IL-13 were also reached using human BEC cultured at air-liquid interface (ALI) [565–569], a model that recapitulates the characteristics of the bronchial epithelium in vivo. Additionally, long-term activation of human BEC in ALI cultures with IL-13 increases MUC5AC production [570].

In addition to augmenting mucus production by BEC, IL-13, and IL-4 have broader effects on BEC and the bronchial epithelium. For instance, IL-13 can damage the epithelial barrier by inducing apoptosis in human epithelial cells [382] and by impairing tight junctions [571, 572]. IL-13 and IL-4 also increase fibrosis via BEC by enhancing their production of TGF-β2, leading to myofibroblasts activation to produce α-SMA and ECM [378, 379, 570]. Note that these effects from the T2 cytokines on TGF-β2 in BEC are attenuated by IFN-γ [378]. Additionally, one of the main T2-response genes in BEC, periostin, is an ECM protein and a component of subepithelial fibrosis in bronchial asthma [573–575]. Periostin is used as a T2-high asthma biomarker [576, 577] and it predicts lung function decline in asthma [578]. Periostin plays a role in the recruitment of inflammatory cells (eosinophils, neutrophils, macrophages) and, thus, indirectly, lung fibrosis [579–581]. BEC-derived periostin also functions directly on BEC themselves as well as fibroblasts to activate MMP, TGF-β, and consequently type 1 collagen production [582]. In addition, periostin binds type 1 collagen and fibronectin and can then participate in the stiffness of the ECM mass [583].

T17 Adaptive Immunity

In vivo, IL-17A enhances the effects of IL-13 on mucus production [433]. Lung epithelial cells express both the IL-17 and IL-22 receptors [584, 585]. Interestingly, comparable to IL-13, IL-17A increases the expression of MUC5AC by nasal epithelial cells and promotes goblet cell hyperplasia [586]. Conversely, in a study, where all three types of cytokines, T2 (IL-13), T1 (IFN-γ), and T17 (IL-17A), were used on nasal epithelial cells cultured at ALI, it was shown that both IFN-γ and IL-13 increased the number of goblet cells and expression of MUC5AC, but IL-17A only augmented MUC5B production [587]. Conversely, a previous study had demonstrated that IL-17A increased the expression of both mucin genes, MUC5B and MUC5AC, in differentiated human BEC in vitro, partially via an autocrine role of IL-6 and JAK2 signaling [588]. To support that study, Pezzulo et al. more recently reported goblet cell hyperplasia in IL-17-activated human BEC cultured at ALI [569].

Besides IL-17A's role in mucus production, IL-17A has other potential detrimental functions on the epithelium that would lead to profibrotic events. For instance, IL-17A promotes epithelial-mesenchymal transition (EMT) of mouse alveolar epithelial cells and their production of collagen in a TGF-β-dependent manner [421]. In addition, through PIK3CA, IL-17 inhibits autophagy in alveolar epithelial cells, which may cause collagen accumulation, and epithelium death and amplify IL-1β/IL-17-induced lung pathology in a vicious circle [421, 589–591].

Thus, there is evidence of the direct role of IL-17A on BEC to produce gel-forming mucin proteins and to amplify airway fibrosis.

B Cells

Human AEC express CD23 and FcεRI, which activation by IgE leads to the production of ET-1 and the release of 15-hydroxyeicosatetraenoic acid (15-HETE), respectively [592, 593]. Downstream the stimulation of BEC by IgE, ET-1 can activate ASMC and increase bronchoconstriction, whereas arachidonic acid-derived 15-HETE that is typically upregulated by T2 cytokines [594], has anti-inflammatory properties [595, 596] and is associated with airway fibrosis in asthma [597]. 15-HETE reduces AHR to methacholine [598] and, in vitro, 15-HETE reduces the production of MUC5AC in a bronchial epithelial cell line [599]. This latter study, however, disagrees with a previous study where 15-HETE enhanced MUC5AC expression in IL-13-activated primary AEC [600]. Therefore, it is possible that IgE induces mucus production by BEC via 15-HETE, yet the role of IgE on BEC remains ill-defined.

The role of the adaptive immune response on airway mucus production as described above is summarized in Fig. 4.

Adaptive Immunity and Fibrin Clots and Angiogenesis

Fibrin Clots

Fibrin clotting is the end-product of the coagulation cascade, and it requires the activity of numerous proteins that are part of this cascade. Exaggerated pro-coagulation activity or impaired fibrin degradation in the airways has been observed in asthma, particularly in severe asthma, and is associated with the number of inflammatory cells [601–606]. Pro-coagulation factors are increased in the airways after SBP-Ag in patients with asthma and they correlate with eosinophil inflammation [607, 608]. In an allergen challenge rat model, the amount of the pro-fibrin accumulation protein, plasminogen activator inhibitor 1 (PAI-1), is augmented in blood and the airways [609], confirming a pro-coagulation activity after allergen challenge. In vitro, OSM increases the level of fibrinogen while IL-4, IL-13, and IL-10 decrease it [610]. In nasal epithelial cells, the anti-fibrin deposition protein, tissue plasminogen activator (tPA) level is decreased by IL-13 and IFN-γ with an additive effect when IL-13 and IFN-γ are used together, indicating that T1 and T2

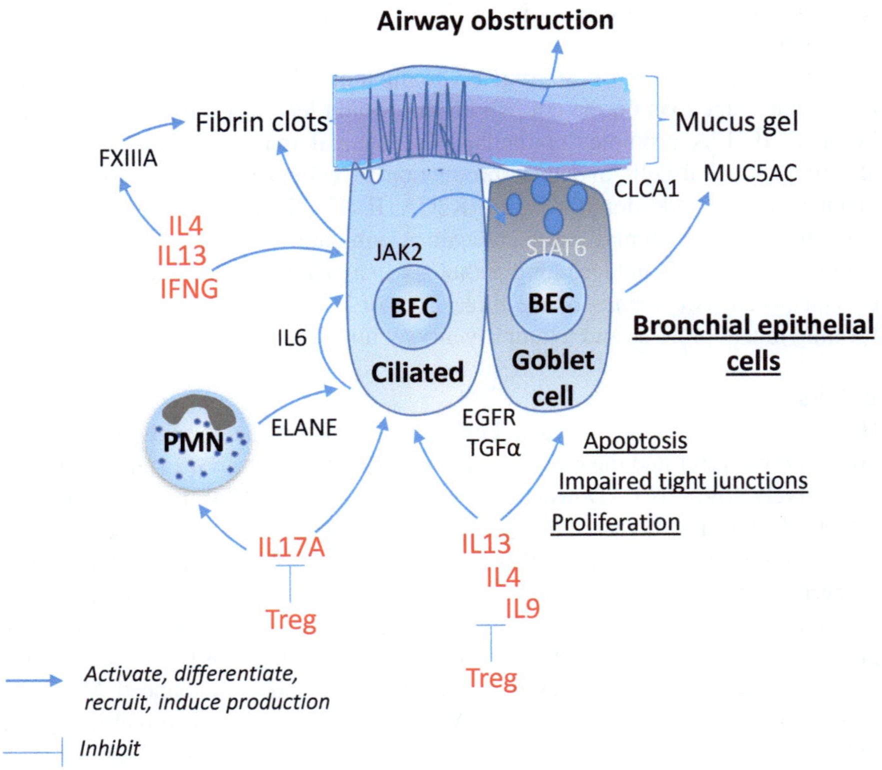

Fig. 4 Adaptive immunity and airway mucus production. Loss of ciliated mucosal bronchial epithelial cells (BEC) and bronchial epithelial goblet cells hyperplasia with mucus accumulation in the airways are features of asthma and airway obstruction. In asthma, increased goblet cells:ciliated cells ratio leads to increase production of gel-forming mucus proteins (e.g., MUC5AC) and lack of mucus gel clearance because of deficiency of cilia movement. IL13, IL4, and IL9 increase mucus production by BEC, in vitro. Yet, in vivo, IL4 and IL9 may also augment MUC5AC production by BEC indirectly by increasing IL13, which remains the essential T2 cytokine directly enhancing mucus production by BEC. IL13 and IL4 also impair BEC tight junctions and increase apoptosis as well as proliferation via an EGF receptor (EGFR) manner and the EGFR ligand, TGFα. Furthermore, IL-13 increases mucus secretion by BEC through enhanced expression of a Ca^{2+} activating chloride channel, CLCA1 and STAT6 signaling. Comparable to IL13, IL17A can increase the expression of MUC5AC by epithelial cells and promotes goblet cells hyperplasia, partially via IL6 and JAK2. IL17A also acts indirectly on mucus production and goblet cell hyperplasia via the production of ELANE by neutrophils (PMN). T-regulatory lymphocytes (Treg) inhibit the adaptive T2 and T17 immune responses. Both T1 (IFNG) and T2 (IL13) cytokines engage epithelial cells into a pro-fibrin deposition function that may further plug the airways with mucus, while T2 cytokines induce the production of the pro-fibrin clot transglutaminase, FXIIIA by antigen-presenting cells

cytokines engage epithelial cells into a pro-fibrin deposition function [611, 612]. In another in vitro model, IFN-γ increases the receptor for TNF-α, which leads to TNF-α-induced production of the pro-fibrin deposition protein, tissue factor, by epithelial cells [613]. In addition to acting on pro- and anti-fibrin formation protein level changes, other proteins such as the transglutaminases have a key role in cross-linking fibrin to form clots. The coagulation factor XIII (FXIIIA) is such a transglutaminase that covalently cross-links fibrin at the end of the coagulation cascade. FXIIIA is produced by the airway cells and correlates with T2 inflammation, markers of DC, and airway obstruction in asthma [614–617]. Finally, the cytokines IL-13, IL-4, and IL-10 enhance the production of FXIII-A by APC [618–620] and thus can enhance fibrin clot formation by raising FXIIIA activity in the airways.

Angiogenesis

Angiogenesis is the process leading to formation of new blood vessels, which is enhanced in the airway walls [63], and it is recognized as an important event part of airway inflammation and remodeling in asthma [31, 621–624]. In inflammatory diseases, angiogenesis may contribute to tissue growth [625], as well as microvascular permeability and edema formation [626]. There is evidence that the increased number and size of blood vessels in asthma contribute to the thickening of the airway wall and narrowing of the airway lumen during ASMC contraction [627]. Several important factors are involved in angiogenesis, including FGF, TGF-α, IL-8, TGF-β, and angiogenin, but VEGF signaling is a critical rate-limiting step of angiogenesis during normal or pathological conditions by promoting the growth of vascular endothelial cells (reviewed in Ref. [628]). In mice, VEGF generated by lung epithelial cells increases airway T2 inflammation, mucus production, angiogenesis, edema, and vascular remodeling [629]. VEGF also induces vascular leakage and enhances thrombin activity in asthmatic airways, explaining its functions on inflammation and fibrin deposition in tissues [630, 631]. In asthma, airway walls display VEGF-dependent increased vascularity [626, 632–635]. The expression level of VEGF and its receptors are increased in the airways [633] and associate with vascularity and reduced pulmonary function [636]. In addition, VEGF level is enhanced in sputum samples and BAL fluids and is associated with airway vascular permeability and asthma severity [637, 638]. In vitro, in addition to TNF-α and IL1-β, the T2 cytokines, IL-4, and IL-13 are inducers of VEGF in ASMC, which is attenuated by IFN-γ [532, 639, 640]. Furthermore, TGF-β1 in presence of IL-4 and IL-13 increases VEGF in bronchial fibroblasts [639–641]; and in mouse, in vivo, overexpression of IL-13 promotes VEGF accumulation in BAL fluids after hyperoxia [642]. These data demonstrate a close interaction between T2 inflammation and VEGF with the potential to lead to airway remodeling. Besides the T2 cytokines, IL17F, another member of the T17 family, augments CD40L-induced VEGF and angiogenin production in PBMC-derived fibroblasts [434]. Finally, as indicated above in the paragraph on mast cells, VEGF can be released from mast cells following the cross-linking of IgE with its cell surface high-affinity receptor, FcεRI [643].

Conclusions

Adaptive immunity plays a critical role in host defense and pathogen elimination. Inappropriate activation of this immune response is a fundamental mechanism leading to airway inflammation in asthma, with eosinophilia as a hallmark of the cellular response. Increased mucus generation, fibroblast activation with matrix protein generation, and new vessel formation are likely the results of inflammatory mediators associated with the response, and collectively they lead to structural airway changes known as airway remodeling which contribute to disease severity and lack of response to therapy. Recent advances in asthma therapeutics have ushered in new biologics aimed at various steps in the inflammatory cascade. These therapies have been approved based on their ability to reduce asthma exacerbations and, in some cases, improve lung functions and quality of life; however, there are no available therapies to date that have been shown to reverse the underlying mechanism of asthma, leading to disease "remission." Beneficial outcomes relative to controlling lung remodeling using biologic therapies targeting one or several specific cytokine/chemokine pathway(s) may first require precisely endotype patients with asthma vis-à-vis their dominant adaptive immune response(s) during asthma exacerbation.

References

1. Laprise C, Laviolette M, Boutet M, Boulet LP. Asymptomatic airway hyperresponsiveness: relationships with airway inflammation and remodelling. Eur Respir J. 1999;14(1):63–73.
2. Chetta A, Foresi A, Del-Donno M, Bertorelli G, Pesci A, Olivieri D. Airways remodeling is a distinctive feature of asthma and is related to severity of disease. Chest. 1997;111:852–7.
3. Goorsenberg AWM, d'Hooghe JNS, Srikanthan K, Ten Hacken NHT, Weersink EJM, Roelofs J, et al. Bronchial thermoplasty induced airway smooth muscle reduction and clinical response in severe asthma. The TASMA Randomized Trial. Am J Respir Crit Care Med. 2021;203(2):175–84.
4. Sumi Y, Hamid Q. Airway remodeling in asthma. Allergol Int. 2007;56(4):341–8.
5. Payne DN, Rogers AV, Adelroth E, Bandi V, Guntupalli KK, Bush A, et al. Early thickening of the reticular basement membrane in children with difficult asthma. Am J Respir Crit Care Med. 2003;167(1):78–82.
6. Redington AE, Howarth PH. Airway wall remodelling in asthma. Thorax. 1997;52:310–2.
7. Nelson HS. Advances in upper airway diseases and allergen immunotherapy. J Allergy Clin Immunol. 2003;111(3 Suppl):S793–8.
8. Bourdin A, Neveu D, Vachier I, Paganin F, Godard P, Chanez P. Specificity of basement membrane thickening in severe asthma. J Allergy Clin Immunol. 2007;119:1367–74.
9. Hinz B, Celetta G, Tomasek JJ, Gabbiani G, Chaponnier C. Alpha-smooth muscle actin expression upregulates fibroblast contractile activity. Mol Biol Cell. 2001;12(9):2730–41.
10. Roy SG, Nozaki Y, Phan SH. Regulation of alpha-smooth muscle actin gene expression in myofibroblast differentiation from rat lung fibroblasts. Int J Biochem Cell Biol. 2001;33(7):723–34.
11. Liu G, Philp AM, Corte T, Travis MA, Schilter H, Hansbro NG, et al. Therapeutic targets in lung tissue remodelling and fibrosis. Pharmacol Ther. 2021;225:107839.
12. Halwani R, Al-Muhsen S, Hamid Q. Airway remodeling in asthma. Curr Opin Pharmacol. 2010;10(3):236–45.
13. Kenyon NJ, Ward RW, McGrew G, Last JA. TGF-beta1 causes airway fibrosis and increased collagen I and III mRNA in mice. Thorax. 2003;58(9):772–7.

14. Nho RS, Ballinger MN, Rojas MM, Ghadiali SN, Horowitz JC. Biomechanical force and cellular stiffness in lung fibrosis. Am J Pathol. 2022;192(5):750–61.
15. Schmidt M, Sun G, Stacey MA, Mori L, Mattoli S. Identification of circulating fibrocytes as precursors of bronchial myofibroblasts in asthma. J Immunol. 2003;171(1):380–9.
16. Cutz E, Levison H, Cooper DM. Ultrastructure of airways in children with asthma. Histopathology. 1978;2:407–21.
17. Davison AG, Haslam PL, Corrin B, Coutts II, Dewar A, Riding WD, et al. Interstitial lung disease and asthma in hard-metal workers: bronchoalveolar lavage, ultrastructural, and analytical findings and results of bronchial provocation tests. Thorax. 1983;38(2):119–28.
18. Roche WR, Beasley R, Williams JH, Holgate ST. Subepithelial fibrosis in the bronchi of asthmatics. Lancet. 1989;1:520–3.
19. Brewster CE, Howarth PH, Djukanovic R, Wilson J, Holgate ST, Roche WR. Myofibroblasts and subepithelial fibrosis in bronchial asthma. Am J Respir Cell Mol Biol. 1990;3(5):507–11.
20. Boulet LP, Laviolette M, Tucotte H, Cartier A, Dugas M, Malo JL, et al. Bronchial subepithelial fibrosis correlates with airway responsiveness to methacholine. Chest. 1997;112:45–52.
21. Minshall EM, Leung DY, Martin RJ, Song YL, Cameron L, Ernst P, et al. Eosinophil-associated TGF-beta1 mRNA expression and airways fibrosis in bronchial asthma. Am J Respir Cell Mol Biol. 1997;17(3):326–33.
22. Haraguchi M, Shimura S, Shirato K. Morphometric analysis of bronchial cartilage in chronic obstructive pulmonary disease and bronchial asthma. Am J Respir Crit Care Med. 1999;159(3):1005–13.
23. Palmans E, Kips JC, Pauwels RA. Prolonged allergen exposure induces structural airway changes in sensitized rats. Am J Respir Crit Care Med. 2000;161(2 Pt 1):627–35.
24. Leigh R, Ellis R, Wattie J, Southam DS, De Hoogh M, Gauldie J, et al. Dysfunction and remodeling of the mouse airway persist after resolution of acute allergen-induced airway inflammation. Am J Respir Cell Mol Biol. 2002;27(5):526–35.
25. Larsen K, Tufvesson E, Malmstrom J, Morgelin M, Wildt M, Andersson A, et al. Presence of activated mobile fibroblasts in bronchoalveolar lavage from patients with mild asthma. Am J Respir Crit Care Med. 2004;170(10):1049–56.
26. Bucala R, Spiegel LA, Chesney J, Hogan M, Cerami A. Circulating fibrocytes define a new leukocyte subpopulation that mediates tissue repair. Mol Med. 1994;1(1):71–81.
27. Abe R, Donnelly SC, Peng T, Bucala R, Metz CN. Peripheral blood fibrocytes: differentiation pathway and migration to wound sites. J Immunol. 2001;166(12):7556–62.
28. Quan TE, Cowper S, Wu SP, Bockenstedt LK, Bucala R. Circulating fibrocytes: collagen-secreting cells of the peripheral blood. Int J Biochem Cell Biol. 2004;36(4):598–606.
29. Chesney J, Bacher M, Bender A, Bucala R. The peripheral blood fibrocyte is a potent antigen-presenting cell capable of priming naive T cells in situ. Proc Natl Acad Sci USA. 1997;94(12):6307–12.
30. Nihlberg K, Larsen K, Hultgardh-Nilsson A, Malmstrom A, Bjermer L, Westergren-Thorsson G. Tissue fibrocytes in patients with mild asthma: a possible link to thickness of reticular basement membrane? Respir Res. 2006;7:50.
31. Chetta A, Zanini A, Foresi A, D'Ippolito R, Tipa A, Castagnaro A, et al. Vascular endothelial growth factor up-regulation and bronchial wall remodelling in asthma. Clin Exp Allergy. 2005;35(11):1437–42.
32. Meerschaert JA, Kelly EA, Mosher DF, Busse WW, Jarjour NN. Segmental antigen challenge increases fibronectin in bronchoalveolar lavage fluid. Am J Respir Crit Care Med. 1999;159(2):619–25.
33. Redington AE, Roche WR, Madden J, Frew AJ, Djukanovic R, Holgate ST, et al. Basic fibroblast growth factor in asthma: measurement in bronchoalveolar lavage fluid basally and following allergen challenge. J Allergy Clin Immunol. 2001;107(2):384–7.
34. Kelly EAB, Busse WW, Jarjour NN. Increased matrix metalloproteinase (MMP)-9 in the airway following allergen challenge. Am J Respir Crit Care Med. 2000;162:1157–61.
35. Kim YH, Kwon HJ, Kim DS. Matrix metalloproteinase 9 (MMP-9)-dependent processing of betaig-h3 protein regulates cell migration, invasion, and adhesion. J Biol Chem. 2012;287(46):38957–69.

36. Engsig MT, Chen QJ, Vu TH, Pedersen AC, Therkidsen B, Lund LR, et al. Matrix metalloproteinase 9 and vascular endothelial growth factor are essential for osteoclast recruitment into developing long bones. J Cell Biol. 2000;151(4):879–89.
37. Ebrahem Q, Chaurasia SS, Vasanji A, Qi JH, Klenotic PA, Cutler A, et al. Cross-talk between vascular endothelial growth factor and matrix metalloproteinases in the induction of neovascularization in vivo. Am J Pathol. 2010;176(1):496–503.
38. Kelly EA, Jarjour NN. Review: Role of MMPs in asthma. Cur Opin Pulm Med. 2003;9:28–33.
39. Tan TK, Zheng G, Hsu TT, Wang Y, Lee VW, Tian X, et al. Macrophage matrix metalloproteinase-9 mediates epithelial-mesenchymal transition in vitro in murine renal tubular cells. Am J Pathol. 2010;176(3):1256–70.
40. Lin CY, Tsai PH, Kandaswami CC, Lee PP, Huang CJ, Hwang JJ, et al. Matrix metalloproteinase-9 cooperates with transcription factor Snail to induce epithelial-mesenchymal transition. Cancer Sci. 2011;102(4):815–27.
41. Huber HL, Koessler KK. The pathology of bronchial asthma. Arch Intern Med. 1922;30(6):689–760.
42. Bento AM, Hershenson MB. Airway remodeling: potential contributions of subepithelial fibrosis and airway smooth muscle hypertrophy/hyperplasia to airway narrowing in asthma. Allergy Asthma Proc. 1998;19(6):353–8.
43. Pepe C, Foley S, Shannon J, Lemiere C, Olivenstein R, Ernst P, et al. Differences in airway remodeling between subjects with severe and moderate asthma. J Allergy Clin Immunol. 2005;116(3):544–9.
44. Heard BE, Hossain S. Hyperplasia of bronchial muscle in asthma. J Pathol. 1973;110(4):319.
45. Ebina M, Takahashi T, Chiba T, Motomiya M. Cellular hypertrophy and hyperplasia of airway smooth muscles underlying bronchial asthma. A 3-D morphometric study. Am Rev Respir Dis. 1993;148(3):720–6.
46. Ebina M, Yaegashi H, Chiba R, Takahashi T, Motomiya M, Tanemura M. Hyperreactive site in the airway tree of asthmatic patients revealed by thickening of bronchial muscles. Am Rev Respir Dis. 1990;141:1327–32.
47. Thomson RJ, Bramley AM, Schellenberg RR. Airway muscle stereology: implications for increased shortening in asthma. Am J Respir Crit Care Med. 1996;154(3 Pt 1):749–57.
48. James AL, Hogg JC. The mechanics of airway narrowing in asthma. Am Rev Respir Dis. 1989;139:242–6.
49. Takizawa T, Thurlbeck WM. Muscle and mucous gland size in the major bronchi of patients with chronic bronchitis, asthma, and asthmatic bronchitis. Am Rev Respir Dis. 1971;104(3):331–6.
50. Kuwano K, Bosken CH, Pare PD, Bai TR, Wiggs BR, Hogg JC. Small airways dimensions in asthma and in chronic obstructive pulmonary disease. Am Rev Respir Dis. 1993;148(5):1220–5.
51. Johnson PR, Roth M, Tamm M, Hughes M, Ge Q, King G, et al. Airway smooth muscle cell proliferation is increased in asthma. Am J Respir Crit Care Med. 2001;164(3):474–7.
52. Tsurikisawa N, Oshikata C, Tsuburai T, Saito H, Sekiya K, Tanimoto H, et al. Bronchial hyperresponsiveness to histamine correlates with airway remodelling in adults with asthma. Respir Med. 2010;104(9):1271–7.
53. Martin JG, Duguet A, Eidelman DH. The contribution of airway smooth muscle to airway narrowing and airway hyperresponsiveness in disease. Eur Respir J. 2000;16(2):349–54.
54. Wiggs BR, Bosken C, Pare PD, James A, Hogg JC. A model of airway narrowing in asthma and in chronic obstructive pulmonary disease. Am Rev Respir Dis. 1992;145(6):1251–8.
55. Johnson PR, Black JL, Carlin S, Ge Q, Underwood PA. The production of extracellular matrix proteins by human passively sensitized airway smooth-muscle cells in culture: the effect of beclomethasone. Am J Respir Crit Care Med. 2000;162(6):2145–51.
56. Coutts A, Chen G, Stephens N, Hirst S, Douglas D, Eichholtz T, et al. Release of biologically active TGF-beta from airway smooth muscle cells induces autocrine synthesis of collagen. Am J Physiol Lung Cell Mol Physiol. 2001;280(5):L999–1008.

57. Chen G, Khalil N. In vitro wounding of airway smooth muscle cell monolayers increases expression of TGF-beta receptors. Respir Physiol Neurobiol. 2002;132(3):341–6.
58. Black PN, Young PG, Skinner SJ. Response of airway smooth muscle cells to TGF-beta 1: effects on growth and synthesis of glycosaminoglycans. Am J Phys. 1996;271(6 Pt 1):L910–7.
59. Fong CY, Pang L, Holland E, Knox AJ. TGF-beta1 stimulates IL-8 release, COX-2 expression, and PGE(2) release in human airway smooth muscle cells. Am J Physiol Lung Cell Mol Physiol. 2000;279(1):L201–7.
60. Tschumperlin DJ, Shively JD, Kikuchi T, Drazen JM. Mechanical stress triggers selective release of fibrotic mediators from bronchial epithelium. Am J Respir Cell Mol Biol. 2003;28(2):142–9.
61. Fagan KA, McMurtry IF, Rodman DM. Role of endothelin-1 in lung disease. Respir Res. 2001;2(2):90–101.
62. McWhinnie R, Pechkovsky DV, Zhou D, Lane D, Halayko AJ, Knight DA, et al. Endothelin-1 induces hypertrophy and inhibits apoptosis in human airway smooth muscle cells. Am J Physiol Lung Cell Mol Physiol. 2007;292(1):L278–86.
63. Dunnill MS. The pathology of asthma, with special reference to changes in the bronchial mucosa. J Clin Pathol. 1960;13:27–33.
64. Dunnill MS, Massarella GR, Anderson JA. A comparison of the quantitative anatomy of the bronchi in normal subjects, in status asthmaticus, in chronic bronchitis, and in emphysema. Thorax. 1969;24:176–9.
65. Bateman JR, Pavia D, Sheahan NF, Agnew JE, Clarke SW. Impaired tracheobronchial clearance in patients with mild stable asthma. Thorax. 1983;38(6):463–7.
66. Aikawa T, Shimura S, Sasaki H, Ebina M, Takishima T. Marked goblet cell hyperplasia with mucus accumulation in the airways of patients who died of severe acute asthma attack. Chest. 1992;101(4):916–21.
67. Dunican EM, Elicker BM, Gierada DS, Nagle SK, Schiebler ML, Newell JD, et al. Mucus plugs in patients with asthma linked to eosinophilia and airflow obstruction. J Clin Invest. 2018;128(3):997–1009.
68. Dickinson JD, Alevy Y, Malvin NP, Patel KK, Gunsten SP, Holtzman MJ, et al. IL13 activates autophagy to regulate secretion in airway epithelial cells. Autophagy. 2016;12(2):397–409.
69. Shim JJ, Dabbagh K, Ueki IF, Dao-Pick T, Burgel PR, Takeyama K, et al. IL-13 induces mucin production by stimulating epidermal growth factor receptors and by activating neutrophils. Am J Physiol Lung Cell Mol Physiol. 2001;280(1):L134–40.
70. Burgel PR, Lazarus SC, Tam DC, Ueki IF, Atabai K, Birch M, et al. Human eosinophils induce mucin production in airway epithelial cells via epidermal growth factor receptor activation. J Immunol. 2001;167(10):5948–54.
71. Xu Q, Chen LX, Ran DH, Xie WY, Li Q, Zhou XD. Bombesin receptor-activated protein regulates neutrophil elastase-induced mucin5AC hypersecretion in human bronchial epithelial cells. Exp Cell Res. 2017;357(2):145–54.
72. Panchabhai TS, Mukhopadhyay S, Sehgal S, Bandyopadhyay D, Erzurum SC, Mehta AC. Plugs of the air passages: a clinicopathologic review. Chest. 2016;150(5):1141–57.
73. Wagers SS, Norton RJ, Rinaldi LM, Bates JHT, Sobel BE, Irvin CG. Extravascular fibrin, plasminogen activator, plasminogen activator inhibitors, and airway hyperresponsiveness. J Clin Invest. 2004;114(1):104–11.
74. Pohunek P, Warner JO, Turzikova J, Kudrmann J, Roche WR. Markers of eosinophilic inflammation and tissue re-modelling in children before clinically diagnosed bronchial asthma. Pediatr Allergy Immunol. 2005;16(1):43–51.
75. Phipps S, Benyahia F, Ou TT, Barkans J, Robinson DS, Kay AB. Acute allergen-induced airway remodeling in atopic asthma. Am J Respir Cell Mol Biol. 2004;31(6):626–32.
76. Lambrecht BN, Carro-Muino I, Vermaelen K, Pauwels RA. Allergen-induced changes in bone-marrow progenitor and airway dendritic cells in sensitized rats. Am J Respir Cell Mol Biol. 1999;20(6):1165–74.

77. Bousquet J, Jeffery PK, Busse WW, Johnson M, Vignola AM. Asthma. From bronchoconstriction to airways inflammation and remodeling. Am J Respir Crit Care Med. 2000;161(5):1720–45.
78. Zhang SL, Smartt H, Holgate ST, Roche WR. Growth factors secreted by bronchial epithelial cells control myofibroblast proliferation: an in vitro co-culture model of airway remodeling in asthma. Lab Investig. 1999;79(4):395–405.
79. Korfhagen TR, Swantz RJ, Wert SE, McCarty JM, Kerlakian CB, Glasser SW, et al. Respiratory epithelial cell expression of human transforming growth factor-alpha induces lung fibrosis in transgenic mice. J Clin Invest. 1994;93(4):1691–9.
80. Holgate ST, Lackie PM, Davies DE, Roche WR, Walls AF. The bronchial epithelium as a key regulator of airway inflammation and remodelling in asthma. Clin Exp Allergy. 1999;29(Suppl 2):90–5.
81. Magnan A, Frachon I, Rain B, Peuchmaur M, Monti G, Lenot B, et al. Transforming growth factor beta in normal human lung: preferential location in bronchial epithelial cells. Thorax. 1994;49(8):789–92.
82. Morishima Y, Nomura A, Uchida Y, Noguchi Y, Sakamoto T, Ishii Y, et al. Triggering the induction of myofibroblast and fibrogenesis by airway epithelial shedding. Am J Respir Cell Mol Biol. 2001;24(1):1–11.
83. Zhang HY, Phan SH. Inhibition of myofibroblast apoptosis by transforming growth factor beta(1). Am J Respir Cell Mol Biol. 1999;21(6):658–65.
84. Xu J, Benyon RC, Leir SH, Zhang S, Holgate ST, Lackie PM. Matrix metalloproteinase-2 from bronchial epithelial cells induces the proliferation of subepithelial fibroblasts. Clin Exp Allergy. 2002;32(6):881–8.
85. Bonser LR, Erle DJ. Airway mucus and asthma: the role of MUC5AC and MUC5B. J Clin Med. 2017;6(12):112.
86. Vliagoftis H, Befus AD, Hollenberg MD, Moqbel R. Airway epithelial cells release eosinophil survival-promoting factors (GM-CSF) after stimulation of proteinase-activated receptor 2. J Allergy Clin Immunol. 2001;107(4):679–85.
87. Vliagoftis H, Schwingshackl A, Milne CD, Duszyk M, Hollenberg MD, Wallace JL, et al. Proteinase-activated receptor-2-mediated matrix metalloproteinase-9 release from airway epithelial cells. J Allergy Clin Immunol. 2000;106(3):537–45.
88. Davies RJ, Wang JH, Trigg CJ, Devalia JL. Expression of granulocyte/macrophage-colony-stimulating factor, interleukin-8 and RANTES in the bronchial epithelium of mild asthmatics is down-regulated by inhaled beclomethasone dipropionate. Int Arch Allergy Immunol. 1995;107(1–3):428–9.
89. Sousa AR, Poston RN, Lane SJ, Nakhosteen JA, Lee TH. Detection of GM-CSF in asthmatic bronchial epithelium and decrease by inhaled corticosteroids. Am Rev Respir Dis. 1993;147(6 Pt 1):1557–61.
90. Kolaczkowska E, Kubes P. Neutrophil recruitment and function in health and inflammation. Nat Rev Immunol. 2013;13(3):159–75.
91. Kaur M, Singh D. Neutrophil chemotaxis caused by chronic obstructive pulmonary disease alveolar macrophages: the role of CXCL8 and the receptors CXCR1/CXCR2. J Pharmacol Exp Ther. 2013;347(1):173–80.
92. Esnault S, Malter JS. Minute quantities of granulocyte-macrophage colony-stimulating factor prolong eosinophil survival. J Interf Cytokine Res. 2001;21(2):117–24.
93. Wang K, Liu CT, Wu YH, Feng YL, Bai HL. Budesonide/formoterol decreases expression of vascular endothelial growth factor (VEGF) and VEGF receptor 1 within airway remodelling in asthma. Adv Ther. 2008;25(4):342–54.
94. Bartemes KR, Kita H. Roles of innate lymphoid cells (ILCs) in allergic diseases: The 10-year anniversary for ILC2s. J Allergy Clin Immunol. 2021;147(5):1531–47.
95. Pulendran B, Artis D. New paradigms in type 2 immunity. Science. 2012;337(6093):431–5.
96. Halim TY, Steer CA, Matha L, Gold MJ, Martinez-Gonzalez I, McNagny KM, et al. Group 2 innate lymphoid cells are critical for the initiation of adaptive T helper 2 cell-mediated allergic lung inflammation. Immunity. 2014;40(3):425–35.

97. Mirchandani AS, Besnard AG, Yip E, Scott C, Bain CC, Cerovic V, et al. Type 2 innate lymphoid cells drive CD4+ Th2 cell responses. J Immunol. 2014;192(5):2442–8.
98. Oliphant CJ, Hwang YY, Walker JA, Salimi M, Wong SH, Brewer JM, et al. MHCII-mediated dialog between group 2 innate lymphoid cells and CD4(+) T cells potentiates type 2 immunity and promotes parasitic helminth expulsion. Immunity. 2014;41(2):283–95.
99. Siracusa MC, Saenz SA, Wojno ED, Kim BS, Osborne LC, Ziegler CG, et al. Thymic stromal lymphopoietin-mediated extramedullary hematopoiesis promotes allergic inflammation. Immunity. 2013;39(6):1158–70.
100. Yao W, Zhang Y, Jabeen R, Nguyen ET, Wilkes DS, Tepper RS, et al. Interleukin-9 is required for allergic airway inflammation mediated by the cytokine TSLP. Immunity. 2013;38(2):360–72.
101. Chu DK, Llop-Guevara A, Walker TD, Flader K, Goncharova S, Boudreau JE, et al. IL-33, but not thymic stromal lymphopoietin or IL-25, is central to mite and peanut allergic sensitization. J Allergy Clin Immunol. 2013;131(1):187–200.e1–8.
102. Sousa AR, Lane SJ, Nakhosteen JA, Yoshimura T, Lee TH, Poston RN. Increased expression of the monocyte chemoattractant protein-1 in bronchial tissue from asthmatic subjects. Am J Respir Cell Mol Biol. 1994;10(2):142–7.
103. Carr MW, Roth SJ, Luther E, Rose SS, Springer TA. Monocyte chemoattractant protein 1 acts as a T-lymphocyte chemoattractant. Proc Natl Acad Sci U S A. 1994;91(9):3652–6.
104. Gu L, Tseng S, Horner RM, Tam C, Loda M, Rollins BJ. Control of TH2 polarization by the chemokine monocyte chemoattractant protein-1. Nature. 2000;404(6776):407–11.
105. Conti P, Boucher W, Letourneau R, Feliciani C, Reale M, Barbacane RC, et al. Monocyte chemotactic protein-1 provokes mast cell aggregation and [3H]5HT release. Immunology. 1995;86(3):434–40.
106. Lee JJ, Jacobsen EA, McGarry MP, Schleimer RP, Lee NA. Eosinophils in health and disease: the LIAR hypothesis. Clin Exp Allergy. 2010;40(4):563–75.
107. Han Z, Junxu, Zhong N. Expression of matrix metalloproteinases MMP-9 within the airways in asthma. Respir Med. 2003;97(5):563–7.
108. Bousquet J, Chanez P, Lacoste JY, Barneon G, Ghavanian N, Enander I, et al. Eosinophilic inflammation in asthma. N Engl J Med. 1990;323(15):1033–9.
109. Walker C, Kaegi MK, Braun P, Blaser K. Activated T cells and eosinophilia in bronchoalveolar lavages from subjects with asthma correlated with disease severity. J Allergy Clin Immunol. 1991;88:935–42.
110. Azzawi M, Johnston PW, Majumdar S, Kay AB, Jeffery PK. T lymphocytes and activated eosinophils in airway mucosa in fatal asthma and cystic fibrosis. Am Rev Respir Dis. 1992;145:1477–82.
111. Filley WV, Holley KE, Kephat GM, Gleich GJ. Identification by immunofluorescence of eosinophil granule major basic protein in lung tissues of patients with bronchial asthma. Lancet. 1982;II:11–5.
112. Lu Y, Huang Y, Li J, Huang J, Zhang L, Feng J, et al. Eosinophil extracellular traps drive asthma progression through neuro-immune signals. Nat Cell Biol. 2021;23(10):1060–72.
113. Mould AW, Ramsay AJ, Matthaei KI, Young IG, Rothenberg ME, Foster PS. The effect of IL-5 and eotaxin expression in the lung on eosinophil trafficking and degranulation and the induction of bronchial hyperreactivity. J Immunol. 2000;164(4):2142–50.
114. Nakajima H, Iwamoto I, Tomoe S, Matsumura R, Tomioka H, Takatsu K, et al. CD4+ T lymphocytes and interleukin-5 mediate antigen-induced eosinophil infiltration into the mouse trachea. Am Rev Respir Dis. 1992;156:374–7.
115. Mattes J, Yang M, Mahalingam S, Kuehr J, Webb DC, Simson L, et al. Intrinsic defect in T cell production of interleukin (IL)-13 in the absence of both IL-5 and eotaxin precludes the development of eosinophilia and airways hyperreactivity in experimental asthma. J Exp Med. 2002;195(11):1433–44.
116. Liu LY, Bates ME, Jarjour NN, Busse WW, Bertics PJ, Kelly EA. Generation of Th1 and Th2 chemokines by human eosinophils: evidence for a critical role of TNF-alpha. J Immunol. 2007;179(7):4840–8.

117. Esnault S, Kelly EA, Nettenstrom LM, Cook EB, Seroogy CM, Jarjour NN. Human eosinophils release IL-1β and increase expression of IL-17A in activated CD4(+) T lymphocytes. Clin Exp Allergy. 2012;42(12):1756–64.
118. Spencer LA, Szela CT, Perez SA, Kirchhoffer CL, Neves JS, Radke AL, et al. Human eosinophils constitutively express multiple Th1, Th2, and immunoregulatory cytokines that are secreted rapidly and differentially. J Leukoc Biol. 2009;85(1):117–23.
119. Moqbel R, Ying S, Barkans J, Newman TT, Kimmitt P, Wakelin M, et al. Identification of messenger RNA for IL-4 in human eosinophils with granule localization and release of the translated product. J Immunol. 1995;155:4939–47.
120. Bernau K, Leet JP, Floerke H, Bruhn EM, Noll AL, McDermott IS, et al. Interleukin-1alpha is a critical mediator of the response of human bronchial fibroblasts to eosinophilic inflammation. Cells. 2021;10(3):528.
121. Gomes I, Mathur SK, Espenshade BM, Mori Y, Varga J, Ackerman SJ. Eosinophil-fibroblast interactions induce fibroblast IL-6 secretion and extracellular matrix gene expression: implications in fibrogenesis. J Allergy Clin Immunol. 2005;116(4):796–804.
122. Esnault S, Bernau K, Torr EE, Bochkov YA, Jarjour NN, Sandbo N. RNA-sequencing analysis of lung primary fibroblast response to eosinophil-degranulation products predicts downstream effects on inflammation, tissue remodeling and lipid metabolism. Respir Res. 2017;18(1):188.
123. Bernau K, Leet JP, Esnault S, Noll AL, Evans MD, Jarjour NN, et al. Eosinophil-degranulation products drive a proinflammatory fibroblast phenotype. J Allergy Clin Immunol. 2018;142(4):1360–3. e3
124. De S, Zelazny ET, Souhrada JF, Souhrada M. IL-1 beta and IL-6 induce hyperplasia and hypertrophy of cultured guinea pig airway smooth muscle cells. J Appl Physiol (1985). 1995;78(4):1555–63.
125. Ohno I, Nitta Y, Yamauchi K, Hoshi H, Honma M, Woolley K, et al. Transforming growth factor β1 (TGFβ1) gene expression by eosinophils in asthmatic airway inflammation. Am J Respir Crit Care Med. 1996;15:404–9.
126. Zagai U, Dadfar E, Lundahl J, Venge P, Skold CM. Eosinophil cationic protein stimulates TGF-beta1 release by human lung fibroblasts in vitro. Inflammation. 2007;30(5):153–60.
127. Hill DJ, Strain AJ, Elstow SF, Swenne I, Milner RD. Bi-functional action of transforming growth factor-beta on DNA synthesis in early passage human fetal fibroblasts. J Cell Physiol. 1986;128(2):322–8.
128. Ignotz RA, Massague J. Transforming growth factor-beta stimulates the expression of fibronectin and collagen and their incorporation into the extracellular matrix. J Biol Chem. 1986;261(9):4337–45.
129. Dubaybo BA, Thet LA. Effect of transforming growth factor beta on synthesis of glycosaminoglycans by human lung fibroblasts. Exp Lung Res. 1990;16(5):389–403.
130. McGowan SE, Jackson SK, Olson PJ, Parekh T, Gold LI. Exogenous and endogenous transforming growth factors-beta influence elastin gene expression in cultured lung fibroblasts. Am J Respir Cell Mol Biol. 1997;17(1):25–35.
131. Goldsmith AM, Bentley JK, Zhou L, Jia Y, Bitar KN, Fingar DC, et al. Transforming growth factor-beta induces airway smooth muscle hypertrophy. Am J Respir Cell Mol Biol. 2006;34(2):247–54.
132. Fisher SA, Ikebe M, Brozovich F. Endothelin-1 alters the contractile phenotype of cultured embryonic smooth muscle cells. Circ Res. 1997;80(6):885–93.
133. Grainger DJ, Metcalfe JC, Grace AA, Mosedale DE. Transforming growth factor-beta dynamically regulates vascular smooth muscle differentiation in vivo. J Cell Sci. 1998;111(Pt 19):2977–88.
134. Sandbo N, Taurin S, Yau DM, Kregel S, Mitchell R, Dulin NO. Downregulation of smooth muscle alpha-actin expression by bacterial lipopolysaccharide. Cardiovasc Res. 2007;74(2):262–9.

135. McMillan SJ, Xanthou G, Lloyd CM. Manipulation of allergen-induced airway remodeling by treatment with anti-TGF-beta antibody: effect on the Smad signaling pathway. J Immunol. 2005;174(9):5774–80.
136. Redington AE, Madden J, Frew AJ, Djukanovic R, Roche WR, Holgate ST, et al. Transforming growth factor-β1 in asthma. Measurement in bronchoalveolar lavage fluid. Am J Respir Crit Care Med. 1997;156:642–7.
137. Hoshino M, Nakamura Y, Sim JJ. Expression of growth factors and remodelling of the airway wall in bronchial asthma. Thorax. 1998;53:21–7.
138. Zhang M, Zhang Z, Pan HY, Wang DX, Deng ZT, Ye XL. TGF-beta1 induces human bronchial epithelial cell-to-mesenchymal transition in vitro. Lung. 2009;187(3):187–94.
139. Gong JL, McCarthy KM, Rogers RA, Schneeberger EE. Interstitial lung macrophages interact with dendritic cells to present antigenic peptides derived from particulate antigens to T cells. Immunology. 1994;81(3):343–51.
140. Kawasaki T, Ikegawa M, Kawai T. Antigen presentation in the lung. Front Immunol. 2022;13:860915.
141. Murray PJ, Wynn TA. Obstacles and opportunities for understanding macrophage polarization. J Leukoc Biol. 2011;89(4):557–63.
142. Barron L, Wynn TA. Fibrosis is regulated by Th2 and Th17 responses and by dynamic interactions between fibroblasts and macrophages. Am J Physiol Gastrointest Liver Physiol. 2011;300(5):G723–8.
143. Shimokado K, Raines EW, Madtes DK, Barrett TB, Benditt EP, Ross R. A significant part of macrophage-derived growth factor consists of at least two forms of PDGF. Cell. 1985;43(1):277–86.
144. Sunderkotter C, Steinbrink K, Goebeler M, Bhardwaj R, Sorg C. Macrophages and angiogenesis. J Leukoc Biol. 1994;55(3):410–22.
145. Ferrari-Lacraz S, Nicod LP, Chicheportiche R, Welgus HG, Dayer JM. Human lung tissue macrophages, but not alveolar macrophages, express matrix metalloproteinases after direct contact with activated T lymphocytes. Am J Respir Cell Mol Biol. 2001;24(4):442–51.
146. Wynn TA. Cellular and molecular mechanisms of fibrosis. J Pathol. 2008;214(2):199–210.
147. Sutherland TE, Maizels RM, Allen JE. Chitinases and chitinase-like proteins: potential therapeutic targets for the treatment of T-helper type 2 allergies. Clin Exp Allergy. 2009;39(7):943–55.
148. Nagarkar DR, Bowman ER, Schneider D, Wang Q, Shim J, Zhao Y, et al. Rhinovirus infection of allergen-sensitized and -challenged mice induces eotaxin release from functionally polarized macrophages. J Immunol. 2010;185(4):2525–35.
149. Mantovani A, Sica A, Sozzani S, Allavena P, Vecchi A, Locati M. The chemokine system in diverse forms of macrophage activation and polarization. Trends Immunol. 2004;25(12):677–86.
150. Kim EY, Battaile JT, Patel AC, You Y, Agapov E, Grayson MH, et al. Persistent activation of an innate immune response translates respiratory viral infection into chronic lung disease. Nat Med. 2008;14(6):633–40.
151. Borthwick LA, Barron L, Hart KM, Vannella KM, Thompson RW, Oland S, et al. Macrophages are critical to the maintenance of IL-13-dependent lung inflammation and fibrosis. Mucosal Immunol. 2016;9(1):38–55.
152. Liu H, Rohowsky-Kochan C. Regulation of IL-17 in human CCR6+ effector memory T cells. J Immunol. 2008;180(12):7948–57.
153. Burgler S, Ouaked N, Bassin C, Basinski TM, Mantel PY, Siegmund K, et al. Differentiation and functional analysis of human T(H)17 cells. J Allergy Clin Immunol. 2009;123(3):588–95. 95
154. Wilson MS, Madala SK, Ramalingam TR, Gochuico BR, Rosas IO, Cheever AW, et al. Bleomycin and IL-1beta-mediated pulmonary fibrosis is IL-17A dependent. J Exp Med. 2010;207(3):535–52.
155. Wight TN, Frevert CW, Debley JS, Reeves SR, Parks WC, Ziegler SF. Interplay of extracellular matrix and leukocytes in lung inflammation. Cell Immunol. 2017;312:1–14.

156. Lukic A, Larssen P, Fauland A, Samuelsson B, Wheelock CE, Gabrielsson S, et al. GM-CSF- and M-CSF-primed macrophages present similar resolving but distinct inflammatory lipid mediator signatures. FASEB J. 2017;31(10):4370–81.
157. Jakubzick CV, Randolph GJ, Henson PM. Monocyte differentiation and antigen-presenting functions. Nat Rev Immunol. 2017;17(6):349–62.
158. Eguiluz-Gracia I, Malmstrom K, Dheyauldeen SA, Lohi J, Sajantila A, Aalokken R, et al. Monocytes accumulate in the airways of children with fatal asthma. Clin Exp Allergy. 2018;48(12):1631–9.
159. Chesney J, Metz C, Stavitsky AB, Bacher M, Bucala R. Regulated production of type I collagen and inflammatory cytokines by peripheral blood fibrocytes. J Immunol. 1998;160(1):419–25.
160. Shipe R, Burdick MD, Strieter BA, Liu L, Shim YM, Sung SS, et al. Number, activation, and differentiation of circulating fibrocytes correlate with asthma severity. J Allergy Clin Immunol. 2016;137(3):750–7.e3.
161. Plantinga M, Guilliams M, Vanheerswynghels M, Deswarte K, Branco-Madeira F, Toussaint W, et al. Conventional and monocyte-derived CD11b(+) dendritic cells initiate and maintain T helper 2 cell-mediated immunity to house dust mite allergen. Immunity. 2013;38(2):322–35.
162. Ko HJ, Brady JL, Ryg-Cornejo V, Hansen DS, Vremec D, Shortman K, et al. GM-CSF-responsive monocyte-derived dendritic cells are pivotal in Th17 pathogenesis. J Immunol. 2014;192(5):2202–9.
163. Schleich FN, Manise M, Sele J, Henket M, Seidel L, Louis R. Distribution of sputum cellular phenotype in a large asthma cohort: predicting factors for eosinophilic vs neutrophilic inflammation. BMC Pulm Med. 2013;13:11.
164. Belda J, Leigh R, Parameswaran K, O'Byrne PM, Sears MR, Hargreave FE. Induced sputum cell counts in healthy adults. Am J Respir Crit Care Med. 2000;161(2 Pt 1):475–8.
165. Hastie AT, Moore WC, Meyers DA, Vestal PL, Li H, Peters SP, et al. Analyses of asthma severity phenotypes and inflammatory proteins in subjects stratified by sputum granulocytes. J Allergy Clin Immunol. 2010;125(5):1028–36. e13
166. Moore WC, Hastie AT, Li X, Li H, Busse WW, Jarjour NN, et al. Sputum neutrophil counts are associated with more severe asthma phenotypes using cluster analysis. J Allergy Clin Immunol. 2014;133(6):1557–63.e5.
167. Esnault S, Khosravi M, Kelly EA, Liu LY, Bochkov YA, Tattersall MC, et al. Increased IL-6 and Potential IL-6 trans-signalling in the airways after an allergen challenge. Clin Exp Allergy. 2021;51(4):564–73.
168. Fischer BM, Voynow JA. Neutrophil elastase induces MUC5AC gene expression in airway epithelium via a pathway involving reactive oxygen species. Am J Respir Cell Mol Biol. 2002;26(4):447–52.
169. Stockley RA. The role of proteinases in the pathogenesis of chronic bronchitis. Am J Respir Crit Care Med. 1994;150(6 Pt 2):S109–13.
170. Voynow JA, Fischer BM, Malarkey DE, Burch LH, Wong T, Longphre M, et al. Neutrophil elastase induces mucus cell metaplasia in mouse lung. Am J Physiol Lung Cell Mol Physiol. 2004;287(6):L1293–302.
171. Pothoven KL, Norton JE, Suh LA, Carter RG, Harris KE, Biyasheva A, et al. Neutrophils are a major source of the epithelial barrier disrupting cytokine oncostatin M in patients with mucosal airways disease. J Allergy Clin Immunol. 2017;139(6):1966–78.e9.
172. Chetty A, Davis P, Infeld M. Effect of elastase on the directional migration of lung fibroblasts within a three-dimensional collagen matrix. Exp Lung Res. 1995;21(6):889–99.
173. Pham DL, Ban GY, Kim SH, Shin YS, Ye YM, Chwae YJ, et al. Neutrophil autophagy and extracellular DNA traps contribute to airway inflammation in severe asthma. Clin Exp Allergy. 2017;47(1):57–70.
174. Saffarzadeh M, Juenemann C, Queisser MA, Lochnit G, Barreto G, Galuska SP, et al. Neutrophil extracellular traps directly induce epithelial and endothelial cell death: a predominant role of histones. PLoS One. 2012;7(2):e32366.

175. Chrysanthopoulou A, Mitroulis I, Apostolidou E, Arelaki S, Mikroulis D, Konstantinidis T, et al. Neutrophil extracellular traps promote differentiation and function of fibroblasts. J Pathol. 2014;233(3):294–307.
176. Chen X, Li Y, Qin L, He R, Hu C. Neutrophil Extracellular Trapping Network Promotes the Pathogenesis of Neutrophil-associated Asthma through Macrophages. Immunol Investig. 2021;50(5):544–61.
177. Vargas A, Roux-Dalvai F, Droit A, Lavoie JP. Neutrophil-derived exosomes: a new mechanism contributing to airway smooth muscle remodeling. Am J Respir Cell Mol Biol. 2016;55(3):450–61.
178. Ventura I, Vega A, Chacon P, Chamorro C, Aroca R, Gomez E, et al. Neutrophils from allergic asthmatic patients produce and release metalloproteinase-9 upon direct exposure to allergens. Allergy. 2014;69(7):898–905.
179. Suga M, Iyonaga K, Okamoto T, Gushima Y, Miyakawa H, Akaike T, et al. Characteristic elevation of matrix metalloproteinase activity in idiopathic interstitial pneumonias. Am J Respir Crit Care Med. 2000;162(5):1949–56.
180. Vermaelen KY, Cataldo D, Tournoy K, Maes T, Dhulst A, Louis R, et al. Matrix metalloproteinase-9-mediated dendritic cell recruitment into the airways is a critical step in a mouse model of asthma. J Immunol. 2003;171(2):1016–22.
181. Esnault S, Kelly EA, Johnson SH, DeLain LP, Haedt MJ, Noll AL, et al. Matrix metalloproteinase-9-dependent release of IL-1beta by human eosinophils. Mediat Inflamm. 2019;2019:7479107.
182. Mautino G, Capony F, Bousquet J, Vignola AM. Balance in asthma between matrix metalloproteinases and their inhibitors. J Allergy Clin Immunol. 1999;104(3 Pt 1):530–3.
183. Kelly EA, Jarjour NN. Role of matrix metalloproteinases in asthma. Curr Opin Pulm Med. 2003;9(1):28–33.
184. Madri JA, Graesser D. Cell migration in the immune system: the evolving inter-related roles of adhesion molecules and proteinases. Dev Immunol. 2000;7(2–4):103–16.
185. Ardi VC, Van den Steen PE, Opdenakker G, Schweighofer B, Deryugina EI, Quigley JP. Neutrophil MMP-9 proenzyme, unencumbered by TIMP-1, undergoes efficient activation in vivo and catalytically induces angiogenesis via a basic fibroblast growth factor (FGF-2)/FGFR-2 pathway. J Biol Chem. 2009;284(38):25854–66.
186. Perng DW, Chang KT, Su KC, Wu YC, Chen CS, Hsu WH, et al. Matrix metalloprotease-9 induces transforming growth factor-beta(1) production in airway epithelium via activation of epidermal growth factor receptors. Life Sci. 2011;89(5–6):204–12.
187. Yu Q, Stamenkovic I. Cell surface-localized matrix metalloproteinase-9 proteolytically activates TGF-beta and promotes tumor invasion and angiogenesis. Genes Dev. 2000;14(2):163–76.
188. Manes S, Llorente M, Lacalle RA, Gomez-Mouton C, Kremer L, Mira E, et al. The matrix metalloproteinase-9 regulates the insulin-like growth factor-triggered autocrine response in DU-145 carcinoma cells. J Biol Chem. 1999;274(11):6935–45.
189. Ellis TN, Beaman BL. Murine polymorphonuclear neutrophils produce interferon-gamma in response to pulmonary infection with Nocardia asteroides. J Leukoc Biol. 2002;72(2):373–81.
190. Yin J, Ferguson TA. Identification of an IFN-gamma-producing neutrophil early in the response to Listeria monocytogenes. J Immunol. 2009;182(11):7069–73.
191. Tecchio C, Micheletti A, Cassatella MA. Neutrophil-derived cytokines: facts beyond expression. Front Immunol. 2014;5:508.
192. Duffy D, Perrin H, Abadie V, Benhabiles N, Boissonnas A, Liard C, et al. Neutrophils transport antigen from the dermis to the bone marrow, initiating a source of memory CD8+ T cells. Immunity. 2012;37(5):917–29.
193. Tillack K, Breiden P, Martin R, Sospedra M. T lymphocyte priming by neutrophil extracellular traps links innate and adaptive immune responses. J Immunol. 2012;188(7):3150–9.
194. Puga I, Cols M, Barra CM, He B, Cassis L, Gentile M, et al. B cell-helper neutrophils stimulate the diversification and production of immunoglobulin in the marginal zone of the spleen. Nat Immunol. 2011;13(2):170–80.

195. Amin K, Ludviksdottir D, Janson C, Nettelbladt O, Bjornsson E, Roomans GM, et al. Inflammation and structural changes in the airways of patients with atopic and nonatopic asthma. BHR Group. Am J Respir Crit Care Med. 2000;162(6):2295–301.
196. Liebler JM, Qu Z, Buckner B, Powers MR, Rosenbaum JT. Fibroproliferation and mast cells in the acute respiratory distress syndrome. Thorax. 1998;53(10):823–9.
197. Chlap Z, Jedynak U, Sladek K. Mast cell: it's significance in bronchoalveolar lavage fluid cytologic diagnosis of bronchial asthma and interstitial lung disease. Pneumonol Alergol Pol. 1998;66(5–6):321–9.
198. Hunt LW, Colby TV, Weiler DA, Sur S, Butterfield JH. Immunofluorescent staining for mast cells in idiopathic pulmonary fibrosis: quantification and evidence for extracellular release of mast cell tryptase. Mayo Clin Proc. 1992;67(10):941–8.
199. Inoue Y, King TE Jr, Tinkle SS, Dockstader K, Newman LS. Human mast cell basic fibroblast growth factor in pulmonary fibrotic disorders. Am J Pathol. 1996;149(6):2037–54.
200. Lezmi G, Galmiche-Rolland L, Rioux S, Jaubert F, Tillie-Leblond I, Scheinmann P, et al. Mast cells are associated with exacerbations and eosinophilia in children with severe asthma. Eur Respir J. 2016;48(5):1320–8.
201. Keglowich LF, Borger P. The Three A's in Asthma - Airway Smooth Muscle, Airway Remodeling & Angiogenesis. Open Respir Med J. 2015;9:70–80.
202. Wang G, Baines KJ, Fu JJ, Wood LG, Simpson JL, McDonald VM, et al. Sputum mast cell subtypes relate to eosinophilia and corticosteroid response in asthma. Eur Respir J. 2016;47(4):1123–33.
203. Prussin C, Metcalfe DD. 5. IgE, mast cells, basophils, and eosinophils. J Allergy Clin Immunol. 2006;117(2 Suppl Mini-Primer):S450–6.
204. Andersson CK, Bergqvist A, Mori M, Mauad T, Bjermer L, Erjefalt JS. Mast cell-associated alveolar inflammation in patients with atopic uncontrolled asthma. J Allergy Clin Immunol. 2011;127(4):905–12.e1–7.
205. Krishnaswamy G, Kelley J, Johnson D, Youngberg G, Stone W, Huang SK, et al. The human mast cell: functions in physiology and disease. Front Biosci. 2001;6:D1109–27.
206. Kanbe N, Kurosawa M, Nagata H, Yamashita T, Kurimoto F, Miyachi Y. Production of fibrogenic cytokines by cord blood-derived cultured human mast cells. J Allergy Clin Immunol. 2000;106(1 Pt 2):S85–90.
207. Pereira PJ, Bergner A, Macedo-Ribeiro S, Huber R, Matschiner G, Fritz H, et al. Human beta-tryptase is a ring-like tetramer with active sites facing a central pore. Nature. 1998;392(6673):306–11.
208. Akers IA, Parsons M, Hill MR, Hollenberg MD, Sanjar S, Laurent GJ, et al. Mast cell tryptase stimulates human lung fibroblast proliferation via protease-activated receptor-2. Am J Physiol Lung Cell Mol Physiol. 2000;278(1):L193–201.
209. Gordon JR, Galli SJ. Promotion of mouse fibroblast collagen gene expression by mast cells stimulated via the Fc epsilon RI. Role for mast cell-derived transforming growth factor beta and tumor necrosis factor alpha. J Exp Med. 1994;180(6):2027–37.
210. Alkhouri H, Hollins F, Moir LM, Brightling CE, Armour CL, Hughes JM. Human lung mast cells modulate the functions of airway smooth muscle cells in asthma. Allergy. 2011;66(9):1231–41.
211. Virk H, Arthur G, Bradding P. Mast cells and their activation in lung disease. Transl Res J Lab Clin Med. 2016;174:60–76.
212. Heard BE, Dewar A, Corrin B. Apposition of fibroblasts to mast cells and lymphocytes in normal human lung and in cryptogenic fibrosing alveolitis. Ultrastructure and cell perimeter measurements. J Pathol. 1992;166(3):303–10.
213. Skold CM, Ohkuni Y, Liu XD, Numerof R, Rennard SI. Co-cultured human mast cells stimulate fibroblast-mediated contraction of collagen gels. Inflammation. 2001;25(1):47–51.
214. Okumura S, Sagara H, Fukuda T, Saito H, Okayama Y. FcepsilonRI-mediated amphiregulin production by human mast cells increases mucin gene expression in epithelial cells. J Allergy Clin Immunol. 2005;115(2):272–9.

215. Wang SW, Oh CK, Cho SH, Hu G, Martin R, Demissie-Sanders S, et al. Amphiregulin expression in human mast cells and its effect on the primary human lung fibroblasts. J Allergy Clin Immunol. 2005;115(2):287–94.
216. Kung TT, Stelts D, Zurcher JA, Jones H, Umland SP, Kreutner W, et al. Mast cells modulate allergic pulmonary eosinophilia in mice. Am J Respir Cell Mol Biol. 1995;12:404–9.
217. Elliott KA, Osna NA, Scofield MA, Khan MM. Regulation of IL-13 production by histamine in cloned murine T helper type 2 cells. Int Immunopharmacol. 2001;1(11):1923–37.
218. Holgate ST. Considerations on third generation antihistamines. Clin Exp Allergy. 2002;32(2):179.
219. Charbeneau RP, Peters-Golden M. Eicosanoids: mediators and therapeutic targets in fibrotic lung disease. Clin Sci (Lond). 2005;108(6):479–91.
220. Holgate ST, Peters-Golden M, Panettieri RA, Henderson WR Jr. Roles of cysteinyl leukotrienes in airway inflammation, smooth muscle function, and remodeling. J Allergy Clin Immunol. 2003;111(1 Suppl):S18–34; discussion S-6.
221. Robbiani DF, Finch RA, Jager D, Muller WA, Sartorelli AC, Randolph GJ. The leukotriene C(4) transporter MRP1 regulates CCL19 (MIP-3beta, ELC)-dependent mobilization of dendritic cells to lymph nodes. Cell. 2000;103(5):757–68.
222. Tager AM, Bromley SK, Medoff BD, Islam SA, Bercury SD, Friedrich EB, et al. Leukotriene B4 receptor BLT1 mediates early effector T cell recruitment. Nat Immunol. 2003;4(10):982–90.
223. Nagy L, Lee TH, Goetzi EJ, Picket WC, Kay AB. Complement receptor enhancement and chemotaxis of human neutrophils and eosinophils by leukotrienes and other lipoxygenase products. Clin Exp Immunol. 1982;47:541–7.
224. Lee E, Robertson T, Smith J, Kilfeather S. Leukotriene receptor antagonists and synthesis inhibitors reverse survival in eosinophils of asthmatic individuals. Am J Respir Crit Care Med. 2000;161(6):1881–6.
225. Kane GC, Pollice M, Kim CJ, Cohn J, Dworski RT, Murray JJ, et al. A controlled trial of the effect of the 5-lipoxygenase inhibitor, zileuton, on lung inflammation produced by segmental antigen challenge in human beings. J Allergy Clin Immunol. 1996;97:646–54.
226. Hojo M, Suzuki M, Maghni K, Hamid Q, Powell WS, Martin JG. Role of cysteinyl leukotrienes in CD4(+) T cell-driven late allergic airway responses. J Pharmacol Exp Ther. 2000;293(2):410–6.
227. Tohda Y, Nakahara H, Kubo H, Haraguchi R, Fukuoka M, Nakajima S. Effects of ONO-1078 (pranlukast) on cytokine production in peripheral blood mononuclear cells of patients with bronchial asthma. Clin Exp Allergy. 1999;29(11):1532–6.
228. Peters-Golden M, Bailie M, Marshall T, Wilke C, Phan SH, Toews GB, et al. Protection from pulmonary fibrosis in leukotriene-deficient mice. Am J Respir Crit Care Med. 2002;165(2):229–35.
229. Brach MA, de Vos S, Arnold C, Gruss HJ, Mertelsmann R, Herrmann F. Leukotriene B4 transcriptionally activates interleukin-6 expression involving NK-chi B and NF-IL6. Eur J Immunol. 1992;22(10):2705–11.
230. Christman JW, Christman BW, Shepherd VL, Rinaldo JE. Regulation of alveolar macrophage production of chemoattractants by leukotriene B4 and prostaglandin E2. Am J Respir Cell Mol Biol. 1991;5(3):297–304.
231. Phan SH, McGarry BM, Loeffler KM, Kunkel SL. Regulation of macrophage-derived fibroblast growth factor release by arachidonate metabolites. J Leukoc Biol. 1987;42(2):106–13.
232. Rajah R, Nunn SE, Herrick DJ, Grunstein MM, Cohen P. Leukotriene D4 induces MMP-1, which functions as an IGFBP protease in human airway smooth muscle cells. Am J Physiol. 1996;271(6 Pt 1):L1014–L22.
233. Parameswaran K, Cox G, Radford K, Janssen LJ, Sehmi R, O'Byrne PM. Cysteinyl leukotrienes promote human airway smooth muscle migration. Am J Respir Crit Care Med. 2002;166(5):738–42.

234. Davidson AB, Lee TH, Scanlon PD, Solway J, McFadden ER, Ingram RH, et al. Bronchoconstrictor effects of leukotriene E4 in normal and asthmatic subjects. Am Rev Respir Dis. 1987;135:333–7.
235. Salmon M, Walsh DA, Huang TJ, Barnes PJ, Leonard TB, Hay DW, et al. Involvement of cysteinyl leukotrienes in airway smooth muscle cell DNA synthesis after repeated allergen exposure in sensitized Brown Norway rats. Br J Pharmacol. 1999;127(5):1151–8.
236. Panettieri RA, Tan EM, Ciocca V, Luttmann MA, Leonard TB, Hay DW. Effects of LTD4 on human airway smooth muscle cell proliferation, matrix expression, and contraction In vitro: differential sensitivity to cysteinyl leukotriene receptor antagonists. Am J Respir Cell Mol Biol. 1998;19(3):453–61.
237. Wang CG, Du T, Xu LJ, Martin JG. Role of leukotriene D4 in allergen-induced increases in airway smooth muscle in the rat. Am Rev Respir Dis. 1993;148(2):413–7.
238. Mensing H, Czarnetzki BM. Leukotriene B4 induces in vitro fibroblast chemotaxis. J Invest Dermatol. 1984;82(1):9–12.
239. Baud L, Perez J, Denis M, Ardaillou R. Modulation of fibroblast proliferation by sulfidopeptide leukotrienes: effect of indomethacin. J Immunol. 1987;138(4):1190–5.
240. Fireman E, Schwartz Y, Mann A, Greif J. Effect of montelukast, a cysteinyl receptor antagonist, on myofibroblasts in interstitial lung disease. J Clin Immunol. 2004;24(4):418–25.
241. Phan SH, McGarry BM, Loeffler KM, Kunkel SL. Binding of leukotriene C4 to rat lung fibroblasts and stimulation of collagen synthesis in vitro. Biochemistry. 1988;27(8):2846–53.
242. Eap R, Jacques E, Semlali A, Plante S, Chakir J. Cysteinyl leukotrienes regulate TGF-beta(1) and collagen production by bronchial fibroblasts obtained from asthmatic subjects. Prostaglandins Leukot Essent Fatty Acids. 2012;86(3):127–33.
243. Asakura T, Ishii Y, Chibana K, Fukuda T. Leukotriene D4 stimulates collagen production from myofibroblasts transformed by TGF-beta. J Allergy Clin Immunol. 2004;114(2):310–5.
244. Pizzichini MM, Pizzichini E, Clelland L, Efthimiadis A, Pavord I, Dolovich J, et al. Prednisone-dependent asthma: inflammatory indices in induced sputum. Eur Respir J. 1999;13(1):15–21.
245. Henderson WR Jr, Tang LO, Chu SJ, Tsao SM, Chiang GK, Jones F, et al. A role for cysteinyl leukotrienes in airway remodeling in a mouse asthma model. Am J Respir Crit Care Med. 2002;165(1):108–16.
246. Parameswaran K, Watson R, Gauvreau GM, Sehmi R, O'Byrne PM. The effect of pranlukast on allergen-induced bone marrow eosinophilopoiesis in subjects with asthma. Am J Respir Crit Care Med. 2004;169(8):915–20.
247. Virchow JC Jr, Prasse A, Naya I, Summerton L, Harris A. Zafirlukast improves asthma control in patients receiving high-dose inhaled corticosteroids. Am J Respir Crit Care Med. 2000;162(2 Pt 1):578–85.
248. Lofdahl CG, Reiss TF, Leff JA, Israel E, Noonan MJ, Finn AF, et al. Randomised, placebo controlled trial of effect of a leukotriene receptor antagonist, montelukast, on tapering inhaled corticosteroids in asthmatic patients. BMJ. 1999;319(7202):87–90.
249. Mosmann TR, Cherwinski H, Bond MW, Giedlin MA, Coffman RH. Two types of murine helper T cell clone. Definition according to profiles of lymphokine activities and secreted proteins. J Immunol. 1986;136:2348–57.
250. Romagnani S. Human TH1 and TH2 subsets: doubt no more. Immunol Today. 1991;12(8):256–7.
251. Annunziato F, Cosmi L, Santarlasci V, Maggi L, Liotta F, Mazzinghi B, et al. Phenotypic and functional features of human Th17 cells. J Exp Med. 2007;204(8):1849–61.
252. Liu T, Song CH, Liu AM, Xie C, Zhao F, Chen X, et al. Forkhead box P3+ T cells express interleukin-17 in nasal mucosa of patients with both allergic rhinitis and polyposis. Clin Exp Immunol. 2011;163(1):59–64.
253. Cosmi L, Maggi L, Santarlasci V, Capone M, Cardilicchia E, Frosali F, et al. Identification of a novel subset of human circulating memory CD4(+) T cells that produce both IL-17A and IL-4. J Allergy Clin Immunol. 2010;125(1):222–30.
254. Romagnani S. The role of lymphocytes in allergic disease. J Allergy Clin Immunol. 2000;105(3):399–408.

255. Hurst SD, Seymour BW, Muchamuel T, Kurup VP, Coffman RL. Modulation of inhaled antigen-induced IgE tolerance by ongoing Th2 responses in the lung. J Immunol. 2001;166(8):4922–30.
256. Romagnani S, Maggi E, Liotta F, Cosmi L, Annunziato F. Properties and origin of human Th17 cells. Mol Immunol. 2009;47(1):3–7.
257. Seder RA, Boulay JL, Finkelman F, Barbier S, Ben-Sasson SZ, Le Gros G, et al. CD8+ T cells can be primed in vitro to produce IL-4. J Immunol. 1992;148(6):1652–6.
258. Erard F, Wild MT, Garcia-Sanz JA, LeGros G. Switch of CD8 T cells to noncytolytic CD8-CD4- cells that make Th2 cytokines and help B cells. Science. 1993;260:1802–5.
259. Kelso A, Groves P. A single peripheral CD8+ T cell can give rise to progeny expressing type 1 and/or type 2 cytokine genes and can retain its multipotentiality through many cell divisions. Proc Natl Acad Sci U S A. 1997;94(15):8070–5.
260. Cerwenka A, Carter LL, Reome JB, Swain SL, Dutton RW. In vivo persistence of CD8 polarized T cell subsets producing type 1 or type 2 cytokines. J Immunol. 1998;161(1):97–105.
261. Lavenu-Bombled C, Trainor CD, Makeh I, Romeo PH, Max-Audit I. Interleukin-13 gene expression is regulated by GATA-3 in T cells: role of a critical association of a GATA and two GATG motifs. J Biol Chem. 2002;277(21):18313–21.
262. Cho SH, Stanciu LA, Holgate ST, Johnston SL. Increased interleukin-4, interleukin-5, and interferon-gamma in airway CD4+ and CD8+ T cells in atopic asthma. Am J Respir Crit Care Med. 2005;171(3):224–30.
263. Fox A, Harland KL, Kedzierska K, Kelso A. Exposure of human CD8(+) T cells to type-2 cytokines impairs division and differentiation and induces limited polarization. Front Immunol. 2018;9:1141.
264. Camiolo MJ, Zhou X, Oriss TB, Yan Q, Gorry M, Horne W, et al. High-dimensional profiling clusters asthma severity by lymphoid and non-lymphoid status. Cell Rep. 2021;35(2):108974.
265. Muehling LM, Heymann PW, Wright PW, Eccles JD, Agrawal R, Carper HT, et al. Human T(H)1 and T(H)2 cells targeting rhinovirus and allergen coordinately promote allergic asthma. J Allergy Clin Immunol. 2020;146(3):555–70.
266. Raundhal M, Morse C, Khare A, Oriss TB, Milosevic J, Trudeau J, et al. High IFN-gamma and low SLPI mark severe asthma in mice and humans. J Clin Invest. 2015;125(8):3037–50.
267. Wisniewski JA, Muehling LM, Eccles JD, Capaldo BJ, Agrawal R, Shirley DA, et al. TH1 signatures are present in the lower airways of children with severe asthma, regardless of allergic status. J Allergy Clin Immunol. 2018;141(6):2048–60. e13
268. Krug N, Erpenbeck VJ, Balke K, Petschallies J, Tschernig T, Hohlfeld JM, et al. Cytokine profile of bronchoalveolar lavage-derived CD4(+), CD8(+), and gammadelta T cells in people with asthma after segmental allergen challenge. Am J Respir Cell Mol Biol. 2001;25(1):125–31.
269. Cembrzynska-Nowak M, Szklarz E, Inglot AD, Teodorczyk-Injeyan JA. Elevated release of tumor necrosis factor-alpha and interferon- gamma by bronchoalveolar leukocytes from patients with bronchial asthma. Am Rev Respir Dis. 1993;147(2):291–5.
270. Corrigan CJ, Hartnell A, Kay AB. T lymphocyte activation in acute severe asthma. Lancet. 1988;331:1129–31.
271. Coyle A, Tsuyuki S, Bertrand C, Huang S, Aguet M, Alkan S, et al. Mice lacking the IFN-g receptor have an impaired ability to resolve a lung eosinophilic inflammatory response associated with a prolonged capacity of T cells to exhibit a Th2 cytokine profile. J Immunol. 1996;156:2680–5.
272. Hansen G, Berry G, Dekruyff RH, Umetsu DT. Allergen-specific Th1 cells fail to counterbalance Th2 cell-induced airway hyperreactivity but cause severe airway inflammation. J Clin Invest. 1999;103(2):175–83.
273. Randolph DA, Carruthers CJ, Szabo SJ, Murphy KM, Chaplin DD. Modulation of airway inflammation by passive transfer of allergen- specific Th1 and Th2 cells in a mouse model of asthma. J Immunol. 1999;162(4):2375–83.
274. Jackson DJ, Gern JE. Rhinovirus infections and their roles in asthma: etiology and exacerbations. J Allergy Clin Immunol Pract. 2022;10(3):673–81.

275. Denlinger LC, Heymann P, Lutter R, Gern JE. Exacerbation-prone asthma. J Allergy Clin Immunol Pract. 2020;8(2):474–82.
276. Wang Y, Shi J, Yan J, Xiao Z, Hou X, Lu P, et al. Germinal-center development of memory B cells driven by IL-9 from follicular helper T cells. Nat Immunol. 2017;18(8):921–30.
277. Takatsuka S, Yamada H, Haniuda K, Saruwatari H, Ichihashi M, Renauld JC, et al. IL-9 receptor signaling in memory B cells regulates humoral recall responses. Nat Immunol. 2018;19(9):1025–34.
278. Cassese G, Arce S, Hauser AE, Lehnert K, Moewes B, Mostarac M, et al. Plasma cell survival is mediated by synergistic effects of cytokines and adhesion-dependent signals. J Immunol. 2003;171(4):1684–90.
279. Laidlaw BJ, Cyster JG. Transcriptional regulation of memory B cell differentiation. Nat Rev Immunol. 2021;21(4):209–20.
280. Geha RS, Jabara HH, Brodeur SR. The regulation of immunoglobulin E class-switch recombination. Nat Rev Immunol. 2003;3(9):721–32.
281. Yang Z, Robinson MJ, Allen CD. Regulatory constraints in the generation and differentiation of IgE-expressing B cells. Curr Opin Immunol. 2014;28:64–70.
282. Koenig JFE, Bruton K, Phelps A, Grydziuszko E, Jimenez-Saiz R, Jordana M. Memory generation and re-activation in food allergy. Immunotargets Ther. 2021;10:171–84.
283. Ueno H, Klechevsky E, Morita R, Aspord C, Cao T, Matsui T, et al. Dendritic cell subsets in health and disease. Immunol Rev. 2007;219:118–42.
284. van Rijt LS, Jung S, Kleinjan A, Vos N, Willart M, Duez C, et al. In vivo depletion of lung CD11c+ dendritic cells during allergen challenge abrogates the characteristic features of asthma. J Exp Med. 2005;201(6):981–91.
285. Banchereau J, Steinman RM. Dendritic cells and the control of immunity. Nature. 1998;392(6673):245–52.
286. Lee CC, Lin CL, Leu SJ, Lee YL. Overexpression of Notch ligand Delta-like-1 by dendritic cells enhances their immunoregulatory capacity and exerts antiallergic effects on Th2-mediated allergic asthma in mice. Clin Immunol. 2018;187:58–67.
287. Huh JC, Strickland DH, Jahnsen FL, Turner DJ, Thomas JA, Napoli S, et al. Bidirectional interactions between antigen-bearing respiratory tract dendritic cells (DCs) and T cells precede the late phase reaction in experimental asthma: DC activation occurs in the airway mucosa but not in the lung parenchyma. J Exp Med. 2003;198(1):19–30.
288. Romagnani S. T-cell subsets (Th1 versus Th2). Ann Allergy Asthma Immunol. 2000;85(1):9–18; quiz, 21.
289. Liu YJ. Dendritic cell subsets and lineages, and their functions in innate and adaptive immunity. Cell. 2001;106(3):259–62.
290. Izumi G, Nakano H, Nakano K, Whitehead GS, Grimm SA, Fessler MB, et al. CD11b(+) lung dendritic cells at different stages of maturation induce Th17 or Th2 differentiation. Nat Commun. 2021;12(1):5029.
291. Constant SL, Brogdon JL, Piggott DA, Herrick CA, Visintin I, Ruddle NH, et al. Resident lung antigen-presenting cells have the capacity to promote Th2 T cell differentiation in situ. J Clin Invest. 2002;110(10):1441–8.
292. Salek-Ardakani S, Song J, Halteman BS, Jember AG, Akiba H, Yagita H, et al. OX40 (CD134) controls memory T helper 2 cells that drive lung inflammation. J Exp Med. 2003;198(2):315–24.
293. Lu P, Zhou X, Chen SJ, Moorman M, Morris SC, Finkelman FD, et al. CTLA-4 ligands are required to induce an in vivo interleukin 4 response to a gastrointestinal nematode parasite. J Exp Med. 1994;180(2):693–8.
294. Kuchroo VK, Das MP, Brown JA, Ranger AM, Zamvil SS, Sobel RA, et al. B7-1 and B7-2 costimulatory molecules activate differentially the Th1/Th2 developmental pathways: application to autoimmune disease therapy. Cell. 1995;80(5):707–18.
295. Stumbles PA. Regulation of T helper cell differentiation by respiratory tract dendritic cells. Immunol Cell Biol. 1999;77(5):428–33.

296. Lambrecht BN, De Veerman M, Coyle AJ, Gutierrez-Ramos JC, Thielemans K, Pauwels RA. Myeloid dendritic cells induce Th2 responses to inhaled antigen, leading to eosinophilic airway inflammation. J Clin Invest. 2000;106(4):551–9.
297. Julia V, Hessel EM, Malherbe L, Glaichenhaus N, O'Garra A, Coffman RL. A restricted subset of dendritic cells captures airborne antigens and remains able to activate specific T cells long after antigen exposure. Immunity. 2002;16(2):271–83.
298. Watts TH. TNF/TNFR family members in costimulation of T cell responses. Annu Rev Immunol. 2005;23:23–68.
299. Binder C, Cvetkovski F, Sellberg F, Berg S, Paternina Visbal H, Sachs DH, et al. CD2 immunobiology. Front Immunol. 2020;11:1090.
300. Moller GM, Overbeek SE, Van Helden-Meeuwsen CG, Van HJM, Prens EP, Mulder PG, et al. Increased numbers of dendritic cells in the bronchial mucosa of atopic asthmatic patients: downregulation by inhaled corticosteroids. Clin Exp Allergy. 1996;26(5):517–24.
301. Tunon-De-Lara JM, Redington AE, Bradding P, Church MK, Hartley JA, Semper AE, et al. Dendritic cells in normal and asthmatic airways: expression of the alpha subunit of the high affinity immunoglobulin E receptor (Fc epsilon RI -alpha). Clin Exp Allergy. 1996;26(6):648–55.
302. Greer AM, Matthay MA, Kukreja J, Bhakta NR, Nguyen CP, Wolters PJ, et al. Accumulation of BDCA1(+) dendritic cells in interstitial fibrotic lung diseases and Th2-high asthma. PLoS One. 2014;9(6):e99084.
303. Broide DH, Lotz M, Cuomo AJ, Coburn DA, Federman EC, Wasserman SI. Cytokines in symptomatic asthma airways. J Allergy Clin Immunol. 1992;89(5):958–67.
304. Cella M, Scheidegger D, Palmer-Lehmann K, Lane P, Lanzavecchia A, Alber G. Ligation of CD40 on dendritic cells triggers production of high levels of interleukin-12 and enhances T cell stimulatory capacity: T-T help via APC activation. J Exp Med. 1996;184(2):747–52.
305. Akbari O, DeKruyff RH, Umetsu DT. Pulmonary dendritic cells producing IL-10 mediate tolerance induced by respiratory exposure to antigen. Nat Immunol. 2001;2(8):725–31.
306. Santos ES, de Aragao-Franca LS, Meira CS, Cerqueira JV, Vasconcelos JF, Nonaka CKV, et al. Tolerogenic dendritic cells reduce cardiac inflammation and fibrosis in chronic chagas disease. Front Immunol. 2020;11:488.
307. Nagai T, Honda S, Sugano Y, Matsuyama TA, Ohta-Ogo K, Asaumi Y, et al. Decreased myocardial dendritic cells is associated with impaired reparative fibrosis and development of cardiac rupture after myocardial infarction in humans. J Am Heart Assoc. 2014;3(3):e000839.
308. Marchal-Somme J, Uzunhan Y, Marchand-Adam S, Kambouchner M, Valeyre D, Crestani B, et al. Dendritic cells accumulate in human fibrotic interstitial lung disease. Am J Respir Crit Care Med. 2007;176(10):1007–14.
309. Tsoumakidou M, Karagiannis KP, Bouloukaki I, Zakynthinos S, Tzanakis N, Siafakas NM. Increased bronchoalveolar lavage fluid CD1c expressing dendritic cells in idiopathic pulmonary fibrosis. Respiration. 2009;78(4):446–52.
310. Peters MC, Mekonnen ZK, Yuan S, Bhakta NR, Woodruff PG, Fahy JV. Measures of gene expression in sputum cells can identify TH2-high and TH2-low subtypes of asthma. J Allergy Clin Immunol. 2014;133(2):388–94.
311. Graber P, Gretener D, Herren S, Aubry JP, Elson G, Poudrier J, et al. The distribution of IL-13 receptor alpha1 expression on B cells, T cells and monocytes and its regulation by IL-13 and IL-4. Eur J Immunol. 1998;28(12):4286–98.
312. Ying S, Durham SR, Corrigan CJ, Hamid Q, Kay AB. Phenotype of cells expressing mRNA for TH2-type (IL-4 and IL-5) and TH1-type (IL-2 and interferon gamma) cytokines in bronchoalveolar lavage and bronchial biopsies from atopic asthmatic and normal control subjects. Am J Respir Cell Mol Biol. 1995;12:477–87.
313. Ferrick DA, Schrenzel MD, Mulvania T, Hselh B, Ferlin WG, Lepper H. Differential production of interferon gamma and IL-4 in response to Th1-and Th2-stimulating pathogens by gamma/delta T cells in vivo. Nature. 1995;373:255–8.

314. Hoshino T, Winkler-Pickett RT, Mason AT, Ortaldo JR, Young HA. IL-13 production by NK cells: IL-13-producing NK and T cells are present in vivo in the absence of IFN-gamma. J Immunol. 1999;162(1):51–9.
315. Li P, Yang QZ, Wang W, Zhang GQ, Yang J. Increased IL-4- and IL-17-producing CD8(+) cells are related to decreased CD39(+)CD4(+)Foxp3(+) cells in allergic asthma. J Asthma. 2018;55(1):8–14.
316. Schaller MA, Lundy SK, Huffnagle GB, Lukacs NW. CD8+ T cell contributions to allergen induced pulmonary inflammation and airway hyperreactivity. Eur J Immunol. 2005;35(7):2061–70.
317. Zuany-Amorim C, Ruffie C, Haile S, Vargaftig BB, Pereira P, Pretolani M. Requirement for gammadelta T cells in allergic airway inflammation. Science. 1998;280(5367):1265–7.
318. Akbari O, Stock P, Meyer E, Kronenberg M, Sidobre S, Nakayama T, et al. Essential role of NKT cells producing IL-4 and IL-13 in the development of allergen-induced airway hyperreactivity. Nat Med. 2003;9(5):582–8.
319. Elyaman W, Bradshaw EM, Uyttenhove C, Dardalhon V, Awasthi A, Imitola J, et al. IL-9 induces differentiation of TH17 cells and enhances function of FoxP3+ natural regulatory T cells. Proc Natl Acad Sci U S A. 2009;106(31):12885–90.
320. Dugas B, Renauld JC, Pene J, Bonnefoy JY, Peti-Frere C, Braquet P, et al. Interleukin-9 potentiates the interleukin-4-induced immunoglobulin (IgG, IgM and IgE) production by normal human B lymphocytes. Eur J Immunol. 1993;23(7):1687–92.
321. Renz H, Lack G, Saloga J, Schwinzer R, Bradley K, Loader J, et al. Inhibition of IgE production and normalization of airways responsiveness by sensitized CD8 T cells in a mouse model of allergen-induced sensitization. J Immunol. 1994;152(1):351–60.
322. McMenamin C, Pimm C, McKersey M, Holt PG. Regulation of IgE responses to inhaled antigen in mice by antigen-specific gamma delta T cells. Science. 1994;265(5180):1869–71.
323. McMenamin C, Holt PG. The natural immune response to inhaled soluble protein antigens involves major histocompatability complex (MHC) class I-restricted CD8+ T-cell dependent immune deviation resulting in selective supression of immunoglobulin E production. J Exp Med. 1993;178(3):889–99.
324. Kaya Z, Dohmen KM, Wang Y, Schlichting J, Afanasyeva M, Leuschner F, et al. Cutting edge: a critical role for IL-10 in induction of nasal tolerance in experimental autoimmune myocarditis. J Immunol. 2002;168(4):1552–6.
325. Levings MK, Roncarolo MG. T-regulatory 1 cells: a novel subset of CD4 T cells with immunoregulatory properties. J Allergy Clin Immunol. 2000;106(1 Pt 2):S109–12.
326. Gardner LM, Thien FC, Douglass JA, Rolland JM, O'Hehir RE. Induction of T 'regulatory' cells by standardized house dust mite immunotherapy: an increase in CD4+ CD25+ interleukin-10+ T cells expressing peripheral tissue trafficking markers. Clin Exp Allergy. 2004;34(8):1209–19.
327. Bohle B, Kinaciyan T, Gerstmayr M, Radakovics A, Jahn-Schmid B, Ebner C. Sublingual immunotherapy induces IL-10-producing T regulatory cells, allergen-specific T-cell tolerance, and immune deviation. J Allergy Clin Immunol. 2007;120(3):707–13.
328. Simonian PL, Wehrmann F, Roark CL, Born WK, O'Brien RL, Fontenot AP. Gammadelta T cells protect against lung fibrosis via IL-22. J Exp Med. 2010;207(10):2239–53.
329. Murakami T, Hatano S, Yamada H, Iwakura Y, Yoshikai Y. Two types of interleukin 17A-producing gammadelta T cells in protection against pulmonary infection with Klebsiella pneumoniae. J Infect Dis. 2016;214(11):1752–61.
330. Vultaggio A, Nencini F, Pratesi S, Petroni G, Romagnani S, Maggi E. Poly(I:C) promotes the production of IL-17A by murine CD1d-driven invariant NKT cells in airway inflammation. Allergy. 2012;67(10):1223–32.
331. Montaldo E, Juelke K, Romagnani C. Group 3 innate lymphoid cells (ILC3s): Origin, differentiation, and plasticity in humans and mice. Eur J Immunol. 2015;45(8):2171–82.
332. Lezmi G, Leite-de-Moraes M. Invariant natural killer T and mucosal-associated invariant T cells in asthmatic patients. Front Immunol. 2018;9:1766.

333. Vijayanand P, Seumois G, Pickard C, Powell RM, Angco G, Sammut D, et al. Invariant natural killer T cells in asthma and chronic obstructive pulmonary disease. N Engl J Med. 2007;356(14):1410–22.
334. Doisne JM, Soulard V, Becourt C, Amniai L, Henrot P, Havenar-Daughton C, et al. Cutting edge: crucial role of IL-1 and IL-23 in the innate IL-17 response of peripheral lymph node NK1.1- invariant NKT cells to bacteria. J Immunol. 2011;186(2):662–6.
335. Jin Z, Sun R, Wei H, Gao X, Chen Y, Tian Z. Accelerated liver fibrosis in hepatitis B virus transgenic mice: involvement of natural killer T cells. Hepatology. 2011;53(1):219–29.
336. Nagata Y, Kamijuku H, Taniguchi M, Ziegler S, Seino K. Differential role of thymic stromal lymphopoietin in the induction of airway hyperreactivity and Th2 immune response in antigen-induced asthma with respect to natural killer T cell function. Int Arch Allergy Immunol. 2007;144(4):305–14.
337. Terashima A, Watarai H, Inoue S, Sekine E, Nakagawa R, Hase K, et al. A novel subset of mouse NKT cells bearing the IL-17 receptor B responds to IL-25 and contributes to airway hyperreactivity. J Exp Med. 2008;205(12):2727–33.
338. Bourgeois E, Van LP, Samson M, Diem S, Barra A, Roga S, et al. The pro-Th2 cytokine IL-33 directly interacts with invariant NKT and NK cells to induce IFN-gamma production. Eur J Immunol. 2009;39(4):1046–55.
339. Camelo A, Barlow JL, Drynan LF, Neill DR, Ballantyne SJ, Wong SH, et al. Blocking IL-25 signalling protects against gut inflammation in a type-2 model of colitis by suppressing nuocyte and NKT derived IL-13. J Gastroenterol. 2012;47(11):1198–211.
340. Ong S, Ligons DL, Barin JG, Wu L, Talor MV, Diny N, et al. Natural killer cells limit cardiac inflammation and fibrosis by halting eosinophil infiltration. Am J Pathol. 2015;185(3):847–61.
341. Mehrotra P, Collett JA, McKinney SD, Stevens J, Ivancic CM, Basile DP. IL-17 mediates neutrophil infiltration and renal fibrosis following recovery from ischemia reperfusion: compensatory role of natural killer cells in athymic rats. Am J Physiol Renal Physiol. 2017;312(3):F385–F97.
342. Lee OJ, Cho YN, Kee SJ, Kim MJ, Jin HM, Lee SJ, et al. Circulating mucosal-associated invariant T cell levels and their cytokine levels in healthy adults. Exp Gerontol. 2014;49:47–54.
343. Kurioka A, Jahun AS, Hannaway RF, Walker LJ, Fergusson JR, Sverremark-Ekstrom E, et al. Shared and distinct phenotypes and functions of human CD161++ Valpha7.2+ T cell subsets. Front Immunol. 2017;8:1031.
344. Dusseaux M, Martin E, Serriari N, Peguillet I, Premel V, Louis D, et al. Human MAIT cells are xenobiotic-resistant, tissue-targeted, CD161hi IL-17-secreting T cells. Blood. 2011;117(4):1250–9.
345. Salio M, Gasser O, Gonzalez-Lopez C, Martens A, Veerapen N, Gileadi U, et al. Activation of human mucosal-associated invariant T cells induces CD40L-dependent maturation of monocyte-derived and primary dendritic cells. J Immunol. 2017;199(8):2631–8.
346. Hinks TS, Zhou X, Staples KJ, Dimitrov BD, Manta A, Petrossian T, et al. Innate and adaptive T cells in asthmatic patients: relationship to severity and disease mechanisms. J Allergy Clin Immunol. 2015;136(2):323–33.
347. Lezmi G, Abou Taam R, Dietrich C, Chatenoud L, de Blic J, Leite-de-Moraes M. Circulating IL-17-producing mucosal-associated invariant T cells (MAIT) are associated with symptoms in children with asthma. Clin Immunol. 2018;188:7–11.
348. Khalil N, Bereznay O, Sporn M, Greenberg AH. Macrophage production of transforming growth factor beta and fibroblast collagen synthesis in chronic pulmonary inflammation. J Exp Med. 1989;170(3):727–37.
349. Shelhamer JH, Levine SJ, Wu T, Jacoby DB, Kaliner MA, Rennard SI. NIH conference. Airway inflammation. Ann Intern Med. 1995;123(4):288–304.
350. Robinson DS, Hamid Q, Ying S, Tsicopoulos A, Barkans J, Bentley AM, et al. Predominant TH2-like bronchoalveolar T-lymphocyte population in atopic asthma. N Eng J Med. 1992;326:298–304.

351. Holgate ST, Davies DE, Lackie PM, Wilson SJ, Puddicombe SM, Lordan JL. Epithelial-mesenchymal interactions in the pathogenesis of asthma. J Allergy Clin Immunol. 2000;105(2 Pt 1):193–204.
352. Kimura T, Ishii Y, Yoh K, Morishima Y, Iizuka T, Kiwamoto T, et al. Overexpression of the transcription factor GATA-3 enhances the development of pulmonary fibrosis. Am J Pathol. 2006;169(1):96–104.
353. Zhu Z, Homer RJ, Wang Z, Chen Q, Geba GP, Wang J, et al. Pulmonary expression of interleukin-13 causes inflammation, mucus hypersecretion, subepithelial fibrosis, physiologic abnormalities, and eotaxin production. J Clin Invest. 1999;103(6):779–88.
354. Huang SK, Xiao HQ, Kleine-Tebbe J, Paciotti G, Marsh DG, Lichtenstein LM, et al. IL-13 expression at the sites of allergen challenge in patients with asthma. J Immunol. 1995;155(5):2688–94.
355. Humbert M, Durham SR, Kimmitt P, Powell N, Assoufi B, Pfister R, et al. Elevated expression of messenger ribonucleic acid encoding IL-13 in the bronchial mucosa of atopic and nonatopic subjects with asthma. J Allergy Clin Immunol. 1997;99(5):657–65.
356. Ying S, Humbert M, Barkans J, Corrigan CJ, Pfister R, Menz G, et al. Expression of IL-4 and IL-5 mRNA and protein product by CD4+ and CD8+ T cells, eosinophils, and mast cells in bronchial biopsies obtained from atopic and nonatopic (intrinsic) asthmatics. J Immunol. 1997;158(7):3539–44.
357. Ohshima Y, Yasutomi M, Omata N, Yamada A, Fujisawa K, Kasuga K, et al. Dysregulation of IL-13 production by cord blood CD4+ T cells is associated with the subsequent development of atopic disease in infants. Pediatr Res. 2002;51(2):195–200.
358. Akaiwa M, Yu B, Umeshita-Suyama R, Terada N, Suto H, Koga T, et al. Localization of human interleukin 13 receptor in non-haematopoietic cells. Cytokine. 2001;13(2):75–84.
359. Hirst SJ, Hallsworth MP, Peng Q, Lee TH. Selective induction of eotaxin release by interleukin-13 or interleukin-4 in human airway smooth muscle cells is synergistic with interleukin-1beta and is mediated by the interleukin-4 receptor alpha-chain. Am J Respir Crit Care Med. 2002;165(8):1161–71.
360. Zheng T, Zhu Z, Wang Z, Homer RJ, Ma B, Riese RJ Jr, et al. Inducible targeting of IL-13 to the adult lung causes matrix metalloproteinase- and cathepsin-dependent emphysema. J Clin Invest. 2000;106(9):1081–93.
361. Firszt R, Francisco D, Church TD, Thomas JM, Ingram JL, Kraft M. Interleukin-13 induces collagen type-1 expression through matrix metalloproteinase-2 and transforming growth factor-beta1 in airway fibroblasts in asthma. Eur Respir J. 2014;43(2):464–73.
362. Zhou X, Hu H, Huynh ML, Kotaru C, Balzar S, Trudeau JB, et al. Mechanisms of tissue inhibitor of metalloproteinase 1 augmentation by IL-13 on TGF-beta 1-stimulated primary human fibroblasts. J Allergy Clin Immunol. 2007;119(6):1388–97.
363. Fulkerson PC, Fischetti CA, Hassman LM, Nikolaidis NM, Rothenberg ME. Persistent effects induced by IL-13 in the lung. Am J Respir Cell Mol Biol. 2006;35(3):337–46.
364. Yang G, Volk A, Petley T, Emmell E, Giles-Komar J, Shang X, et al. Anti-IL-13 monoclonal antibody inhibits airway hyperresponsiveness, inflammation and airway remodeling. Cytokine. 2004;28(6):224–32.
365. McAnulty RJ, Hernandez-Rodriguez NA, Mutsaers SE, Coker RK, Laurent GJ. Indomethacin suppresses the anti-proliferative effects of transforming growth factor-beta isoforms on fibroblast cell cultures. Biochem J. 1997;321(Pt 3):639–43.
366. Liu X, Kohyama T, Wang H, Zhu YK, Wen FQ, Kim HJ, et al. Th2 cytokine regulation of type I collagen gel contraction mediated by human lung mesenchymal cells. Am J Physiol Lung Cell Mol Physiol. 2002;282(5):L1049–56.
367. Elias JA, Lee CG, Zheng T, Ma B, Homer RJ, Zhu Z. New insights into the pathogenesis of asthma. J Clin Invest. 2003;111(3):291–7.
368. Wills-Karp M, Chiaramonte M. Interleukin-13 in asthma. Curr Opin Pulm Med. 2003;9(1):21–7.
369. Ingram JL, Rice A, Geisenhoffer K, Madtes DK, Bonner JC. Interleukin-13 stimulates the proliferation of lung myofibroblasts via a signal transducer and activator of transcription-6-

dependent mechanism: a possible mechanism for the development of airway fibrosis in asthma. Chest. 2003;123(3 Suppl):422S–4S.
370. Lee CG, Homer RJ, Zhu Z, Lanone S, Wang X, Koteliansky V, et al. Interleukin-13 induces tissue fibrosis by selectively stimulating and activating transforming growth factor beta(1). J Exp Med. 2001;194(6):809–21.
371. Stolarski B, Kurowska-Stolarska M, Kewin P, Xu D, Liew FY. IL-33 exacerbates eosinophil-mediated airway inflammation. J Immunol. 2010;185(6):3472–80.
372. Kurowska-Stolarska M, Stolarski B, Kewin P, Murphy G, Corrigan CJ, Ying S, et al. IL-33 amplifies the polarization of alternatively activated macrophages that contribute to airway inflammation. J Immunol. 2009;183(10):6469–77.
373. Mann JS, Holgate ST, Renwick AG, Cushley MJ. Airway effects of purine nucleosides and nucleotides and release with bronchial provocation in asthma. J Appl Physiol. 1986;61:1667–76.
374. Driver AG, Kukoly CA, Ali S, Mustafa SJ. Adenosine in bronchoalveolar lavage fluid in asthma. Am Rev Respir Dis. 1993;148(1):91–7.
375. Blackburn MR, Lee CG, Young HW, Zhu Z, Chunn JL, Kang MJ, et al. Adenosine mediates IL-13-induced inflammation and remodeling in the lung and interacts in an IL-13-adenosine amplification pathway. J Clin Invest. 2003;112(3):332–44.
376. Tiggelman AM, Boers W, Linthorst C, Sala M, Chamuleau RA. Collagen synthesis by human liver (myo)fibroblasts in culture: evidence for a regulatory role of IL-1 beta, IL-4, TGF beta and IFN gamma. J Hepatol. 1995;23(3):307–17.
377. Sempowski GD, Derdak S, Phipps RP. Interleukin-4 and interferon-gamma discordantly regulate collagen biosynthesis by functionally distinct lung fibroblast subsets. J Cell Physiol. 1996;167(2):290–6.
378. Wen FQ, Kohyama T, Liu X, Zhu YK, Wang H, Kim HJ, et al. Interleukin-4- and interleukin-13-enhanced transforming growth factor-beta2 production in cultured human bronchial epithelial cells is attenuated by interferon-gamma. Am J Respir Cell Mol Biol. 2002;26(4):484–90.
379. Richter A, Puddicombe SM, Lordan JL, Bucchieri F, Wilson SJ, Djukanovic R, et al. The contribution of interleukin (IL)-4 and IL-13 to the epithelial-mesenchymal trophic unit in asthma. Am J Respir Cell Mol Biol. 2001;25(3):385–91.
380. Nagahama KY, Togo S, Holz O, Magnussen H, Liu X, Seyama K, et al. Oncostatin M modulates fibroblast function via signal transducers and activators of transcription proteins-3. Am J Respir Cell Mol Biol. 2013;49(4):582–91.
381. Hashimoto S, Gon Y, Takeshita I, Maruoka S, Horie T. IL-4 and IL-13 induce myofibroblastic phenotype of human lung fibroblasts through c-Jun NH2-terminal kinase-dependent pathway. J Allergy Clin Immunol. 2001;107(6):1001–8.
382. Borowski A, Kuepper M, Horn U, Knupfer U, Zissel G, Hohne K, et al. Interleukin-13 acts as an apoptotic effector on lung epithelial cells and induces pro-fibrotic gene expression in lung fibroblasts. Clin Exp Allergy. 2008;38(4):619–28.
383. Doucet C, Brouty-Boye D, Pottin-Clemenceau C, Canonica GW, Jasmin C, Azzarone B. Interleukin (IL) 4 and IL-13 act on human lung fibroblasts. Implication in asthma. J Clin Invest. 1998;101(10):2129–39.
384. Doucet C, Brouty-Boye D, Pottin-Clemenceau C, Jasmin C, Canonica GW, Azzarone B. IL-4 and IL-13 specifically increase adhesion molecule and inflammatory cytokine expression in human lung fibroblasts. Int Immunol. 1998;10(10):1421–33.
385. Kraft M, Lewis C, Pham D, Chu HW. IL-4, IL-13, and dexamethasone augment fibroblast proliferation in asthma. J Allergy Clin Immunol. 2001;107(4):602–6.
386. Roufosse F. Targeting the interleukin-5 pathway for treatment of eosinophilic conditions other than asthma. Front Med. 2018;5:49.
387. Esnault S, Kelly EA. Essential mechanisms of differential activation of eosinophils by IL-3 compared to GM-CSF and IL-5. Crit Rev Immunol. 2016;36(5):429–44.

388. Gavett SH, Chen X, Finkelman F, Wills-Karp M. Depletion of murine CD4+ T lymphocytes prevents antigen-induced airway hyperreactivity and pulmonary eosinophilia. Am J Respir Cell Mol Biol. 1994;10:587–93.
389. Gerblich AA, Salik H, Schuyler MR. Dynamic T-cell changes in peripheral blood and bronchoalveolar lavage after antigen bronchoprovocation in asthmatics. Am Rev Respir Dis. 1991;143(3):533–7.
390. Kelly EA, Rodriguez RR, Busse WW, Jarjour NN. The effect of segmental bronchoprovocation with allergen on airway lymphocyte function. Am J Respir Crit Care Med. 1997;156(5):1421–8.
391. Walker C, Bode E, Boer L, Hansel TT, Blaser K, Virchow JC. Allergic and nonallergic asthmatics have distinct patterns of T-cell activation and cytokine production in peripheral blood and bronchoalveolar lavage. Am Rev Respir Dis. 1992;146:109–15.
392. Flood-Page P, Menzies-Gow A, Phipps S, Ying S, Wangoo A, Ludwig MS, et al. Anti-IL-5 treatment reduces deposition of ECM proteins in the bronchial subepithelial basement membrane of mild atopic asthmatics. J Clin Invest. 2003;112(7):1029–36.
393. Esnault S, Kelly EA, Johansson MW, Liu LY, Han S-H, Akhtar M, et al. Semaphorin 7A is expressed on airway eosinophils and upregulated by IL-5 family cytokines. Clin Immunol. 2014;150(1):90–100.
394. Kang HR, Lee CG, Homer RJ, Elias JA. Semaphorin 7A plays a critical role in TGF-beta1-induced pulmonary fibrosis. J Exp Med. 2007;204(5):1083–93.
395. De Minicis S, Rychlicki C, Agostinelli L, Saccomanno S, Trozzi L, Candelaresi C, et al. Semaphorin 7A contributes to TGF-beta-mediated liver fibrogenesis. Am J Pathol. 2013;183(3):820–30.
396. Schmitt E, Germann T, Goedert S, Hoehn P, Huels C, Koelsch S, et al. IL-9 production of naive CD4+ T cells depends on IL-2, is synergistically enhanced by a combination of TGF-beta and IL-4, and is inhibited by IFN-gamma. J Immunol. 1994;153(9):3989–96.
397. Kaplan MH, Hufford MM, Olson MR. The development and in vivo function of T helper 9 cells. Nat Rev Immunol. 2015;15(5):295–307.
398. Li J, Chen S, Xiao X, Zhao Y, Ding W, Li XC. IL-9 and Th9 cells in health and diseases-From tolerance to immunopathology. Cytokine Growth Factor Rev. 2017;37:47–55.
399. Noelle RJ, Nowak EC. Cellular sources and immune functions of interleukin-9. Nat Rev Immunol. 2010;10(10):683–7.
400. Schmitt E, Klein M, Bopp T. Th9 cells, new players in adaptive immunity. Trends Immunol. 2014;35(2):61–8.
401. Stassen M, Schmitt E, Bopp T. From interleukin-9 to T helper 9 cells. Ann N Y Acad Sci. 2012;1247:56–68.
402. Nowak EC, Weaver CT, Turner H, Begum-Haque S, Becher B, Schreiner B, et al. IL-9 as a mediator of Th17-driven inflammatory disease. J Exp Med. 2009;206(8):1653–60.
403. Abdelilah S, Latifa K, Esra N, Cameron L, Bouchaib L, Nicolaides N, et al. Functional expression of IL-9 receptor by human neutrophils from asthmatic donors: role in IL-8 release. J Immunol. 2001;166(4):2768–74.
404. Kearley J, Erjefalt JS, Andersson C, Benjamin E, Jones CP, Robichaud A, et al. IL-9 governs allergen-induced mast cell numbers in the lung and chronic remodeling of the airways. Am J Respir Crit Care Med. 2011;183(7):865–75.
405. Wiener Z, Falus A, Toth S. IL-9 increases the expression of several cytokines in activated mast cells, while the IL-9-induced IL-9 production is inhibited in mast cells of histamine-free transgenic mice. Cytokine. 2004;26(3):122–30.
406. Temann UA, Geba GP, Rankin JA, Flavell RA. Expression of interleukin 9 in the lungs of transgenic mice causes airway inflammation, mast cell hyperplasia, and bronchial hyperresponsiveness. J Exp Med. 1998;188(7):1307–20.
407. Temann UA, Ray P, Flavell RA. Pulmonary overexpression of IL-9 induces Th2 cytokine expression, leading to immune pathology. J Clin Invest. 2002;109(1):29–39.
408. Ying S, Meng Q, Kay AB, Robinson DS. Elevated expression of interleukin-9 mRNA in the bronchial mucosa of atopic asthmatics and allergen-induced cutaneous late-phase

reaction: relationships to eosinophils, mast cells and T lymphocytes. Clin Exp Allergy. 2002;32(6):866–71.
409. Sitkauskiene B, Radinger M, Bossios A, Johansson AK, Sakalauskas R, Lotvall J. Airway allergen exposure stimulates bone marrow eosinophilia partly via IL-9. Respir Res. 2005;6:33.
410. Gounni AS, Gregory B, Nutku E, Aris F, Latifa K, Minshall E, et al. Interleukin-9 enhances interleukin-5 receptor expression, differentiation, and survival of human eosinophils. Blood. 2000;96(6):2163–71.
411. van den Brule S, Heymans J, Havaux X, Renauld JC, Lison D, Huaux F, et al. Profibrotic effect of IL-9 overexpression in a model of airway remodeling. Am J Respir Cell Mol Biol. 2007;37(2):202–9.
412. Yao Z, Painter SL, Fanslow WC, Ulrich D, Macduff BM, Spriggs MK, et al. Human IL-17: a novel cytokine derived from T cells. J Immunol. 1995;155(12):5483–6.
413. Fossiez F, Djossou O, Chomarat P, Flores-Romo L, Ait-Yahia S, Maat C, et al. T cell interleukin-17 induces stromal cells to produce proinflammatory and hematopoietic cytokines. J Exp Med. 1996;183(6):2593–603.
414. Shin HC, Benbernou N, Esnault S, Guenounou M. Expression of IL-17 in human memory CD45RO+ T lymphocytes and its regulation by protein kinase A pathway. Cytokine. 1999;11(4):257–66.
415. Hwang SY, Kim JY, Kim KW, Park MK, Moon Y, Kim WU, et al. IL-17 induces production of IL-6 and IL-8 in rheumatoid arthritis synovial fibroblasts via NF-kappaB- and PI3-kinase/Akt-dependent pathways. Arthritis Res Ther. 2004;6(2):R120–8.
416. Zhu L, Wu Y, Wei H, Xing X, Zhan N, Xiong H, et al. IL-17R activation of human periodontal ligament fibroblasts induces IL-23 p19 production: Differential involvement of NF-kappaB versus JNK/AP-1 pathways. Mol Immunol. 2011;48(4):647–56.
417. Ye P, Rodriguez FH, Kanaly S, Stocking KL, Schurr J, Schwarzenberger P, et al. Requirement of interleukin 17 receptor signaling for lung CXC chemokine and granulocyte colony-stimulating factor expression, neutrophil recruitment, and host defense. J Exp Med. 2001;194(4):519–27.
418. Ye P, Garvey PB, Zhang P, Nelson S, Bagby G, Summer WR, et al. Interleukin-17 and lung host defense against Klebsiella pneumoniae infection. Am J Respir Cell Mol Biol. 2001;25(3):335–40.
419. Li Q, Gu Y, Tu Q, Wang K, Gu X, Ren T. Blockade of interleukin-17 restrains the development of acute lung injury. Scand J Immunol. 2016;83(3):203–11.
420. Yan Z, Xiaoyu Z, Zhixin S, Di Q, Xinyu D, Jing X, et al. Rapamycin attenuates acute lung injury induced by LPS through inhibition of Th17 cell proliferation in mice. Sci Rep. 2016;6:20156.
421. Mi S, Li Z, Yang HZ, Liu H, Wang JP, Ma YG, et al. Blocking IL-17A promotes the resolution of pulmonary inflammation and fibrosis via TGF-beta1-dependent and -independent mechanisms. J Immunol. 2011;187(6):3003–14.
422. Chakir J, Shannon J, Molet S, Fukakusa M, Elias J, Laviolette M, et al. Airway remodeling-associated mediators in moderate to severe asthma: Effect of steroids on TGF-beta, IL-11, IL-17, and type I and type III collagen expression. J Allergy Clin Immunol. 2003;111(6):1293–8.
423. Cosmi L, Liotta F, Annunziato F. Th17 regulating lower airway disease. Curr Opin Allergy Clin Immunol. 2016;16(1):1–6.
424. Molet S, Hamid Q, Davoine F, Nutku E, Taha R, Page N, et al. IL-17 is increased in asthmatic airways and induces human bronchial fibroblasts to produce cytokines. J Allergy Clin Immunol. 2001;108(3):430–8.
425. Bullens DM, Truyen E, Coteur L, Dilissen E, Hellings PW, Dupont LJ, et al. IL-17 mRNA in sputum of asthmatic patients: linking T cell driven inflammation and granulocytic influx? Respir Res. 2006;7:135.
426. Evans MD, Esnault S, Denlinger LC, Jarjour NN. Sputum cell IL-1 receptor expression level is a marker of airway neutrophilia and airflow obstruction in asthmatic patients. J Allergy Clin Immunol. 2018;142(2):415–23.

427. Ciprandi G, De AM, Murdaca G, Fenoglio D, Ricciardolo F, Marseglia G, et al. Serum interleukin-17 levels are related to clinical severity in allergic rhinitis. Allergy. 2009;64(9):1375–8.
428. Semik-Orzech A, Barczyk A, Wiaderkiewicz R, Pierzchala W. Interleukin 17 and RANTES levels in induced sputum of patients with allergic rhinitis after a single nasal allergen challenge. Ann Allergy Asthma Immunol. 2009;103(5):418–24.
429. Hasegawa T, Uga H, Mori A, Kurata H. Increased serum IL-17A and Th2 cytokine levels in patients with severe uncontrolled asthma. Eur Cytokine Netw. 2017;28(1):8–18.
430. Al-Muhsen S, Letuve S, Vazquez-Tello A, Pureza MA, Al-Jahdali H, Bahammam AS, et al. Th17 cytokines induce pro-fibrotic cytokines release from human eosinophils. Respir Res. 2013;14(1):34.
431. Peters M, Kohler-Bachmann S, Lenz-Habijan T, Bufe A. Influence of an Allergen-Specific Th17 Response on Remodeling of the Airways. Am J Respir Cell Mol Biol. 2016;54(3):350–8.
432. Zhou X, Loomis-King H, Gurczynski SJ, Wilke CA, Konopka KE, Ptaschinski C, et al. Bone marrow transplantation alters lung antigen-presenting cells to promote TH17 response and the development of pneumonitis and fibrosis following gammaherpesvirus infection. Mucosal Immunol. 2016;9(3):610–20.
433. Hall SL, Baker T, Lajoie S, Richgels PK, Yang Y, McAlees JW, et al. IL-17A enhances IL-13 activity by enhancing IL-13-induced signal transducer and activator of transcription 6 activation. J Allergy Clin Immunol. 2017;139(2):462–71. e14
434. Hayashi H, Kawakita A, Okazaki S, Yasutomi M, Murai H, Ohshima Y. IL-17A/F modulates fibrocyte functions in cooperation with CD40-mediated signaling. Inflammation. 2013;36(4):830–8.
435. Annunziato F, Cosmi L, Liotta F, Maggi E, Romagnani S. Defining the human T helper 17 cell phenotype. Trends Immunol. 2012;33(10):505–12.
436. Antony PA, Restifo NP. CD4+CD25+ T regulatory cells, immunotherapy of cancer, and interleukin-2. J Immunother. 2005;28(2):120–8.
437. Sakaguchi S, Yamaguchi T, Nomura T, Ono M. Regulatory T cells and immune tolerance. Cell. 2008;133(5):775–87.
438. Raimondi G, Turner MS, Thomson AW, Morel PA. Naturally occurring regulatory T cells: recent insights in health and disease. Crit Rev Immunol. 2007;27(1):61–95.
439. Frydrychowicz M, Boruczkowski M, Kolecka-Bednarczyk A, Dworacki G. The dual role of treg in cancer. Scand J Immunol. 2017;86(6):436–43.
440. Bergmann C, Strauss L, Wang Y, Szczepanski MJ, Lang S, Johnson JT, et al. T regulatory type 1 cells in squamous cell carcinoma of the head and neck: mechanisms of suppression and expansion in advanced disease. Clin Cancer Res. 2008;14(12):3706–15.
441. Nakamura K, Kitani A, Strober W. Cell contact-dependent immunosuppression by CD4(+) CD25(+) regulatory T cells is mediated by cell surface-bound transforming growth factor beta. J Exp Med. 2001;194(5):629–44.
442. Green EA, Gorelik L, McGregor CM, Tran EH, Flavell RA. CD4+CD25+ T regulatory cells control anti-islet CD8+ T cells through TGF-beta-TGF-beta receptor interactions in type 1 diabetes. Proc Natl Acad Sci U S A. 2003;100(19):10878–83.
443. Roncarolo MG, Bacchetta R, Bordignon C, Narula S, Levings MK. Type 1 T regulatory cells. Immunol Rev. 2001;182:68–79.
444. Wan YY, Flavell RA. 'Yin-Yang' functions of transforming growth factor-beta and T regulatory cells in immune regulation. Immunol Rev. 2007;220:199–213.
445. Miyara M, Yoshioka Y, Kitoh A, Shima T, Wing K, Niwa A, et al. Functional delineation and differentiation dynamics of human CD4+ T cells expressing the FoxP3 transcription factor. Immunity. 2009;30(6):899–911.
446. Magnan A, Retornaz F, Tsicopoulos A, Brisse J, Van Pee D, Gosset P, et al. Altered compartmentalization of transforming growth factor-a in asthmatic airways. Clin Exp Allergy. 1997;27:389–95.
447. Baumgartner RA, Deramo VA, Beaven MA. Constitutive and inducible mechanisms for synthesis and release of cytokines in immune cell lines. J Immunol. 1996;157(9):4087–93.

448. Tsunawaki S, Sporn M, Ding A, Nathan C. Deactivation of macrophages by transforming growth factor-beta. Nature. 1988;334(6179):260–2.
449. Fontana A, Frei K, Bodmer S, Hofer E, Schreier MH, Palladino MA Jr, et al. Transforming growth factor-beta inhibits the generation of cytotoxic T cells in virus-infected mice. J Immunol. 1989;143(10):3230–4.
450. Fargeas C, Wu CY, Nakajima T, Cox D, Nutman T, Deleespesse G. Differential effect of transforming growth factor beta on the synthesis of Th1- and Th2-like lymphokines by human T lymphocytes. Eur J Immunol. 1992;22:2173–6.
451. Bogdan C, Vodovotz Y, Nathan C. Macrophage deactivation by interleukin 10. J Exp Med. 1991;174(6):1549–55.
452. de Waal Malefyt R, Abrams J, Bennett B, Figdor CG, de Vries JE. Interleukin 10(IL-10) inhibits cytokine synthesis by human monocytes: an autoregulatory role of IL-10 produced by monocytes. J Exp Med. 1991;174(5):1209–20.
453. de Waal Malefyt R, Haanen J, Spits H, Roncarolo MG, te Velde A, Figdor C, et al. Interleukin 10 (IL-10) and viral IL-10 strongly reduce antigen-specific human T cell proliferation by diminishing the antigen-presenting capacity of monocytes via downregulation of class II major histocompatibility complex expression. J Exp Med. 1991;174(4):915–24.
454. D'Andrea A, Aste-Amezaga M, Valiante NM, Ma X, Kubin M, Trinchieri G. Interleukin 10 (IL-10) inhibits human lymphocyte interferon gamma-production by suppressing natural killer cell stimulatory factor/IL-12 synthesis in accessory cells. J Exp Med. 1993;178(3):1041–8.
455. Creery WD, Diaz-Mitoma F, Filion L, Kumar A. Differential modulation of B7-1 and B7-2 isoform expression on human monocytes by cytokines which influence the development of T helper cell phenotype. Eur J Immunol. 1996;26(6):1273–7.
456. Groux H, Bigler M, de Vries JE, Roncarolo MG. Inhibitory and stimulatory effects of IL-10 on human CD8+ T cells. J Immunol. 1998;160(7):3188–93.
457. Coomes SM, Kannan Y, Pelly VS, Entwistle LJ, Guidi R, Perez-Lloret J, et al. CD4(+) Th2 cells are directly regulated by IL-10 during allergic airway inflammation. Mucosal Immunol. 2017;10(1):150–61.
458. Galati D, De Martino M, Trotta A, Rea G, Bruzzese D, Cicchitto G, et al. Peripheral depletion of NK cells and imbalance of the Treg/Th17 axis in idiopathic pulmonary fibrosis patients. Cytokine. 2014;66(2):119–26.
459. Moye S, Bormann T, Maus R, Sparwasser T, Sandrock I, Prinz I, et al. Regulatory T cells limit pneumococcus-induced exacerbation of lung fibrosis in mice. J Immunol. 2020;204(9):2429–38.
460. do Valle Duraes F, Lafont A, Beibel M, Martin K, Darribat K, Cuttat R, et al. Immune cell landscaping reveals a protective role for regulatory T cells during kidney injury and fibrosis. JCI Insight. 2020;5(3):e130651.
461. Kalekar LA, Cohen JN, Prevel N, Sandoval PM, Mathur AN, Moreau JM, et al. Regulatory T cells in skin are uniquely poised to suppress profibrotic immune responses. Sci Immunol. 2019;4(39):eaaw2910.
462. Peng X, Moore MW, Peng H, Sun H, Gan Y, Homer RJ, et al. CD4+CD25+FoxP3+ Regulatory Tregs inhibit fibrocyte recruitment and fibrosis via suppression of FGF-9 production in the TGF-beta1 exposed murine lung. Front Pharmacol. 2014;5:80.
463. Garibaldi BT, D'Alessio FR, Mock JR, Files DC, Chau E, Eto Y, et al. Regulatory T cells reduce acute lung injury fibroproliferation by decreasing fibrocyte recruitment. Am J Respir Cell Mol Biol. 2013;48(1):35–43.
464. Provoost S, Maes T, Van Durme YM, Gevaert P, Bachert C, Schmidt-Weber CB, et al. Decreased FOXP3 protein expression in patients with asthma. Allergy. 2009;64(10):1539–46.
465. Hartl D, Koller B, Mehlhorn AT, Reinhardt D, Nicolai T, Schendel DJ, et al. Quantitative and functional impairment of pulmonary CD4+CD25hi regulatory T cells in pediatric asthma. J Allergy Clin Immunol. 2007;119(5):1258–66.
466. Nguyen KD, Vanichsarn C, Fohner A, Nadeau KC. Selective deregulation in chemokine signaling pathways of CD4+CD25(hi)CD127(lo)/(−) regulatory T cells in human allergic asthma. J Allergy Clin Immunol. 2009;123(4):933–9. e10

467. Boonpiyathad T, Capova G, Duchna HW, Croxford AL, Farine H, Dreher A, et al. Impact of high-altitude therapy on type-2 immune responses in asthma patients. Allergy. 2020;75(1):84–94.
468. Bohm L, Maxeiner J, Meyer-Martin H, Reuter S, Finotto S, Klein M, et al. IL-10 and regulatory T cells cooperate in allergen-specific immunotherapy to ameliorate allergic asthma. J Immunol. 2015;194(3):887–97.
469. Lo Re S, Lecocq M, Uwambayinema F, Yakoub Y, Delos M, Demoulin JB, et al. Platelet-derived growth factor-producing CD4+ Foxp3+ regulatory T lymphocytes promote lung fibrosis. Am J Respir Crit Care Med. 2011;184(11):1270–81.
470. Leigh R, Ellis R, Wattie J, Donaldson DD, Inman MD. Is interleukin-13 critical in maintaining airway hyperresponsiveness in allergen-challenged mice? Am J Respir Crit Care Med. 2004;170(8):851–6.
471. Wills-Karp M. Interleukin-13 in asthma pathogenesis. Immunol Rev. 2004;202:175–90.
472. Kudo F, Ikutani M, Seki Y, Otsubo T, Kawamura YI, Dohi T, et al. Interferon-gamma constrains cytokine production of group 2 innate lymphoid cells. Immunology. 2016;147(1):21–9.
473. Giri SN, Hyde DM, Marafino BJ Jr. Ameliorating effect of murine interferon gamma on bleomycin-induced lung collagen fibrosis in mice. Biochem Med Metab Biol. 1986;36(2):194–7.
474. Gurujeyalakshmi G, Giri SN. Molecular mechanisms of antifibrotic effect of interferon gamma in bleomycin-mouse model of lung fibrosis: downregulation of TGF-beta and procollagen I and III gene expression. Exp Lung Res. 1995;21(5):791–808.
475. Clark JG, Dedon TF, Wayner EA, Carter WG. Effects of interferon-gamma on expression of cell surface receptors for collagen and deposition of newly synthesized collagen by cultured human lung fibroblasts. J Clin Invest. 1989;83(5):1505–11.
476. Chen ES, Greenlee BM, Wills-Karp M, Moller DR. Attenuation of lung inflammation and fibrosis in interferon-gamma-deficient mice after intratracheal bleomycin. Am J Respir Cell Mol Biol. 2001;24(5):545–55.
477. Czaja MJ, Weiner FR, Eghbali M, Giambrone MA, Eghbali M, Zern MA. Differential effects of gamma-interferon on collagen and fibronectin gene expression. J Biol Chem. 1987;262(27):13348–51.
478. Shaw RJ, Benedict SH, Clark RA, King TE Jr. Pathogenesis of pulmonary fibrosis in interstitial lung disease. Alveolar macrophage PDGF(B) gene activation and up-regulation by interferon gamma. Am Rev Respir Dis. 1991;143(1):167–73.
479. Crosby JR, Tappan KA, Seifert RA, Bowen-Pope DF. Chimera analysis reveals that fibroblasts and endothelial cells require platelet-derived growth factor receptorbeta expression for participation in reactive connective tissue formation in adults but not during development. Am J Pathol. 1999;154(5):1315–21.
480. Gao Z, Sasaoka T, Fujimori T, Oya T, Ishii Y, Sabit H, et al. Deletion of the PDGFR-beta gene affects key fibroblast functions important for wound healing. J Biol Chem. 2005;280(10):9375–89.
481. Rajkumar VS, Shiwen X, Bostrom M, Leoni P, Muddle J, Ivarsson M, et al. Platelet-derived growth factor-beta receptor activation is essential for fibroblast and pericyte recruitment during cutaneous wound healing. Am J Pathol. 2006;169(6):2254–65.
482. Sikkeland LIB, Qiao SW, Ueland T, Myrdal O, Wyrozemski L, Aukrust P, et al. Lung CD4+ T-cells in patients with lung fibrosis produce pro-fibrotic interleukin-13 together with interferon-gamma. Eur Respir J. 2021;57(3):2000983.
483. Naumov D, Gassan D, Kotova O, Sheludko E, Afanaseva E, Perelman J, et al. Role of interferon-gamma as a marker of asthma severity and control. Eur Respir J. 2019;54:PA4378.
484. Irifune K, Yokoyama A, Sakai K, Watanabe A, Katayama H, Ohnishi H, et al. Adoptive transfer of T-helper cell type 1 clones attenuates an asthmatic phenotype in mice. Eur Respir J. 2005;25(4):653–9.
485. Honore I, Nunes H, Groussard O, Kambouchner M, Chambellan A, Aubier M, et al. Acute respiratory failure after interferon-gamma therapy of end-stage pulmonary fibrosis. Am J Respir Crit Care Med. 2003;167(7):953–7.

486. Kalra S, Utz JP, Ryu JH, Mayo Clinic Interstitial Lung Diseases Group. Interferon gamma-1b therapy for advanced idiopathic pulmonary fibrosis. Mayo Clin Proc. 2003;78(9):1082–7.
487. Raghu G, Brown KK, Bradford WZ, Starko K, Noble PW, Schwartz DA, et al. A placebo-controlled trial of interferon gamma-1b in patients with idiopathic pulmonary fibrosis. N Engl J Med. 2004;350(2):125–33.
488. Komura K, Yanaba K, Horikawa M, Ogawa F, Fujimoto M, Tedder TF, et al. CD19 regulates the development of bleomycin-induced pulmonary fibrosis in a mouse model. Arthritis Rheum. 2008;58(11):3574–84.
489. Matsushita T, Kobayashi T, Mizumaki K, Kano M, Sawada T, Tennichi M, et al. BAFF inhibition attenuates fibrosis in scleroderma by modulating the regulatory and effector B cell balance. Sci Adv. 2018;4(7):eaas9944.
490. Francois A, Chatelus E, Wachsmann D, Sibilia J, Bahram S, Alsaleh G, et al. B lymphocytes and B-cell activating factor promote collagen and profibrotic markers expression by dermal fibroblasts in systemic sclerosis. Arthritis Res Ther. 2013;15(5):R168.
491. Bosello S, De Santis M, Lama G, Spano C, Angelucci C, Tolusso B, et al. B cell depletion in diffuse progressive systemic sclerosis: safety, skin score modification and IL-6 modulation in an up to thirty-six months follow-up open-label trial. Arthritis Res Ther. 2010;12(2):R54.
492. Ma K, Wang X, Shi X, Lin X, Xiao F, Ma X, et al. The expanding functional diversity of plasma cells in immunity and inflammation. Cell Mol Immunol. 2020;17(4):421–2.
493. Jansen K, Satitsuksanoa P, Wirz OF, Schneider SR, van de Veen W, Tan G, et al. T regulatory cells from atopic asthmatic individuals show a Th2-like phenotype. Allergy. 2022;77(4):1320–4.
494. Liu F, Dai W, Li C, Lu X, Chen Y, Weng D, et al. Role of IL-10-producing regulatory B cells in modulating T-helper cell immune responses during silica-induced lung inflammation and fibrosis. Sci Rep. 2016;6:28911.
495. Kaminska M, Foley S, Maghni K, Storness-Bliss C, Coxson H, Ghezzo H, et al. Airway remodeling in subjects with severe asthma with or without chronic persistent airflow obstruction. J Allergy Clin Immunol. 2009;124(1):45–51.e1–4.
496. Ramos-Barbon D, Presley JF, Hamid QA, Fixman ED, Martin JG. Antigen-specific CD4+ T cells drive airway smooth muscle remodeling in experimental asthma. J Clin Invest. 2005;115(6):1580–9.
497. Woodruff PG, Modrek B, Choy DF, Jia G, Abbas AR, Ellwanger A, et al. T-helper type 2-driven inflammation defines major subphenotypes of asthma. Am J Respir Crit Care Med. 2009;180(5):388–95.
498. Walter DM, McIntire JJ, Berry G, McKenzie AN, Donaldson DD, DeKruyff RH, et al. Critical role for IL-13 in the development of allergen-induced airway hyperreactivity. J Immunol. 2001;167(8):4668–75.
499. Wills-Karp M, Luyimbazi J, Xu XY, Schofield B, Neben TY, Karp CL, et al. Interleukin-13: Central mediator of allergic asthma. Science. 1998;282(5397):2258–61.
500. Leigh R, Ellis R, Wattie JN, Hirota JA, Matthaei KI, Foster PS, et al. Type 2 cytokines in the pathogenesis of sustained airway dysfunction and airway remodeling in mice. Am J Respir Crit Care Med. 2004;169(7):860–7.
501. Venkayya R, Lam M, Willkom M, Grunig G, Corry DB, Erle DJ. The Th2 lymphocyte products IL-4 and IL-13 rapidly induce airway hyperresponsiveness through direct effects on resident airway cells. Am J Respir Cell Mol Biol. 2002;26(2):202–8.
502. Yang M, Hogan SP, Henry PJ, Matthaei KI, McKenzie AN, Young IG, et al. Interleukin-13 mediates airways hyperreactivity through the IL-4 receptor-alpha chain and STAT-6 independently of IL-5 and eotaxin. Am J Respir Cell Mol Biol. 2001;25(4):522–30.
503. Laporte JC, Moore PE, Baraldo S, Jouvin MH, Church TL, Schwartzman IN, et al. Direct effects of interleukin-13 on signaling pathways for physiological responses in cultured human airway smooth muscle cells. Am J Respir Crit Care Med. 2001;164(1):141–8.
504. Kuperman DA, Huang X, Koth LL, Chang GH, Dolganov GM, Zhu Z, et al. Direct effects of interleukin-13 on epithelial cells cause airway hyperreactivity and mucus overproduction in asthma. Nat Med. 2002;8:885.

505. Aravamudan B, Thompson M, Pabelick C, Prakash YS. Brain-derived neurotrophic factor induces proliferation of human airway smooth muscle cells. J Cell Mol Med. 2012;16(4):812–23.
506. Espinosa K, Bosse Y, Stankova J, Rola-Pleszczynski M. CysLT1 receptor upregulation by TGF-beta and IL-13 is associated with bronchial smooth muscle cell proliferation in response to LTD4. J Allergy Clin Immunol. 2003;111(5):1032–40.
507. Wardlaw AJ, Hay H, Cromwell O, Collins JV, Kay AB. Leukotrienes, LTC4 and LTB4, in bronchoalveolar lavage in bronchial asthma and other respiratory diseases. J Allergy Clin Immunol. 1989;84(1):19–26.
508. Thivierge M, Stankova J, Rola-Pleszczynski M. IL-13 and IL-4 up-regulate cysteinyl leukotriene 1 receptor expression in human monocytes and macrophages. J Immunol. 2001;167(5):2855–60.
509. Bautsch W, Hoymann HG, Zhang Q, Meier-Wiedenbach I, Raschke U, Ames RS, et al. Cutting edge: Guinea pigs with a natural C3a-receptor defect exhibit decreased bronchoconstriction in allergic airway disease: evidence for an involvement of the C3a anaphylatoxin in the pathogenesis of asthma. J Immunol. 2000;165(10):5401–5.
510. Humbles AA, Lu B, Nilsson CA, Lilly C, Israel E, Fujiwara Y, et al. A role for the C3a anaphylatoxin receptor in the effector phase of asthma. Nature. 2000;406(6799):998–1001.
511. Manson ML, Safholm J, James A, Johnsson AK, Bergman P, Al-Ameri M, et al. IL-13 and IL-4, but not IL-5 nor IL-17A, induce hyperresponsiveness in isolated human small airways. J Allergy Clin Immunol. 2020;145(3):808–17. e2
512. Grunstein MM, Hakonarson H, Leiter J, Chen M, Whelan R, Grunstein JS, et al. IL-13-dependent autocrine signaling mediates altered responsiveness of IgE-sensitized airway smooth muscle. Am J Physiol Lung Cell Mol Physiol. 2002;282(3):L520–8.
513. Zhong H, Belardinelli L, Maa T, Feoktistov I, Biaggioni I, Zeng D. A(2B) adenosine receptors increase cytokine release by bronchial smooth muscle cells. Am J Respir Cell Mol Biol. 2004;30(1):118–25.
514. Cho JY, Miller M, Baek KJ, Han JW, Nayar J, Lee SY, et al. Inhibition of airway remodeling in IL-5-deficient mice. J Clin Invest. 2004;113(4):551–60.
515. Calzetta L, Ritondo BL, Matera MG, Facciolo F, Rogliani P. Targeting IL-5 pathway against airway hyperresponsiveness: a comparison between benralizumab and mepolizumab. Br J Pharmacol. 2020;177(20):4750–65.
516. Gounni AS, Hamid Q, Rahman SM, Hoeck J, Yang J, Shan L. IL-9-mediated induction of eotaxin1/CCL11 in human airway smooth muscle cells. J Immunol. 2004;173(4):2771–9.
517. Baraldo S, Faffe DS, Moore PE, Whitehead T, McKenna M, Silverman ES, et al. Interleukin-9 influences chemokine release in airway smooth muscle: role of ERK. Am J Physiol Lung Cell Mol Physiol. 2003;284(6):L1093–L102.
518. Yamasaki A, Saleh A, Koussih L, Muro S, Halayko AJ, Gounni AS. IL-9 induces CCL11 expression via STAT3 signalling in human airway smooth muscle cells. PLoS One. 2010;5(2):e9178.
519. McKinley L, Alcorn JF, Peterson A, Dupont RB, Kapadia S, Logar A, et al. TH17 cells mediate steroid-resistant airway inflammation and airway hyperresponsiveness in mice. J Immunol. 2008;181(6):4089–97.
520. Rahman MS, Yang J, Shan LY, Unruh H, Yang X, Halayko AJ, et al. IL-17R activation of human airway smooth muscle cells induces CXCL-8 production via a transcriptional-dependent mechanism. Clin Immunol. 2005;115(3):268–76.
521. Vanaudenaerde BM, Wuyts WA, Dupont LJ, Van Raemdonck DE, Demedts MM, Verleden GM. Interleukin-17 stimulates release of interleukin-8 by human airway smooth muscle cells in vitro: a potential role for interleukin-17 and airway smooth muscle cells in bronchiolitis obliterans syndrome. J Heart Lung Transplant. 2003;22(11):1280–3.
522. Kwofie K, Scott M, Rodrigues R, Guerette J, Radford K, Nair P, et al. Regulation of IL-17A responses in human airway smooth muscle cells by Oncostatin M. Respir Res. 2015;16(1):14.

523. Chang Y, Al-Alwan L, Risse PA, Halayko AJ, Martin JG, Baglole CJ, et al. Th17-associated cytokines promote human airway smooth muscle cell proliferation. FASEB J. 2012;26(12):5152–60.
524. Al-Alwan LA, Chang Y, Baglole CJ, Risse PA, Halayko AJ, Martin JG, et al. Autocrine-regulated airway smooth muscle cell migration is dependent on IL-17-induced growth-related oncogenes. J Allergy Clin Immunol. 2012;130(4):977–85. e6
525. Kudo M, Melton AC, Chen C, Engler MB, Huang KE, Ren X, et al. IL-17A produced by alphabeta T cells drives airway hyper-responsiveness in mice and enhances mouse and human airway smooth muscle contraction. Nat Med. 2012;18(4):547–54.
526. Willis CR, Siegel L, Leith A, Mohn D, Escobar S, Wannberg S, et al. IL-17RA Signaling in Airway Inflammation and Bronchial Hyperreactivity in Allergic Asthma. Am J Respir Cell Mol Biol. 2015;53(6):810–21.
527. Chiba Y, Tanoue G, Suto R, Suto W, Hanazaki M, Katayama H, et al. Interleukin-17A directly acts on bronchial smooth muscle cells and augments the contractility. Pharmacol Rep. 2017;69(3):377–85.
528. Kumar RK, Webb DC, Herbert C, Foster PS. Interferon-gamma as a possible target in chronic asthma. Inflamm Allergy Drug Targets. 2006;5(4):253–6.
529. Shimokado K, Yokota T, Kato N, Kosaka C, Sasaguri T, Masuda J, et al. Bidirectional regulation of smooth muscle cell proliferation by IFN-gamma. J Atheroscler Thromb. 1994;1(Suppl 1):S29–33.
530. Wang W, Chen HJ, Schwartz A, Cannon PJ, Rabbani LE. T cell lymphokines modulate bFGF-induced smooth muscle cell fibrinolysis and migration. Am J Phys. 1997;272(2 Pt 1):C392–8.
531. Amrani Y, Tliba O, Choubey D, Huang CD, Krymskaya VP, Eszterhas A, et al. IFN-gamma inhibits human airway smooth muscle cell proliferation by modulating the E2F-1/Rb pathway. Am J Physiol Lung Cell Mol Physiol. 2003;284(6):L1063–71.
532. Wen FQ, Liu X, Manda W, Terasaki Y, Kobayashi T, Abe S, et al. TH2 Cytokine-enhanced and TGF-beta-enhanced vascular endothelial growth factor production by cultured human airway smooth muscle cells is attenuated by IFN-gamma and corticosteroids. J Allergy Clin Immunol. 2003;111(6):1307–18.
533. Ford CL, Wang Y, Morgan K, Boktor M, Jordan P, Castor TP, et al. Interferon-gamma depresses human intestinal smooth muscle cell contractility: Relevance to inflammatory gut motility disturbances. Life Sci. 2019;222:69–77.
534. Peat JK, Toelle BG, Dermand J, van den Berg R, Britton WJ, Woolcock AJ. Serum IgE levels, atopy, and asthma in young adults: results from a longitudinal cohort study. Allergy. 1996;51(11):804–10.
535. Gounni AS, Wellemans V, Yang J, Bellesort F, Kassiri K, Gangloff S, et al. Human airway smooth muscle cells express the high affinity receptor for IgE (Fc epsilon RI): a critical role of Fc epsilon RI in human airway smooth muscle cell function. J Immunol. 2005;175(4):2613–21.
536. Belleau JT, Gandhi RK, McPherson HM, Lew DB. Research upregulation of CD23 (FcepsilonRII) expression in human airway smooth muscle cells (huASMC) in response to IL-4, GM-CSF, and IL-4/GM-CSF. Clin Mol Allergy. 2005;3:6.
537. Redhu NS, Saleh A, Shan L, Gerthoffer WT, Kung SK, Halayko AJ, et al. Proinflammatory and Th2 cytokines regulate the high affinity IgE receptor (FcepsilonRI) and IgE-dependant activation of human airway smooth muscle cells. PLoS One. 2009;4(7):e6153.
538. Redhu NS, Shan L, Al-Subait D, Ashdown HL, Movassagh H, Lamkhioued B, et al. IgE induces proliferation in human airway smooth muscle cells: role of MAPK and STAT3 pathways. Allergy Asthma Clin Immunol. 2013;9(1):41.
539. Roth M, Zhong J, Zumkeller C, S'Ng CT, Goulet S, Tamm M. The role of IgE-receptors in IgE-dependent airway smooth muscle cell remodelling. PLoS One. 2013;8(2):e56015.
540. Xia YC, Schuliga M, Shepherd M, Powell M, Harris T, Langenbach SY, et al. Functional expression of IgG-Fc receptors in human airway smooth muscle cells. Am J Respir Cell Mol Biol. 2011;44(5):665–72.

541. Nemoto K, Okamura T. Intracellular signals in IgG-mediated anaphylactic contraction of single smooth muscle cells. Arerugi. 1992;41(2 Pt 1):125–34.
542. Habener A, Happle C, Grychtol R, Skuljec J, Busse M, Daluge K, et al. Regulatory B cells control airway hyperreactivity and lung remodeling in a murine asthma model. J Allergy Clin Immunol. 2021;147(6):2281–94. e7
543. Houtmeyers E, Gosselink R, Gayan-Ramirez G, Decramer M. Regulation of mucociliary clearance in health and disease. Eur Respir J. 1999;13(5):1177–88.
544. Thornton DJ, Rousseau K, McGuckin MA. Structure and function of the polymeric mucins in airways mucus. Annu Rev Physiol. 2008;70:459–86.
545. Hovenberg HW, Davies JR, Carlstedt I. Different mucins are produced by the surface epithelium and the submucosa in human trachea: identification of MUC5AC as a major mucin from the goblet cells. Biochem J. 1996;318(Pt 1):319–24.
546. Welsh KG, Rousseau K, Fisher G, Bonser LR, Bradding P, Brightling CE, et al. MUC5AC and a glycosylated variant of MUC5B alter mucin composition in children with acute asthma. Chest. 2017;152(4):771–9.
547. Lachowicz-Scroggins ME, Yuan S, Kerr SC, Dunican EM, Yu M, Carrington SD, et al. Abnormalities in MUC5AC and MUC5B protein in airway mucus in asthma. Am J Respir Crit Care Med. 2016;194(10):1296–9.
548. Kirkham S, Sheehan JK, Knight D, Richardson PS, Thornton DJ. Heterogeneity of airways mucus: variations in the amounts and glycoforms of the major oligomeric mucins MUC5AC and MUC5B. Biochem J. 2002;361(Pt 3):537–46.
549. Longphre M, Li D, Gallup M, Drori E, Ordonez CL, Redman T, et al. Allergen-induced IL-9 directly stimulates mucin transcription in respiratory epithelial cells. J Clin Invest. 1999;104(10):1375–82.
550. Dabbagh K, Takeyama K, Lee HM, Ueki IF, Lausier JA, Nadel JA. IL-4 induces mucin gene expression and goblet cell metaplasia in vitro and in vivo. J Immunol. 1999;162(10):6233–7.
551. Vermeer PD, Harson R, Einwalter LA, Moninger T, Zabner J. Interleukin-9 induces goblet cell hyperplasia during repair of human airway epithelia. Am J Respir Cell Mol Biol. 2003;28(3):286–95.
552. Wills-Karp M. Trophic slime, allergic slime. Am J Respir Cell Mol Biol. 2000;22(6):637–9.
553. Temann UA, Prasad B, Gallup MW, Basbaum C, Ho SB, Flavell RA, et al. A novel role for murine IL-4 in vivo: induction of MUC5AC gene expression and mucin hypersecretion. Am J Respir Cell Mol Biol. 1997;16(4):471–8.
554. Louahed J, Toda M, Jen J, Hamid Q, Renauld JC, Levitt RC, et al. Interleukin-9 upregulates mucus expression in the airways. Am J Respir Cell Mol Biol. 2000;22(6):649–56.
555. Cohn L, Homer RJ, MacLeod H, Mohrs M, Brombacher F, Bottomly K. Th2-induced airway mucus production is dependent on IL-4Ralpha, but not on eosinophils. J Immunol. 1999;162(10):6178–83.
556. Grunig G, Warnock M, Wakil AE, Venkayya R, Brombacher F, Rennick DM, et al. Requirement for IL-13 independently of IL-4 in experimental asthma. Science. 1998;282(5397):2261–3.
557. Blease K, Jakubzick C, Westwick J, Lukacs N, Kunkel SL, Hogaboam CM. Therapeutic effect of IL-13 immunoneutralization during chronic experimental fungal asthma. J Immunol. 2001;166(8):5219–24.
558. McKenzie GJ, Emson CL, Bell SE, Anderson S, Fallon P, Zurawski G, et al. Impaired development of Th2 cells in IL-13-deficient mice. Immunity. 1998;9(3):423–32.
559. Gavett SH, O'Hearn DJ, Karp CL, Patel EA, Schofield BH, Finkelman FD, et al. Interleukin-4 receptor blockade prevents airway responses induced by antigen challenge in mice. Am J Phys. 1997;272(2 Pt 1):L253–61.
560. Kuperman D, Schofield B, Wills-Karp M, Grusby MJ. Signal transducer and activator of transcription factor 6 (Stat6)-deficient mice are protected from antigen-induced airway hyperresponsiveness and mucus production. J Exp Med. 1998;187(6):939–48.
561. Laoukili J, Perret E, Willems T, Minty A, Parthoens E, Houcine O, et al. IL-13 alters mucociliary differentiation and ciliary beating of human respiratory epithelial cells. J Clin Invest. 2001;108(12):1817–24.

562. Booth BW, Adler KB, Bonner JC, Tournier F, Martin LD. Interleukin-13 induces proliferation of human airway epithelial cells in vitro via a mechanism mediated by transforming growth factor-alpha. Am J Respir Cell Mol Biol. 2001;25(6):739–43.
563. Taniguchi K, Yamamoto S, Aoki S, Toda S, Izuhara K, Hamasaki Y. Epigen is induced during the interleukin-13-stimulated cell proliferation in murine primary airway epithelial cells. Exp Lung Res. 2011;37(8):461–70.
564. Zhou Y, Dong Q, Louahed J, Dragwa C, Savio D, Huang M, et al. Characterization of a calcium-activated chloride channel as a shared target of Th2 cytokine pathways and its potential involvement in asthma. Am J Respir Cell Mol Biol. 2001;25(4):486–91.
565. Zhao J, Minami Y, Etling E, Coleman JM, Lauder SN, Tyrrell V, et al. Preferential generation of 15-HETE-PE induced by IL-13 regulates goblet cell differentiation in human airway epithelial cells. Am J Respir Cell Mol Biol. 2017;57(6):692–701.
566. Harrop CA, Gore RB, Evans CM, Thornton DJ, Herrick SE. TGF-beta(2) decreases baseline and IL-13-stimulated mucin production by primary human bronchial epithelial cells. Exp Lung Res. 2013;39(1):39–47.
567. Kanoh S, Tanabe T, Rubin BK. IL-13-induced MUC5AC production and goblet cell differentiation is steroid resistant in human airway cells. Clin Exp Allergy. 2011;41(12):1747–56.
568. Atherton HC, Jones G, Danahay H. IL-13-induced changes in the goblet cell density of human bronchial epithelial cell cultures: MAP kinase and phosphatidylinositol 3-kinase regulation. Am J Physiol Lung Cell Mol Physiol. 2003;285(3):L730–9.
569. Pezzulo AA, Tudas RA, Stewart CG, Buonfiglio LGV, Lindsay BD, Taft PJ, et al. HSP90 inhibitor geldanamycin reverts IL-13- and IL-17-induced airway goblet cell metaplasia. J Clin Invest. 2019;129(2):744–58.
570. Malavia NK, Mih JD, Raub CB, Dinh BT, George SC. IL-13 induces a bronchial epithelial phenotype that is profibrotic. Respir Res. 2008;9(1):27.
571. Schmidt H, Braubach P, Schilpp C, Lochbaum R, Neuland K, Thompson K, et al. IL-13 impairs tight junctions in airway epithelia. Int J Mol Sci. 2019;20(13):3222.
572. Sugita K, Steer CA, Martinez-Gonzalez I, Altunbulakli C, Morita H, Castro-Giner F, et al. Type 2 innate lymphoid cells disrupt bronchial epithelial barrier integrity by targeting tight junctions through IL-13 in asthmatic patients. J Allergy Clin Immunol. 2018;141(1):300–10. e11
573. Suzaki I, Kawano S, Komiya K, Tanabe T, Akaba T, Asano K, et al. Inhibition of IL-13-induced periostin in airway epithelium attenuates cellular protein expression of MUC5AC. Respirology. 2017;22(1):93–100.
574. Woodruff PG, Boushey HA, Dolganov GM, Barker CS, Yang YH, Donnelly S, et al. Genome-wide profiling identifies epithelial cell genes associated with asthma and with treatment response to corticosteroids. Proc Natl Acad Sci USA. 2007;104(40):15858–63.
575. Takayama G, Arima K, Kanaji T, Toda S, Tanaka H, Shoji S, et al. Periostin: a novel component of subepithelial fibrosis of bronchial asthma downstream of IL-4 and IL-13 signals. J Allergy Clin Immunol. 2006;118(1):98–104.
576. Corren J, Lemanske RF, Hanania NA, Korenblat PE, Parsey MV, Arron JR, et al. Lebrikizumab treatment in adults with asthma. N Engl J Med. 2011;365(12):1088–98.
577. Jia G, Erickson RW, Choy DF, Mosesova S, Wu LC, Solberg OD, et al. Periostin is a systemic biomarker of eosinophilic airway inflammation in asthmatic patients. J Allergy Clin Immunol. 2012;130(3):647–54. e10
578. Kanemitsu Y, Ito I, Niimi A, Izuhara K, Ohta S, Ono J, et al. Osteopontin and periostin are associated with a 20-year decline of pulmonary function in patients with asthma. Am J Respir Crit Care Med. 2014;190(4):472–4.
579. Masuoka M, Shiraishi H, Ohta S, Suzuki S, Arima K, Aoki S, et al. Periostin promotes chronic allergic inflammation in response to Th2 cytokines. J Clin Invest. 2012;122(7):2590–600.
580. Uchida M, Shiraishi H, Ohta S, Arima K, Taniguchi K, Suzuki S, et al. Periostin, a matricellular protein, plays a role in the induction of chemokines in pulmonary fibrosis. Am J Respir Cell Mol Biol. 2012;46(5):677–86.

581. Blanchard C, Mingler MK, McBride M, Putnam PE, Collins MH, Chang G, et al. Periostin facilitates eosinophil tissue infiltration in allergic lung and esophageal responses. Mucosal Immunol. 2008;1(4):289–96.
582. Sidhu SS, Yuan S, Innes AL, Kerr S, Woodruff PG, Hou L, et al. Roles of epithelial cell-derived periostin in TGF-beta activation, collagen production, and collagen gel elasticity in asthma. Proc Natl Acad Sci USA. 2010;107(32):14170–5.
583. O'Dwyer DN, Moore BB. The role of periostin in lung fibrosis and airway remodeling. Cell Mol Life Sci. 2017;74(23):4305–14.
584. Chen K, Eddens T, Trevejo-Nunez G, Way EE, Elsegeiny W, Ricks DM, et al. IL-17 receptor signaling in the lung epithelium is required for mucosal chemokine gradients and pulmonary host defense against K. pneumoniae. Cell Host Microbe. 2016;20(5):596–605.
585. Pociask DA, Scheller EV, Mandalapu S, McHugh KJ, Enelow RI, Fattman CL, et al. IL-22 is essential for lung epithelial repair following influenza infection. Am J Pathol. 2013;182(4):1286–96.
586. Xia W, Bai J, Wu X, Wei Y, Feng S, Li L, et al. Interleukin-17A promotes MUC5AC expression and goblet cell hyperplasia in nasal polyps via the Act1-mediated pathway. PLoS One. 2014;9(6):e98915.
587. Jiao J, Duan S, Meng N, Li Y, Fan E, Zhang L. Role of IFN-gamma, IL-13, and IL-17 on mucociliary differentiation of nasal epithelial cells in chronic rhinosinusitis with nasal polyps. Clin Exp Allergy. 2016;46(3):449–60.
588. Chen Y, Thai P, Zhao YH, Ho YS, DeSouza MM, Wu R. Stimulation of airway mucin gene expression by interleukin (IL)-17 through IL-6 paracrine/autocrine loop. J Biol Chem. 2003;278(19):17036–43.
589. Hu M, Yang S, Yang L, Cheng Y, Zhang H. Interleukin-22 alleviated palmitate-induced endoplasmic reticulum stress in INS-1 cells through activation of autophagy. PLoS One. 2016;11(1):e0146818.
590. Liu H, Mi S, Li Z, Hua F, Hu ZW. Interleukin 17A inhibits autophagy through activation of PIK3CA to interrupt the GSK3B-mediated degradation of BCL2 in lung epithelial cells. Autophagy. 2013;9(5):730–42.
591. Reed M, Morris SH, Owczarczyk AB, Lukacs NW. Deficiency of autophagy protein Map1-LC3b mediates IL-17-dependent lung pathology during respiratory viral infection via ER stress-associated IL-1. Mucosal Immunol. 2015;8(5):1118–30.
592. Campbell AM, Vignola AM, Chanez P, Godard P, Bousquet J. Low-affinity receptor for IgE on human bronchial epithelial cells in asthma. Immunology. 1994;82(4):506–8.
593. Campbell AM, Vachier I, Chanez P, Vignola AM, Lebel B, Kochan J, et al. Expression of the high-affinity receptor for IgE on bronchial epithelial cells of asthmatics. Am J Respir Cell Mol Biol. 1998;19(1):92–7.
594. Jayawickreme SP, Gray T, Nettesheim P, Eling T. Regulation of 15-lipoxygenase expression and mucus secretion by IL-4 in human bronchial epithelial cells. Am J Phys. 1999;276(4):L596–603.
595. Takata S, Matsubara M, Allen PG, Janmey PA, Serhan CN, Brady HR. Remodeling of neutrophil phospholipids with 15(S)-hydroxyeicosatetraenoic acid inhibits leukotriene B4-induced neutrophil migration across endothelium. J Clin Invest. 1994;93(2):499–508.
596. van Dijk AP, McCafferty DM, Wilson JH, Zijlstra FJ. 15-Hydroxy-eicosatetraenoic acid has minor anti-inflammatory properties in colitis. Agents Actions. 1993;38:C120–1.
597. Chu HW, Balzar S, Westcott JY, Trudeau JB, Sun Y, Conrad DJ, et al. Expression and activation of 15-lipoxygenase pathway in severe asthma: relationship to eosinophilic phenotype and collagen deposition. Clin Exp Allergy. 2002;32(11):1558–65.
598. Lai CK, Phillips GD, Jenkins JR, Holgate ST. The effect of inhaled 15-(s)-hydroxyeicosatetraenoic acid (15-HETE) on airway calibre and non-specific responsiveness in normal and asthmatic human subjects. Eur Respir J. 1990;3(1):38–45.
599. Song YS, Kim MS, Lee DH, Oh DK, Yoon DY. 15-Hydroxyeicosatetraenoic Acid Inhibits Phorbol-12-Myristate-13-Acetate-Induced MUC5AC Expression in NCI-H292 Respiratory Epithelial Cells. J Microbiol Biotechnol. 2015;25(5):589–97.

600. Zhao J, Maskrey B, Balzar S, Chibana K, Mustovich A, Hu H, et al. Interleukin-13-induced MUC5AC is regulated by 15-lipoxygenase 1 pathway in human bronchial epithelial cells. Am J Respir Crit Care Med. 2009;179(9):782–90.
601. Brims FJ, Chauhan AJ, Higgins B, Shute JK. Coagulation factors in the airways in moderate and severe asthma and the effect of inhaled steroids. Thorax. 2009;64(12):1037–43.
602. Tomasiak-Lozowska MM, Misztal T, Rusak T, Branska-Januszewska J, Bodzenta-Lukaszyk A, Tomasiak M. Asthma is associated with reduced fibrinolytic activity, abnormal clot architecture, and decreased clot retraction rate. Allergy. 2017;72(2):314–9.
603. Bazan-Socha S, Mastalerz L, Cybulska A, Zareba L, Kremers R, Zabczyk M, et al. Asthma is associated with enhanced thrombin formation and impaired fibrinolysis. Clin Exp Allergy. 2016;46(7):932–44.
604. Chambers RC. Procoagulant signalling mechanisms in lung inflammation and fibrosis: novel opportunities for pharmacological intervention? Br J Pharmacol. 2008;153(Suppl 1):S367–78.
605. de Boer JD, Majoor CJ, van't Veer C, Bel EH, van der Poll T. Asthma and coagulation. Blood. 2012;119(14):3236–44.
606. Gabazza EC, Taguchi O, Tamaki S, Takeya H, Kobayashi H, Yasui H, et al. Thrombin in the airways of asthmatic patients. Lung. 1999;177(4):253–62.
607. Terada M, Kelly EA, Jarjour NN. Increased thrombin activity after allergen challenge: a potential link to airway remodeling? Am J Respir Crit Care Med. 2004;169(3):373–7.
608. Zhu Y, Esnault S, Ge Y, Jarjour NN, Brasier AR. Airway fibrin formation cascade in allergic asthma exacerbation: implications for inflammation and remodeling. Clin Proteomics. 2022;19(1):15.
609. Kucharewicz I, Mogielnicki A, Kasacka I, Buczko W, Bodzenta-Lukaszyk A. Plasmin system regulation in an ovalbumin-induced rat model of asthma. Int Arch Allergy Immunol. 2008;147(3):190–6.
610. Vasse M, Paysant J, Soria J, Collet JP, Vannier JP, Soria C. Regulation of fibrinogen biosynthesis by cytokines, consequences on the vascular risk. Haemostasis. 1996;26(Suppl 4):331–9.
611. Chen CL, Yao Y, Pan L, Hu ST, Ma J, Wang ZC, et al. Common fibrin deposition and tissue plasminogen activator downregulation in nasal polyps with distinct inflammatory endotypes. J Allergy Clin Immunol. 2020;146(3):677–81.
612. Takabayashi T, Kato A, Peters AT, Hulse KE, Suh LA, Carter R, et al. Increased expression of factor XIII-A in patients with chronic rhinosinusitis with nasal polyps. J Allergy Clin Immunol. 2013;132(3):584–92. e4
613. Bastarache JA, Sebag SC, Grove BS, Ware LB. Interferon-gamma and tumor necrosis factor-alpha act synergistically to up-regulate tissue factor in alveolar epithelial cells. Exp Lung Res. 2011;37(8):509–17.
614. Katona E, Nagy B, Kappelmayer J, Baktai G, Kovacs L, Marialigeti T, et al. Factor XIII in bronchoalveolar lavage fluid from children with chronic bronchoalveolar inflammation. J Thromb Haemost. 2005;3(7):1407–13.
615. Esnault S, Kelly EA, Schwantes EA, Liu LY, Delain LP, Hauer JA, et al. Identification of genes expressed by human airway eosinophils after an in vivo allergen challenge. PLoS One. 2013;8(7):e67560.
616. Esnault S, Kelly EA, Sorkness RL, Evans MD, Busse WW, Jarjour NN. Airway factor XIII associates with type 2 inflammation and airway obstruction in asthmatic patients. J Allergy Clin Immunol. 2016;137(3):767–73. e6
617. Mitchell JL, Mutch NJ. Let's cross-link: diverse functions of the promiscuous cellular transglutaminase factor XIII-A. J Thromb Haemost. 2019;17(1):19–30.
618. Alshehri FSM, Whyte CS, Tuncay A, Williams ML, Wilson HM, Mutch NJ. Monocytes expose factor XIII-A and stabilize thrombi against fibrinolytic degradation. Int J Mol Sci. 2021;22(12):6591.

619. Chaitidis P, O'Donnell V, Kuban RJ, Bermudez-Fajardo A, Ungethuem U, Kuhn H. Gene expression alterations of human peripheral blood monocytes induced by medium-term treatment with the TH2-cytokines interleukin-4 and -13. Cytokine. 2005;30(6):366–77.
620. Torocsik D, Bardos H, Nagy L, Adany R. Identification of factor XIII-A as a marker of alternative macrophage activation. Cell Mol Life Sci. 2005;62(18):2132–9.
621. Walters EH, Soltani A, Reid DW, Ward C. Vascular remodelling in asthma. Curr Opin Allergy Clin Immunol. 2008;8(1):39–43.
622. McDonald DM. Angiogenesis and remodeling of airway vasculature in chronic inflammation. Am J Respir Crit Care Med. 2001;164(10):S39–45.
623. Wilson J. The bronchial microcirculation in asthma. Clin Exp Allergy. 2000;30(Suppl 1):51–3.
624. Baluk P, Tammela T, Ator E, Lyubynska N, Achen MG, Hicklin DJ, et al. Pathogenesis of persistent lymphatic vessel hyperplasia in chronic airway inflammation. J Clin Invest. 2005;115(2):247–57.
625. Walsh DA, Pearson CI. Angiogenesis in the pathogenesis of inflammatory joint and lung diseases. Arthritis Res. 2001;3(3):147–53.
626. Li X, Wilson JW. Increased vascularity of the bronchial mucosa in mild asthma. Am J Respir Crit Care Med. 1997;156(1):229–33.
627. Moreno RH, Hogg JC, Pare PD. Mechanics of airway narrowing. Am Rev Respir Dis. 1986;133(6):1171–80.
628. Ferrara N, Gerber HP, LeCouter J. The biology of VEGF and its receptors. Nat Med. 2003;9(6):669–76.
629. Lee CG, Link H, Baluk P, Homer RJ, Chapoval S, Bhandari V, et al. Vascular endothelial growth factor (VEGF) induces remodeling and enhances TH2-mediated sensitization and inflammation in the lung. Nat Med. 2004;10(10):1095–103.
630. Dvorak HF, Brown LF, Detmar M, Dvorak AM. Vascular permeability factor/vascular endothelial growth factor, microvascular hyperpermeability, and angiogenesis. Am J Pathol. 1995;146(5):1029–39.
631. Kanazawa H, Yoshikawa T. Up-regulation of thrombin activity induced by vascular endothelial growth factor in asthmatic airways. Chest. 2007;132(4):1169–74.
632. Vrugt B, Wilson S, Bron A, Holgate ST, Djukanovic R, Aalbers R. Bronchial angiogenesis in severe glucocorticoid-dependent asthma. Eur Respir J. 2000;15(6):1014–21.
633. Hoshino M, Takahashi M, Aoike N. Expression of vascular endothelial growth factor, basic fibroblast growth factor, and angiogenin immunoreactivity in asthmatic airways and its relationship to angiogenesis. J Allergy Clin Immunol. 2001;107(2):295–301.
634. Ribatti D. The crucial role of vascular permeability factor/vascular endothelial growth factor in angiogenesis: a historical review. Br J Haematol. 2005;128(3):303–9.
635. Psarras S, Volonaki E, Skevaki CL, Xatzipsalti M, Bossios A, Pratsinis H, et al. Vascular endothelial growth factor-mediated induction of angiogenesis by human rhinoviruses. J Allergy Clin Immunol. 2006;117(2):291–7.
636. Hoshino M, Nakamura Y, Hamid QA. Gene expression of vascular endothelial growth factor and its receptors and angiogenesis in bronchial asthma. J Allergy Clin Immunol. 2001;107(6):1034–8.
637. Asai K, Kanazawa H, Otani K, Shiraishi S, Hirata K, Yoshikawa J. Imbalance between vascular endothelial growth factor and endostatin levels in induced sputum from asthmatic subjects. J Allergy Clin Immunol. 2002;110(4):571–5.
638. Siddiqui S, Sutcliffe A, Shikotra A, Woodman L, Doe C, McKenna S, et al. Vascular remodeling is a feature of asthma and nonasthmatic eosinophilic bronchitis. J Allergy Clin Immunol. 2007;120(4):813–9.
639. Alagappan VK, McKay S, Widyastuti A, Garrelds IM, Bogers AJ, Hoogsteden HC, et al. Proinflammatory cytokines upregulate mRNA expression and secretion of vascular endothelial growth factor in cultured human airway smooth muscle cells. Cell Biochem Biophys. 2005;43(1):119–29.

640. Faffe DS, Flynt L, Bourgeois K, Panettieri RA Jr, Shore SA. Interleukin-13 and interleukin-4 induce vascular endothelial growth factor release from airway smooth muscle cells: role of vascular endothelial growth factor genotype. Am J Respir Cell Mol Biol. 2006;34(2):213–8.
641. Lee YC, Kwak YG, Song CH. Contribution of vascular endothelial growth factor to airway hyperresponsiveness and inflammation in a murine model of toluene diisocyanate-induced asthma. J Immunol. 2002;168(7):3595–600.
642. Corne J, Chupp G, Lee CG, Homer RJ, Zhu Z, Chen Q, et al. IL-13 stimulates vascular endothelial cell growth factor and protects against hyperoxic acute lung injury. J Clin Invest. 2000;106(6):783–91.
643. Boesiger J, Tsai M, Maurer M, Yamaguchi M, Brown LF, Claffey KP, et al. Mast cells can secrete vascular permeability factor/vascular endothelial cell growth factor and exhibit enhanced release after immunoglobulin E-dependent upregulation of Fc epsilon receptor I expression. J Exp Med. 1998;188(6):1135–45.

Part IV

Influence of Heterogeneity on Approaches to Management

Approaches to Management of Asthma: Guidelines for Stepped Care and Self-Monitoring

Daniel Rosenberg and Philip Palmon

Abstract

While the twentieth century brought rapid changes to the diagnosis and management of asthma, consensus guidelines did not exist until the mid-1980s. Over the subsequent four decades, guidelines evolved from expert-based recommendations to comprehensive, evidenced-based references. The main two guidelines to be discussed in this chapter are the expert panel reports (EPR) published by the United States National Heart, Lung, and Blood Institute (NHLBI) and the Global Strategy for Asthma Management and Prevention Report (GINA) published by the NHLBI and World Health Organization (WHO). While these guidelines both focus on evidence-based approaches to the diagnosis and management of asthma, there are significant differences in both organization and recommendations. Just as the introduction of evidence-based guidelines was revolutionary to asthma management over the last 20 years, we anticipate further development of recommendations specific to precision medicine based on ongoing cutting-edge clinical research.

Keywords

Evidence-based guidelines · Diagnosis · Management · Prevention · Exacerbations · Expert panel report · Implementation guidelines · Precision medicine

---

The original version of this chapter was revised. The correction to this chapter is available at https://doi.org/10.1007/978-3-031-32259-4_18

D. Rosenberg (✉)
University of Wisconsin School of Medicine and Public Health, Department of Medicine, Division of Allergy, Pulmonary and Critical Care Medicine, Madison, WI, USA
e-mail: drosenberg2@wisc.edu

P. Palmon
University of Wisconsin School of Medicine and Public Health, Madison, WI, USA

A. R. Brasier, N. N. Jarjour (eds.), *Precision Approaches to Heterogeneity in Asthma*, Advances in Experimental Medicine and Biology 1426,
https://doi.org/10.1007/978-3-031-32259-4_15

Evolution of Guidelines

The twentieth century brought rapid changes to the diagnosis and management of asthma, particularly with the advent of inhaled corticosteroid and beta-agonist therapy. However, there was little consensus guidance available to treating physicians until the mid-1980s. The importance of guideline-directed therapy came to the forefront, given both the increasing prevalence of asthma and fatal asthma attacks throughout the "Western" world. From 1984 to 1994, a prevalence increase of around 20-fold was seen in Australian children [1]. Even more concerning was a fourfold increase in asthma deaths from age 15 to 64 seen in New Zealand starting in 1976 compared to previous years [2].

As such, it is not surprising that the first guidelines in asthma were developed in Australia and New Zealand. These early guidelines were composed of expert-based recommendations based on the severity of disease, with a particular focus on preventing fatal asthma attacks [3]. Due to the rapidly evolving nature of diagnosis and treatment of asthma, guideline-based therapy continued to evolve throughout the 1990s. In 1989 the United States National Heart, Lung, and Blood Institute (NHLBI) and the National Asthma Education and Prevention Program (NAEPP) began to develop the first Expert Panel Report (EPR-1), which was published in 1991 [4]. In 1995, a joint effort between the NHLBI and WHO led to the first Global Strategy for Asthma Management and Prevention Report (GINA) [5]. While far more comprehensive with respect to epidemiology, diagnosis, and management than the earlier guidelines, the EPR and GINA reports were still primarily based on expert opinion during the late 1990s. The NHLBI subsequently published a second Expert Panel Report (EPR-2) in 1997, followed by an update of selected topics in 2002 [4].

With the increasing number of randomized control trials assessing pharmacologic interventions in asthma, it became clear that guidelines must be updated to properly assess rapidly changing scientific literature. In 2004, the NHLBI expert panel laid the framework to include evidence-based medicine in the guidelines for the first time. Four areas of focus were chosen for the new guidelines, including assessment and monitoring of asthma, patient education, environmental factors, and pharmacological treatments. In addition, grades were provided to assess the levels of evidence for each recommendation. These advances culminated in the third Expert Panel Report published in 2007 [6].

The most recent Expert Panel Report was published almost 13 years later, in 2020 [7]. The experts and stakeholders behind EPR-4 sought to focus on six a priori topics for systematic review. These included fractional excretion of nitric oxide (FeNO), indoor allergen mitigation, adjustable medication dosing, long-acting antimuscarinic agents (LAMA), allergen immunotherapy, and bronchial thermoplasty (BT). Of particular importance to this report is the methodologic framework of "Grading of Recommendations Assessment, Development and Evaluation" (GRADE). This internationally endorsed methodology provides a framework for consistency in assessing the quality of evidence behind all recommendations. Another novel feature of EPR-4 was an "implementation guidance" section that provided more straightforward guidance to clinicians on how to apply discussed recommendations.

Starting in 2002, the GINA guidelines similarly began to include evidence-based recommendations, mirroring the paradigm shift seen with EPR-3 [8]. Unlike the expert panel reports, the GINA guidelines have been published on an annual basis since 2002. Each year, a comprehensive literature review is performed in PubMed encompassing conventional randomized control trials as well as pragmatic and observational studies. Each article is screened by a medical librarian and at least two nonconflicted members of the Science Committee. The committee performs a three-part process to assess each publication including assessment of quality and relevance, decision about the inclusion of the evidence, and a discussion of related changes to the GINA report. As all topics are systemically reviewed twice per year rather than focusing on several a priori questions, the GINA yearly updates are far more comprehensive than the aforementioned EPR reports. The first major updates occurred in 2006, with subsequent extensive revisions in 2014 and 2019 [9].

GINA Guidelines

The Global Initiative for Asthma (GINA) provides a global strategy for asthma management and prevention. While a comprehensive review of the GINA Guideline is outside the scope of this chapter, we will provide a brief summarization of asthma and severity definitions, and a general review of pharmacologic and nonpharmacologic recommendations.

GINA describes asthma as a heterogeneous disease characterized by chronic airway inflammation, respiratory symptoms (wheeze, shortness of breath, chest tightness, and cough) that vary over time and intensity, and variable expiratory airflow limitation that may become persistent. GINA recommends that a diagnosis of asthma is based on the history of symptom patterns and evidence of variable expiratory airflow limitation documented from bronchodilator reversibility [10, 11]. Diagnosis in patients less than 5 years of age may be challenging given that intermittent wheezing and cough are common in children without asthma, and performing bronchodilator studies on children may not always be possible [12, 13]. An underlying diagnosis of asthma is more likely if the following are present: wheezing or coughing that occurs with exercise, laughing, crying, or in the absence of respiratory infection, a history of other atopic diseases (eczema or allergic rhinitis), allergen sensitization, or asthma in first-degree relatives, clinical improvement during 2–3 months of controller treatment, and worsening after cessation [10].

GINA recommends assessment of asthma control in two domains, symptom control and risk of adverse outcomes [10]. Symptom control can be assessed over the course of 4 weeks using the GINA consensus control tool incorporating the following questions: Daytime asthma symptoms more than twice per week, any night waking due to asthma, short-acting beta agonist use (SABA) reliever for symptoms more than twice per week, and any activity limitation due to asthma. For adults and adolescents, other options include categorical control tools such as the Royal

College of Physicians Three Questions' tool or Asthma APGAR tool, and numerical asthma control tools such as the Asthma Control Questionnaire and Asthma Control Test [14–17]. A similar symptom control assessment tool is available for children aged 5 years or younger [10].

Uncontrolled asthma symptoms are an important risk factor for exacerbations [18]. The GINA report lists the following as risk factors for exacerbations: medication use (high SABA use, inadequate inhaled corticosteroids), comorbid medical conditions (obesity, chronic rhinosinusitis, gastroesophageal reflux disease, food allergy, and pregnancy), exposures such as smoking, major psychological, or socioeconomic problems, low FEV_1, high bronchodilator responsiveness, and elevated type 2 inflammatory markers such as blood eosinophils and the fraction of exhaled nitric oxide (FeNO) [19–28].

Assessment of asthma severity is determined by the concept of 'difficulty to treat' and is separated into three categories including mild, Moderate, and severe. Severity is assessed retrospectively from the level of treatment required to control symptoms and exacerbations. Mild asthma is well-controlled with as-needed inhaled corticosteroids (ICS)-formoterol or with low-dose ICS plus as-needed SABA. Moderate asthma is well-controlled with low or medium-dose maintenance ICS-formoterol. Severe asthma is defined as asthma that remains uncontrolled despite optimized treatment with high-dose inhaled corticosteroid and long-acting beta agonist (ICS/LABA), or that requires high-dose ICS-LABA to prevent it from being uncontrolled. Severe asthma is a subset of difficult-to-treat asthma as all modifiable risk factors such as incorrect inhaler technique, poor adherence, smoking, or comorbidities have been addressed and remains asthma uncontrolled [29–31].

GINA recommends a step-wise approach for reliever and controller therapies for asthma control for the following age groups: adults and adolescents 12+ years in age, children aged six to 11, and children aged five or younger.

For adults and adolescents, the following steps are recommended (see Fig. 1):

Track 1 with controller and preferred reliever of low-dose ICS-formoterol

- Steps 1 and 2 are combined with as-needed low-dose ICS-formoterol
- Step 3 recommends low-dose maintenance ICS-formoterol
- Step 4 medium dose maintenance ICS-formoterol
- Step 5 add-on Long-acting muscarinic antagonist, referral for assessment of phenotype, consider high-dose maintenance ICS-formoterol, anti-IgE, anti-IL5/5R, anti-IL4R, and anti-thymic stromal lymphopoietin (TSLP)

Track 2 outlines the controller with alternative reliever of SABA

- Step 1 Take ICS whenever SABA taken
- Step 2 Low-dose maintenance ICS
- Step 3 recommends low-dose maintenance ICS-formoterol
- Step 4 medium dose maintenance ICS-formoterol

- Step 5 add-on Long-acting muscarinic antagonist, referral for assessment of phenotype, consider high-dose maintenance ICS-formoterol, anti-IgE, anti-IL5/5R, anti-IL4R, and anti-TSLP

For Children aged 6–11, step-up therapy is as follows (see Fig. 2):

- Step 1 Low-dose ICS taken whenever SABA taken
- Step 2 Daily low-dose ICS

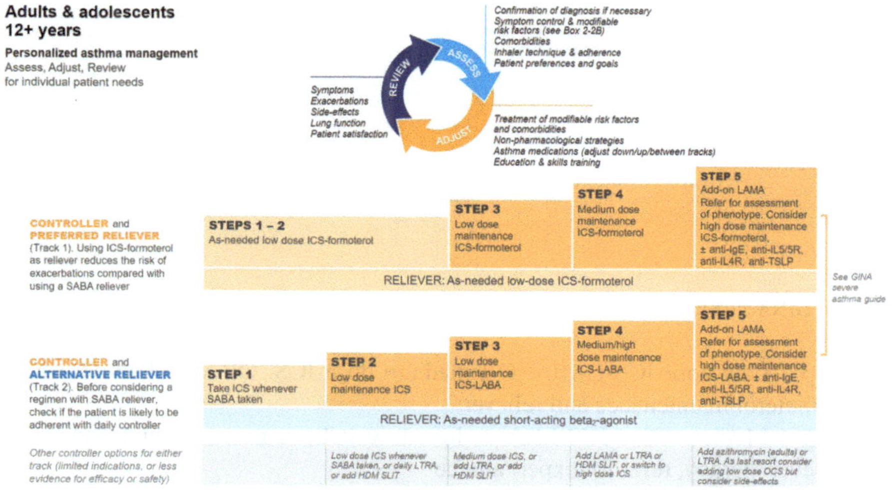

Fig. 1 GINA step-up adolescents and adults

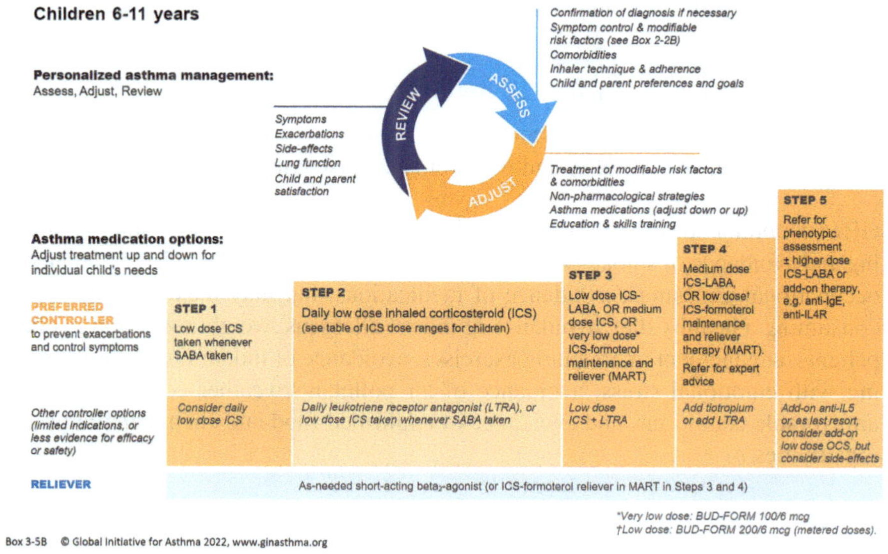

Fig. 2 GINA step-up 6–11

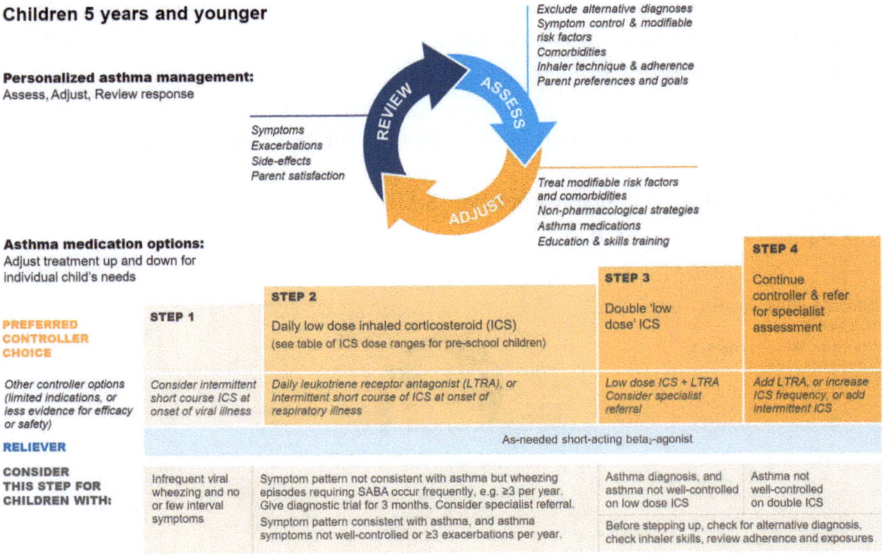

Fig. 3 GINA step-up ages 5 and younger

- Step 3 Low-dose ICS-LABA, OR medium dose ICS, OR very low-dose ICS-formoterol maintenance and reliever
- Step 4 Medium dose ICS-LBA or low-dose ICS-formoterol maintenance and reliever therapy, refer for expert advice
- Step 5 Refer for phenotypic assessment, higher-dose ICS-LABA or add-on therapy including anti-IgE, anti-IL4R

For children aged 5 or younger step-up therapy is as follows (see Fig. 3):

- Step 1 as-needed SABA
- Step 2 Daily low-dose ICS
- Step 3 Double 'low-dose' ICS
- Step 4 Continue controller and refer for specialist assessment

GINA recommends the following nonpharmacologic strategies: cessation of smoking and avoidance of smoke exposure, engagement in physical activity, avoidance of occupational exposures, avoidance of medications that may make asthma worse, consuming a healthy diet, avoidance of indoor allergens, weight reduction in obese patients, engagement in breathing exercises, avoidance of indoor air pollution, dealing with emotional stress, avoidance of air pollutants/weather conditions during unfavorable conditions, and avoidance of confirmed food and chemical allergy and sensitivities.

National Heart, Lung, and Blood Institute Focused 2020 Update

Here we will review the six topics the NHLBI reviewed along with their recommendations regarding each topic. Additionally, the strength of each recommendation will be included. While a comprehensive review is outside the scope of this chapter, we will to introduce each major topic reviewed by the EPR-4.

FeNO Use

Exhaled nitric oxide can be used to measure the level of airway inflammation. There are four recommendations by the NHLBI with regard to FeNO. The first recommendation is that for patients who are aged 5 years and older, the addition of FeNO can be used as an adjunct measure when the diagnosis of asthma is uncertain despite history, clinical findings, clinical course, and spirometry. This is a conditional recommendation with moderate certainty of evidence. The second recommendation is for patients who are 5 years and older with persistent allergic asthma for whom there is uncertainty in choosing, monitoring, or adjusting anti-inflammatory therapies. For these patients, the addition of FeNO can be used as a part of ongoing asthma monitoring and management strategy, including frequent assessments. This is a conditional recommendation with a low certainty of evidence.

The third recommendation is that in patients aged 5 years or older, the expert panel recommends against the use of FeNO in isolation to assess asthma control, predict future exacerbations, or assess exacerbation severity. This is a strong recommendation with low certainty of evidence. The fourth recommendation is that in patients aged 4 years or younger with recurrent episodes of wheezing, FeNO measurement does not reliably predict the future development of asthma.

Indoor Allergen Mitigation

Environmental control was an essential part of asthma management in ERP-3 [32]. There are four recommendations regarding indoor allergen mitigation. The first is that for individuals with asthma who do not have sensitization to specific indoor allergens or who do not have symptoms related to exposure to specific indoor allergens, the expert panel recommends against allergen mitigation interventions. This is a conditional recommendation with low certainty of evidence. The second recommendation is that in individuals with asthma who have symptoms with exposure to specific indoor allergens confirmed by history or allergy testing, a multicomponent allergen-specific mitigation intervention is recommended. This is a conditional recommendation with low certainty of evidence. The third recommendation is that in individuals with asthma with sensitization or symptoms related to exposure to pests (cockroaches and roaches), the panel recommends integrated pest management alone or as part of a multicomponent allergen-specific mitigation intervention. This is a conditional recommendation with a low certainty of evidence. The last recommendation is in patients with asthma who have sensitization or symptoms related to exposure to dust mites, the panel recommends impermeable pillow and mattress covers only as part of a multicomponent allergen mitigation intervention.

Intermittent ICS Use

The use of scheduled daily ICS was the recommended controller therapy for persistent asthma at the time of EPR-3, though there was a suggestion that intermittent dosing was potentially useful per expert opinion [32]. There are five recommendations in EPR-4 regarding the use of intermittent ICS. The first recommendation is in children between ages 0 and 4 with recurrent wheezing secondary to respiratory tract infections, and the expert panel recommends a short course of daily ICS with as-needed SABA at the onset of symptoms compared to SABA monotherapy. Budesonide suspension 1 mg BID for 7 days was noted to be effective at the first sign of infectious symptoms [33]. This is a conditional recommendation with a high certainty of evidence. The second recommendation is that in patients age 12 years or older with mild persistent asthma, the expert panel conditionally recommends either daily low-dose ICS and as-needed SABA for quick relief therapy, or as-needed ICS and SABA used concomitantly. The third recommendation is that in patients ages 4 years or older with mild to moderate persistent asthma who are adherent to daily ICS treatment, the expert panel recommends against short-term increase in the ICS dose for increased symptoms or decreased peak flow.

The fourth recommendation is that in individuals ages 4 years and older with moderate to severe persistent asthma, the expert panel recommends ICS-formoterol in a single inhaler used as both daily controller and reliever therapy rather than either a higher-dose ICS as daily controller therapy and SABA for quick relief, or the same dose ICS-LABA as daily controller therapy and SABA for quick relief therapy. This is a strong recommendation with certainty for ages 4–11 years. The last recommendation is that in individuals ages 12 and older with moderate to severe persistent asthma, the expert panel conditionally recommends ICS-formoterol in a single inhaler used as both daily controller and reliever therapy compared to higher-dose ICS-LABA as daily controller therapy and SABA for quick relief therapy.

Use of LAMA for Asthma

The use of long-acting muscarinic antagonists as bronchodilators for asthma was not addressed in EPR-3, but multiple studies have been published since then [32]. There are three recommendations regarding LAMA use for asthma. In patients aged 12 years or older with uncontrolled persistent asthma, the expert panel recommends against adding LAMA to ICS compared to adding LABA. This is a conditional recommendation with moderate certainty of evidence. The second recommendation is that if a LABA cannot be used in an individual aged 12 years and older with uncontrolled persistent asthma, the expert panel conditionally recommends adding LAMA to ICS compared to continuing the same dose of ICS alone. This is also a conditional recommendation with a moderate certainty of evidence. The final recommendation is for patients 12 years and older with uncontrolled persistent asthma, the addition of LAMA to ICS and LABA compared to continuing with the same dose of ICS and LABA. This is a conditional recommendation with moderate certainty of evidence.

Subcutaneous and Sublingual Immunotherapy

Two recommendations were made regarding immunotherapy in patients with allergic asthma. The first recommendation is in individuals ages 5 and older with milder to moderate allergic asthma, and the expert panel conditionally recommends the use of subcutaneous immunotherapy as an adjunct therapy as long as asthma remains well-controlled. This is a conditional recommendation with moderate certainty of evidence. The second recommendation is that in patients with persistent allergic asthma, the expert panel recommends against the use of sublingual immunotherapy in asthma treatment.

Bronchial Thermoplasty

The expert panel made one recommendation regarding bronchial thermoplasty. In individuals ages 18 years and older with persistent asthma it is conditionally recommended against to perform bronchial thermoplasty. This is noted to have low certainty of evidence.

Differences Between the Guidelines

There are differences in recommended asthma treatment guidance between the GINA Main Report and ERP-4. We will spend the next portion discussing these differences.

There are differences initial and step-up therapy between GINA and EPR-4, particularly in ages 12 and older and ages 5/6–11. Per GINA, in patients aged 12 and older initial step-up therapy begins with as-needed ICS-formoterol as a combined step 1 and 2. If further increase in therapy is indicated, GINA recommends ICS-formoterol as both a schedule controller and as-needed reliever. This contrasts with EPR-4 which recommends the initial use of SABA for rescue therapy as step 1 therapy, daily low-dose ICS and as-needed SABA vs as-needed ICS and SABA as step 2, then in step 3 recommends ICS- formoterol for both daily controller therapy and as a reliever. Please see Figs. 4 and 5 for full EPR-4 step-up recommendations for ages 12 and older and ages 5–11.

GINA notes that in a large double-blind study and an open-label study, there is evidence of risk reduction for severe exacerbations with the use of ICS-formoterol compared to SABA alone [34, 35]. Additionally, GINA recommends against SABA alone as such patients are at an increased risk of asthma-related death and urgent

Figure I.d: Stepwise Approach for Management of Asthma in Individuals Ages 12 Years and Older

	Intermittent Asthma	Management of Persistent Asthma in Individuals Ages 12+ Years				
Treatment	STEP 1	STEP 2	STEP 3	STEP 4	STEP 5	STEP 6■
Preferred	PRN SABA	Daily low-dose ICS and PRN SABA or PRN concomitant ICS and SABA▲	Daily and PRN combination low-dose ICS-formoterol▲	Daily and PRN combination medium-dose ICS-formoterol▲	Daily medium-high dose ICS-LABA + LAMA and PRN SABA▲	Daily high-dose ICS-LABA + oral systemic corticosteroids + PRN SABA
Alternative		Daily LTRA* and PRN SABA or Cromolyn,* or Nedocromil,* or Zileuton,* or Theophylline,* and PRN SABA	Daily medium-dose ICS and PRN SABA or Daily low-dose ICS-LABA, or daily low-dose ICS + LAMA,▲ or daily low-dose ICS + LTRA,* and PRN SABA or Daily low-dose ICS + Theophylline* or Zileuton,* and PRN SABA	Daily medium-dose ICS-LABA or daily medium-dose ICS + LAMA, and PRN SABA▲ or Daily medium-dose ICS + LTRA,* or daily medium-dose ICS + Theophylline,* or daily medium-dose ICS + Zileuton,* and PRN SABA	Daily medium-high dose ICS-LABA or daily high-dose ICS + LTRA,* and PRN SABA	
		Steps 2-4: Conditionally recommend the use of subcutaneous immunotherapy as an adjunct treatment to standard pharmacotherapy in individuals ≥ 5 years of age whose asthma is controlled at the initiation, build up, and maintenance phases of immunotherapy▲			Consider adding Asthma Biologics (e.g., anti-IgE, anti-IL5, anti-IL5R, anti-IL4/IL13)**	

Fig. 4 EPR-4 step-up therapy for 12 and older

Figure I.c: Stepwise Approach for Management of Asthma in Individuals Ages 5–11 Years

	Intermittent Asthma	Management of Persistent Asthma in Individuals Ages 5–11 Years				
Treatment	STEP 1	STEP 2	STEP 3	STEP 4	STEP 5	STEP 6
Preferred	PRN SABA	Daily low-dose ICS and PRN SABA	Daily and PRN combination low-dose ICS-formoterol▲	Daily and PRN combination medium-dose ICS-formoterol▲	Daily high-dose ICS-LABA and PRN SABA	Daily high-dose ICS-LABA + oral systemic corticosteroid and PRN SABA
Alternative		Daily LTRA,* or Cromolyn,* or Nedocromil,* or Theophylline,* and PRN SABA	Daily medium-dose ICS and PRN SABA or Daily low-dose ICS-LABA, or daily low-dose ICS + LTRA,* or daily low-dose ICS +Theophylline,* and PRN SABA	Daily medium-dose ICS-LABA and PRN SABA or Daily medium-dose ICS + LTRA* or daily medium-dose ICS + Theophylline,* and PRN SABA	Daily high-dose ICS + LTRA* or daily high-dose ICS + Theophylline,* and PRN SABA	Daily high-dose ICS + LTRA* + oral systemic corticosteroid or daily high-dose ICS + Theophylline* + oral systemic corticosteroid, and PRN SABA
		Steps 2-4: Conditionally recommend the use of subcutaneous immunotherapy as an adjunct treatment to standard pharmacotherapy in individuals ≥ 5 years of age whose asthma is controlled at the initiation, build up, and maintenance phases of immunotherapy▲			Consider Omalizumab**▲	

Fig. 5 EPR-4 step-up therapy for ages 5–11

asthma-related health care [36–38]. In fact, there is evidence regular use of SABA in patients with newly diagnosed asthma leads to lower lung function and worse outcomes compared to those on daily controller ICS [39]. GINA notes that

Figure I.b: Stepwise Approach for Management of Asthma in Individuals Ages 0-4 Years

	Intermittent Asthma	Management of Persistent Asthma in Individuals Ages 0-4 Years				
Treatment	STEP 1	STEP 2	STEP 3	STEP 4	STEP 5	STEP 6
Preferred	PRN SABA and At the start of RTI: Add short course daily ICS▲	Daily low-dose ICS and PRN SABA	Daily medium-dose ICS and PRN SABA	Daily medium-dose ICS-LABA and PRN SABA	Daily high-dose ICS-LABA and PRN SABA	Daily high-dose ICS-LABA + oral systemic corticosteroid and PRN SABA
Alternative		Daily montelukast* or Cromolyn,* and PRN SABA		Daily medium-dose ICS + montelukast* and PRN SABA	Daily high-dose ICS + montelukast* and PRN SABA	Daily high-dose ICS + montelukast*+ oral systemic corticosteroid and PRN SABA
			For children age 4 years only, see Step 3 and Step 4 on Management of Persistent Asthma in Individuals Ages 5-11 Years diagram.			

Fig. 6 EPR-4 step-up therapy for ages 0–4

adherence to daily ICS in patients with intermittent symptoms is low, thereby increasing the risk of SABA monotherapy [40].

For patients aged 6–11, GINA recommends as-needed ICS and SABA concomitantly as step 1 therapy, and for step 2, increasing ICS to daily controller therapy. This contrasts with EPR-4 for children aged 5–11, which recommends SABA alone for step 1 and then the addition of daily controller ICS with the addition of as-needed SABA. GINA notes that there are significantly fewer exacerbations with the use of intermittent ICS when SABA is taken compared to SABA alone [41]. GINA also notes there are no studies of as-needed only ICS-formoterol in children aged 6–11 at this point in time.

GINA and EPR-4 are in agreement for initial step-up therapy for children aged 0–4/5, with step 1 therapy incorporating the use of as-needed SABA and step 2 incorporating the use of daily ICS plus an as-needed SABA. There is the support for a short course of ICS at the onset of viral-induced wheezing by both GINA and EPR-4, particularly since the risk of exacerbation between regular low-dose ICS and episodic high-dose ICS is similar [42] (Fig. 6).

Immunotherapy

GINA and EPR-4 have different recommendations regarding immunotherapy, particularly in the case of the sublingual formulation. GINA recommends sublingual immunotherapy to dust mites in patients with allergic rhinitis with dust mite sensitization and persistent symptoms despite low-medium dose ICS, provided FEV1 is greater than 70%. EPR-4 recommends against the use of SLIT for asthma specifically but notes that therapy may be indicated for allergic rhinoconjunctivitis. GINA notes that in a trial of house dust mite SLIT, patients with asthma were able to reduce their ICS dose [43]. Another study showed that the addition of house dust mite SLIT in patients with uncontrolled allergic asthma had been shown to increase time to exacerbation during a trial of ICS reduction [44]. EPR-4 recognizes these studies but also notes that there is minimal benefit for the following outcomes identified as critical: exacerbations, asthma control, and quality of life [43–50].

With regards to subcutaneous immunotherapy (SCIT), both GINA and EPR-4 recommend the use in patients with allergic asthma. GINA states it plans to review the evidence regarding allergen immunotherapy, but if the benefits outweigh the risks SCIT can be considered. NHLBI has more specific recommendations, noting that for patients 5 years or older with mild to moderate asthma, SCIT can be used as adjunct therapy, provided asthma is controlled. There is potentially a small benefit to the critical outcomes of exacerbations, quality of life, and asthma control leading to the EPR-4 recommendation.

Allergen Mitigation

There are differences in recommendations regarding allergen mitigation between GINA and EPR-4. GINA notes that allergen avoidance, in general, is not a recommended strategy. It notes that for sensitized patients, there is a limited benefit for adults with single avoidance strategies; in children, there is a similarly limited benefit for multicomponent avoidance strategies. While there may be some benefits, they are complicated, and expensive, without validated methods to identify those that would benefit [10]. EPR-4 notes that for sensitized patients with symptoms related to exposure, a multicomponent allergen-specific mitigation intervention is recommended. Similarly, for pests and dust mites, integrated pest management or a multicomponent allergen-specific mitigation intervention is conditionally recommended. EPR-4 recognizes that the literature regarding allergen mitigation is quite heterogeneous and that for most interventions, the certainty of the evidence is low, and the benefits are. For those who have symptoms when exposed, consideration for the use of the four potential interventions (mattress and pillow covers, integrated pest management, HEPA vacuum cleaners, and mold mitigation) is recommended.

Bronchial Thermoplasty

GINA notes that bronchial thermoplasty (BT) is a potential step 5 therapy for adult patients whose asthma remains uncontrolled despite optimized treatments and referral to an asthma specialty center. Per GINA, BT should be performed in patients with severe asthma only in the context of an Institutional Review Board (IRB) approved systematic registry or clinical study. GINA notes that a large placebo effect gives no beneficial effect on lung function or asthma symptoms when compared to patients who underwent sham bronchoscopy procedures [51]. EPR-4 recommends against the use of BT for adults with persistent asthma as a part of routine asthma care. EPR-4 agrees with GINA that BT should only be performed in the context of IRB-approved registry and clinical studies so that long-term efficacy and safety can be assessed in BT. EPR-4 notes that the three randomized clinical studies of BT were funded by the company that markets the BT device. The Research in Severe Asthma Study and the Asthma Interventional Research Study compared BT to the standard of care treatment and found a modest improvement in exacerbations not requiring corticosteroids, a reduction in number of emergency department visits, improved Asthma Control Questionnaire scores, and less rescue medication use [52, 53]. Despite these improvements, therapy was associated with adverse outcomes including ongoing or increased dyspnea, chest discomfort, bronchial irritation, wheezing, and cough. Additionally, hospitalizations during and after the treatment period were increased in patients undergoing BT [51–53]. Overall, EPR-4 notes that the benefits of BT are small, and the harms are moderate, leading to their recommendation. It is noted that further randomized and registry studies may be useful in determining which patient population may benefit the most from BT, such as patients who had unsuccessfully trialed different biologic medications.

FeNO

The use of fraction of exhaled nitric oxide (FeNO) is discussed by both the GINA and NHLBI guidelines. In general, the EPR-4 recommends that in patients aged 5 years or older FeNO can be used as an adjunct test to help with the diagnosis, initiation of treatment, and ongoing monitoring of response for asthma, but should not be used in isolation. There is also insufficient evidence to use FeNO in children to predict the development of asthma.

With regards to diagnosis, GINA notes that FeNO has not been established for the diagnosis of asthma, given the variability of FeNO correlation with an underlying diagnosis of asthma. For example, FeNO may be elevated in patients with type 2 high inflammation but low in patients with other phenotypes [54]. FeNO is also elevated in nonasthma diseases such as eosinophilic bronchitis and allergic rhinitis as well as tobacco use.

GINA and EPR-4 both note there is data supporting the use of FeNO for monitoring including evidence that FeNO-guided treatment is not associated with a significant reduction in exacerbation rates when compared to guideline treatments in both children and young adults [55, 56]. GINA notes there are problems with the design of the intervention and control algorithms that make comparisons and conclusions difficult. EPR-4 similarly notes issues with the algorithms in the literature and acknowledges its potential lack of relevance for clinical practice.

For initiation of treatment, GINA notes that an elevated FeNO, particularly FeNO >50 parts per billion, is associated with a favorable short-term response to ICS [57, 58]. However, a low FeNO does not preclude the need for ICS. EPR-4 notes that FeNO levels have been responsive to certain anti-inflammatory controller medications including ICS, montelukast, and omalizumab. Given the costs, availability of testing, and increased burden on patients, EPR-4 recommends shared decision-making when assessing the utility of frequent FeNO monitoring as a part of an individual's asthma care plan.

Long-Acting Muscarinic Antagonists

With regards to long-acting muscarinic antagonists, GINA and EPR-4 overall are in agreement. Both recognize that LAMAs can be additional therapy to ICS and LABA, particularly when the ICS is at least at a moderate dose. Both note that the addition of LAMA is associated with modest improvements in lung function and reduced rates of exacerbations, but symptoms often remain unchanged. As stated above, EPR-4 notes that ICS and LAMA combination therapy should not be used in patients instead of LABA unless contraindications to LABA exist. It is preferred to add-on LAMA rather than continue with high-dose ICS and LABA if symptoms are poorly controlled.

Inhaled Corticosteroids

GINA and EPR-4 have differing recommendations for ICS use during exacerbations. GINA recommends that for patients already on ICS controller therapies, providers should increase doses for 2–4 weeks. The rationale for this recommendation includes studies showing increasing ICS dose reduces health care utilization, prevents progression to severe exacerbations, and reduces oral corticosteroid use [59–62]. GINA does acknowledge there are negative studies that do not support this recommendation. EPR-4 takes the opposite stand in their most recent update. As noted earlier, for patients aged 4 years or older with mild to moderate asthma who are likely to be adherent to daily ICS, the expert panel recommends against a short-term increase in ICS for worsening asthma control. EPR-4 notes the recent work of Jackson et al. in which there were no difference in rates of OCS use for exacerbations in children whose ICS dose was quintupled [63]. The work of McKeever et al. is also referenced, but EPR-4 notes that this was not a blinded, placebo controlled trial and baseline adherence to controller therapy was noted to be low [62]. This suggests it is unclear if the benefit was from increasing ICS or simply improving adherence to ICS independent of dosing.

The Use of Biologics

GINA makes general recommendations regarding the use of biologics as add-on therapy for patients uncontrolled on steps 4–5 therapy. This is based on biomarkers of IgE, eosinophil count, and OCS use. The NHLBI notes that their review of the literature did not include studies examining the role of biologics in the treatment of asthma. Subsequently, the report does not include specific recommendations other than the consideration of adding a biologic in steps 5 and 6.

Areas Needing Further Exploration

While both guidelines are quite comprehensive with both the breadth and depth of the literature review, there are areas that still require further investigation. One such area is further guidance for stepping-down therapy. This was not one of the areas chosen for EPR-4. EPR-3 does note that step-down therapy can be considered after 3 months of asthma stability. The GINA main report has a section on stepping-down therapy with varying levels of evidence. It recommends that patients with well-controlled symptoms for 3 months or more and stable lung function tolerate step-down in therapy. Current evidence suggests that discontinuation of LABA may lead to deterioration, but most patients will likely tolerate reductions of ICS of 25–50% in 3-month increments [64, 65]. A targeted approach with the use of biomarkers is a potential avenue. A randomized control trial of patients with well-controlled asthma on high-dose ICS-LABA comparing a strategy with type 2 biomarkers versus ACQ-7 and history of recent exacerbations was unfortunately inconclusive given many patients did not follow recommendations [66].

Another area needing further exploration is what to do if ICS-formoterol is not specifically covered by a patient's insurance. Given the variability of insurance formularies, many patients may find themselves with ICS/LABA combinations that are not applicable to the ICS/Formoterol recommendations. Similarly, the relevant literature supporting as-needed and single inhaler ICS-formoterol combinations specifically used budesonide-formoterol [34, 35]. While the GINA guidelines suggest mometasone-formoterol could potentially be used for single maintenance and reliever therapy (SMART) therapy, this specific formulation has not been studied specifically.

A related question to SMART therapy is the optimal dosing of ICS-formoterol. For instance, EPR-4 makes a recommendation of one to two puffs administered one to two times per day. This potentially leads to large variability in reliever and maintenance therapy, with some patients likely to be over-treated and others with lower-than-needed maintenance dosing. While there are clear upper limits for the formoterol component for both adults and children, EPR-4 provides the same maximum daily dosing guidance regardless of low versus medium steroid dose.

There are some practical considerations with regard to intermittent therapy as well. Both guidelines have as-needed ICS/SABA as an option for patients, but given that this formulation is not yet available in the United States, it requires patients to carry two inhalers for rescue therapy. While a recent study did show a reduction in severe exacerbations in patients with moderate to severe asthma with the use of ICS/SABA compared to SABA alone, the required use of multiple inhalers likely limits the external validity of this data [67]. With regard to the recommendations made to use a short course of daily ICS at the start of a viral infection in children, clinicians must carefully keep track of the frequency of use. In addition, this approach may not be ideal for children who experience rapid clinical deterioration resulting in intensive care unit admissions.

The role of immunotherapy per the guidelines is limited to patients with well-controlled mild to moderate asthma. There is an outstanding question regarding the role of immunotherapy for patients with severe asthma. A recent study by Chow et al. reviewed the rate of severe systemic reactions to subcutaneous immunotherapy [68]. Their group found that there was no difference in rates of grade 3 and 4 severe reactions in patients with severe asthma compared to those without asthma or those with less severe asthma. This is an encouraging finding suggesting that patients with well-controlled severe asthma may potentially benefit from SCIT, though the absolute number of such patients in the study was relatively small.

Further Directions in Search of Precision Medicine

While there has been a tremendous evolution in guideline-based diagnosis and management of asthma, clinicians are still somewhat limited in providing an approach that is truly "personalized." As asthma results from a complex interplay between genetics and early-in-life exposures, targeting one particular pathway can only provide so much benefit. This is clear from the relevant clinical trials of biologic medications for two reasons. With respect to symptoms, current data suggest that biologics are more likely to provide more modest benefits, such as small increases in lung volume. Greater efficacy has been seen in the prevention of symptomatic flares and the use of oral steroids, but interestingly reductions in exacerbations do not seem to decrease by more than 50% in the relevant trials [69–71].

Furthermore, it is often difficult to choose between type 2 high-focused biologics given similar efficacy in preventing exacerbations and increased response to similar biomarkers. Head-to-head trials between these therapies in the literature are lacking, and one must use caution in comparing different studies with varying inclusion criteria. An illustrative example is the use of an eosinophil cut-off of 300 in the benralizumab trials vs. 400 in the reslizumab trials, which makes comparison of exacerbation rates challenging [69, 71]. Some potential considerations include the existence of comorbid disease. As multiple asthma biologics have been FDA-approved for alternative conditions, one might lean toward dupilumab for concomitant nasal polyposis, while omalizumab would be an ideal option in a patient with chronic urticaria. However, biomarkers in these other allergic conditions are even more poorly understood than in asthma, which causes difficulty when multiple medications are approved for the same disease (such as in nasal polyposis).

The introduction of tezepelumab, a monoclonal antibody that blocks thymic stromal lymphopoietin (TSLP), is illustrative of the limitations of a type 2 high-focused approach. As TSLP is a more "upstream" mediator with wide-ranging effects on inflammation, it was plausible that blockade would provide more profound benefits in both type 2 high and low asthma phenotypes. However, the absolute reduction in exacerbations was found to be around 50%, with the majority of the reduction seen in patients with higher baselines of exacerbations [72]. In addition, reduction in exacerbations was more profound in patients with type 2 biomarkers such as eosinophil count >150, similar to what was found in the dupilumab asthma trials [70]. However, in spite of the decreased efficacy in type 2 low phenotypes, clinicians now have access to a biologic medication that can be approved without absolute minimums of peripheral eosinophils or FeNO.

The difficulty of treating patients with type 2 low phenotypes has been well illustrated in the relevant trials of biologic medications. Type 2 low asthma has historically been difficult to classify given its unclear pathophysiology and lack of clinically relevant biomarkers. Some theories for pathogenesis include TH1 cytokines, TH17 cytokines, IL-6, IL-17, and tumor necrosis factor. Of particular note is the association between asthma and an IL-17-mediated "psoriasis-like" phenotype [73]. However, clinical trials of anti-IL-17A monoclonal antibodies failed to find significant efficacy, which illustrates the difficulties in implementing precision medicine [74]. One promising area of investigation is the identification of possible IL17a biomarkers, particularly urinary 11-dehydrothromboxane B2 [75]. Such a targeted proposed approach bears some similarities to the identification of peripheral eosinophilia as an actionable target in anti-IL-5 therapy.

One particular phenotype that continues to confound our understanding is the so-called "neutrophilic phenotype." The Severe Asthma Research Program (SARP) has identified a subset of patients with moderate-severe asthma and high healthcare utilization despite high doses of inhaled or oral glucocorticoids with significant neutrophilia (with and without concomitant eosinophilia) [76, 77]. In addition, sputum neutrophilia has been associated with a relatively high proportion of sudden-onset fatal asthma attacks, up to 23% in the relevant literature [78]. We should note, however, that this phenotype is controversial given possible confounding from air pollution, respiratory infection, gastroesophageal disease, and other plausible etiologies of increased neutrophil counts. A wide variety of interventions such as monoclonal antibodies, phosphodiesterase inhibitors, macrolide antibiotics, and bronchial thermoplasty have been evaluated, but without clear biomarkers, results have often been inconclusive. Another potential therapeutic target identified by SARP is the association between plasma IL-6 concentrations, metabolic dysfunction, and obesity [79]. As anti-IL-6 therapy is well-described in the Rheumatology literature, this provides yet another possibility for personalized medicine in severe asthma.

A systematic approach to personalized medicine is embodied by the ongoing Precision Interventions for Severe and/or Exacerbation-Prone Asthma Network Study (PRECISE) [80]. This phase 2 study seeks to study the efficacy of several distinct biologic therapies in biomarker-selected groups of patients. The patient population recruited for this study have moderate-severe asthma with lack of improvement on conventional therapies, and undergoes testing to identify potentially actionable biomarkers. Interventions in this study include novel therapies such as anti-IL-6 therapy, CFTR stabilizer therapy, tyrosine kinase inhibitors, dietary supplementation, and bacterial lysate. The PRECISE study will be described in detail in the third section of Part IV.

References

1. Peat JK, Berg RHVD, Green WF, Mellis CM, Leeder SR, Wolcock AJ. Changing prevalence of asthma in Australian children. BMJ. 1994;308(6944):1591.
2. Sears MR, Rea HH, Fenwick J, Beaglehole R, Gillies AJ, Holst PE, et al. Deaths from asthma in New Zealand. Arch Dis Child. 1986;61(1):6.

3. Woolcock A, Rubinfeld AR, Seale JP, Landau LL, Antic R, Mitchell C, et al. Thoracic Society of Australia and New Zealand. Asthma management plan, 1989. Med J Aust. 1989;151(11–12):650–3.
4. Berg AO, Moy JG. Guidelines for the diagnosis and management of asthma. J Am Board Fam Pract. 1992;5(6):629–34.
5. Pauwels RA. Global strategy for asthma management and prevention. Jpn J Allergol. 1996;45(8–9):792.
6. Morosco G, Kiley J. Expert Panel Report 3 (EPR-3): guidelines for the diagnosis and management of asthma-summary report 2007. J Allergy Clin Immunol. 2007;120(5):S93.
7. (NAEPPCC) EPWG of the NH Lung, and Blood Institute (NHLBI) administered and coordinated National Asthma Education and Prevention Program Coordinating Committee, Cloutier MM, Baptist AP, Blake KV, Brooks EG, Bryant-Stephens T, et al. 2020 focused updates to the asthma management guidelines: a report from the National Asthma Education and Prevention Program Coordinating Committee Expert Panel Working Group. J Allergy Clin Immunol. 2020;146(6):1217–70.
8. Bateman ED, Hurd SS, Barnes PJ, Bousquet J, Drazen JM, FitzGerald M, et al. Global strategy for asthma management and prevention: GINA executive summary. Eur Respir J. 2008;31(1):143–78.
9. Boulet LP, Reddel HK, Bateman E, Pedersen S, FitzGerald JM, O'Byrne PM. The Global Initiative for Asthma (GINA): 25 years later. Eur Respir J. 2019;54(2):1900598.
10. GINA. Global strategy for asthma management and prevention, 2021. 2022. Available from: https://ginaasthma.org/.
11. Levy ML, Fletcher M, Price DB, Hausen T, Halbert RJ, Yawn BP. International Primary Care Respiratory Group (IPCRG) guidelines: diagnosis of respiratory diseases in primary care. Prim Care Respir J. 2006;15(1):20–34.
12. Pedersen S. Preschool asthma–not so easy to diagnose. Prim Care Resp J. 2007;16(1):4–6.
13. Doherty G, Bush A. Diagnosing respiratory problems in young children. Practitioner. 2007;251(20):22–5.
14. Pinnock H, Burton C, Campbell S, Gruffydd-Jones K, Hannon K, Hoskins G, et al. Clinical implications of the Royal College of Physicians three questions in routine asthma care: a real-life validation study. Prim Care Respir J. 2012;21(3):288–94.
15. Yawn BP, Wollan PC, Rank MA, Bertram SL, Juhn Y, Pace W. Use of asthma APGAR tools in primary care practices: a cluster-randomized controlled trial. Ann Fam Med. 2018;16(2):100–10.
16. Juniper EF, Bousquet J, Abetz L, Bateman ED. Identifying "well-controlled" and "not well-controlled" asthma using the Asthma Control Questionnaire. Respir Med. 2006;100(4):616–21.
17. Nathan RA, Sorkness CA, Kosinski M, Schatz M, Li JT, Marcus P, et al. Development of the asthma control test: a survey for assessing asthma control. J Allergy Clin Immunol. 2004;113(1):59–65.
18. Haselkorn T, Fish JE, Zeiger RS, Szefler SJ, Miller DP, Chipps BE, et al. Consistently very poorly controlled asthma, as defined by the impairment domain of the Expert Panel Report 3 guidelines, increases risk for future severe asthma exacerbations in The Epidemiology and Natural History of Asthma: Outcomes and Treatment Regimens (TENOR) study. J Allergy Clin Immunol. 2009;124(5):895–902.e1–4.
19. Fitzpatrick S, Joks R, Silverberg JI. Obesity is associated with increased asthma severity and exacerbations, and increased serum immunoglobulin E in inner-city adults. Clin Exp Allergy. 2012;42(5):747–59.
20. Denlinger LC, Phillips BR, Ramratnam S, Ross K, Bhakta NR, Cardet JC, et al. Inflammatory and comorbid features of patients with severe asthma and frequent exacerbations. Am J Respir Crit Care Med. 2017;195(3):302–13.
21. Burks AW, Tang M, Sicherer S, Muraro A, Eigenmann PA, Ebisawa M, et al. ICON: food allergy. J Allergy Clin Immunol. 2012;129(4):906–20.
22. Murphy VE, Jensen ME, Gibson PG. Asthma during pregnancy: exacerbations, management, and health outcomes for mother and infant. Semin Respir Crit Care Med. 2017;38(02):160–73.

23. Osborne ML, Pedula KL, O'Hollaren M, Ettinger KM, Stibolt T, Buist AS, et al. Assessing future need for acute care in adult asthmatics: the Profile of Asthma Risk Study: a prospective health maintenance organization-based study. Chest. 2007;132(4):1151–61.
24. Sturdy PM, Victor CR, Anderson HR, Bland JM, Butland BK, Harrison BD, et al. Psychological, social and health behaviour risk factors for deaths certified as asthma: a national case-control study. Thorax. 2002;57(12):1034–9.
25. Zeiger RS, Schatz M, Zhang F, Crawford WW, Kaplan MS, Roth RM, et al. Elevated exhaled nitric oxide is a clinical indicator of future uncontrolled asthma in asthmatic patients on inhaled corticosteroids. J Allergy Clin Immunol. 2011;128(2):412–4.
26. Patel M, Pilcher J, Reddel HK, Pritchard A, Corin A, Helm C, et al. Metrics of salbutamol use as predictors of future adverse outcomes in asthma. Clin Exp Allergy. 2013;43(10):1144–51.
27. Ernst P, Spitzer WO, Suissa S, Cockcroft D, Habbick B, Horwitz RI, et al. Risk of fatal and near-fatal asthma in relation to inhaled corticosteroid use. JAMA. 1992;268:3462–4.
28. Raissy HH, Kelly HW, Harkins M, Szefler SJ. Inhaled corticosteroids in lung diseases. Am J Respir Crit Care Med. 2013;187(8):798–803.
29. Chung KF, Wenzel SE, Brozek JL, Bush A, Castro M, Sterk PJ, et al. International ERS/ATS guidelines on definition, evaluation and treatment of severe asthma. Eur Respir J [Internet]. 2014;43(2):343–73. Available from: https://doi.org/10.1183/09031936.00202013.
30. Taylor DR, Bateman ED, Boulet LP, Boushey HA, Busse WW, Casale TB, et al. A new perspective on concepts of asthma severity and control. Eur Respir J. 2008;32(3):545–54.
31. Reddel HK, Taylor DR, Bateman ED, Boulet LP, Boushey HA, Busse WW, et al. An Official American Thoracic Society/European Respiratory Society statement: asthma control and exacerbations: standardizing endpoints for clinical asthma trials and clinical practice. Am J Respir Crit Care Med. 2009;180(1):59–99.
32. National Asthma Education and Prevention Program, Third Expert Panel on the Diagnosis and Management of Asthma. Expert Panel Report 3: Guidelines for the diagnosis and management of asthma. 2007. Available from: https://www.ncbi.nlm.nih.gov/books/NBK7232/.
33. Zeiger RS, Mauger D, Bacharier LB, Guilbert TW, Martinez FD, Lemanske RF Jr, et al. Daily or intermittent budesonide in preschool children with recurrent wheezing. N Engl J Med. 2011;365(21):1990–2001.
34. O'Byrne PM, FitzGerald JM, Bateman ED, Barnes PJ, Zhong N, Keen C, et al. Inhaled combined budesonide-formoterol as needed in mild asthma. N Engl J Med. 2018;378(20):1865–76.
35. Beasley R, Holliday M, Reddel HK, Braithwaite I, Ebmeier S, Hancox RJ, et al. Controlled trial of budesonide-formoterol as needed for mild asthma. N Engl J Med. 2019;380(21):2020–30.
36. Nwaru BI, Ekström M, Hasvold P, Wiklund F, Telg G, Janson C. Overuse of short-acting β(2)-agonists in asthma is associated with increased risk of exacerbation and mortality: a nationwide cohort study of the global SABINA programme. Eur Respir J. 2020;55(4):1901872.
37. Suissa S, Ernst P, Benayoun S, Baltzan M, Cai B. Low-dose inhaled corticosteroids and the prevention of death from asthma. N Engl J Med. 2000;343(5):332–6.
38. Suissa S, Ernst P, Kezouh A. Regular use of inhaled corticosteroids and the long term prevention of hospitalisation for asthma. Thorax. 2002;57(10):880–4.
39. Haahtela T, Järvinen M, Kava T, Kiviranta K, Koskinen S, Lehtonen K, et al. Comparison of a beta 2-agonist, terbutaline, with an inhaled corticosteroid, budesonide, in newly detected asthma. N Engl J Med. 1991;325(6):388–92.
40. Bårnes CB, Ulrik CS. Asthma and adherence to inhaled corticosteroids: current status and future perspectives. Respir Care. 2015;60(3):455–68.
41. Martinez FD, Chinchilli VM, Morgan WJ, Boehmer SJ, Lemanske RF Jr, Mauger DT, et al. Use of beclomethasone dipropionate as rescue treatment for children with mild persistent asthma (TREXA): a randomised, double-blind, placebo-controlled trial. Lancet. 2011;377(9766):650–7.
42. Kaiser SV, Huynh T, Bacharier LB, Rosenthal JL, Bakel LA, Parkin PC, et al. Preventing exacerbations in preschoolers with recurrent wheeze: a meta-analysis. Pediatrics. 2016;137(6):e20154496.

43. Mosbech H, Deckelmann R, de Blay F, Pastorello EA, Trebas-Pietras E, Andres LP, et al. Standardized quality (SQ) house dust mite sublingual immunotherapy tablet (ALK) reduces inhaled corticosteroid use while maintaining asthma control: a randomized, double-blind, placebo-controlled trial. J Allergy Clin Immunol. 2014;134(3):568–575.e7.
44. Virchow JC, Backer V, Kuna P, Prieto L, Nolte H, Villesen HH, et al. Efficacy of a house dust mite sublingual allergen immunotherapy tablet in adults with allergic asthma: a randomized clinical trial. JAMA. 2016;315(16):1715–25.
45. Wang L, Yin J, Fadel R, Montagut A, de Beaumont O, Devillier P. House dust mite sublingual immunotherapy is safe and appears to be effective in moderate, persistent asthma. Allergy. 2014;69(9):1181–8.
46. Mosbech H, Canonica GW, Backer V, de Blay F, Klimek L, Broge L, et al. SQ house dust mite sublingually administered immunotherapy tablet (ALK) improves allergic rhinitis in patients with house dust mite allergic asthma and rhinitis symptoms. Ann Allergy Asthma Immunol. 2015;114(2):134–40.
47. de Blay F, Kuna P, Prieto L, Ginko T, Seitzberg D, Riis B, et al. SQ HDM SLIT-tablet (ALK) in treatment of asthma--post hoc results from a randomised trial. Respir Med. 2014;108(10):1430–7.
48. Devillier P, Fadel R, de Beaumont O. House dust mite sublingual immunotherapy is safe in patients with mild-to-moderate, persistent asthma: a clinical trial. Allergy. 2016;71(2):249–57.
49. Marogna M, Braidi C, Bruno ME, Colombo C, Colombo F, Massolo A, et al. The contribution of sublingual immunotherapy to the achievement of control in birch-related mild persistent asthma: a real-life randomised trial. Allergol Immunopathol. 2013;41(4):216–24.
50. Vera JG, Sandoval GF, Solano MO, Tiro JL, Saab NJ. Safety and efficacy of specific sublingual immunotherapy in patients with asthma and allergy to Dermatophagoides pteronyssinus. Rev Alerg Mex. 2005;52:231–6.
51. Castro M, Rubin AS, Laviolette M, Fiterman J, Lima MDA, Shah PL, et al. Effectiveness and safety of bronchial thermoplasty in the treatment of severe asthma: a multicenter, randomized, double-blind, sham-controlled clinical trial. Am J Respir Crit Care Med. 2010;181(2):116–24.
52. Cox G, Thomson NC, Rubin AS, Niven RM, Corris PA, Siersted HC, et al. Asthma control during the year after bronchial thermoplasty. N Engl J Med. 2007;356(13):1327–37.
53. Pavord ID, Cox G, Thomson NC, Rubin AS, Corris PA, Niven RM, et al. Safety and efficacy of bronchial thermoplasty in symptomatic, severe asthma. Am J Respir Crit Care Med. 2007;176(12):1185–91.
54. Fahy JV. Type 2 inflammation in asthma--present in most, absent in many. Nat Rev Immunol. 2015;15(1):57–65.
55. Petsky HL, Kew KM, Turner C, Chang AB. Exhaled nitric oxide levels to guide treatment for adults with asthma. Cochrane Database Syst Rev. 2016;9(9):CD011440.
56. Petsky HL, Kew KM, Chang AB. Exhaled nitric oxide levels to guide treatment for children with asthma. Cochrane Database Syst Rev. 2016;11(11):CD011439.
57. Dweik RA, Boggs PB, Erzurum SC, Irvin CG, Leigh MW, Lundberg JO, et al. An official ATS clinical practice guideline: interpretation of exhaled nitric oxide levels (FENO) for clinical applications. Am J Respir Crit Care Med. 2011;184(5):602–15.
58. Price DB, Buhl R, Chan A, Freeman D, Gardener E, Godley C, et al. Fractional exhaled nitric oxide as a predictor of response to inhaled corticosteroids in patients with non-specific respiratory symptoms and insignificant bronchodilator reversibility: a randomised controlled trial. Lancet Respir Med. 2018;6(1):29–39.
59. Gibson PG, Powell H. Written action plans for asthma: an evidence-based review of the key components. Thorax. 2004;59(2):94–9.
60. Reddel HK, Barnes DJ. Pharmacological strategies for self-management of asthma exacerbations. Eur Respir J. 2006;28(1):182–99.
61. Ducharme FM, Lemire C, Noya FJ, Davis GM, Alos N, Leblond H, et al. Preemptive use of high-dose fluticasone for virus-induced wheezing in young children. N Engl J Med. 2009;360(4):339–53.

62. McKeever T, Mortimer K, Wilson A, Walker S, Brightling C, Skeggs A, et al. Quadrupling inhaled glucocorticoid dose to abort asthma exacerbations. N Engl J Med. 2018;378(10):902–10.
63. Jackson DJ, Bacharier LB, Mauger DT, Boehmer S, Beigelman A, Chmiel JF, et al. Quintupling inhaled glucocorticoids to prevent childhood asthma exacerbations. N Engl J Med. 2018;378(10):891–901.
64. Hagan JB, Samant SA, Volcheck GW, Li JT, Hagan CR, Erwin PJ, et al. The risk of asthma exacerbation after reducing inhaled corticosteroids: a systematic review and meta-analysis of randomized controlled trials. Allergy. 2014;69(4):510–6.
65. Ahmad S, Kew KM, Normansell R. Stopping long-acting beta2-agonists (LABA) for adults with asthma well controlled by LABA and inhaled corticosteroids. Cochrane Database Syst Rev. 2015;(6):CD011306.
66. Heaney LG, Busby J, Hanratty CE, Djukanovic R, Woodcock A, Walker SM, et al. Composite type-2 biomarker strategy versus a symptom-risk-based algorithm to adjust corticosteroid dose in patients with severe asthma: a multicentre, single-blind, parallel group, randomised controlled trial. Lancet Respir Med. 2021;9(1):57–68.
67. Papi A, Chipps BE, Beasley R, Panettieri RA Jr, Israel E, Cooper M, et al. Albuterol-budesonide fixed-dose combination rescue inhaler for asthma. N Engl J Med. 2022;386(22):2071–83.
68. Chow TG, Palka JM, Stone B, Trojan T, Epstein TG, Khan DA. Safety of subcutaneous immunotherapy in patients with severe asthma. Ann Allergy Asthma Immunol. 2022;129:585.
69. Castro M, Zangrilli J, Wechsler ME, Bateman ED, Brusselle GG, Bardin P, et al. Reslizumab for inadequately controlled asthma with elevated blood eosinophil counts: results from two multicentre, parallel, double-blind, randomised, placebo-controlled, phase 3 trials. Lancet Respir Med. 2015;3(5):355–66.
70. Castro M, Corren J, Pavord ID, Maspero J, Wenzel S, Rabe KF, et al. Dupilumab efficacy and safety in moderate-to-severe uncontrolled asthma. N Engl J Med. 2018;378(26):2486–96.
71. FitzGerald JM, Bleecker ER, Nair P, Korn S, Ohta K, Lommatzsch M, et al. Benralizumab, an anti-interleukin-5 receptor α monoclonal antibody, as add-on treatment for patients with severe, uncontrolled, eosinophilic asthma (CALIMA): a randomised, double-blind, placebo-controlled phase 3 trial. Lancet. 2016;388(10056):2128–41.
72. Menzies-Gow A, Corren J, Bourdin A, Chupp G, Israel E, Wechsler ME, et al. Tezepelumab in adults and adolescents with severe, uncontrolled asthma. N Engl J Med. 2021;384(19):1800–9.
73. Östling J, van Geest M, Schofield JPR, Jevnikar Z, Wilson S, Ward J, et al. IL-17–high asthma with features of a psoriasis immunophenotype. J Allergy Clin Immunol. 2019;144(5):1198–213.
74. Busse WW, Holgate S, Kerwin E, Chon Y, Feng J, Lin J, et al. Randomized, double-blind, placebo-controlled study of brodalumab, a human anti–IL-17 receptor monoclonal antibody, in moderate to severe asthma. Am J Respir Crit Care Med. 2013;188(11):1294–302.
75. Busse WW. Asthma and psoriasis – what do they have in common? IL-17A! J Allergy Clin Immunol. 2019;144(5):1169–71.
76. Hastie AT, Moore WC, Meyers DA, Vestal PL, Li H, Peters SP, et al. Analyses of asthma severity phenotypes and inflammatory proteins in subjects stratified by sputum granulocytes. J Allergy Clin Immunol. 2010;125(5):1028–1036.e13.
77. Hastie AT, Mauger DT, Denlinger LC, Coverstone A, Castro M, Erzurum S, et al. Baseline sputum eosinophil+neutrophil subgroups' clinical characteristics and longitudinal trajectories for NHLBI Severe Asthma Research Program (SARP 3) Cohort. J Allergy Clin Immunol. 2020;146(1):222–6.
78. Sur S, Crotty TB, Kephart GM, Hyma BA, Colby TV, Reed CE, et al. Sudden-onset fatal asthma: a distinct entity with few eosinophils and relatively more neutrophils in the airway submucosa? Am Rev Respir Dis. 2012;148(3):713–9.
79. Peters MC, McGrath KW, Hawkins GA, Hastie AT, Levy BD, Israel E, et al. Plasma interleukin-6 concentrations, metabolic dysfunction, and asthma severity: a cross-sectional analysis of two cohorts. Lancet Respir Med. 2016;4(7):574–84.
80. Israel E, Denlinger LC, Bacharier LB, LaVange LM, Moore WC, Peters MC, et al. PrecISE: Precision Medicine in Severe Asthma: an adaptive platform trial with biomarker ascertainment. J Allergy Clin Immunol. 2021;147(5):1594–601.

Global Considerations in Asthma Treatment: Management in Low Resource Settings

Neha T. Agnihotri and Sharmilee M. Nyenhuis

Abstract

Low-resource settings have a disproportionately higher burden of asthma due to factors that include environmental triggers, access to healthcare, availability of medications, and uncoordinated health systems. The application of guideline-based management can vary, which further impacts the treatment delivered. This chapter aims to outline the global landscape of asthma management, including cultural and social factors, with suggestions for interventions.

Keywords

Prevalence · Low- and middle-income countries (LMIC) · Survey · Diagnosis · Wheezing · School-age children · Physician-diagnosed asthma · Precision medicine · Phenotypes · Environmental factors · Triggers

N. T. Agnihotri
University of Illinois at Chicago, Department of Medicine, Division of Pulmonary, Critical Care, Sleep and Allergy, Chicago, IL, USA

S. M. Nyenhuis (✉)
University of Chicago, Pediatrics, Section of Allergy and Immunology, Chicago, IL, USA
e-mail: snyenhuis@bsd.uchicago.edu

A. R. Brasier, N. N. Jarjour (eds.), *Precision Approaches to Heterogeneity in Asthma*, Advances in Experimental Medicine and Biology 1426,
https://doi.org/10.1007/978-3-031-32259-4_16

Asthma Prevalence and Severity in Low-Resource Settings (Fig. 1)

Asthma is a common chronic respiratory disease and affects over 339 million people worldwide, most of whom live in low-resource settings, particularly in low- and middle-income countries (LMICs) [1]. While low-resource settings can occur in both developed and developing nations, for the purposes of this chapter, we will focus on low-resource settings of LMICs. The impact of asthma is substantial globally, and includes societal, financial, and health care consequences for many countries.

The global prevalence of asthma is poorly documented. Some of the reasons for this include the lack of a single test to definitively diagnose asthma, healthcare provider underdiagnosis, and the lack of standardized surveys. The prevalence of asthma in LMICs was not evaluated until the 1990s. It was generally thought that LMICs had much lower rates of asthma than high-income countries, yet recent efforts have found the asthma prevalence to be higher than previously thought. Efforts to accurately determine the prevalence of asthma globally include the International Study for Asthma and Allergy in Childhood (ISAAC), European Community Respiratory Health Survey (ECRHS), International Study of Wheezing in Infants (EISL), Global Asthma Network (GAN), and World Health Survey (WHS). These studies have not only filled our knowledge gaps in global asthma prevalence but have provided insight into the disease burden of asthma globally.

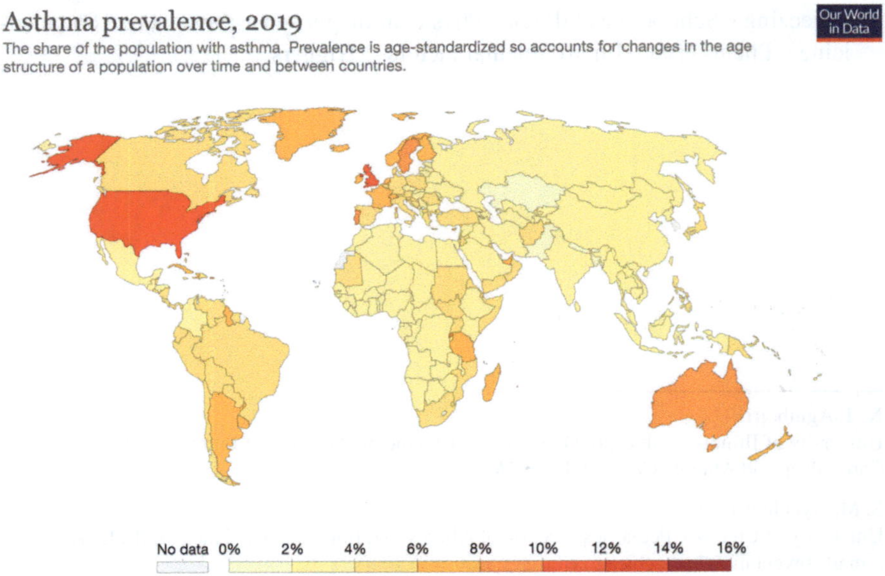

Fig. 1 Global Asthma Prevalence 2019. (From https://ourworldindata.org/grapher/asthma-prevalence?tab=table, Accessed August 31, 2022. Source: IHME, Global Burden of Disease (2019))

The most recent published data on the prevalence of asthma in school-aged children is from the 2021 Global Asthma Network [2]. This study used the same methods as ISAAC Phase III to conduct a global surveillance of the prevalence of asthma symptoms including trends over a 27-year period [3]. Almost 120,000 participants from 14 countries were included, and 10% of participants indicated they had a wheezing episode in the preceding year. Over the 27-year period, there was an increase in the prevalence of asthma symptoms, particularly in children and adolescents in LMICs. The prevalence was stable in upper-middle-income and high-income countries. The global prevalence of asthma in adults has also been examined. The World Health Organization (WHO) conducted the World Health Survey (WHS) in 2002–2003 among adults in 70 countries. A secondary analysis of the WHS reported the global prevalence rate of physician-diagnosed asthma in adults to be 4.3%, but the proportion of participants reporting asthma symptoms was 8.6%, suggesting an underdiagnosis of asthma by physicians [4].

While the asthma prevalence in LMICs is rising, it is still lower than what is seen in developed countries. Interestingly, despite the lower asthma prevalence in LMICs, the burden of asthma appears to be greater in these low-resource settings. Findings from the ISAAC Phase III database showed the presentation of asthma symptoms is more severe in LMICs [3]. Further, similar to what is seen in affluent countries, rates of uncontrolled asthma are much higher in medically disadvantaged communities of developing countries as well. A secondary analysis of the WHS found symptoms of asthma were more frequent in the extremes: low-income and high-income countries. This U-shaped distribution suggests the disease has multiple phenotypes that are driven by different mechanisms.

Contextualizing the Global Landscape

Local Environmental Factors for Risk of Asthma and Triggering Symptoms

There is heterogeneity in the prevalence and severity of asthma worldwide, which is likely due to variations in genetic and environmental exposures. Phase I of The International Study for Asthma and Allergy in Childhood (ISAAC) involved over 700,000 children aged 6–7 and 13–14 years from 156 centers in 56 countries [5, 6]. The study found significant geographic variations in the prevalence of asthma symptoms even within genetically similar groups, supporting the contribution of environmental factors in the development of asthma. Some of the environmental factors that contribute to poor asthma control and poor asthma outcomes in LMICs include urbanization, exposure to air pollution, unhygienic living conditions, environmental tobacco smoke, and allergens (outdoor/indoor). We still do not know all the environmental factors that contribute to asthma or how they interact with each other and with genes. Further, these environmental factors may act differently in different settings worldwide.

Urbanization

The rise of asthma in LMICs may be due to increasing urbanization. This is supported by a study conducted in children from rural Ecuador where a dose-response effect was shown with increasing urbanization associated with increased frequency of asthma symptoms [7]. Additional studies in Peru and Macedonia have shown that urban populations have a greater prevalence of asthma compared to rural populations in those countries [8, 9]. The impact of urbanization on asthma control has also been examined in LMICs. In Brazil, an evaluation of urbanization and asthma morbidity and mortality found that municipalities with an increasingly urban population were less likely to reduce hospital admissions and death rates for asthma even after correcting for confounders [10]. The biologic mechanisms of urbanization in asthma include a lack of natural regulation of the immune system. Sensitization to inhalant allergens has been shown to be in urban areas, as compared to rural areas, suggesting certain environmental exposures in urban areas deviate the immune system toward a Th2 phenotype. Additionally, rural environments can provide exposure to microorganisms that provide adequately stimulate to physiological immunomodulatory pathways, deviating the immune system toward a Th2 phenotype, which is important in the development of asthma [11].

Air Pollution

Urbanization is often associated with an increase in air pollution. Air pollution from motor vehicles, very common in urban areas, is associated with persistent respiratory symptoms and a higher prevalence of asthma as well [12, 13]. In particular, diesel exposure is linked with poor asthma outcomes, which may be especially relevant in LMICs. Trucks are often poorly maintained, use unregulated fuel, and drive near residential areas [14–16]. Air pollution seems to induce unspecific airway inflammation, which exacerbates airway symptoms in people with asthma.

Other environmental air exposures, more commonly seen in rural areas of LMICs, include indoor combustion of solid fuels for heating or cooking. This can contribute to the development of asthma and worsening asthma symptoms and are not related to urbanization [17]. In a study of 5539 adults in Colombia, indoor air pollution was found to be associated with increased prevalence and severity of asthma [18].

Unhygienic Living Conditions

Helminths are frequent pathogens in rural areas in LMICs. These pathogens may downregulate the immune responses related to atopy [19]. Ponte et al. conducted a longitudinal ecological study in Brazil and found that communities reducing helminths-related hospitalizations were less likely to reduce hospital admission and death rates from asthma compared to communities that have never been endemic for helminth infection or communities that maintained the same rate of hospitalizations because of helminths [20]. Dampness is also a potentially modifiable risk factor for asthma worldwide, though its association with asthma is stronger in LMICs compared to high-income countries. This risk has been shown to be independent of allergic sensitization to house dust mites, which is more common in damp homes.

Environmental Tobacco Smoke

Rates of tobacco smoking are increasing in LMICs, and of the 1.1 billion smokers in the world, 80% live in LMICs [21]. Environmental tobacco smoke (ETS) is a well-documented risk factor for asthma, prenatally and in children and adults. The relationship between ETS and asthma is causal, as asthma prevalence and severity would reduce if exposure to tobacco smoke was reduced/eliminated [11].

Allergens (Outdoor/Indoor)

In LMICs, the proportion of people with nonallergic asthma is greater than in high-income countries, but environmental allergens may still contribute to uncontrolled or worsening asthma symptoms [1]. Sensitization to environmental allergens should be evaluated, if possible, but access to allergy testing is often limited in LMICs. Further, if sensitization to allergen(s) is found, avoidance of the allergen(s) may be more difficult in LMICs. Allergen avoidance education is discussed elsewhere in this chapter.

Economic Burden

The economic impact of asthma goes beyond the individual and impacts families, health systems, and communities through financial and productivity consequences. The costs of asthma include but are not limited to expenses with visits to providers, medications, emergency room visits, hospital admissions, and missed school/workdays. At the individual and family level, asthma may cause a significant reduction in quality of life due to lethargy, sleep disturbances, decreased activity, and poor concentration [22, 23]. This may lead to school and work absenteeism, and financial and productivity consequences in the individual, family, and even the wider community. These economic impacts are huge and have been examined primarily in high-income countries with data from LMICs limited. In 2008, it was estimated that asthma causes economic losses exceeding billions of dollars in Germany, Switzerland, and the United States [22, 24]. In the United States alone, in 2013, the total annual cost of asthma reached $81.9 billion [25, 26]. The economic burden of asthma is closely associated with disease severity [24]. In LMICs, patients with uncontrolled asthma account for approximately 50% of all direct and indirect costs, even though they represent a small proportion (10–20%) of asthma sufferers

[27–29]. Globally, the economic burden of asthma exceeds those of tuberculosis and human immunodeficiency virus/acquired immune deficiency syndrome [HIV/AIDS] combined [30].

Optimizing Diagnosis of Asthma in Low-Resource Settings

The Global Initiative for Asthma (GINA) describes asthma as "...a heterogeneous disease, usually characterized by chronic airway inflammation [31]. Asthma is defined by the history of respiratory symptoms such as wheezing, shortness of breath, chest tightness, and cough that vary over time and in intensity, together with variable expiratory airflow limitation." While this is not a strict definition, this description captures the essential features for clinical purposes. In low-resource countries, clinicians often rely on clinical findings and a syndromic approach to diagnosis and management [1]. This can decrease precision and lead to underdiagnosis and undertreatment of asthma. Additionally, making a diagnosis of asthma in LMICs may be more difficult due to the myriad of other endemic respiratory diseases such as tuberculosis, HIV/AIDS-associated lung diseases, and parasitic or fungal lung diseases that must be considered in the differential diagnosis. Multiple healthcare provider tools are available to help improve the diagnosis of respiratory disease in primary care in LMICs [32]. Further, in many primary care settings in LMICs, there is a lack of detailed understanding or resistance to the implementation of National or International Guidelines for asthma diagnosis and management [33, 34].

The measurement of expiratory airflow limitation may be difficult, particularly in LMICs where access to lung function testing may be limited and, even when available, underutilized. Reasons for underutilization include affordability for the patient or health system, time-consuming to perform, and lack of appropriately trained staff to perform the procedure [35]. Where access to accurate lung function testing in low-resourced countries is limited, GINA guidelines suggest the utilization of peak expiratory flow (PEF) meters which are low-cost, more readily available, and do not require extensive staff training [36]. Measuring PEF rate (PEFR) can be used to support a clinical diagnosis of asthma [31]. GINA recommends twice-daily measurements of PEFR over a 2-week period [31]. If excessive variability is found (Children: >13%; Adults: >10%), this is suggestive of asthma. Additionally, the use of PEF measurements before and after administration of short-acting beta-agonist (e.g., albuterol and salbutamol) can be utilized to assist in a diagnosis of asthma. If there is a ≥20% improvement in PEFR after two puffs of a bronchodilator, this increases the likelihood of asthma over other lung diseases such as COPD [37]. The WHO Package of Essential Noncommunicable disease interventions supports the use of this testing method to complement a clinical diagnosis of asthma. GINA guideline also recommends considering a 4-week trial of anti-inflammatory therapy (including one week of oral corticosteroids if necessary) with symptoms and performing PEFR monitoring to help confirm a diagnosis of asthma prior to starting long-term controller therapy [31]. To improve the diagnosis of asthma in LMICs, affordable diagnostic tools such as spirometry and PEF meters, and training on their use, needs to be expanded. National programs in middle- and high-income countries that focus on educating primary care providers on the diagnosis and management of asthma have been shown to have meaningful changes in emergency room visits, hospital stays, and asthma-related disability [1, 38]. Wider dissemination of these programs with necessary adaptations in low-resource settings are needed to help improve the diagnosis of asthma.

Barriers to Management

The heterogeneity of asthma is a major challenge in managing the care of patients with this complex disease. The variation in symptom severity in itself impacts the recognition and delivery of appropriate treatment. In low-resource settings, in particular, this challenge is further exaggerated given limitations in availability and access.

Access to Healthcare

The initial access point for treatment in asthma care varies considerably, particularly in low-resource countries. While preventative management is expected to begin in the outpatient primary care setting, many patients do not encounter a provider until an exacerbation of their respiratory status occurs and frequently present in an emergency department (ED) or hospital setting. More frequent ED visits for asthma have been demonstrated to be increasingly prevalent in patients from low socioeconomic backgrounds [39]. Visits to the ED for asthma are commonly associated with poorly controlled or inadequately treated disease and have been linked to future poor outcomes [40, 41]. However, much of asthma treatment, particularly globally, is focused primarily on exacerbations [33]. Universally, patients from low-resource backgrounds are more likely to delay seeking treatment and neglect preventative self-management techniques resulting in insufficient everyday management and recurrent urgent care presentations [42]. Even in well-developed countries, interruption in insurance coverage precipitates further gaps in treatment.

It is generally thought that access to continuity of care can allow for consistent and longer-term monitoring of symptoms and treatment response [43]. A significant challenge in low-resource settings is finding access to a longitudinal provider. In the global landscape, moreover, there is limited availability of providers who are familiar with recognizing severe asthma and/or guideline-based asthma management. There is considerable research to support the role of subspecialists, such as pulmonologists or allergists, for patients with asthma, particularly with severe disease. The lack of availability of such providers in many low-resource settings, whether due to lack of trained professionals or insurance limitation, further adds to healthcare burden. With a range of uncoordinated health systems, the ports of entry into the healthcare system at various points can further be scattered.

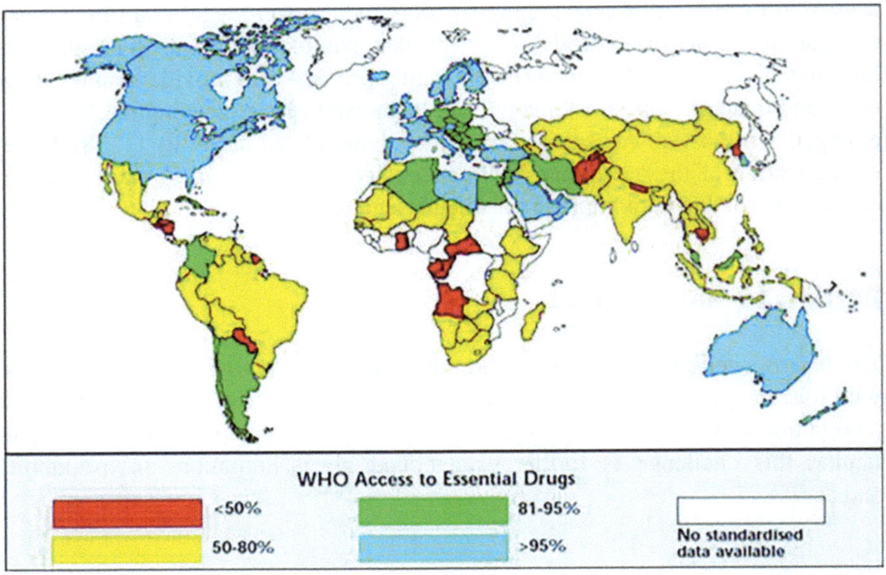

Fig. 2 World map of the proportion of the population with access to essential drugs [47]

Access to Medications (Figs. 2 and 3)

If patients are able to access a provider, access to medications can become a subsequent barrier. An "Essential Medication List" has been developed by the World Health Organization (WHO) to help LMICs identify efficacious and cost-effective medications [44]. Asthma medications specifically include: (1) inhaled corticosteroids (ICS): beclomethasone and budesonide; (2) a combination of ICS with long-acting beta-2 agonist (LABA): budesonide with formoterol; (3) short-acting beta-2 agonist (inhaled, nebulized, and injected): salbutamol. However, further assessment of the actual availability of these identified asthma drugs in low-income countries was around 30.1% in public and 43.1% in private sectors [45, 46]. This assessment does not include extensive regions within many countries that do not have access to any inhalers. Eighty percent of 32 LMICs surveyed did not have any treatment options for steps 4 and 5 of severe asthma guideline management, which includes medium to high dose ICS-LABA [45]. Other studies in similar low-resource settings have also demonstrated scarce availability or interruptions in the supply of essentially identified asthma medications [47]. Moreover, biologics, or monoclonal antibodies, have the potential to reduce asthma burden and possibly mortality, but access is limited primarily to developed countries [48].

The affordability of medications is another leading cause of barriers to adequate asthma care. Asthma is a chronic disease where long-term treatment and, thereby, associated costs must be considered. This is a universal challenge even in well-developed countries where medications may be available, but insurance-related costs can restrict usage. High medication costs are not only determined by

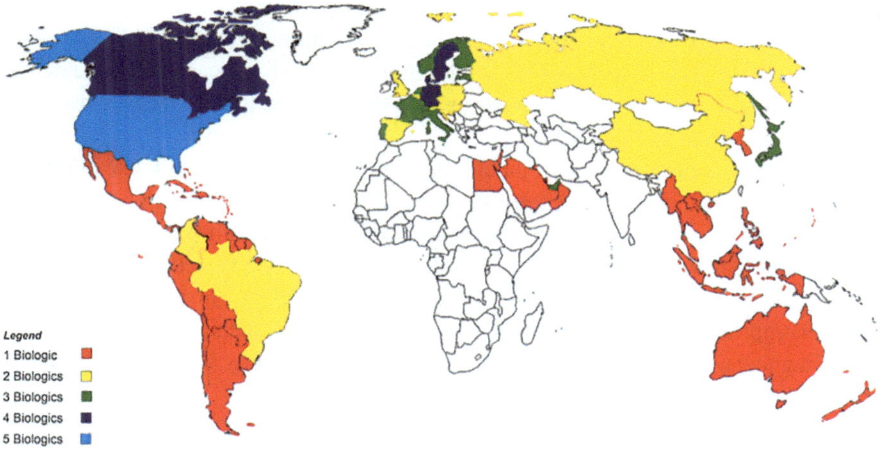

Fig. 3 Worldwide availability of biologics for severe asthma. The different colors identify the number of marketed biologics in each country [48]

manufacturers but add-on costs such as import tariffs, fees, margins, and wholesale mark-ups [47]. The cost of medication for a year in LMICs, however, can ultimately take a large portion of a patient's annual income [47]. As a further example, generic salbutamol can cost 4 days' wages for one inhaler, whereas a yearlong supply of budesonide inhalers for severe asthma can equate to more than 50 days' wages in several LMICs [49]. Furthermore, although on the WHO list of essential medicines and a key part of global guidelines, combination medications like budesonide-formoterol are much more expensive than inhaled corticosteroids and short-acting medications, which limit their use.

Adherence Challenges

Medication adherence remains an ongoing concern among asthma patients worldwide, regardless of socioeconomic status. While financial reasons and access impact proper adherence, social and cultural factors additionally influence asthma perceptions and, thereby, management. Consistent inhaler use has been repeatedly cited in global findings to be hindered by fear of addiction to medications [50]. Qualitative studies that consolidate patient views commonly express concern for medication dependency, hesitation surrounding side effects, and assumption of negative impact on the immune system and ability to fight infections [42, 51, 52]. This sentiment is echoed in minority populations in the United States, where lower medication adherence was correlated with negative asthma beliefs and minority status [50, 53].

Moreover, a part of patient uncertainty comes from a disconnect during the diagnosis and treatment initiation process. Following physician visits, patients still report unfamiliarity that asthma is a chronic, inflammatory condition and fail to grasp the importance of routine follow-up or the use of medications [51]. There have been reported knowledge gaps in understanding when to use a quick reliever medication as opposed to a controller medication. Stemming from beliefs surrounding asthma or a lack of awareness of the course of the disease, studies situated in low-resource settings often demonstrate self-discontinuation of inhaler use with symptom improvement [51].

Social stigma associated with inhaler use can also dissuade public use during an exacerbation [54]. In a school setting, children and adolescents with asthma have had to use medications in front of their peers or miss certain physical activities. Research has shown that children and adolescents have been teased or bullied about inhaler use and report anxiety, embarrassment, and feelings of isolation that can hinder compliance [55]. These feelings or experiences can cause individuals to hide their diagnosis, prompt delays in seeking care, reduce activity level, and ultimately negatively impact disease control.

Role of Culture in Asthma Management

Cultural beliefs have been shown to impact the perception and self-management of asthma. Limited studies focus on the cultural role in asthma management. Cultural background, family or community beliefs, and rituals play a central role in many ethnic groups globally; for some, there is pride in maintaining these aspects including in the management of health [56]. While conventional asthma management stems from the avoidance of triggers along with anti-inflammatory medication, many patients unfamiliar with this approach or rooted in traditional approaches to medicine as a whole may favor complementary or alternative approaches to management. Perception of asthma triggers also can vary globally. One study was done in a low-resource setting identified factors like exercise, being in the cold, foods, and changes in temperature and emotions to trigger asthma, and by a significantly less margin, triggers like dust, animals, or viral illnesses [42]. A study in Mozambique found that patients believed that asthma was contagious in 11% of cases and curable in 43% of cases by alternative treatment compared to 7% by medical treatment [57]. This is echoed in other studies, including African and Hispanic populations, where alternative medicine including folk remedies were overwhelmingly favored for treating asthma [42, 58].

Alternative approaches to asthma management most commonly including breathing exercises, inclusive of yoga, acupuncture, as well as herbal medications and vitamins [59]. In a qualitative Malaysian study, as well as a separate study done in a Hispanic population, patients thought that uncontrolled asthma was an imbalance between "hot or cold" in the body [42, 60]. There was a strong inclination to treat asthma with foods or plant ingredients that had cooling or warming properties on the body to aid in balancing the system and ease the asthma burden. Likewise, in East and South Asian communities, drugs that derive from plants or herbs are particularly favored for management. While these remedies can contain pharmacological ingredients that may lend to benefits, they likewise may have side effects or medication interactions, in part due to variability in preparation and lack of standardization in dosing [61].

The evidence overall to support the use of complementary or alternative medicine is limited. Regardless, cultural perceptions form a large foundation of nonallopathic approaches. These have been widely practiced for centuries with patient satisfaction. Understanding an individual's beliefs and experiences, which can be influenced by advice from family or members of the community, may allow opportunities for education and collaboration. At a minimum, it is recommended to schedule closer interval follow-up to strengthen the patient-physician relationship, acknowledge common ground, improve mutual understanding, and navigate recommendations that reconcile various perspectives.

Interventions for Improvement (Table 1)

Given asthma is a public health concern, effective interventions integrate individual interaction with community and policy engagement. A more coordinated approach on large system levels can have a greater impact on asthma management. In LMICs, increased government recognition of the enduring socioeconomic burden and mortality resulting from improper diagnosis or lack of effective treatment may allow for resource delegation and policy implementation [62]. Ideal policy change goes beyond medical care; it also involves social and environmental factors including housing conditions, general public awareness, school systems, insurance coverage, and health delivery systems [63]. There have been large organizational attempts to create a blueprint for changes, understanding while these may be gradual, the commitment to reform is essential [62, 63].

Availability of Medications

One of the foremost interventions involves expanding the availability of medications. Widespread and proper distribution of simple, low-cost quality asthma medications can have a lasting impact. In a global setting, prioritizing access to salbutamol and corticosteroid inhalers, along with a spacer that can even be made from a 500 mL plastic bottle [64], and oral prednisolone for emergencies, can form the basic requirement for asthma care. Ensuring they are affordable in relation to the national minimum wage allows accessibility. Where universal health coverage is not available, there should be regard toward developing methods of payment to attain essential medications. Although short-acting bronchodilator availability is becoming more widespread in developing countries, there should be just as much emphasis on inhaled corticosteroids as the latter works toward addressing airway inflammation. An example of a step toward a practical solution was the Asthma Drug Facility that was available from 2006 to 2013. It provided quality-assured medications and lower pricing through a "revolving drug fund" that started with initial capital and was replenished with sales from medications.

Table 1 Barriers to asthma control in low-resource settings and suggested Interventions

Barrier	Suggested intervention
Diagnostic limitations	Adapt global asthma guidelines in a systemic manner Continued medical education in asthma for primary care providers and other asthma providers Standardize elements including the use of peak flow meters and asthma assessment questionnaires
Limited access to primary care and specialist providers	Train more asthma-specific providers to create a large network (including physicians, physician extenders, nurses, medical assistants, teachers, and patient navigators)
Limited availability of asthma medications	Widespread distribution Prioritized access to medications on WHO "essential medication list"
Affordability of medications	Make medications low-cost (in relation to the national minimum wage) Create programs that develop methods of payment to attain medications or funds that lower pricing
Lack of knowledge regarding asthma	Literacy and language-appropriate patient education (verbal counseling, handouts, videos) Patient engagement with motivational interviewing and shared decision making Create school-based programs that promote asthma education and stock inhaler, and partner with school health teams Community engagement through religious and social groups, social media, and learning modules
Medication compliance/ adherence	Simplify medication regimens Demonstrate proper inhaler technique Create individualized asthma action plans Facilitate the development of self-management skills Encouraged continuity of care and regular asthma assessment with providers outpatient
Use of alternative therapies	Closer interval follow-up Recognize patient experience and express understanding Avoid biases and assumptions Acknowledge common ground Navigate recommendations that reconcile various perspectives
Weak health care infrastructure and low public health priority	Increase government recognition of the socioeconomic burden, morbidity, and mortality due to asthma Policy change advocacy that integrates social and environmental factors (housing, clean air, etc.) School-based asthma programs

Education

Communication between patient and provider forms a large part of the foundation for improvement in education on an individual level. Patient education that is language and literacy appropriate, while also cognizant of the cultural context, is an important part of the process. Numerous studies performed in low-resource settings highlight patient favorability in receiving their asthma diagnosis in a more objective and etiological manner [51, 65, 66]. Education that is not through information alone, but further involves patient engagement in shared decision-making may notably improve outcomes [50, 67]. Motivational interviewing, a guided style of counseling that assesses patients' beliefs or ambivalences and assists them in finding their intention for committing to a behavioral change, has been shown to be helpful in overcoming barriers and promoting adherence [68]. Sometimes a simplified regimen is needed to start. Specific tools like standardized asthma assessment questionnaires, peak flow meters, and written action plans have been well correlated with patient adherence. Demonstration of proper inhaler technique is likewise imperative when prescribing treatment for asthma; even after training, a large portion of patients with poor technique are likely to revert to using incorrectly; thus regular check-ins are needed [51].

Interventions for the improvement of medical care should be coupled with the avoidance of key environmental exposures. Several trials have demonstrated the benefits of indoor home remediation strategies, specifically dust mites and pest control like cockroach and mouse. While beneficial, the cost of services such as pest or mold removal or high-efficiency particulate air filtration devices can be limiting. Simple interventions include keeping food sealed, sealing cracks that allow pest access into the home, washing bedding in hot water weekly, and dust-mite proof bedding covers if available. Where feasible, increased ventilation, central heating, and optimizing homes to decrease moisture are also beneficial. Education on smoke avoidance, including secondhand smoke, is also critical.

The opportunity for education best presents with routine and structured follow-up to facilitate the development of self-management skills. This gradually allows a shift from an "acutely" seeking care approach to a more "routinely" seeking care approach. A preventative care model also depends on the availability of health professionals skilled in the diagnosis and management of asthma. An effective network of trained providers, with adherence to guidelines that are cognizant of medications available in that country, and with support for long-term management can have lasting implications for asthma health.

Community Engagement

Increasing public awareness of asthma can begin in the community setting such as church, school, or local groups focused on asthma support. Where the Internet is available, social media or active learning modules have been a successful means of mitigating stigma associated with the diagnosis of this disease and promoting environmental awareness, medication adherence, and symptom recognition. The Global Initiative for Asthma (GINA) offers a helpful patient guide online in multiple languages (https://ginasthma.org/gina-patient-guide-you-can-control-your-asthma/), as well as useful links to advocacy groups (https://ginasthma.org/useful-links/). The Global Allergy and Airways Patient Platform (https://gaapp.org/) also contains helpful online content including severe asthma patient education and breathing exercises and techniques for asthma. Finally, community engagement that embraces cultural differences and offers programs or activities that cater to a mutual sense of respect and reconciliation has the potential for improvement.

School-based programs that promote recognition, treatment, and support of asthma are another key aspect of practice and policy development. Children spend a sizable portion of their days in school; hence, policies targeting the school setting are important to successful asthma management. Simple, but effective, steps include written, standardized school asthma care plans and availability of quick-relief inhaler onsite. There are barriers to leaving the onus to children to carry their inhalers to school, including developmentally appropriate ages to carry, forgetting their inhaler, carrying an empty or expired inhaler, social stigma, or in many global settings in particular limited access to any inhaler at all [69]. In the United States, several states have enacted laws authorizing schools to stock asthma rescue medication. A study of an urban, low-income area in Arizona, one year after piloting stock albuterol, showed 20% fewer emergency calls and 40% less emergency medical transport [70]. Stock availability can alleviate anxiety, improve access, and even provide therapeutic intervention for those that previously had undiagnosed asthma but experience concerning symptoms, often noted during exercise routines or other encounters with triggers [70]. Furthermore, creating or increasing support for an onsite school health team, primarily a nurse, has been shown to reduce missed school days, ED visits, and costs [71]. Such programs provide opportunities to identify asthmatics at the beginning of the school year, check in daily with those with poor control, observe inhaler techniques, and educate students, staff, and caregivers. Partnering with community providers enhances the effectiveness of these school-based asthma programs.

There are some successful examples of global programs implemented to improve asthma care. One example is the National Asthma Programme in Finland, undertaken from 1994 to 2004. Their focus was to disseminate knowledge, especially in primary care, through regional education sessions regarding early diagnosis and guided self-management, implemented through an effective network of trained providers. Primary care physicians and nurses were able to initiate guideline-based treatment quickly and provide action plans with more regular follow-up, along with routine peak flow and spirometry measurements; severe cases were referred to specialists, and pharmacists followed up on written and oral information regarding proper technique and differentiating rescue and preventative inhalers. This intervention demonstrated a 54% reduction in hospital days over 10 years, annual cost per patient reduction by 50%, and improvement in patient satisfaction [38]. Similarly, the Wheezy Child Program in Belo Horizante, Brazil resulted in significant pediatric emergency department visits and hospitalizations, with a 300% increase in inhaler use and at least 50% sustained medication adherence [72]. This was achieved through the education of pediatricians and other health care workers by pulmonologists of adapted GINA guidelines and free dispersion of inhaled beclomethasone with spacers and albuterol.

Conclusion

Despite lower rates of asthma in LMICs compared to high-income countries, LMICs carry a greater burden of asthma with more severe disease and higher asthma-related death rates. Multiple factors contribute to this greater burden in LMICs including lack of appropriate diagnosis, availability of guideline-recommended treatment options, weak infrastructure of health services, persistent exposure to risk factors/triggers, and social/cultural barriers. Despite these barriers, case studies of the implementation of asthma education programs show improvements in asthma care are attainable but need to be prioritized by governments, health systems, and providers (Table 1). Finally, prioritizing affordable access to guideline-based asthma treatments (inhaled steroids, short-acting beta-agonists, and spacers), especially in low-resource settings, is necessary to improve asthma outcomes worldwide.

References

1. The global asthma report. Auckland: Global Asthma Network; 2018.
2. Asher MI, Rutter CE, Bissell K, Chiang CY, El Sony A, Ellwood E, et al. Worldwide trends in the burden of asthma symptoms in school-aged children: Global Asthma Network Phase I cross-sectional study. Lancet. 2021;398(10311):1569–80.
3. Pearce N, Aït-Khaled N, Beasley R, Mallol J, Keil U, Mitchell E, et al. Worldwide trends in the prevalence of asthma symptoms: phase III of the International Study of Asthma and Allergies in Childhood (ISAAC). Thorax. 2007;62(9):758–66.
4. To T, Stanojevic S, Moores G, Gershon AS, Bateman ED, Cruz AA, et al. Global asthma prevalence in adults: findings from the cross-sectional world health survey. BMC Public Health. 2012;12(1):204.
5. Worldwide variation in prevalence of symptoms of asthma, allergic rhinoconjunctivitis, and atopic eczema: ISAAC. The International Study of Asthma and Allergies in Childhood (ISAAC) Steering Committee. Lancet. 1998;351(9111):1225–32.
6. Worldwide variations in the prevalence of asthma symptoms. the International Study of Asthma and Allergies in Childhood (ISAAC). Eur Respir J. 1998;12(2):315–35.
7. Rodriguez A, Vaca M, Oviedo G, Erazo S, Chico ME, Teles C, et al. Urbanisation is associated with prevalence of childhood asthma in diverse, small rural communities in Ecuador. Thorax. 2011;66(12):1043–50.
8. Vlaski E, Lawson JA. Urban-rural differences in asthma prevalence among young adolescents: the role of behavioural and environmental factors. Allergol Immunopathol (Madr). 2015;43(2):131–41.
9. Gaviola C, Miele CH, Wise RA, Gilman RH, Jaganath D, Miranda JJ, et al. Urbanisation but not biomass fuel smoke exposure is associated with asthma prevalence in four resource-limited settings. Thorax. 2016;71(2):154–60.
10. Ponte EV, Cruz AA, Athanazio R, Carvalho-Pinto R, Fernandes FLA, Barreto ML, et al. Urbanization is associated with increased asthma morbidity and mortality in Brazil. Clin Respir J. 2018;12(2):410–7.
11. Cruz ÁA, Stelmach R, Ponte EV. Asthma prevalence and severity in low-resource communities. Curr Opin Allergy Clin Immunol. 2017;17(3):188–93.
12. Hasunuma H, Sato T, Iwata T, Kohno Y, Nitta H, Odajima H, et al. Association between traffic-related air pollution and asthma in preschool children in a national Japanese nested case-control study. BMJ Open. 2016;6(2):e010410.

13. Singh S, Sharma BB, Sharma SK, Sabir M, Singh V. Prevalence and severity of asthma among Indian school children aged between 6 and 14 years: associations with parental smoking and traffic pollution. J Asthma. 2016;53(3):238–44.
14. Balmes JR. How does diesel exhaust impact asthma? Thorax. 2011;66(1):4–6.
15. Brunekreef B, Stewart AW, Anderson HR, Lai CK, Strachan DP, Pearce N. Self-reported truck traffic on the street of residence and symptoms of asthma and allergic disease: a global relationship in ISAAC phase 3. Environ Health Perspect. 2009;117(11):1791–8.
16. Pandya RJ, Solomon G, Kinner A, Balmes JR. Diesel exhaust and asthma: hypotheses and molecular mechanisms of action. Environ Health Perspect. 2002;110 Suppl 1(Suppl 1):103–12.
17. Wong GW, Brunekreef B, Ellwood P, Anderson HR, Asher MI, Crane J, et al. Cooking fuels and prevalence of asthma: a global analysis of phase three of the International Study of Asthma and Allergies in Childhood (ISAAC). Lancet Respir Med. 2013;1(5):386–94.
18. Gonzalez-Garcia M, Caballero A, Jaramillo C, Maldonado D, Torres-Duque CA. Prevalence, risk factors and underdiagnosis of asthma and wheezing in adults 40 years and older: a population-based study. J Asthma. 2015;52(8):823–30.
19. de Almeida T, Fernandes JS, Lopes DM, Andrade LS, Oliveira SC, Carvalho EM, et al. Schistosoma mansoni antigens alter activation markers and cytokine profile in lymphocytes of patients with asthma. Acta Trop. 2017;166:268–79.
20. Ponte EV, Rasella D, Souza-Machado C, Stelmach R, Barreto ML, Cruz AA. Reduced asthma morbidity in endemic areas for helminth infections: a longitudinal ecological study in Brazil. J Asthma. 2014;51(10):1022–7.
21. Health AoSa. Tobacco and the developing world 2019. Available from: http://ash.org.uk/category/information-and-resources/.
22. Bahadori K, Doyle-Waters MM, Marra C, Lynd L, Alasaly K, Swiston J, et al. Economic burden of asthma: a systematic review. BMC Pulm Med. 2009;9:24.
23. WAO White book on allergy. Milwaukee: World Allergy Organization; 2011. pp. 1–216.
24. Godard P, Chanez P, Siraudin L, Nicoloyannis N, Duru G. Costs of asthma are correlated with severity: a 1-yr prospective study. Eur Respir J. 2002;19(1):61–7.
25. Nurmagambetov T, Kuwahara R, Garbe P. The economic burden of asthma in the United States, 2008-2013. Ann Am Thorac Soc. 2018;15(3):348–56.
26. Beasley R. The burden of asthma with specific reference to the United States. J Allergy Clin Immunol. 2002;109(5 Suppl):S482–9.
27. Soto-Martínez ME, Soto-Quiros ME, Custovic A. Childhood asthma: low and middle-income countries perspective. Acta Med Acad. 2020;49(2):181–90.
28. Serebrisky D, Wiznia A. Pediatric Asthma: A Global Epidemic. Ann Glob Health. 2019;85(1):6.
29. Asher I, Haahtela T, Selroos O, Ellwood P, Ellwood E. Global Asthma Network survey suggests more national asthma strategies could reduce burden of asthma. Allergol Immunopathol (Madr). 2017;45(2):105–14.
30. WHO factsheet 206: Bronchial Asthma: World Health Organization; Available from: www.who.int/mediacentre/factsheets/fs206/en.
31. Global Initiatve for Asthma. Global Strategy for Asthma MAnagement and Prevention, 2022.
32. Mortimer K, Reddel HK, Pitrez PM, Bateman ED. Asthma management in low- and middle-income countries: case for change. Eur Respir J. 2022;60:2103179.
33. Sánchez-Borges M, Capriles-Hulett A, Caballero-Fonseca F. Asthma care in resource-poor settings. World Allergy Organ J. 2011;4(4):68–72.
34. Dubaybo BA. The care of asthma patients in communities with limited resources. Res Rep Trop Med. 2021;12:33–8.
35. Masekela R, Zurba L, Gray D. Dealing with access to spirometry in Africa: a commentary on challenges and solutions. Int J Environ Res Public Health. 2018;16(1):62.
36. Huang WC, Fox GJ, Pham NY, Nguyen TA, Vu VG, Ngo QC, et al. A syndromic approach to assess diagnosis and management of patients presenting with respiratory symptoms to healthcare facilities in Vietnam. ERJ Open Res. 2021;7(1):00572.
37. WHO package of essential noncommunicable (PEN) disease interventions for primary health care. Geneva: World Health Organization; 2020.

38. Haahtela T, Tuomisto LE, Pietinalho A, Klaukka T, Erhola M, Kaila M, et al. A 10 year asthma programme in Finland: major change for the better. Thorax. 2006;61(8):663–70.
39. Gushue C, Miller R, Sheikh S, Allen ED, Tobias JD, Hayes D Jr, et al. Gaps in health insurance coverage and emergency department use among children with asthma. J Asthma. 2019;56(10):1070–8.
40. Mohammed F, Bootoor S, Panday A, Ramdass A, Reemaul J, Sharma A, et al. Predictors of repeat visits to the emergency room by asthmatic children in primary care. J Natl Med Assoc. 2006;98(8):1278–85.
41. Emerman CL. Relapse following treatment of acute asthma in the emergency department. J Asthma. 2000;37(8):701–8.
42. Bearison DJ, Minian N, Granowetter L. Medical management of asthma and folk medicine in a Hispanic community. J Pediatr Psychol. 2002;27(4):385–92.
43. DeVoe JE, Tillotson CJ, Wallace LS. Children's receipt of health care services and family health insurance patterns. Ann Fam Med. 2009;7(5):406–13.
44. Bazargani YT, de Boer A, Leufkens HG, Mantel-Teeuwisse AK. Essential medicines for COPD and asthma in low and middle-income countries. Thorax. 2014;69(12):1149–51.
45. Bissell K, Perrin C, Beran D. Access to essential medicines to treat chronic respiratory disease in low-income countries. Int J Tuberc Lung Dis. 2016;20(6):717–28.
46. Cameron A, Roubos I, Ewen M, Mantel-Teeuwisse AK, Leufkens HG, Laing RO. Differences in the availability of medicines for chronic and acute conditions in the public and private sectors of developing countries. Bull World Health Organ. 2011;89(6):412–21.
47. Mendis S, Fukino K, Cameron A, Laing R, Filipe A Jr, Khatib O, et al. The availability and affordability of selected essential medicines for chronic diseases in six low- and middle-income countries. Bull World Health Organ. 2007;85(4):279–88.
48. Caminati M, Morais-Almeida M, Bleecker E, Ansotegui I, Canonica GW, Bovo C, et al. Biologics and global burden of asthma: a worldwide portrait and a call for action. World Allergy Organ J. 2021;14(2):100502.
49. Babar ZU, Lessing C, Mace C, Bissell K. The availability, pricing and affordability of three essential asthma medicines in 52 low- and middle-income countries. PharmacoEconomics. 2013;31(11):1063–82.
50. McQuaid EL. Barriers to medication adherence in asthma: the importance of culture and context. Ann Allergy Asthma Immunol. 2018;121(1):37–42.
51. Hannane A, Misane L, Devouassoux G, Colin C, Letrilliart L. Asthma patients' perception on their care pathway: a qualitative study. NPJ Prim Care Respir Med. 2019;29(1):9.
52. Apter AJ, Boston RC, George M, Norfleet AL, Tenhave T, Coyne JC, et al. Modifiable barriers to adherence to inhaled steroids among adults with asthma: it's not just black and white. J Allergy Clin Immunol. 2003;111(6):1219–26.
53. Le TT, Bilderback A, Bender B, Wamboldt FS, Turner CF, Rand CS, et al. Do asthma medication beliefs mediate the relationship between minority status and adherence to therapy? J Asthma. 2008;45(1):33–7.
54. Kolbe J, Vamos M, Fergusson W. Socio-economic disadvantage, quality of medical care and admission for acute severe asthma. Aust NZ J Med. 1997;27(3):294–300.
55. Kansra S, Calvert R, Jones S. Stigma from medication use: an under recognised burden of care. Breathe. 2021;17(1):210002.
56. Koinis-Mitchell D, McQuaid EL, Jandasek B, Kopel SJ, Seifer R, Klein RB, et al. Identifying individual, cultural and asthma-related risk and protective factors associated with resilient asthma outcomes in urban children and families. J Pediatr Psychol. 2012;37(4):424–37.
57. Mavale-Manuel S, Duarte N, Alexandre F, Albuquerque O, Scheinmann P, Poisson-Salomon AS, et al. Knowledge, attitudes, and behavior of the parents of asthmatic children in Maputo. J Asthma. 2004;41(5):533–8.
58. Shimwela M, Mwita JC, Mwandri M, Rwegerera GM, Mashalla Y, Mugusi F. Asthma prevalence, knowledge, and perceptions among secondary school pupils in rural and urban coastal districts in Tanzania. BMC Public Health. 2014;14:387.

59. Chen W, Fitzgerald JM, Rousseau R, Lynd LD, Tan WC, Sadatsafavi M. Complementary and alternative asthma treatments and their association with asthma control: a population-based study. BMJ Open. 2013;3(9):e003360.
60. Koh WM, Abu Bakar AI, Hussein N, Pinnock H, Liew SM, Hanafi NS, et al. Sociocultural influences on asthma self-management in a multicultural society: a qualitative study amongst Malaysian adults. Health Expect. 2021;24(6):2078–86.
61. Passalacqua G, Bousquet PJ, Carlsen KH, Kemp J, Lockey RF, Niggemann B, et al. ARIA update: I--Systematic review of complementary and alternative medicine for rhinitis and asthma. J Allergy Clin Immunol. 2006;117(5):1054–62.
62. Bateman ED, Hurd SS, Barnes PJ, Bousquet J, Drazen JM, FitzGerald JM, Gibson P, Ohta K, O'Byrne P, Pedersen SE, Pizzichini E, Sullivan SD, Wenzel SE, Zar HJ. Global strategy for asthma management and prevention: GINA executive summary. Eur Respir J. 2008;31:143–78. Eur Respir J. 2018;51(2).
63. Lara M, Rosenbaum S, Rachelefsky G, Nicholas W, Morton SC, Emont S, et al. Improving childhood asthma outcomes in the United States: a blueprint for policy action. Pediatrics. 2002;109(5):919–30.
64. Zar HJ, Asmus MJ, Weinberg EG. A 500-ml plastic bottle: an effective spacer for children with asthma. Pediatr Allergy Immunol. 2002;13(3):217–22.
65. Pelaez S, Lamontagne AJ, Collin J, Gauthier A, Grad RM, Blais L, et al. Patients' perspective of barriers and facilitators to taking long-term controller medication for asthma: a novel taxonomy. BMC Pulm Med. 2015;15:42.
66. Shum M, Poureslami I, Liu J, FitzGerald J. Perceived barriers to asthma therapy in ethno-cultural communities: the role of culture, beliefs and social support. Health. 2017;9:1029–46.
67. Trivedi M, Pappalardo AA, Udoko M, Garg A, Phipatanakul W, Szefler SJ, et al. Social determinants of health in asthma through the life course. J Allergy Clin Immunol Pract. 2022;10(4):953–61.
68. Young HN, Kanchanasuwan S, Cox ED, Moreno MM, Havican NS. Barriers to medication use in rural underserved patients with asthma. Res Social Adm Pharm. 2015;11(6):909–14.
69. O'Rourke A, Zimmerman A, Platt H, Pappalardo AA. Preventing asthma emergencies in schools. Pediatrics. 2020;145(4):e20191995.
70. Gerald LB, Snyder A, Disney J, Gerald JK, Thomas A, Wilcox G, et al. Implementation and evaluation of a Stock Albuterol Program for students with asthma. Ann Am Thorac Soc. 2016;13(2):295–6.
71. Rodriguez E, Rivera DA, Perlroth D, Becker E, Wang NE, Landau M. School nurses' role in asthma management, school absenteeism, and cost savings: a demonstration project. J Sch Health. 2013;83(12):842–50.
72. Lasmar L, Fontes MJ, Mohallen MT, Fonseca AC, Camargos P. Wheezy child program: the experience of the belo horizonte pediatric asthma management program. World Allergy Organ J. 2009;2(12):289–95.

Clinical Trial Design Innovations for Precision Medicine in Asthma

Salman Siddiqui, Elin Haf Davies, Majid Afshar, and Loren C. Denlinger

Abstract

Severe asthma is a spectrum disorder with numerous subsets, many of which are defined by clinical history and a general predisposition for T2 inflammation. Most of the approved therapies for severe asthma have required clinical trial designs with population enrichment for exacerbation frequency and/or elevation of blood eosinophils. Moving beyond this framework will require trial designs that increase efficiency for studying nondominant subsets and continue to improve upon biomarker signatures. In addition to reviewing the current literature on biomarker-informed trials for severe asthma, this chapter will also review the advantages of master protocols and adaptive design methods for establishing the efficacy of new interventions in prospectively defined subsets of patients. The incorporation of methods that allow for data collection outside of traditional study visits at academic centers, called remote decentralized trial design, is a growing trend that may increase diversity in study participation and allow for enhanced resiliency during the COVID-19 pandemic. Finally, reaching the goals of precision medicine in asthma will require increased emphasis on effectiveness studies. Recent advances in real-world data utilization from electronic health records are also discussed with a view toward pragmatic trial designs that could also incorporate the evaluation of biomarker signatures.

S. Siddiqui
National Heart and Lung Institute, Imperial College, London, England, UK

E. Haf Davies
Aparito Limited, Wrexham, Wales, UK

M. Afshar · L. C. Denlinger (✉)
Division of Allergy, Pulmonary and Critical Care, Department of Medicine, University of Wisconsin School of Medicine and Public Health, Madison, WI, USA
e-mail: lcd@medicine.wisc.edu

A. R. Brasier, N. N. Jarjour (eds.), *Precision Approaches to Heterogeneity in Asthma*, Advances in Experimental Medicine and Biology 1426,
https://doi.org/10.1007/978-3-031-32259-4_17

Keywords

Phenotypes · Endotypes · Longitudinal analysis · Cohort · Lung function · Biomarkers · Risk estimate · Placebo-controlled trials · Exacerbation · Intervention · Stratification

Introduction

Despite recent advances in therapy, approximately 5% of patients with asthma have severe disease that is refractory to current maintenance treatments and is associated with exacerbations, loss of lung function, increased health care utilization, and diminished quality of life [1]. The underlying mechanisms are thought to be diverse. Additionally, there is a growing recognition by patients that this diversity affects responsiveness to current treatment options [2]. This chapter will attempt to integrate the latest information on severe asthma phenotypes with ongoing trial approaches. In addition to a focus on biomarker-stratified design efficiency, decentralized and pragmatic designs will also be discussed as the need for trial adaptation has grown during the COVID-19 pandemic.

Pheno-Endotyping for the Purposes of Trials

Longitudinal Clinical Phenotypes

Longitudinal cohorts of patients with severe asthma continue to demonstrate clinical heterogeneity. The largest of these with data beyond 2 years of follow-up is the NHLBI Severe Asthma Research Program-3. The structure of the cohort at the baseline visit included 715 participants, 25% of whom were children and 60% in both age strata met ATS/ERS definitions for severe disease [3]. Longitudinal patterns of adult lung function and exacerbation trajectories demonstrated a few surprising aspects that challenge convention. First, only 43% of the cohort with a baseline and at least two follow-up visits had evidence of progressive loss of lung function. Severe asthma status, exacerbations, and blood eosinophils at baseline had no impact on a participant's longitudinal lung function classification [4]. A second surprising observation is that 41% of the adult cohort remained free of exacerbation for at least 3 years beyond the baseline visit [5]. Again, severe asthma status had no relationship to the patterns of longitudinal exacerbations. The clinical history of past exacerbations was the dominant predictor of future events. Without clinical history, the probability of having at least one exacerbation in the next year was approximately 40%. This probability declines to approximately 15% for participants lacking exacerbation for 2 years in a row, and climbs 80% for those with two consecutive past years of at least one exacerbation. Baseline blood eosinophils and plasma IL-6 levels are independent predictors of future exacerbations, but only after removing clinical history from the models [5]. Collectively, these observations regarding patterns of lung function and exacerbation suggest that there are likely to be composite biomarker signatures that remain to be defined and that are more informative than their individual constituents.

Lessons from Blood Eosinophils, IgE, FeNO, and Periostin in Clinical Trials

Severe asthma trials have focused predominantly on asthma exacerbations as the primary endpoint, for trials evaluating biologic therapies. Enrichment of participants into these trials requires biomarkers that are themselves predictive of asthma exacerbation risk – to ensure that the right phenotype(s) of trial participants are enrolled. Figure 1 reviews current and developing treatment options in the framework of the type 2 inflammatory target.

Type-2 asthma biomarkers readily available for enrichment in clinical trials, include blood eosinophils counts (BEC), fractional exhaled nitric oxide (FeNO), and immunoglobulin E levels (IgE). Real-world studies have demonstrated that BEC and FeNO levels are associated with increased asthma exacerbation risk in primary care [6]. The highest risk estimates were seen in patients with both an

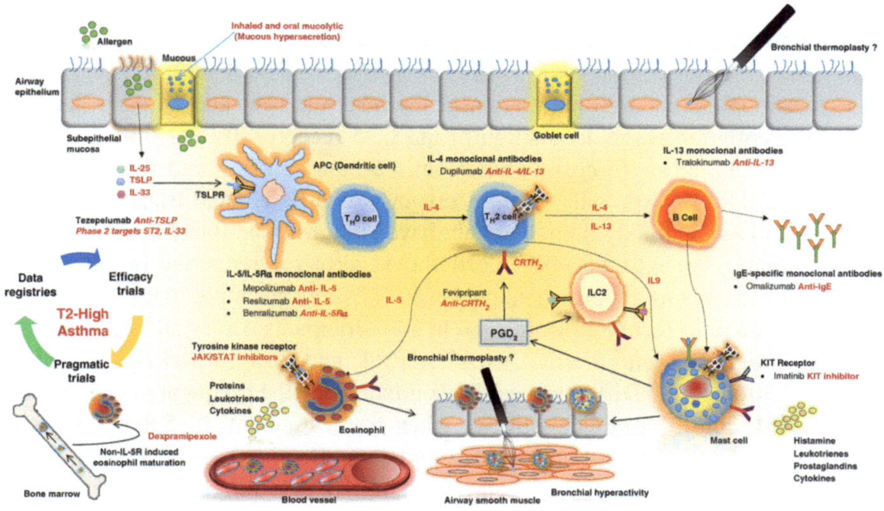

Fig. 1 Biomarker paired interventions for severe asthma targeting components of type 2 inflammation. Molecular targets of approved and developing interventions for severe asthma are identified in red. Bronchial thermoplasty stands out as an exception in that the molecular target remains unclear for this treatment. Each intervention was approved primarily on efficacy studies. Incorporation of pragmatic trials and real-world data from registries should lead to better understanding of effectiveness. (Reprinted with permission of the American Thoracic Society. Copyright © 2019, American Thoracic Society, all rights reserved. Cite: Siddiqui et al. [1]. *The American Journal of Respiratory & Critical Care Medicine* is an official journal of the American Thoracic Society)

elevated BEC ≥ 0.30×10^9/L and FeNO ≥ 50 ppb, yielding an unadjusted exacerbation rate of approximately 3.7 events per year. More recently, a prototype asthma exacerbation risk scoring tool developed from post hoc analyses of clinical trial data acquired across the spectrum of asthma severity, has demonstrated utility in providing estimates of prospective asthma exacerbation risk in patients stratified by both global initiative for asthma (GINA) treatment intensity and according to their prior exacerbation rate [7].

In contrast, data from multiple pooled placebo-controlled phase II/III biologic trials failed to identify any association between serum IgE level and asthma exacerbation rate [8]. Furthermore, the epithelial cell-derived, matricellular protein periostin has been proposed to offer further additive risk estimates for asthma exacerbations based upon results from phase II clinical trials of the anti-IL-13 biologics [9]; however, the additive risk (i.e., additive above BEC and FeNO in combination) was relatively small. Furthermore, a subsequent clinical trial in severe asthma designed to taper maintenance oral steroids according to a composite biomarker score combining serum periostin, BEC, and FeNO compared to standard of care failed to meet its primary endpoint [10]. These exemplars underscore the association between biomarkers of type-2 asthma and exacerbation risk and are cognate with the role of pragmatic biomarkers BEC and FeNO in clinical trial stratification in severe asthma.

T2 Biomarkers and Clinical Trial Stratification

Blood eosinophils undergo maturation in the bone marrow, and an early stage in their differentiation is the acquisition of the IL-5 receptor. IL-5 is critical for the maturation, and egress of eosinophils from the bone marrow into the circulating blood pool [11]. Pooled analyses from pivotal phase III clinical trials of the anti-IL-5 monoclonal antibody mepolizumab, have demonstrated that the effect size of mepolizumab relative to placebo on asthma exacerbation reduction increases as a function of increasing blood eosinophil level [12]. For example, the exacerbation reduction rate ratio increased from 47% in all participants across all BEC levels to 52% in participants with a BEC ≥ 0.15×10^9/L and 70% in participants with a BEC ≥ 0.50×10^9/L. Different health economies have subsequently selected the optimal BEC cutoff for clinical practice when prescribing anti-IL-5 therapeutics in severe asthma based upon both efficacy and cost-effectiveness. For example, the UK regulator NICE, has set a BEC threshold of ≥ 0.30×10^9/L for the prescribing of anti-IL-5 therapeutics – which represents a tradeoff between biologic efficacy, cost-effectiveness and the relative population likely to be able to access a biologic given the selected threshold for stratification.

FeNO is generated from the airway mucosal barrier after activation of inducible nitric oxide by type-2 cytokines interleukin 4 (IL-4) and interleukin 13 (IL-13) [13]. Consequently, biologic therapy targeting the IL-4Rα, via which both IL-14 and IL-13 signal – would be predicted to be effective in patient enriched for FeNO. The phase III clinical trial of the IL-4Rα targeting biologic Dupilumab, has evaluated this concept via subgroup analysis of trial primary outcome data. In these analyses a baseline FeNO ≥ 50 ppb was associated with a relative risk reduction of exacerbations of approximately 70%, compared to a 20% reduction in trial participants with a FeNO < 25 ppb and 48% reduction in the overall trial population [14]. In contrast, post hoc analysis has failed to identify that serum total IgE is a useful biomarker for anti-IgE therapy [15], with the majority of the evidence suggesting that blood eosinophilia is not a biomarker of anti-IgE response either [16, 17].

All in all, these observations highlight the important of trial enrichment using biomarkers that are both proximal to the mechanistic pathway of interest of the drug target and also predictive of the clinical outcome of interest that the drug target is seeking to modify. Finally, it must be noted that despite these promising advances in biomarker guided precision medicine, recent trials targeting upstream epithelially derived alarmins, such as thymic stromal lymphopoietin (TSLP), have demonstrated a treatment effect irrespective of BEC and FeNO levels on asthma exacerbations [18], highlighting the need for potentially more targeted enrichment tools for upstream type-2 biological targets such as pharmacogenomic methods [19].

Examples of T2 Low Trials Here That Have Failed Potentially Due to Inadequate Stratification

Several key mechanisms may drive disease expression in severe asthma that is beyond type-2 immune biology (Fig. 2). These include alternative immune programs including Th17 immunity, which has an important role in epithelial barrier antimicrobial defense. Key effector cytokines and effector cells driving Th17 responses are IL-21, IL-23, IL-6, and neutrophils, which may be putative targets in severe asthma. However, targeting these pathways has been unsuccessful in severe asthma due to inadequate stratification of patients and an incomplete understanding of how these pathways relate to asthma outcome measures. For example, the monoclonal antibody targeting IL-17 brodalumab, failed to meet its primary endpoint of asthma control [20]. In addition, targeting IL-23 with rizankizumab in an unstratified population also failed to demonstrate a signal against the primary endpoint of time to first asthma worsening [21]. Finally, a stratified design enriching for blood neutrophil levels failed to demonstrate a signal with neutrophil chemokine (CXCR2) inhibition in severe asthma [22]. Another potential approach is to target the airway microbiome in severe asthma. Several studies have reported airway dysbiosis with proteobacteria such as *Haemophilus influenzae* in severe asthma [23]. A recent post hoc analysis identified that the macrolide antibiotic – azithromycin selectively reduced *H. influenzae* bacterial load compared to placebo in severe asthma [24], it would be tempting to speculate that future trials enriching for the dysbiosis phenotype of severe asthma would identify the optimal responder population. Another potential approach in microbiome directed therapeutics would be to target the downstream NLRP3-IL-1β inflammasome activation axis, however this has yet to be explored in human trials [25].

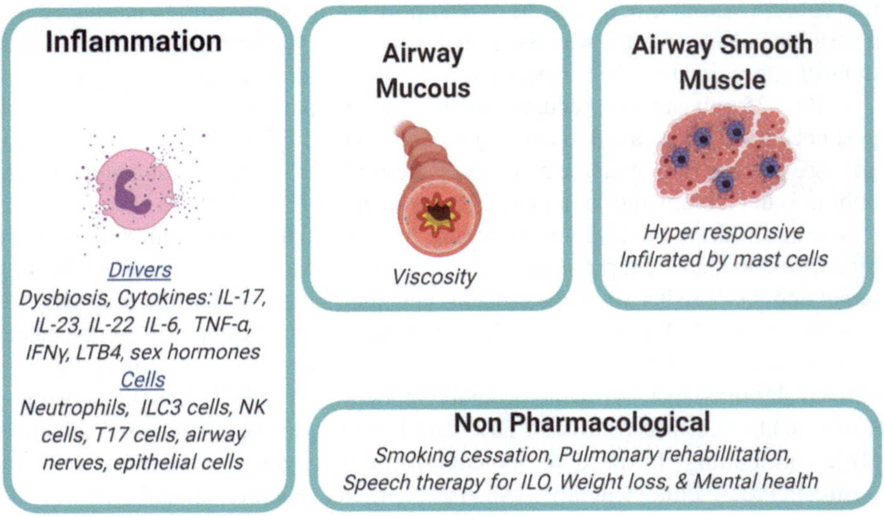

Fig. 2 Biomarker observations in severe asthma without current therapy. Many patients with severe asthma have ongoing symptoms and risk, while lacking evidence of persistently elevated T2-inflammation if studied in the setting of existing therapy. This figure categorizes various molecular targets, behaviors, and/or comorbid conditions that could lead to future trial designs. Created with BioRender.com

Implementation Challenges for Molecular Signatures in Standard Trial Designs

The need to link mechanisms to more tightly coupled asthma immune-pathological programs has led to the development of several initiatives that have reported multi-marker signatures in severe asthma. For example, the European Union, Innovative Medicines Initiative (EU-IMI) funded UBIOPRED program [26], recently reported multiple phenotypes of severe asthma using sputum-derived transcriptomics biomarkers [27]. Patient segmentation using these transcriptional biomarkers offers the potential for much deeper insight into asthma immune pathology. For example, whilst one transcriptome-associated patient cluster (TAC) was predominantly T2-driven (characterized by immune receptors *IL33R*, *CCR3*, and *TSLPR*), and would likely be detected with BEC and FeNO as surrogates, other TACs would not anchor to current biomarkers with enrichment of metabolic and mitochondrial genes in one TAC and interferon and inflammasome associated genes in another TAC. It is tempting to speculate that the stratification of patients using this approach might be a future route for novel metabolic therapies in severe asthma or, indeed inflammasome-directed therapies. However, several limitations and barriers prevent the implementation of this approach in trials. Firstly, the patient assignment to TAC clusters varies over time [28] in a significant proportion of patients (>40%) potentially due to the variability of driving mechanisms in asthma. Secondly, the TAC clusters reported would require further validation using qPCR to develop robust multiplexed diagnostics for clinical trial implementation with robust thresholds for each component of a multi-marker signature. Nonetheless, notwithstanding these potential limitations, these results underscore the potential for multi-marker signatures to provide more precise segmentations of patient populations for highly stratified precision medicine trials in the future.

Master Protocols and Adaptive Design Methods

Summary of the Master Protocol Features

Master protocols are the emerging standard for adaptive design trials. There are numerous advantages and a few disadvantages, reviewed by Woodcock and LaVange [29]. Three basic types of master protocols include umbrella, allowing multiple interventions for a single disease; basket, using a single intervention for multiple diseases; and platform, a perpetual version of an umbrella protocol with a prespecified decision algorithm to stop existing and/or start new interventions during the conduct of the trial. Gains in the efficiency of the trial infrastructure are the main advantage to master protocols, particularly with respect to the governance of a single protocol as opposed to multiple smaller protocols. Common methods for participant screening, data capture, and quality control are also very helpful. The main disadvantage is the increased time needed at the beginning of the trial and complexity in negotiations with sponsors to provide assurances on proposed futility decision-making and the presence or absence of comparisons across interventions.

Brief Overview of Adaptive Methods

Adaptive trial designs allow for the modification of prespecified variables during the course of the trial based on the outcomes of interim analyses of accrued and/or simulated trial data. These data update the overall model for treatment responses, and the trial continues to enrollment until a prespecified sample size or posterior probability has been reached. An overview of key steps in the workflow for any adaptive trial and guidelines for best practices are reviewed by the Adaptive Platform Trial Coalition [30]. Several reviews of the advantages and disadvantages of different types of adaptive trials have also been written (e.g. [31, 32]). Prior to the COVID-19 pandemic, the most common designs in the literature were Seamless phase II/III, Group Sequential, and Biomarker Adaptive [33]. Examples of these design types are discussed in the next section.

Learning from Adaptive Trials in Oncology and Infectious Diseases

Clinical trials with adaptive designs have become more common in Oncology and Infectious Diseases. The FOCUS4 trials are a good example of using this method to promote early results from phase 2 stages seamlessly into a phase 3 trial [34]. This collaborative group used an umbrella protocol of trials for the management of patients with metastatic colorectal cancer after standard induction chemotherapy. Four molecularly stratified trials were launched along with a fifth nonstratified trial for patients without a biomarker match and/or logistical issues precluding participation in a stratified trial. After passing an interim analysis for futility, trials were promoted with sample sizes and designs to support phase 3 regulatory needs. Two of these interim analyses are presented by Seligman et al. and Adams et al. [35, 36].

The early phase of the COVID-19 pandemic presented clinicians and trialists with multiple unknowns regarding disease natural history and reliable trial endpoints, making traditional phase 2 designs challenging. The RECOVERY collaborative group in the United Kingdom mobilized quickly to promote numerous trials using a group sequential adaptive design, again with provision for early termination of trials that did not pass interim futility analysis. The RECOVERY trials of dexamethasone and continuous positive airway pressure (CPAP) are good examples of the success of this method [37, 38]. Recent simulation analyses suggest that the RECOVERY trial efficiency might be further enhanced by the use of response adaptive randomization [39].

The I-SPY2 trials for breast cancer are a widely cited example of biomarker adaptive design that also utilizes response adaptive randomization and adaptation on the biomarker signature(s) during the course of the trial. Early examples of this success are the trials described by Park et al. and Rugo et al. [40, 41]. In this design, patients are stratified according to predictive biomarker signatures and randomized to interventions thought to be the best match for this biomarker. Interim analysis is continuous with an analysis plan using Bayesian probabilities. These event rates modify the randomization probabilities during the course of the trial to favor the more successful intervention. This approach also allows for the refinement of the predictive biomarker signature during the trial. The strengths and weaknesses of this approach are reviewed by Gyawali and Niraula [42], and a revised method for biomarker signature refinement is discussed by Wolf et al. [43].

PrecISE Design Features

The Precision Medicine in Severe and/or Exacerbation Prone (PrecISE) trial is sponsored by the National Heart Lung and Blood Institute and is currently midway through its enrollment period. It uses a platform master protocol to test five interventions in adults and adolescents ages 12 years and older with severe asthma. The clinical and statistical aspects of the protocol have been published, as has a more detailed report on the rationale for the interventions, the governance structure of the trial, and adaptations made for the COVID-19 pandemic [44–46]. The phenotypes targeted by the interventions in the trial include mast cell activation, metabolic dysfunction, oxidant stress, and frequent exacerbations. Each intervention is associated with a primary predictive biomarker. A placebo-controlled crossover design is in use with 16-week treatment periods interspersed with 8- to 16-week washout periods. The trial was originally designed with three primary endpoints (prebronchodilator FEV_1, ACQ-6, and CompEx events, a composite measure of loss of control events and exacerbations, [47]) with an analysis plan structured to determine improvement in any of the three endpoints.

During a common screening and run-in period, the primary predictive biomarkers are assessed; these include blood eosinophils, FeNO levels, plasma IL-6 concentrations, and genotypes for S-nitrosoglutathione reductase. Secondary predictive biomarkers are also collected at this point using several types of samples including sputum, peripheral blood mononuclear cells, high-resolution chest CT imaging, both urine and stool samples. Participants are randomized to an intervention on a 2:1 basis favoring the biomarker-positive status group (Fig. 3). The initial treatment period has a second randomization regarding the sequence for placebo or active drug. Participants may continue on in the trial after completing the first two treatment periods to receive other interventions and have a reduced probability of assignment to placebo. An interim analysis is planned to evaluate the futility of each intervention after accruing 40% of the biomarker-positive population needed for the final efficacy analysis. Interventions passing the futility assessment will undergo revision of the biomarker signature with the addition of up to four of the other primary or secondary predictive biomarkers prior to enrolling another 30% of the needed sample size. A second revision of the biomarker signature will be completed prior to enrolling the final 30% of the sample. The primary analysis focuses on

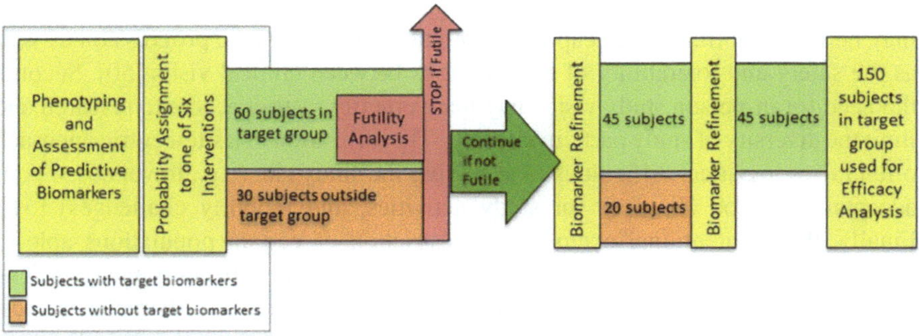

Fig. 3 Schematic of the biomarker adaptation planned for the Precision Interventions for Severe and/or Exacerbation prone asthma (PrecISE) trial. Predictive biomarkers assessed during the run-in period are used for assignments to paired interventions on a 2:1 basis, including patients with the biomarker-negative profile as controls. The interim futility and overall efficacy analyses will be performed only on the biomarker-positive group. Interventions proceeding beyond the futility analysis will also have the opportunity to revise the biomarker signature based on an interim responder analysis. (This figure has been reproduced from Israel et al. [45], Copyright (2021). With permission from Elsevier)

comparisons to placebo in the biomarker-positive efficacy population. A secondary precision medicine analysis is planned, also including the biomarker-negative participants to evaluate the associated cut points and begin to assess best treatment options for composite biomarker profiles.

Digitally Enabled Clinical Research Modalities

Biomarker-informed adaptive trials like PrecISE are in many ways quintessential efficacy studies, designed to recruit highly defined study populations and use state-of-the-art diagnostic methods at academic research centers. The COVID-19 pandemic forced reevaluation and innovation in both clinical care and research. The following section will review how investments in digital technology can provide resiliency to these types of trials and may lead to new opportunities. Although hybrid approaches are more likely to become standard, pure remote, decentralized trial designs, have also been envisioned (Fig. 4), with an example of a pure design recently published testing the role of hydroxychloroquine for postexposure prophylaxis for COVID-19 [48].

Limitations of Centralized Clinical Trials

Clinical research is often site-centric, adhering to schedules largely driven by tradition and operational convenience rather than the natural history of diseases and the diversity of target populations [49]. Several limitations of centralized trials exist. First, the approach mostly provides snapshots of patient data which are often determined by and correspond to the in-person clinical visits where patients' biological samples and vital statistics are measured and recorded. The implication is that there may be lost opportunities to capture information about disease progression as well as the safety and tolerability of interventions between clinical visits [50]. Second, the need for in-person study visits may inadvertently restrict access to research and limit the diversity of study participants [50]. The utilization of a site-centric approach often raises logistical difficulties in accessing the study site (e.g., travel costs, non-acceptance of job absences for study activities, and mobility challenges) [51]. Finally, the unintentional exclusion or restriction of certain populations able to enroll in clinical research limits the applicability of treatment benefits and risks. The implication is that there is limited evidence to support the use of therapies in individuals from underserved groups leading to potentially suboptimal care [52].

Benefits of Digitally Enabled Decentralized Clinical Trials or Hybrid Approaches

The rapidly advancing field of digital health technologies (DHTs) provides an opportunity to address some of the issues associated with site-centric clinical

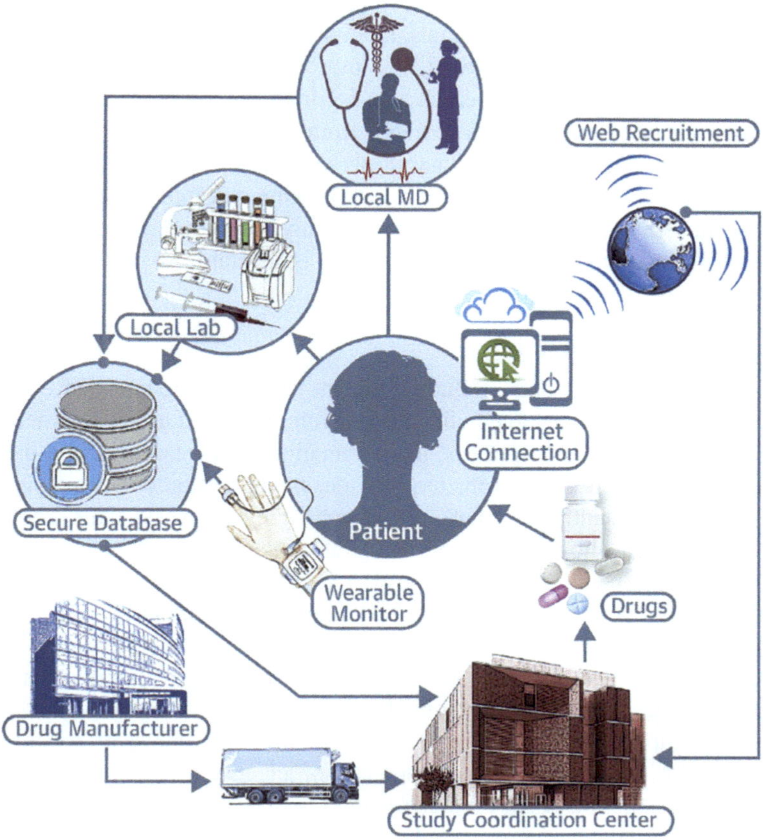

Fig. 4 Idealized workflow and participation in a remote decentralized clinical trial. In response to social media or electronic health record-embedded study advertisements, interested participants can learn about trial opportunities and interact with study coordinators online, leading to an electronic consent process. Additional screening could be done by local clinical care teams and labs with support for simplified case report forms and access to a secure central database. Confirmation of eligibility would trigger randomization at the study coordination center with the shipment of drug supplies and digital health monitors to the patient. Data collection could be in real-time from the health monitor, coupled with reminders to submit patient-reported outcome information via electronic portals. (This figure has been reproduced from Van Norman [51], Copyright (2021), with permission from Elsevier)

research and transform the way clinical research is conducted [50]. For the purposes of this review, decentralized clinical trials are referred to as those utilizing DHT to promote recruitment, retention, and ease of data collection.

Facilitating Participant Recruitment and Retention

The use of digital tools including database-screening tools (e.g., Clinical Practice Research Datalink [CPRD]) and recruitment methods via social media (e.g., Facebook and Twitter), text messaging, and online project websites could reduce the screening of ineligible participants, improve recruitment and retention rates by widen the potential pool of participants, facilitating quicker identification of eligible participants, reducing screening time, and improving record keeping [53].

Digitally enabled clinical research also facilitates recruitment and retention by allowing patients to engage with research from their homes or any place of their choice rather than attend conventional in-person study visits, thus removing travel costs and the need to take time off work. By making research more accessible and ensuring that study cohorts are more representative of the target population, study findings are enriched, and better-quality data is available to inform the care of diverse patients.

Facilitating Data Collection

Patient-generated health data, including patient-reported outcomes (PROs), are an integral part of digitally enabled clinical research. Increased use of internet access, home computing, and smartphones, in combination with wearables, and biosensor technology in inhalers and spirometry, provides the opportunity to remotely collect patient-generated health data that are closer to the real-world daily contexts of patients while still allowing for rigorous control, support, and management [50]. Such technology enables participants to provide data at a time that best suits them, thus potentially maintaining participant motivation and retention.

PROs capture patient's perspectives, contributing to a more holistic and comprehensive assessment of the safety, efficacy, and tolerability of interventions under investigation [52]. PROs have traditionally been measured using paper-based questionnaires completed by participants while at site visits. However, technological advances have facilitated the development of electronic PROs (ePROs) [54]. Advantages of utilizing ePROs include reduction of administrative and respondent burden, reduced data entry errors, and real-time data collection, which could allow patients to remain at home. Furthermore, ePROs can be configured to trigger notifications that alert research and clinical teams in case of important clinical symptoms. These alerts can prompt more timely clinical intervention and onward referral to specialist care if required.

The digital biomarkers captured by DHTs have become an exciting new tool for advancing precision medicine and supporting clinical trials. While many digital biomarkers are still being validated, in the future they may provide detailed information on physiological processes and explain, influence, and/or predict health-related outcomes [55]. They may also inform diagnostics and dosing titration and serve as endpoints for clinical trials [56]. Asthma-specific examples of DHTs include the use of digital inhalers and the use of remote spirometry. Several companies have developed sensors that can be attached to inhalers to monitor the frequency of use and technique (reviewed in [57]). These systems often come with a smartphone dashboard that also has an interface to collect PRO data. Several proof-of-concept studies have also incorporated measurement of lung function with portable spirometers [58–62]. These methods have promise for decentralized trials, particularly if coupled with a video-health method to facilitate session coaching by a study coordinator and/or a dashboard that allows for overreading to maximize quality control.

Regulatory Position and Guidance for Digital Methods

There is strong regulatory buy-in for digitally enabled clinical research. In the United Kingdom, the Medicines and Healthcare Products Regulatory Agency's (MHRA) Delivery Plan 2021–2023 outlines a commitment to innovative clinical trials for the immediate future [63]. The European Medicine Agency (EMA) Network Strategy to 2025 also encourages the use of digital tools in clinical research and the use of real-world data (RWD) [64]. The US Food and Drug Administration has also issued draft guidance regarding digital health technologies [65].

Ensuring the integrity, reliability, and robustness of data generated by digitally enabled decentralized research is essential. The European Medicines Agency (EMA) has produced guidance on data access, source data verification, and principles to be followed regarding remote access to patients' data [66]. In the United Kingdom, guidance for sponsors, contract research organizations (CROs), and investigator sites when considering the management of personal data processed in relation to research has been jointly developed by the Heath Research Authority (HRA) and the Medicines and Healthcare products Regulatory Agency (MHRA) [67]. Co-development of systems with patient partners and usability testing with representative patients will help ensure that systems meet participant needs.

Caveats

Whilst decentralized digitally enabled research brings many benefits; access to the technology is not yet universal, and pockets of broadband-famine and social groups refusing technology will remain for years to come. Therefore, care must be taken to consider potential issues of digital inclusivity and diversity. Digital literacy and digital exclusion (both devices and connectivity) are issues that could inadvertently lead to the lack of diversity that digitally enabled clinical research seeks to address. There is a risk that digital exclusion due to the use of digitally enabled clinical research may widen health inequalities and worsen health data poverty for specific groups.

Pragmatic Designs With Informatics Support

The Salford Lung Study for chronic obstructive pulmonary disease (COPD) and asthma is frequently cited as a real-world comparative effectiveness study, taking advantage of an integrated health system and its electronic medical record [68]. In the growing digital era of healthcare, pragmatic designs in adaptive trials can also be achieved using informatics solutions via electronic health record (EHR) systems from diverse health systems. Best practices in clinical trial design include frequent oversight and regulation of data quality with strict protocols, specialized clinical trial management systems, and specific case report forms. While these measures allow for strong internal validity in efficacy studies, they threaten the generalizability and external validity. Pragmatic designs are intended to integrate the existing infrastructure and health system with the electronic health record (EHR) as the core for patient tracking and data capture and may be better suited for questions of effectiveness. Interoperability enforced by The Office of the National Coordination for Health Information Technology (ONC) and the adoption of common data models (i.e., Observational Medical Outcomes Partnership – OMOP) has increased the capabilities of using clinical data for research within EHRs [69]. Computable phenotypes, data from the EHR to identify diseases and conditions, can automatically identify eligible patients across an adaptive design. Clinical decision support systems facilitate pragmatic intervention delivery because it gets incorporated into existing clinical workflows and health systems operations [70]. Overall, pragmatic designs with informatics solutions can provide better generalizability and external validity.

Data quality and data capture in a multi-center setting are challenging to adapt across different instances of EHRs between health systems. Interoperable tools and data exchange have helped overcome the unique data operations at each center. Increased adoption of Health Level 7 (HL7) Fast Healthcare Interoperability Resource (FHIR) application programming interfaces (APIs) in recent years has increased the availability of patient data for external applications such as REDCap and other third-party applications [71]. Data from the EHR can interface directly with REDCap to automatically populate case report forms, which are routinely collected from enterprise data warehouses that store the data from EHRs. FHIR resources reduce the additional builds routinely enforced in multi-site clinical trials. These features allow for standardized and automated data readiness to ensure adaptive designs can be executed across sites [72].

Determining whether health interventions are effective in real-world settings requires a pragmatic design at scale and are the goals of many learning health systems [73]. EHRs can support the delivery of interventions or new clinical practices using several clinical decision support features (e.g., best practice alerts). Standards-based clinical decision support systems based on HL7 FHIR can reduce variability and assist providers in diagnostic and therapeutic interventions [74]. Leveraging standards-based clinical decision support tools like CDS Hooks to overcome the highly customized EHR installations across health systems, including site-specific features. Therefore, clinical coordinating centers should invest in hospital IT staff prior to trial initiation to take existing clinical decision support standards and deploy them in the EHR. Sometimes, the data critically needed for a trial, such as patient-reported outcomes (PROs), are not collected during routine care in the EHR. Researchers must develop or enhance systems to collect these data and interact with operations staff to ensure alignment with workflows. Federal funding agencies like the Patient-Centered Outcomes Research Institute (PCORI) provide guidance to integrate PROs in EHRs (users guide website).

Increasingly, informatics professionals must engage with research teams and health system leaders to support pragmatic research, underscoring the inevitable union between clinical research and health care operations. As the sophistication of pragmatic trials evolves with adaptive designs and multiple interim analyses, leveraging informatics tools and learning about the flexibility of EHRs are imperative to effective and efficient data analytics needed in adaptive designs.

Lessons Learned and Conclusions

To push the frontier of therapeutic development for severe asthma beyond T2 focused interventions, trial designs will need to include strategies that increase efficiency for testing in uncommon subsets. Biomarker stratified trials will likely need to retain traditional elements of study execution with centralized laboratories in order to support the associated molecular diagnostics. That said, the disruptions to all things considered normal during the COVID-19 pandemic may also spring new trial methods that encourage patient participation. Greater use of remote, decentralized, and/or pragmatic trial designs may be key to developing treatments that matter most to patients.

References

1. Siddiqui S, et al. Unmet needs in severe asthma subtyping and precision medicine trials. Bridging clinical and patient perspectives. Am J Respir Crit Care Med. 2019;199(7):823–829.
2. Coleman C, et al. Narrative review to capture patients' perceptions and opinions about non-response and response to biological therapy for severe asthma. Eur Respir J. 2022;61:2200837.
3. Teague WG, et al. Baseline features of the severe asthma research program (SARP III) cohort: differences with age. J Allergy Clin Immunol Pract. 2018;6(2):545–554 e4.
4. Denlinger LC, et al. Responsiveness to parenteral corticosteroids and lung function trajectory in adults with moderate-to-severe asthma. Am J Respir Crit Care Med. 2021;203(7):841–52.
5. Peters MC, et al. Evidence for exacerbation-prone asthma and predictive biomarkers of exacerbation frequency. Am J Respir Crit Care Med. 2020;202(7):973–82.
6. Price DB, et al. Association of elevated fractional exhaled nitric oxide concentration and blood eosinophil count with severe asthma exacerbations. Clin Transl Allergy. 2019;9:41.
7. Couillard S, et al. Derivation of a prototype asthma attack risk scale centred on blood eosinophils and exhaled nitric oxide. Thorax. 2022;77(2):199–202.
8. Kraft M, et al. Patient characteristics, biomarkers and exacerbation risk in severe, uncontrolled asthma. Eur Respir J. 2021;58(6):2100413.
9. Hanratty CE, et al. A randomised pragmatic trial of corticosteroid optimization in severe asthma using a composite biomarker algorithm to adjust corticosteroid dose versus standard care: study protocol for a randomised trial. Trials. 2018;19(1):5.
10. Heaney LG, et al. Composite type-2 biomarker strategy versus a symptom-risk-based algorithm to adjust corticosteroid dose in patients with severe asthma: a multicentre, single-blind, parallel group, randomised controlled trial. Lancet Respir Med. 2021;9(1):57–68.
11. Pelaia C, et al. Interleukin-5 in the pathophysiology of severe asthma. Front Physiol. 2019;10:1514.
12. Ortega HG, et al. Severe eosinophilic asthma treated with mepolizumab stratified by baseline eosinophil thresholds: a secondary analysis of the DREAM and MENSA studies. Lancet Respir Med. 2016;4(7):549–56.
13. Escamilla-Gil JM, Fernandez-Nieto M, Acevedo N. Understanding the cellular sources of the fractional exhaled nitric oxide (FeNO) and its role as a biomarker of type 2 inflammation in asthma. Biomed Res Int. 2022;2022:5753524.

14. Castro M, et al. Dupilumab efficacy and safety in moderate-to-severe uncontrolled asthma. N Engl J Med. 2018;378(26):2486–96.
15. Bousquet J, et al. Predicting and evaluating response to omalizumab in patients with severe allergic asthma. Respir Med. 2007;101(7):1483–92.
16. Casale TB, et al. Omalizumab effectiveness by biomarker status in patients with asthma: evidence from PROSPERO, a prospective real-world study. J Allergy Clin Immunol Pract. 2019;7(1):156–164.e1.
17. Humbert M, et al. Omalizumab effectiveness in patients with severe allergic asthma according to blood eosinophil count: the STELLAIR study. Eur Respir J. 2018;51(5):1702523.
18. Menzies-Gow A, et al. Tezepelumab in adults and adolescents with severe, uncontrolled asthma. N Engl J Med. 2021;384(19):1800–9.
19. Rabe KF, et al. Safety and efficacy of itepekimab in patients with moderate-to-severe COPD: a genetic association study and randomised, double-blind, phase 2a trial. Lancet Respir Med. 2021;9(11):1288–98.
20. Busse WW, et al. Randomized, double-blind, placebo-controlled study of brodalumab, a human anti-IL-17 receptor monoclonal antibody, in moderate to severe asthma. Am J Respir Crit Care Med. 2013;188(11):1294–302.
21. Brightling CE, et al. Risankizumab in severe asthma – a phase 2a, placebo-controlled trial. N Engl J Med. 2021;385(18):1669–79.
22. O'Byrne PM, et al. Efficacy and safety of a CXCR2 antagonist, AZD5069, in patients with uncontrolled persistent asthma: a randomised, double-blind, placebo-controlled trial. Lancet Respir Med. 2016;4(10):797–806.
23. Diver S, et al. Sputum microbiomic clustering in asthma and chronic obstructive pulmonary disease reveals a Haemophilus-predominant subgroup. Allergy. 2020;75(4):808–17.
24. Taylor SL, et al. Long-term azithromycin reduces haemophilus influenzae and increases antibiotic resistance in severe asthma. Am J Respir Crit Care Med. 2019;200(3):309–17.
25. Kim RY, et al. Role for NLRP3 inflammasome-mediated, IL-1β-dependent responses in severe, steroid-resistant asthma. Am J Respir Crit Care Med. 2017;196(3):283–97.
26. Shaw DE, et al. Clinical and inflammatory characteristics of the European U-BIOPRED adult severe asthma cohort. Eur Respir J. 2015;46(5):1308–21.
27. Kuo CS, et al. T-helper cell type 2 (Th2) and non-Th2 molecular phenotypes of asthma using sputum transcriptomics in U-BIOPRED. Eur Respir J. 2017;49(2):1602135.
28. Kermani NZ, et al. Instability of sputum molecular phenotypes in U-BIOPRED severe asthma. Eur Respir J. 2021;57(2):2001836.
29. Woodcock J, LaVange LM. Master protocols to study multiple therapies, multiple diseases, or both. N Engl J Med. 2017;377(1):62–70.
30. Adaptive Platform Trials Coalition. Adaptive platform trials: definition, design, conduct and reporting considerations. Nat Rev Drug Discov. 2019;18(10):797–807.
31. Chow SC. Adaptive clinical trial design. Annu Rev Med. 2014;65:405–15.
32. Bhatt DL, Mehta C. Adaptive designs for clinical trials. N Engl J Med. 2016;375(1):65–74.
33. Bothwell LE, et al. Adaptive design clinical trials: a review of the literature and ClinicalTrials.gov. BMJ Open. 2018;8(2):e018320.
34. Brown LC, et al. Experiences of running a stratified medicine adaptive platform trial: challenges and lessons learned from 10 years of the FOCUS4 trial in metastatic colorectal cancer. Clin Trials. 2022;19(2):146–57.
35. Seligmann JF, et al. Inhibition of WEE1 Is effective in TP53- and RAS-mutant metastatic colorectal cancer: a randomized trial (FOCUS4-C) comparing adavosertib (AZD1775) with active monitoring. J Clin Oncol. 2021;39(33):3705–15.
36. Adams RA, et al. Capecitabine versus active monitoring in stable or responding metastatic colorectal cancer after 16 weeks of first-line therapy: results of the randomized FOCUS4-N trial. J Clin Oncol. 2021;39(33):3693–704.
37. Group, R.C, et al. Dexamethasone in hospitalized patients with Covid-19. N Engl J Med. 2021;384(8):693–704.

38. Perkins GD, et al. Effect of noninvasive respiratory strategies on intubation or mortality among patients with acute hypoxemic respiratory failure and COVID-19: the RECOVERY-RS randomized clinical trial. JAMA. 2022;327(6):546–58.
39. Sirkis T, Jones B, Bowden J. Should RECOVERY have used response adaptive randomisation? Evidence from a simulation study. BMC Med Res Methodol. 2022;22(1):216.
40. Park JW, et al. Adaptive randomization of neratinib in early breast cancer. N Engl J Med. 2016;375(1):11–22.
41. Rugo HS, et al. Adaptive randomization of veliparib-carboplatin treatment in breast cancer. N Engl J Med. 2016;375(1):23–34.
42. Gyawali B, Niraula S. Lessons from adaptive randomization: spying the I-SPY2 trial in breast cancer. J Natl Compr Cancer Netw. 2020;18(11):1441–4.
43. Wolf DM, et al. Redefining breast cancer subtypes to guide treatment prioritization and maximize response: predictive biomarkers across 10 cancer therapies. Cancer Cell. 2022;40(6):609–623.e6.
44. Ivanova A, et al. The precision interventions for severe and/or exacerbation-prone asthma (PrecISE) adaptive platform trial: statistical considerations. J Biopharm Stat. 2020;30(6):1026–37.
45. Israel E, et al. PrecISE: precision medicine in severe asthma: an adaptive platform trial with biomarker ascertainment. J Allergy Clin Immunol. 2021;147(5):1594–601.
46. Georas SN, et al. The Precision Interventions for Severe and/or Exacerbation-Prone (PrecISE) Asthma Network: an overview of network organization, procedures, and interventions. J Allergy Clin Immunol. 2022;149(2):488–516 e9.
47. Fuhlbrigge AL, et al. A novel endpoint for exacerbations in asthma to accelerate clinical development: a post-hoc analysis of randomised controlled trials. Lancet Respir Med. 2017;5(7):577–90.
48. Boulware DR, et al. A randomized trial of hydroxychloroquine as postexposure prophylaxis for Covid-19. N Engl J Med. 2020;383(6):517–25.
49. De Brouwer W, et al. Empowering clinical research in a decentralized world. NPJ Digit Med. 2021;4(1):102.
50. Dockendorf MF, et al. Digitally enabled, patient-centric clinical trials: shifting the drug development paradigm. Clin Transl Sci. 2021;14(2):445–59.
51. Van Norman GA. Decentralized clinical trials: the future of medical product development?∗. JACC Basic Transl Sci. 2021;6(4):384–7.
52. Calvert MJ, et al. Patient reported outcome assessment must be inclusive and equitable. Nat Med. 2022;28:1120.
53. Blatch-Jones A, et al. Using digital tools in the recruitment and retention in randomised controlled trials: survey of UK Clinical Trial Units and a qualitative study. Trials. 2020;21(1):304.
54. Hughes SE, et al. Development and validation of the symptom burden questionnaire for long covid (SBQ-LC): Rasch analysis. BMJ. 2022;377:e070230.
55. Marra C, et al. Quantifying the use of connected digital products in clinical research. NPJ Digit Med. 2020;3(1):50.
56. Inan OT, et al. Digitizing clinical trials. NPJ Digit Med. 2020;3(1):101.
57. Mosnaim GS, et al. Digital inhalers and remote patient monitoring for asthma. J Allergy Clin Immunol Pract. 2022;10(10):2525–33.
58. Fossati A, et al. Telehome monitoring of symptoms and lung function in children with asthma. Healthcare (Basel). 2022;10(6):1131.
59. Huang C, et al. Remote FEV1 monitoring in asthma patients: a pilot study. Clin Transl Sci. 2021;14(2):529–35.
60. Kruizinga MD, et al. Technical validity and usability of a novel smartphone-connected spirometry device for pediatric patients with asthma and cystic fibrosis. Pediatr Pulmonol. 2020;55(9):2463–70.
61. Kupczyk M, et al. Home self-monitoring in patients with asthma using a mobile spirometry system. J Asthma. 2021;58(4):505–11.

62. Ring B, et al. Validation of an app-based portable spirometer in adolescents with asthma. J Asthma. 2021;58(4):497–504.
63. MHRA. Putting patients first: a new era for our agency. Delivery Plan 2021–2023. 2021.
64. EMA. European medicines agencies network strategy to 2025. 2020.
65. Digital Health Technologies for Remote Data Acquisition in Clinical Investigations, F.a.D.A. U.S. Department of Health and Human Services, Editor. 2021. Federal Register.
66. EMA. Guideline for good clinical practice E6(R2). 2016.
67. MHRA. Access to electronic health records by sponsor representatives in clinical trials. 2020. [08/09/2021, 03/05/2022]. Available from: https://www.gov.uk/guidance/on-site-access-to-electronic-health-records-by-sponsor-representatives-in-clinical-trials.
68. Albertson TE, et al. The Salford Lung Study: a pioneering comparative effectiveness approach to COPD and asthma in clinical trials. Pragmat Obs Res. 2017;8:175–81.
69. Sun Y, et al. Building an OMOP common data model-compliant annotated corpus for COVID-19 clinical trials. J Biomed Inform. 2021;118:103790.
70. Richesson RL, et al. Enhancing the use of EHR systems for pragmatic embedded research: lessons from the NIH Health Care Systems Research Collaboratory. J Am Med Inform Assoc. 2021;28(12):2626–40.
71. Cheng AC, et al. REDCap on FHIR: clinical data interoperability services. J Biomed Inform. 2021;121:103871.
72. Yamamoto K, et al. A pragmatic method for transforming clinical research data from the research electronic data capture "REDCap" to Clinical Data Interchange Standards Consortium (CDISC) Study Data Tabulation Model (SDTM): development and evaluation of REDCap2SDTM. J Biomed Inform. 2017;70:65–76.
73. Murdoch TB, Detsky AS. The inevitable application of big data to health care. JAMA. 2013;309(13):1351–2.
74. Strasberg HR, et al. Contemporary clinical decision support standards using Health Level Seven International Fast Healthcare Interoperability Resources. J Am Med Inform Assoc. 2021;28(8):1796–806.

Correction to: Approaches to Management of Asthma: Guidelines for Stepped Care and Self-Monitoring

Daniel Rosenberg and Philip Palmon

Correction to:
Chapter 15 in: A. R. Brasier, N. N. Jarjour (eds.), *Precision Approaches to Heterogeneity in Asthma*, Advances in Experimental Medicine and Biology 1426, https://doi.org/10.1007/978-3-031-32259-4_15

In Chapter 15 of the original version of this book, the name of co-author "Philip Palmon" was missed out. This has now been updated.

The updated original version of this chapter can be found at
https://doi.org/10.1007/978-3-031-32259-4_15

A. R. Brasier, N. N. Jarjour (eds.), *Precision Approaches to Heterogeneity in Asthma*, Advances in Experimental Medicine and Biology 1426,
https://doi.org/10.1007/978-3-031-32259-4_18

Conclusion: Precision Approaches to Heterogeneity in Asthma

The past few decades have ushered in significant advances in understanding of the mechanisms of asthma, available therapies and important practical guidelines informing the use of these advances to improve disease control, enhance patients' quality of life, and reduce the burden of asthma on patients, families, and communities.

Improved understanding of the etiology of disease complexity in asthma has led to greater focus on patient characterization to guide therapy, moving the field of asthma clinical care closer toward precision medicine. Therefore, there has been a welcome transition from using one approach for all patients to greater implementation of patient phenotyping to select best treatment plans that target specific patient's endotype. The increased recognition of disease heterogeneity has paved the way for patient-centered approaches with greater efficacy and improved outcomes. However, there remain many unmet needs because the current recommendations regarding patient characterization continue to rely on defining the inflammatory profile to help guide the likely effective therapy. Following these recommendations, patient with type 2 (T2) inflammation who do not respond to glucocorticoid therapy can be provided one of several approved biologic therapies that target T2 inflammation and can generally expect a good chance of response with reduction in exacerbations and improved quality of lie. However, more limited advances have been made in management of patient who do not exhibit a T2 high endotype or in patients with T2 endotype who do not respond to the therapies targeting these pathways.

In this second edition of "Heterogeneity in Asthma," we aimed to address recent developments in asthma etiology, genomics, epidemiology, and biomarker-informed treatment approaches. We have retained the topics of several previously published chapters, however, most of them have been addressed from new angles by new authors. We have also included a number of chapters that address important pathophysiological aspects of the disease (e.g. adaptive immunity and imaging) and its clinical features (e.g. evolution across the life span and exacerbations). The revised edition includes more emphasis on the systemic aspects of asthma including its potential contributions to cardiovascular and neurological diseases. We are delighted to have contributions from a highly accomplished group of asthma experts

A. R. Brasier, N. N. Jarjour (eds.), *Precision Approaches to Heterogeneity in Asthma*, Advances in Experimental Medicine and Biology 1426,
https://doi.org/10.1007/978-3-031-32259-4

addressing disease epidemiology (including social determinants of health effect of climate change and evolving asthma manifestation over the lifespan) mechanistic pathways (cellular inflammation, proteomics, metabolomics, and transcriptomics), and the effect of asthma-related inflammation on other organs (e.g. brain, cardiovascular system). The chapters in this book have nicely addressed the various aspects of asthma heterogeneity, including its phenotypes and response to treatment. The authors provided excellent overview of current guidelines for disease management and recent and expected advances in understanding asthma pathophysiology and approaches to its management. Because a large share of asthma-related burden is related to disease severity and impact of exacerbations, the advances in these areas are discussed in detail. Specific focus in this book has been placed on the social determinants of health which are of high importance to asthma prevalence, severity, and overall burden.

The authors in this book cover the spectrum of asthma expertise from the bench to the bedside, from the clinic to population health and from North America to Europe. We are also pleased to have several chapters written by a joint effort of an early career physician who is guided and mentored by an established expert in the field.

This book has laid the foundation for future work in the precision approach to asthma. More work on asthma endotypes is needed, particularly in drivers of disease outside of the adaptive immune system including, but not limited to, innate inflammation, etiologies of frequent exacerbations and structural remodeling. We are excited about the future developments in these areas.

The authors paid special attention to including illustrations, tables, and references which we hope to provide added value to our readers. Each chapter has undergone peer review and editorial input; however, we recognize that there could still be errors that these processes did not identify, and for that we apologize to our readers and ask for their understanding and feedback. We are grateful to the wonderful colleagues who have generously contributed to this book and hope that you find it a valuable addition to the literature.

On behalf of the contributors,

Nizar N. Jarjour, MD
Allan R. Brasier, MD
Department of Medicine and Institute for Clinical and Translational Research, University of Wisconsin-Madison, Madison, WI, USA.

Index

A. R. Brasier, N. N. Jarjour (eds.), *Precision Approaches to Heterogeneity in Asthma*, Advances in Experimental Medicine and Biology 1426, https://doi.org/10.1007/978-3-031-32259-4

MIX
Papier aus verantwortungsvollen Quellen
Paper from responsible sources
FSC® C105338

If you have any concerns about our products,
you can contact us on
ProductSafety@springernature.com

In case Publisher is established outside the EU,
the EU authorized representative is:
Springer Nature Customer Service Center GmbH
Europaplatz 3, 69115 Heidelberg, Germany

Printed by Libri Plureos GmbH
in Hamburg, Germany